For All Practical Purposes

PROJECT DIRECTOR

Solomon Garfunkel, *Consortium for Mathematics and Its Applications*

COORDINATING EDITOR, FIRST EDITION

Lynn A. Steen, *St. Olaf College*

CONTRIBUTING AUTHORS

PART I MANAGEMENT SCIENCE

Joseph Malkevitch, *York College, CUNY*
Rochelle Meyer, *Nassau Community College*
Walter Meyer, *Adelphi University*

PART II STATISTICS: THE SCIENCE OF DATA

David S. Moore, Purdue University

PART III SOCIAL CHOICE AND DECISION MAKING

William F. Lucas, *Claremont Graduate School*

PART IV ON SIZE AND SHAPE

Donald Albers, *Menlo College*
Paul J. Campbell, *Beloit College*
Donald Crowe, *University of Wisconsin*
Seymour Schuster, *Carleton College*
Maynard Thompson, *Indiana University*

PART V COMPUTERS

Wayne Carlson, *Cranston/Csuri Productions, Inc.*
Zaven Karian, *Denison University*
Sartaj Sahni, *University of Minnesota*
Paul Wang, *Kent State University*

Joseph Blatt, *Chedd-Angier Production Company*

For All Practical Purposes

INTRODUCTION TO

CONTEMPORARY MATHEMATICS

SECOND EDITION

W. H. FREEMAN AND COMPANY
NEW YORK

Front Cover: Wassily Kandinsky, *On Points,* 1928. Musée National d'Arte Moderne, Centre Georges Pompidou, Paris. [Giraudon/Art Resource, N.Y. Copyright 1990 ARS N.Y./ADAGP.]

Library of Congress Cataloging-in-Publication Data

For all practical purposes: introduction to contemporary mathematics / [project director, Solomon Garfunkel; coordinating editor, Lynn A. Steen; contributing authors, Joseph Malkevitch . . . et al.]. — 2nd ed.
    p.   cm.
    A project of Consortium for Mathematics and Its Applications, Arlington, Mass.
    Includes index.
    ISBN 0-7167-2115-5
    1. Mathematics.   I. Garfunkel, Solomon A., 1943– .   II. Steen, Lynn Arthur, 1941– .
III. Malkevitch, Joseph, 1942– . IV. Consortium for Mathematics and Its Applications (U.S.)
    QA7.F68   1991
    510 — dc20   90-46527
                                                            CIP

Printed in the United States of America

1 2 3 4 5 6 7 8 9 0   RRD   9 9 8 7 6 5 4 3 2 1 0

◆ Contents ◆

vi   **Contents**

✦ Preface to the Second Edition ✦

In a very real sense, this, the second edition of *For All Practical Purposes,* is a users' edition. All of us at COMAP, Inc., and W. H. Freeman and Company were pleasantly surprised and extremely grateful for the broad and immediate success of the first edition. Back in what now seems ancient history when the idea for our book and accompanying telecourse first emerged, most colleagues (and publishers) were dubious about the future of both the text and the liberal arts mathematics course. It is clear in retrospect that our idea was one whose time had truly come.

The first edition has been used at over 300 schools in just a three-year period, and it has not only been adopted by committed COMAPers but has become the basis of a standard course at many two-year and four-year colleges and universities. Because the text was so different, the content so new, and the connection with video such a new idea for many mathematics faculty, we were able to collect numerous suggestions, critical reviews, ideas, demands, and even pleas from faculty teaching from the first edition. We are extremely grateful to our users for their careful, detailed feedback.

With the exception of typos, wrong answers, and a list of favorite or least liked sections, reviewers' comments were remarkably consistent on one point: More exercises! More examples! You strongly felt that the idea and approach of the text were on target, but that it was a difficult text to teach from due to an insufficient number of exercises and worked examples. Our response is an edition with approximately 200 additional pages of exercises and examples — *double* the original number. In our second edition you will find that:

◆ all chapters have been reworked to provide clearer exposition and clarify the concepts and ideas presented in the first edition. In all chapters, this has involved providing new and more extensive exercise sets and substantially increasing the use of labelled examples.

◆ essential mathematical procedures that appeared in boxes in the first edition have been integrated into the text. Human-interest "Spotlights" appear in well-placed, clearly marked boxes. These historical anecdotes and present-day conversations provide the book with a stronger cultural component.

◆ the spirit of the first edition has been retained by achieving a good balance between the expository and conceptual approach unique to *FAPP* and the necessary skill-building features required of any introductory mathematics course.

◆ optional sections in chapters (along with clearly marked exercises that go with these sections) give instructors more flexibility in organizing their courses.

◆ suggested readings now follow each chapter and many have been annotated.

◆ answers are provided to all exercises (odd-numbered in the text; even-numbered in the Instructor's Guide).

◆ the significantly improved Instructor's Guide provides chapter outlines, thorough chapter summaries, examples, teaching strategies, suggestions for coordinating the videos with the text, and additional exercises and answers for every chapter.

◆ the Telecourse Guide for instructors who use the videos has been updated.

◆ 62 transparency masters, with enlarged figures and tables from the text, are available for the convenience of those instructors who want to use overhead transparencies.

These changes and additions represent our efforts to create a text that is worthy of the adventuresome math department faculty who have made *For All Practical Purposes* a fixture on the mathematics education scene. We only hope (as we stated in the first edition preface) what these materials will continue to provide "a rich and exciting environment in which to learn more about the power and centrality of mathematics in our world."

ACKNOWLEDGMENTS

In addition to the heartfelt acknowledgments of the first edition, there are new people to be thanked. First of course, the authors:

Part I   Management Science
Joseph Malkevitch, York College, CUNY

Rochelle Meyer, Nassau Community College
Walter Meyer, Adelphi University

Part II   Statistics: The Science of Data
David S. Moore, Purdue University

Part III   Social Choice and Decision Making
William F. Lucas, Claremont Graduate School

Part IV   On Size and Shape
Paul J. Campbell, Beloit College

Part V   Computers
Zaven Karian, Denison University

As indicated above, the tireless efforts of our reviewers were crucial to the design and concepts of this edition. Specifically, we want to thank:

Carol A. Bouma, The Park School, Maryland

Ronald Czochor, Glassboro State College, New Jersey

Duane E. Deal, Ball State University, Indiana

Arthur Gittelman, California State University, Long Beach, California

Dale T. Hoffman, Bellevue Community College, Washington

Frederick Hoffman, Florida Atlantic University, Florida

Robert W. Hunt, Humboldt State University, California

Charles V. Jones, Ball State University, Indiana

Kay I. Meeks, Ball State University, Indiana

Thomas E. Moore, Bridgewater State College, Massachusetts

James Osterburg, University of Cincinnati, Ohio

Ron Palcic, Washburn University of Topeka, Kansas

Eli Passow, Temple University, Pennsylvania

Delene Perley, The College of Wooster, Ohio

Michael Petricig, Saddleback College, California

Sandra H. Savage, Orange Coast College, California

Stephen J. Willson, Iowa State University, Iowa

The production of the second edition was a superb team effort. Once again the editors of W. H. Freeman and Company somehow managed to bring order from chaos. Special thanks go to publisher Jerry Lyons, whose enthusiasm and faith in the book have been unflagging; to the development editorial staff: Linda Baron

Davis and Natalie Zabrocky, who started development of the second edition, Kay Ueno and Elisa Adams, who saw the development of the book through to completion, and editorial assistant Larry Marcus; to Sonia DiVittorio, project editor, whose tireless attention to editorial questions on both text and art has been invaluable, and to project editorial assistant Jodi Creditor.

We greatly appreciate the efforts of the production staff: Susan Stetzer, who, as production coordinator for both the first *and* second editions, diligently juggled the schedule to keep the project moving; Alice Fernandes-Brown, who designed the text and the outstanding cover; production artists John Hatzakis and Howard Johnson, who achieved a layout that is both handsome and readable; and Mara Kasler, who, as illustration coordinator, kept all the pieces of the illustration program straight.

And once again, the final plaudits must go to the COMAP staff: Laurie Aragon, Phil McGaw, Roland Cheyney, Annemarie Morgan, Laurie Holbrook, Dale Jolliffe, and Kathy Hynes, who manage, despite my unbelievable personal disorganization, to make all of COMAP (even me) look good.

September 1990   *Sol Garfunkel*
                                                   Arlington, MA

◆ Preface to the First Edition ◆

Every mathematician at some time has been called upon to answer the innocent question, "Just what is mathematics used for?" With understandable frequency, usually at social gatherings, the question is raised in similar ways: "What do mathematicians do, practice, or believe in?" At a time when success in our society depends heavily on satisfying the need for developing quantitative skills and reasoning ability, the mystique surrounding mathematics persists. *For All Practical Purposes: Introduction to Contemporary Mathematics* is our response to these questions and our attempt to fill this need.

For All Practical Purposes represents our effort to bring the excitement of contemporary mathematical thinking to the nonspecialist, as well as help him or her develop the capacity to engage in logical thinking and to read critically the technical information with which our contemporary society abounds. We attempt to implement for the study of mathematics Thomas Jefferson's notion of an "enlightened citizenry," in which individuals having acquired a broad knowledge of topics exercise sound judgment in making personal and political decisions. Environmental and economic issues dominate modern life, and behind these issues are complex matters of science, technology, and mathematics that call for an awareness of fundamental principles.

To encourage achievement of these goals, *For All Practical Purposes* stresses the connections between contemporary mathematics and modern society. Since the technological explosion that followed World War II, mathematics has become a cluster of mathematical sciences encompassing statistics, computer science, operations research, and decision science, as well as the more traditional areas. In science and industry mathematical models are the tools par excellence for solving complex problems. In this book our goal is to convey the power of mathematics as illustrated by the great variety of problems that can be modeled and solved by quantitative means.

This book is designed for use in a one-term course in liberal arts mathematics or in courses that survey mathematical ideas. The assumed background of our audience is varied. We expect some ability in arithmetic, geometry, and elementary algebra. Even though a few of the topics included here are traditionally thought of as part of advanced mathematics, we aim to develop in the reader strong conceptual understanding and appreciation, not computational expertise. Our stated bias throughout the book is toward presenting the subject through its contemporary applications.

In determining the selection of topics for this book, the authors posed a question to leading mathematicians and educators nationwide: "What would you teach students if they took only one semester of math during their entire college career?" Their answers are best demonstrated by the main topic selections for the text: management science, statistics, social choice, the geometry of size and shape, and mathematics for computer science. These topics were chosen both for their basic mathematical importance and for the critical role their applications play in a person's economic, political, and personal life.

Because readers will approach this book with a great range of expertise in using mathematical symbols, the material has been designed in layers to accommodate different objectives and diverse backgrounds. *For All Practical Purposes* concentrates on discussions about mathematics — about its nature, its content, its applications. At various places, in separate boxes or optional sections, certain mathematical issues are pursued with explicit use of equations and appropriate mathematical symbols. Problems are also divided by type, with some emphasizing descriptive matters and others providing practice in calculation and symbolic manipulation.

Additionally, *For All Practical Purposes* includes interviews with practitioners — the people who put mathematics to work. Each major topic section is complete and self-contained so that instructors can adjust the topic ordering to suit their particular needs. A large number of photographs and 69 full-color illustrations (in 24 pages of inserts) have been incorporated to emphasize the fact that contemporary mathematics is visually alive. Many of the color images are computer-generated and represent the accomplishments at the forefront of some mathematical research.

Perhaps the most distinctive new feature of this text is its relation to the 26 half-hour television programs that constitute the series "For All Practical Purposes." In 1983, the Consortium for Mathematics and Its Applications (COMAP) received a grant from The Annenberg/CPB Project, with co-funding from the Carnegie Corporation of New York, to produce a telecourse in mathematics for public broadcast on PBS. Development of the textbook was funded by the Alfred P. Sloan Foundation. An impressive array of people committed to educational excellence in mathematics as well as experienced television producers and computer graphics specialists were gathered to work on this project from the original outlines to final shows and text. Though this book stands alone from the video series, the authors worked simultaneously on developing television scripts and book chapters.

Some instructors will use the text and television shows in concert. Some will use the text alone in a traditional lecture course. Still others will make the videos available to students on tape for enrichment or special credit. The choice of how best to use this mix of resources will depend on each individual's particular preference and course requirements. An Instructor's Manual that relates the video programs with the text is available. Information about television-course licensing, off-air taping rights, and prerecorded video cassettes can be obtained by calling (202) 879-9655 (collect), or by writing The Annenberg/CPB Project at 901 E Street NW, Washington, D.C. 20004.

Students enrolled in the telecourse will view the videos and read the text on their own. To aid them a Course Guide has been developed by the text authors that provides an overview of each program, skill objectives, and a sample short-answer examination. We hope that these materials will provide a rich and exciting environment in which to learn more about the power and centrality of mathematics in our world.

ACKNOWLEDGMENTS

First, it is a pleasure to acknowledge the support of the Alfred P. Sloan Foundation, which provided a generous grant to help fund the production of this book. Over the almost four years of this undertaking, a remarkable number of people have made significant contributions. It is difficult to find words expressing our gratitude and appreciation. For the authors, this was no ordinary writing task. The revision process was enriched but certainly made a great deal more complicated by the relationship with the television series. With so many contributors, coordination was of necessity a key factor. The cluster leaders — Donald Albers, Zaven Karian, William F. Lucas, David S. Moore, and Joseph Malkevitch — did yeoman duty.

Professor Malkevitch truly deserves a very special recognition. To a great extent the underlying philosophy of this book is a reflection of his ideas, beliefs, and dedication.

We also owe a huge debt to Stephanie Stewart, John Rubin, and David Gifford, the writers, producers, and directors of the Chedd-Angier Production Company, who created "For All Practical Purposes" as a television series. The selection of locations, the interviews, and the scripts were all their domain. In particular, we acknowledge the tireless efforts of Joseph Blatt, the series's producer. His ingenuity and creativity, along with those of producer Olga Rakich, permeate this book. Their words and their spirit are found throughout the text.

We are indebted to the many instructors at colleges and universities around the country who offered us their critical comments of the manuscript during the development and production of *For All Practical Purposes.* Their efforts helped improve the conceptual and pedagogical quality of the text.

Larry A. Curnutt, Bellevue Community College, Washington

Jane Edgar, Brevard County Community College, Florida

Marjorie A. Fitting, San Jose State University, California

Jerome A. Goldstein, Tulane University, Louisiana

Elizabeth Hodes, Santa Barbara Community College District, California

Dale Hoffman, Bellevue Community College, Washington

Frederick Hoffman, Florida Atlantic University, Florida

Barbara Juister, Elgin Community College, Illinois

Peter Lindstrom, North Lake College, Texas

Francis Masat, Glassboro State College, New Jersey

Charles Nelson, University of Florida, Florida

Virginia Taylor, University of Lowell, Massachusetts

Special thanks is extended to Larry A. Curnutt and Dale Hoffman of Bellevue Community College and Frederick Hoffman of Florida Atlantic University who class tested prepublication portions of this book.

Many of the outstanding full color images were obtained from research scientists. We thank them for allowing us to reproduce some of their important work: H. E. Benzinger, S. A. Burns, and J. Palmore, University of Illinois (chaotic basins of attraction); Douglas Dunham, University of Minnesota (hyperbolic patterns); David Hoffman, University of Massachusetts (minimal surfaces); Benoit Mandelbrot, IBM (fractal images); Heinz-Otto Peitgen, University of Santa Cruz (Mandelbrot sets); Roger Penrose, Oxford University (Penrose tilings).

Other images appearing in the color plates are courtesy of Philip Zucco (computer graphics) and Cordon Art (Escher patterns). Color renderings were done by Tom Moore.

The writing of this book by our author team took place simultaneously with the production of the television series, and the complications and deadlines were numerous. The manuscript and complex illustrations would not have come together without the expert guidance of the W. H. Freeman and Company staff. In particular we wish to thank Susan Moran, project editor, Susan Stetzer, production coordinator, Mike Suh, art director, and Bill Page, art coordinator. Lloyd Black, senior development editor, arrived on the scene just in time to pull the words and figures together and guide the book through the production process. We send a special note of thanks to Jeremiah Lyons, senior mathematics editor. His faith in the project and efforts over two plus years have brought order out of chaos.

The final acknowledgments must go to the COMAP staff. To the production and administrative staff—Philip McGaw, Nancy Hawley, and Annemarie Morgan—go all our thanks. And finally, we recognize the contribution of Laurie Aragon, the COMAP business, development, personnel, etc. manager, who kept this project, as she does all of COMAP, running smoothly and efficiently. To everyone who helped make our purposes practical, we offer our appreciation for an exciting, exhausting, and exhilarating time.

Solomon Garfunkel
COMAP

Lynn A. Steen
St. Olaf College

For All Practical Purposes

◆ I ◆

Management Science

Those who were watching live television one night in July 1969 will never forget the spectacle of seeing the first person walk on the moon. The element of danger and uncertainty, heightened by a history of trial and error, added to the suspense of the lunar landing. Before the 1960s, no one really knew if rockets would ever be able to carry humans into space.

Neil Armstrong's first step onto the moon's surface was a triumph for American science and technology and the culmination of a national quest that had begun in the office of President John F. Kennedy. It was Kennedy's goal to put a man on the moon before the decade was out, a goal realized in the Nixon administration.

In the eight years from 1961 to 1969, we moved from a president's vision to the reality of a lunar landing. Most of us think of this achievement in terms of the tremendous scientific advances it represented — in physics, engineering, chemistry, and associated technologies. But there was another side to this far-reaching project. Someone had to set the objectives, commission the work, suffer the setbacks, overcome unforeseen obstacles, and tie together a project with thousands of disparate components. The kind of science responsible for such details is a branch of mathematics called **management science.**

NASA administrators faced many new problems in the Apollo moon project: they had to choose the best design for the spacecraft, design realistic ground simulations, and weigh the priorities of conducting experiments with immediate returns against carrying out tests that would serve long-term goals. When NASA

The first steps on the moon represented a great leap for management science. Here, Buzz Aldrin poses for Neil Armstrong. [NASA.]

commissioned the Apollo module, it was asking several hundred companies to design, build, test, and deliver components and systems that had never been built before.

Supporting these space age goals, however, were the nuts and bolts issues that make up the major concerns of management science, namely, finding ways to make the operations as productive and economical as possible; details of this support are in the Spotlight on the Apollo 11 launch. Issues of efficiency are important in all organizations. In business, industry, or government, operating efficiently is hardly a novel idea. Directors of large corporations and those in the high ranks of the military have always pursued efficiency. What is new about management science is that it distinguishes between trial-and-error — "seat of the pants" — approaches and new ideas guided by systematic mathematical analysis.

SPOTLIGHT Apollo 11 Launch Owes Success to Management Science

Captain Robert F. Freitag, who later became director of Policy and Plans for NASA's space station, in 1969 headed the team responsible for landing the Apollo 11 safely on the moon. The success of the lunar mission can be traced to management-science techniques that ensured that thousands of small tasks would come together to meet a single giant objective. Freitag shares his observations about that historic event:

Captain Robert F. Freitag, NASA.

I think the feeling most of us in NASA shared was, "My gosh, now we really have to do it." When you think that the enterprise we were about to undertake was ten times larger than any that had ever been undertaken, including the Manhattan Project, it was a pretty awesome event. But we knew it was the kind of thing that could be broken down into manageable pieces and that if we could get the right people and the right arrangement of these people, it would be possible.

In the case of the Apollo program, it was very important that we take a comprehensive system-engineering approach. We had to analyze in a very strict sense exactly what the mission was going to be, what each piece of equipment needed was and how it would perform, and all the elements of the system from the concept on through to the execution of the mission, to its recovery back on earth.

We started out, in a very logical way, by having a space station in earth orbit. We would then take the lunar spacecraft and build it in orbit, and then send it off to the moon and bring it back. It turned out that this approach was probably a little more risky

The need to establish scientific principles for operations management arose in World War II. The founders of management science were mathematicians and industrial technicians associated with the armed services who worked together to improve military operations. In applying quantitative techniques to project planning, these pioneers founded a new science whose impact reaches beyond the military to many corners of our lives.

Management science, or **operations research** as it sometimes is called, turned out to be a powerful notion with wide-ranging applications such as cleaning streets and planning multistop vacations. It enabled mathematicians to bring a long history of pure research to bear on practical problems. We will explore some of these problems and solutions in the chapters ahead.

and took a lot longer, so with the analyses we made, we shifted our whole operation to building a rocket that would go all the way to the moon after it took off from Cape Canaveral. It would then go into orbit around the moon rather than landing on the moon, and from orbit around the moon would descend to the surface of the moon and perform its exploration. Then it would return to its orbit around the moon and come back home.

Well, that was a very comprehensive analysis job. It was probably more deep-seated than the kind of job one would do for building an airplane or a dam because there were so many variables involved. What you do is break it down into pieces: the launch site, the launch vehicles, the spacecraft, the lunar module, and worldwide tracking networks, for example. Then, once these pieces are broken down, you assign them to one organization or another. They, in turn, take those small pieces, like the rocket, and break it down into engines or structures or guidance equipment. And this breakdown, or "tree," is the really tough part about managing.

In the Apollo program, it was decided that three NASA centers would do the work. One was Huntsville, where Dr. von Braun and his team built the rocket. The other was Houston, where Dr. Gerous and his team built the spacecraft and controlled the flight operations. The third was Cape Canaveral, were Dr. Debries and his team did the launching and the preparation of the rocket.

Those three centers were pieces, and they could break their pieces down into about 10 or 20 major industrial contractors who would build pieces of the rocket. And then each of those industrial contractors would break them down into maybe 20 to 30 or 50 subcontractors—and they, in turn, would break them down into perhaps 300,000 or 400,000 pieces, each of which would end up being the job of one person. But you need to be sure that the pieces come together at the right time, and that they work when put together. Management science helps with that. The total number of people who worked on the Apollo was about 400,000 to 500,000, all working toward a single objective. But that objective was clear when President Kennedy said, "I want to land a man on the moon and have him safely returned to the earth, and to do so within the decade." Of course, Congress set aside $20 billion. So you had cost, performance, and schedule, and you knew what the job was in one simple sentence. It took a lot of effort to make that happen.

·1·

Street Networks

The underlying theme of management science is finding the best method for solving some problem — what mathematicians call the **optimal solution.** In some cases, it may be to finish a job as quickly as possible. In other situations, the goal might be to maximize profit or minimize cost. What we define as "optimal" depends on the nature of the goal.

Whatever the aim, the optimal solution is directly linked to the goal of efficiency in managing a complex activity. Complex activities arise in many places, and management-science techniques are not limited to multi-million-dollar corporations. They also turn out to be useful in planning the public services we depend on. In a city, for example, such techniques might benefit the public works department, the sanitation authority, or the parking department.

Let's concentrate on the parking department. Most cities and many small towns have parking meters that must be regularly checked for parking violations or emptied of coins. We will use an imaginary town to show how management-science techniques can help to make parking control more efficient.

EULER CIRCUITS

The street map in Figure 1.1 is typical of many towns across the United States, with streets, residential blocks, and a village green. There are almost unlimited possibilities for parking-control routes. Our job, or that of the commissioner of parking, is to find the optimal solution, the most efficient route for the parking-control officers, who travel on foot, checking the meters in an area.

The commissioner has two goals in mind: (1) The parking-control officer must cover all the sidewalks that have parking meters without retracing steps any more than is necessary;

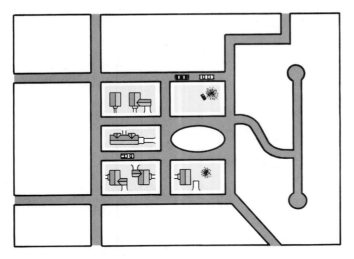

Figure 1.1 A street map for part of a town.

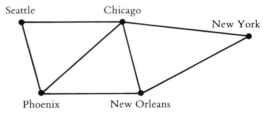

Figure 1.2 The edges of this graph show nonstop routes that an airline might offer.

(2) the route should end at the same point it began, perhaps where the officer's patrol car is parked.

We can think of this problem in terms of a structure called a graph, one of the many mathematical models that can help to simplify complex problems. A **graph** is a finite set of dots and connecting links. The dots are called **vertices** (a single dot is called a **vertex**), and the links are called **edges.** Each edge must connect two different vertices. A **path** is a connected sequence of edges showing a route on the graph that starts at a vertex and ends at a vertex; a path is usually described by naming in turn the vertices visited in traversing it. A path that starts and ends at the same vertex is called a **circuit.** A graph can represent our city map, a communications network, or even a system of air routes.

EXAMPLE: Parts of a Graph. In Figure 1.2, the vertices represent cities and the edges represent nonstop airline routes between them. We see that there is a nonstop flight between New Orleans and Phoenix but no such flight between Seattle and New York.

There are several paths that describe how a person might travel with this airline from Seattle to New York: the path that seems most direct is Seattle, Chicago, New York, but Seattle, Phoenix, New Orleans, New York is also such a path. An example of a circuit is Phoenix, New Orleans, Chicago, Phoenix. It is a circuit because the path starts and ends at the same vertex. In this chapter we are especially interested in circuits just as we are in real life; most of us end our day in the same location where we start it — at home!

In the case of parking control, we can use a graph to represent the whole territory to be patrolled: we can think of each street intersection as a vertex and each sidewalk that contains a meter as an edge, as in Figure 1.3. Notice in Figure 1.3b that the street separating the blocks is not explicitly represented; it has been shrunk to nothing. In effect, we are simplifying our problem by ignoring any distance traveled in crossing streets at the street corners.

The numbered sequence of edges in Figure 1.4a shows one circuit that covers all the meters (note that it is a circuit because its path returns to its starting point). But Figure 1.4b shows another solution that is better because its circuit covers every edge (sidewalk) exactly once. In Figure 1.4b there is no reuse of an edge already covered, or *deadheading* (a term borrowed from shipping, which means making a return trip without a load). Circuits that cover every edge only once are called **Euler**

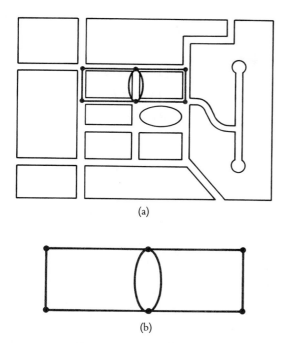

(a)

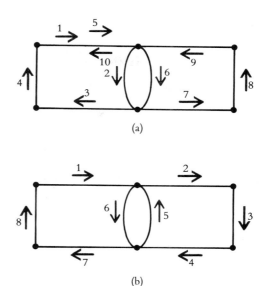

(a)

(b)

Figure 1.4 (a) A circuit and (b) an Euler circuit.

(b)

Figure 1.3 (a) A graph superimposed upon a street map. The edges show which sidewalks have parking meters. (b) The same graph enlarged.

circuits, after the great eighteenth-century mathematician Leonhard Euler (pronounced oy'lur), who first studied them (see Spotlight 1.1, p. 10). Figure 1.4b shows an Euler circuit.

Euler was the founder of the theory of graphs. One of his first discoveries was that some graphs have no Euler circuits at all. For example, in the graph in Figure 1.5b, it would be impossible to start at one point and cover all the edges without retracing some steps: if we try to start a circuit at the leftmost vertex, we discover that once we have left the vertex, we have "used up" the only edge meeting it. We have no way to return to our starting point except to reuse that edge. But this is not allowed in an Euler circuit. If we try to start a circuit at one of the other two vertices, we likewise can't complete it to form an Euler circuit.

Since we are interested in finding circuits, and Euler circuits are the most efficient circuits, we will want to know how to find Euler circuits. If a graph has no Euler circuit, we will want to develop the next best circuits, those having minimum deadheading. These topics are the focus of the rest of this chapter.

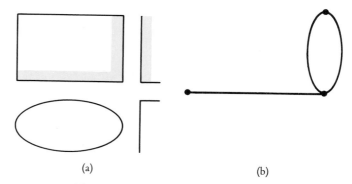

(a)   (b)

Figure 1.5 (a) The three shaded sidewalks cannot be covered by an Euler circuit. (b) The graph of the shaded sidewalks in part (a).

FINDING EULER CIRCUITS

Now that we know what an Euler circuit is, we have two obvious questions to ask about Euler circuits:

1. Is there a way to tell by calculation, not by trial and error, if a particular graph has an Euler circuit?

2. Is there a method, other than trial and error, for finding an Euler circuit when one exists?

Euler answered these questions in 1735 by using the concepts of valence and connectedness. The **valence** of a vertex in a graph is the number of edges meeting at the point. Figure 1.6 illustrates the concept of valence, with vertices *A* and *D* having valence 3, vertex *B* having valence 2, and vertex *C* having valence 0. (Isolated vertices such as vertex *C* are an annoyance in Euler circuit theory. Because they don't occur in typical applications, we henceforth assume that our graphs have no vertices of valence 0.)

A graph is said to be **connected** if for any pair of its vertices there is at least one path of one or more edges connecting the two vertices. Given a graph, if we can find even one pair of vertices not connected by a path, then we say that the graph is not connected. For

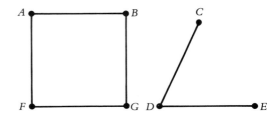

Figure 1.7 A nonconnected graph.

example, the graph in Figure 1.7 is not connected because we are unable to join *A* to *D* with a path of edges. However, the graph does consist of two "pieces" or connected components, one containing the vertices *A, B, F,* and *G,* the other containing *C, D,* and *E.* A connected graph will contain a *single* connected component.

We can now state Euler's theorem, his simple answer to the problem of detecting when a graph G has an Euler circuit:

1. If G is connected and has all valences even, then G has an Euler circuit.

2. Conversely, if G has an Euler circuit, then G must be connected and all its valences must be even numbers.

In the optional section entitled "Proving Euler's Theorem," you will find an outline of a proof of the theorem.

Once we know there is an Euler circuit in a certain graph, how do we find it? The set of rules Euler gave us is of theoretical interest and could be of practical interest if we wanted to program a computer to find Euler circuits mechanically. However, it turns out that most human beings, with a little practice, can find Euler circuits by trial and error and a little ingenuity — even in fairly complex graphs. In Spotlight 1.2 (p. 14) there is a discussion of the different contributions made by humans and computers to problem solving.

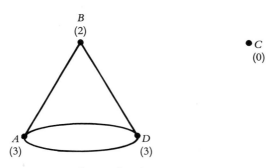

Figure 1.6 Valences of vertices.

EXAMPLE: Finding an Euler Circuit.
Check the valences of the vertices of the graph in Figure 1.8a to verify that the graph does have an Euler circuit. Now try to find an Euler circuit for that graph. You can start at any vertex. Sometimes it is helpful to "use up" all the edges in one section of the graph before moving on to another section, but make sure you leave a way to return to the starting vertex. When you are done, compare your solution with the Euler circuit given in Figure 1.8b. If your path covers each edge exactly once and returns to its original vertex (is a circuit), then it is an Euler circuit, even if it is not the same as the one we give.

Armed with Euler's theorem, we can now take another look at the parking-meter problem first presented in Figure 1.3. In terms of our new vocabulary, the key question becomes this: Is there an Euler circuit in the graph? We can answer this question by checking connec-
tedness and valence, the two conditions in Euler's theorem. First, we observe that the graph is connected, so the first condition is satisfied. Next, we determine the valence, odd or even, of each vertex. Because each vertex does have even valence, the graph meets the second condition, too. The parking commissioner, knowing that the graph has an Euler circuit, can now be assured of designing an efficient patrol route, such as the one in Figure 1.4b.

Optional PROVING EULER'S THEOREM

We begin by defining a **simple circuit** as a graph that we can draw entirely, starting and ending at the same vertex, without revisiting any vertex except the starting point. In Figure 1.9 we show some examples. A simple circuit looks like a polygon, although its edges need not be straight. Clearly a simple circuit has an Euler circuit and is a connected, even-valent graph. We plan to prove Euler's theorem by showing that a graph is built from simple circuits if and only if it has an Euler circuit and

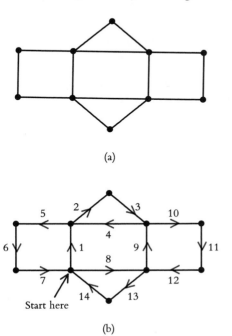

Figure 1.8 (a) A graph having (b) an Euler circuit.

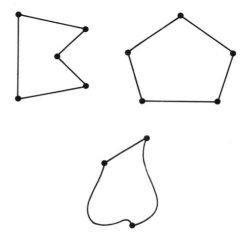

Figure 1.9 Simple circuits.

SPOTLIGHT 1.1 Leonhard Euler

Portrait of Leonhard Euler.

Although born in Switzerland, Leonhard Euler (1707–1783) spent a large part of his life in St. Petersburg, Russia. He was extremely prolific, publishing over 500 works in his lifetime. His collected works comprise 70 volumes. Euler made major contributions to many areas of mathematics, including algebra, geometry, and calculus. A contemporary claimed that Euler could calculate effortlessly, "just as men breathe, as eagles sustain themselves in the air."

Human interest stories about Euler have been handed down through two centuries. He was extremely fond of children and had thirteen of his own, of whom only five survived childhood. It is said that he often wrote difficult mathematical works with a child or two in his lap. He was a prodigy at doing complex mathematical calculations under less than ideal conditions and continued to do them even after he became totally blind later in life. His blindness diminished neither the quantity nor the quality of his output. Throughout his life, he was able to mentally calculate in a short time what would have taken ordinary mathematicians hours of pencil-and-paper work.

Euler's mathematical mind found new mathematics in everyday life. In the old German town of Königsberg, people frequently tried to take a Sunday stroll whose route crossed each of the seven bridges in the town exactly once. Euler analyzed this local pastime using what are now known as Euler circuits.

Euler divided his working life between the St. Petersburg Academy in Russia and the Berlin Academy. He moved to Berlin largely because of difficult political conditions in Russia. When asked in Berlin why he spoke so little, Euler replied, "Madam, I come from a country where if you speak, you are hanged."

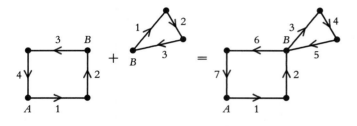

Figure 1.10 Joining two simple circuits.

also if and only if it is connected and even-valent.

We first show that if a graph is built from simple circuits, then it has an Euler circuit. If we put two simple circuits together so that they touch at one or more vertices but not along whole edges, we can build an extended Euler circuit out of the individual Euler circuits. In Figure 1.10 we can begin the Euler circuit for the first circuit at vertex A, interrupt ourselves when we get to vertex B, where the two circuits meet, go around the second circuit, return to vertex B, and then finish the Euler circuit for the first circuit at vertex A. If we hang yet another simple circuit on this graph, we can similarly extend the Euler circuit of the two-circuit graph to an Euler circuit for the three-circuit graph in Figure 1.11. Notice that the four-sided simple circuit is added at two vertices, C and D, not along edges.

This process could be continued if more simple circuits were added. Thus, if a graph is built out of simple circuits, it has an Euler circuit. We also see that if a graph is built in this way, it is connected and even-valent, since the valence of every vertex is twice the number of original circuits to which that vertex belongs.

Having shown one direction of the "if and only if" statements, we next need to look at the other, more difficult, direction. Start with a graph having an Euler circuit; we use it to help us find the simple circuits we need. Begin at any vertex and follow the Euler circuit until it revisits a vertex. The portion of the Euler circuit between the visits is a simple circuit. Cut the simple circuit out of the graph. Now find some vertex in the remaining portion of the graph and pick up the Euler circuit there.

Follow the Euler circuit as before and cut out another simple circuit. As long as some edges have not yet been cut out in forming simple circuits, there will be a vertex and a portion of the Euler circuit that you can follow. Since the Euler circuit contained each edge exactly once, each and every edge will eventually find its way into one of the simple circuits.

Lastly, we show that a connected, even-valent graph can always be built out of simple circuits. Start a path at any vertex and continue it without reusing any edge; stop when you revisit a vertex. There is only one circumstance that would force you to stop before you could make a revisit: going down a dead-end edge (i.e., an edge ending in a one-valent vertex). But since there are no odd valences, you will eventually revisit. When you do, you have found a simple circuit. Cut the simple circuit out of the graph. This leaves a graph that is also even-valent, since the only vertices whose valence has been affected have had their valences reduced by two. So we can repeat the previous process, obtaining a second simple circuit. Continuing, we can totally decompose the original graph into simple circuits. Since the original graph was connected, the simple circuits "hang together." ●

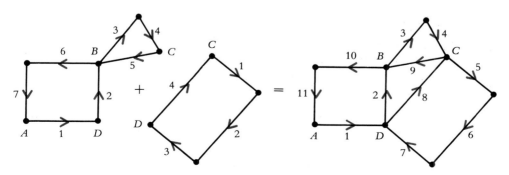

Figure 1.11 Adding a third simple circuit.

CIRCUITS WITH REUSED EDGES

Now let's see what Euler's theorem tells us about the three-block neighborhood with parking meters represented by dots in Figure 1.12a. Figure 1.12b shows the corresponding graph. (Notice that the sidewalk with no meters is not represented by any edge in the graph. We only use edges to represent sidewalks along which the officer must walk.) This graph has two odd valences, so Euler's theorem tells us that there is no Euler circuit for this graph.

Since we must reuse some edges in this graph if we are to cover all edges in a circuit, for efficiency we need to keep the total length of reused edges to a minimum. This type of problem, in which we are interested in minimizing the length of a circuit by carefully choosing which edges to retrace, is often called the **Chinese postman problem**: like parking-control routes, mail routes need to be efficient. The problem was first studied by the Chinese mathematician Meigu Guan in 1962, hence the name. Although the Euler circuit theory doesn't deal directly with reused edges or edges of different lengths, we can extend the theory to help solve the Chinese postman problem.

In a realistic Chinese postman problem, we need to take account of the lengths of the sidewalks, streets, or whatever the edges represent because we want to minimize the total length of the reused edges. However, we will simplify things at the start by supposing that all edges represent the same length. (This is often called the *simplified* Chinese postman problem.) In this case, we need only count reused edges and need not measure their lengths. Thus, we want to find a circuit that covers each edge and that has the minimal number of reuses of edges already covered.

To illustrate the procedure we are going to develop, consider the graph of Figure 1.13a, the same graph as in Figure 1.12b but with labeled vertices. It has no Euler circuit, but there is a circuit that has only one reuse of an edge *(CG)*, namely, *ABCDHGCGFBFEA*. Let's draw this circuit so that when edge *CG* is about to be reused, we install a new, extra edge in the graph for the circuit to use. By duplicating edge *CG*, we can avoid reusing the edge. To duplicate an edge, we must add an edge that joins the two vertices that are already joined by the edge we want to duplicate. We have now created the graph of Figure 1.13b. In this graph the original circuit can be traced as an Euler circuit, using the new edge when needed; the circuit is shown in Figure 1.13c. Our theory will be based on using this idea in reverse, as follows:

1. Take the given graph and add edges, duplicating existing ones until you arrive at a graph that is connected and even-valent. We call this process **eulerizing** a graph because the graph

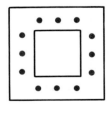
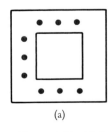
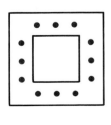

(a)

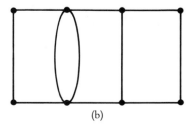

(b)

Figure 1.12 (a) A street network and (b) its graphic representation.

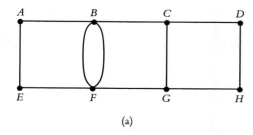

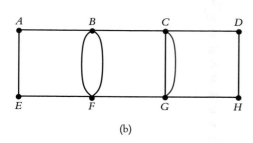

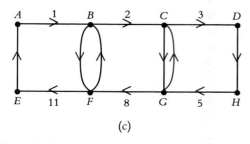

Figure 1.13 Making a circuit by reusing an edge.

we produce will have an Euler circuit. (In our graphs, the edges we add are in color, and thus can be distinguished from the original edges, which are black. You may want to use some system to help you remember which edges are original and which are duplicates.)

2. Find an Euler circuit on the eulerized graph.

3. "Squeeze" this Euler circuit from the eulerized graph onto the original graph by reusing an edge of the original each time the circuit on the eulerized graph uses an added edge.

EXAMPLE: Eulerizing a Graph. When we eulerize a graph, first we locate the vertices with odd valence. The graph in Figure 1.14a has two, B and C. Next, at each such vertex, we add one end of an edge, matching the new edge up with an existing edge in the original graph. Figure 1.14b shows one way to eulerize the graph. Note that B and C have even valence in the second graph. After eulerization, each vertex has even valence. Figure 1.14c shows an Euler circuit on the eulerized graph: simply follow the edges in numerical order and in the direction of the arrows, beginning and ending at vertex A. The final step, in Figure 1.14d, is to squeeze our Euler circuit into the original graph. There are two reuses of previously covered edges. Notice that each

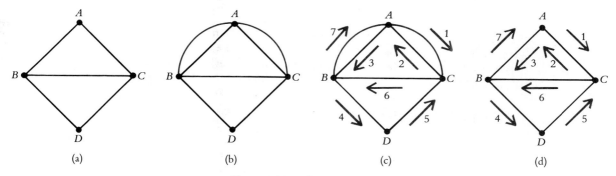

Figure 1.14 Eulerizing a graph.

SPOTLIGHT 1.2 The Human Aspect of Problem Solving

Thomas Magnanti, professor of operations research and management, heads the department of management science at MIT's Sloan School of Management. Here are some of his observations:

Thomas Magnanti, department of management science, Sloan School of Management, MIT.

In applying Euler circuits and other management-science tools and techniques, it's very important to recognize that you need data in order to be effective. In the case of Euler circuits, you must have information on the underlying patterns of the street network. You also need information concerning the demand patterns — what the flow of traffic is like, what the parking situation is, and so on.

But in addition to that, I think one should recognize that the underlying algorithms — the procedures for solving the Euler circuits — are typically only a very small part of a management-science approach to problem solving. We want to be able to collect information and data, and we want to be able to use them effectively. In many cases, collecting, storing, and updating the information is as much of a challenge to mathematicians and computer scientists as are the solution methods themselves. Often, a large-scale data base underlies the actual application of Euler circuits or other management-science techniques.

Typically, a management-science approach has several different ingredients. One is just structuring the problem — understanding that the problem is an Euler-circuit problem or a related management-science problem. After that, one has to develop the solution methods.

But one should also recognize that you don't just push a button and get the answer. In using these underlying mathematical tools, we never want to lose sight of our common sense, of understanding, intuition, and judgment. The computer provides certain kinds of insights. It deals with some of the combinatorial complexities of these problems very nicely. But a model such as an Euler circuit can never capture the full essence of a decision-making problem.

Typically, when we solve the mathematical problem, we see that it doesn't quite correspond to the real problem we want to solve. So we make modifications in the underlying model. It is an interactive approach, using the best of what computers and mathematics have to offer and the best of what we, as human beings, with our own decision-making capabilities, have to offer.

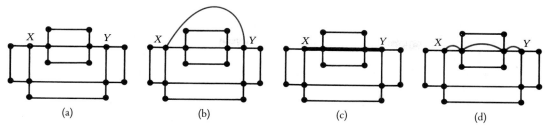

Figure 1.15 Eulerizing when the vertices are more than one edge apart.

reuse of an edge corresponds to an added edge. This is generally true in this type of problem: *the number of reuses of edges equals the number of edges added during eulerization.*

When adding new edges to eulerize a graph, it is desirable to add only edges that are duplicates of existing edges. Adding an edge that doesn't duplicate an existing edge is like saying, for example, that the street map shows a street when in fact it does not. In Figure 1.15a, we need to make X and Y even. Adding one long edge from X to Y (Figure 1.15b) might seem like an attractive idea, but this would not duplicate only one existing edge. (Remember, to duplicate an edge means to add an edge that joins two vertices that are already joined by the edge to be duplicated.)

If we add this long edge and then try to squeeze an Euler circuit on this new graph back into the old graph, this long edge won't count for just one reused edge. This is because the alternative to using the long edge, which we can't use during squeezing, is not one reused edge but a whole series stretching from X to Y (the heavy line in Figure 1.15c). If we don't add the long edge but instead follow the rule for duplicating existing edges, we'll be able to count (in Figure 1.15d) the edges that will be reused according to how many duplicate edges we added (three in this case).

Now that we have learned to eulerize, the next step is to try to get the best eulerization we can. It turns out that there are many ways to eulerize a graph.

EXAMPLE: A Better Eulerization. In Figure 1.16a, we begin with the same graph we used in Figure 1.14, but we eulerize it in a different way—by adding only one edge (see Figure 1.16b). Figure 1.16c shows an Euler circuit on the eulerized graph, and in Figure

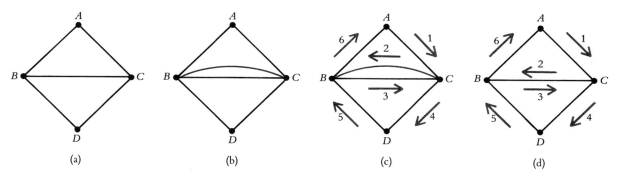

Figure 1.16 A better eulerization of Figure 1.14.

1.16d we see how it is squeezed onto the original graph. There is one reuse of an edge, because we added one edge during eulerization.

The solution in Figure 1.16 is better than the solution in Figure 1.14 because one reuse is better than two. These examples suggest the following addition to our solution procedure: *try to find the eulerization with the smallest number of added edges.* This extra requirement makes the problem both more interesting and more difficult. For large graphs, the best eulerization may not be obvious. We can try out a few and pick the best among the ones we find, but there may be an even better one that our haphazard search does not turn up.

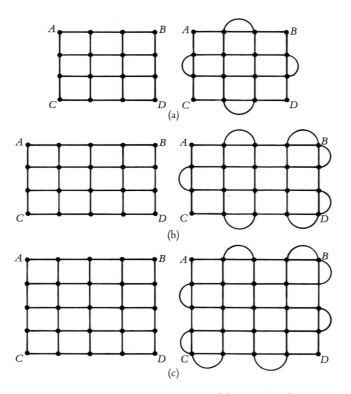

Figure 1.17 Eulerizations of three rectangular networks.

A systematic procedure for finding the best eulerization does exist, but it is complicated. There is an especially easy technique for eulerizing rectangular street networks. We will look at rectangular networks first, then we will comment on how to approach nonrectangular ones.

Many street networks are composed of a series of rectangular blocks that form a large rectangle *m* blocks high by *n* blocks wide. Examples of rectangular street networks are shown in Figure 1.17. The graph on the right of each pair shows the best eulerization for the rectangular street network on the left. There appear to be three different eulerization patterns, depending upon whether in the original graph the rectangle height (*m* blocks) and width (*n* blocks) are odd or even numbers. In Figure 1.17a, both lengths are 3, both odd; in Figure 1.17b, one length is odd (3) and one is even (4); in Figure 1.17c, both lengths are 4, an even number.

Although the patterns appear different, one technique can be used to create all of them. The technique can be thought of as involving four "edge walkers" who work in pairs. To start out, the pairs of walkers are positioned at diagonally opposite corners of the original large rectangle. In all the uneulerized graphs of Figure 1.17 (left column), the walkers start at the upper left and lower right corners, labeled *A* and *D*. Each walker walks along one side of the large rectangle alternately duplicating and not duplicating the edges on that side; the walker finishes when the next corner of the large rectangle is reached. In Figure 1.17, the walkers finish upon reaching either the upper right corner, *B*, or the lower left corner, *C*. If the number of edges in the length and width of the large rectangle are both odd as in Figure 1.17a, or both even as in Figure 1.17c, then all four walkers start out by not duplicating the first edge, then duplicating the second edge, and so on. If the large rectangle is odd by even in dimension as in Figure 1.17b, then one pair of walkers, in our example the ones starting at *D,* switches so that the first edge is duplicated, the second not, and so on.

In a street network that is not rectangular, the eulerization process starts by locating all the vertices with odd valence and then pairing these vertices with each other and finding the length of the shortest path between each pair, since each edge on those paths will be duplicated. The idea is to make the pairings cleverly so that the sum of the lengths of those paths is the smallest it can be. With a little practice, most people can find the best or nearly best eulerization using only this idea, trial and error, and a little ingenuity. For those interested in a further discussion, there is the fol-

lowing optional section of "Finding Good Eulerizations." On a large eulerized graph we know that there is an Euler circuit, but finding one is not necessarily an easy job.

Optional **FINDING GOOD EULERIZATIONS**

Suppose we want a perfect procedure for eulerizing a graph. What theoretical ideas and methods could we use to build such a tool?

One building block we could use is a method for finding the shortest path between two given vertices of a graph. For example, let us focus on vertices *X* and *Y* in Figure 1.18a;

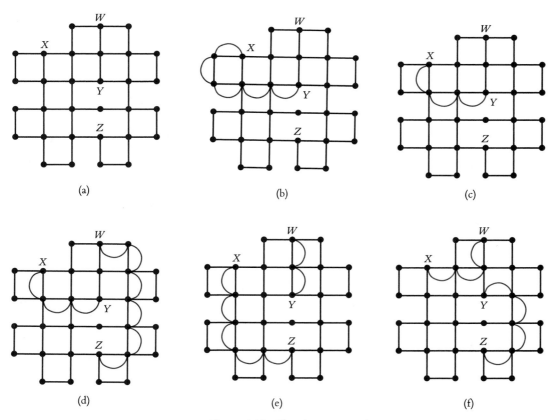

Figure 1.18 Choosing among eulerizations.

both have odd valence. We can connect them with a pattern of duplicate edges, as in Figure 1.18b. The cost of this is the length of the path we duplicated from X to Y. A shorter path from X to Y, such as the one shown in Figure 1.18c, would be better. Fortunately, the *shortest-path problem* has been well studied, and we have many good procedures for solving it exactly, even in large, complex graphs. These procedures are discussed in some of the suggested readings given at the end of this chapter but are beyond the scope of this text.

But there is more to eulerizing the graph in Figure 1.18a than dealing with X and Y: notice that we have odd valences at Z and W. Should we connect X and Y with a path and then connect Z and W, as in Figure 1.18d? Or should we connect X to Z and Y to W, as in Figure 1.18e? Another alternative is to use connections X to W and Y to Z, as in Figure 1.18f. It turns out that the alternatives in both Figure 1.18e and 1.18f are preferable to the one in Figure 1.18d. At the start, it is often not clear which alternatives are best. The problem is how to pair up vertices for connection to get a set of paths whose total length is minimal. This problem, called the *matching problem,* has also been studied and solved. As with the shortest-path problem, refer to the suggested readings for further details. •

Each of these problems has its own special requirements that may call for modifications in the theory. For example, in the case of garbage collection, the edges of our graph will represent streets, not sidewalks. If some of the streets are one-way, we need to put arrows on the corresponding edges, resulting in a directed graph, or **digraph.** The circuits we seek will have to obey these arrows. In the case of salt spreaders and snowplows, each lane of a street needs to be modeled as a directed edge, as shown in Figure 1.19. Note that the arrows on the map and digraph are not in color because these arrows denote restrictions in traversal possibilities, not parts of circuits.

Like salt spreaders, street-sweeping trucks can travel in only one lane at a time and need to obey the direction of traffic. Street sweepers, however, have an additional complication: parked cars. It is very difficult to clean the street if cars are parked along the curb. Yet for overall efficiency, those who are responsible for routing street sweepers want to interfere with parking as little as possible. The common solution is to post signs specifying times when parking is prohibited, such as Thursday between 8 a.m. and 2 p.m. Because the parking-time factor is a constraint on street sweeping, it is important not only to find an Euler circuit, or a circuit with very few duplications, but a

CIRCUITS WITH MORE COMPLICATIONS

Euler circuits and eulerizing have many more practical applications than just checking parking meters. Almost any time services must be delivered along streets or roads, our theory can make the job more efficient. Examples include collecting garbage, salting icy roads, plowing snow, inspecting railroad tracks, and reading electric meters (see Spotlight 1.3, p. 20).

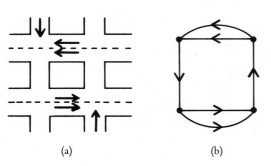

(a)   (b)

Figure 1.19 (a) Salt-spreading route, where each east-west street has a traffic lane for each direction, and (b) an appropriate digraph model.

circuit that can be completed in the time available. Once again, the theory can be modified to handle this constraint.

Finally, because towns and cities of any size will have more than one street sweeper, parking officer, or garbage truck, a single best route will not suffice. Instead, they will have to divide the territory into multiple routes. The general goal is to find optimal solutions while taking into account traffic direction, number of lanes, time restrictions, and divided routes. (See Figure 1.20.)

Management science makes all this possible. For example, a pilot study done in the 1970s in New York City showed that applying these techniques to street sweepers in just one district could save about $30,000 per year. With 57 sanitation districts in New York, this would amount to a savings of more than $1.5 million in a single year. In addition, the same principles could be extended to garbage collection, parking control, and other services carried out on street networks.

But New York City never adopted this plan. Because city services take place in a political context, several other factors come into play. For example, union leaders try to protect the jobs of city workers, bureaucrats might try to keep their departmental budgets high, and elected politicians rarely want to be accused of cutting the jobs of their constituents. Thus political obstacles can overrule management science.

Despite the complications of real-world problems, management-science principles provide ways to understand them by using graphs as models. We can reason about the graphs and then return to the real-world problem with a workable solution. The results we get can have a lasting effect on the efficiency and economic well-being of any organization or community.

(a)

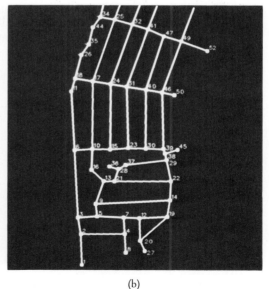

(b)

Figure 1.20 (a) An aerial view of street networks in Seattle, WA. [Peter Arnold, Inc.] Today, finding optimal routes within complex street networks is often done with sophisticated computer-based color graphics systems. (b) A computer-generated street network. [Courtesy of Sidney B. Bowne & Son, Consulting Engineers.]

SPOTLIGHT 1.3 Israel Electric Company Reduces Meter-Reading Task

The Beersheba branch of Israel's major electric company wanted to make the job of meter reading more efficient. When the branch managers decided to minimize the number of people required to read the electric meters in the houses of one particular neighborhood, they set a precedent by applying management science. Formerly, each person's route had been worked out by trial and error and intuition, with no help from mathematics. The whole job required 24 people, each doing a part of the neighborhood in a five-hour shift.

Finding a more efficient way of doing the work sounds a lot like the Chinese postman problem, but there are two important differences. First, the neighborhood was big enough to negate any possibility of having only one route assigned to one person. It was necessary to find a number of routes that, taken together, covered all the edges (sidewalks). Second, a meter reader who was done with a route was allowed to return home directly. Thus, there was no reason for the individual routes to return to their starting points: routes could be paths instead of circuits.

The Beersheba researchers started by doing a partial eulerization in which each odd valence was converted to an even valence, except for two vertices that remained odd. In such a partially eulerized graph, it is possible to find a path, not a circuit, that covers every edge and that starts at one odd-valent vertex and ends at the other.

Next, they chopped the path into five-hour chunks by stepping through the path and marking each time five hours' worth of edges were covered. By following this procedure, researchers managed to cover the neighborhood with 15 five-hour routes, a 40% reduction of the original 24 five-hour routes. Altogether, these routes involve a total of 4338 minutes of walking time, of which 41 minutes (less than 1%) is deadheading.

REVIEW VOCABULARY

Chinese postman problem The problem of finding a circuit on a graph that covers every edge of the graph at least once and that has the shortest possible length.

Circuit A path that starts and ends at the same vertex.

Connected graph A graph is connected if it is possible to reach any vertex from any specified starting vertex by traversing edges.

Digraph A graph in which each edge has an arrow indicating the direction of the edge. Such directed edges are appropriate when the relationship is "one-sided" rather than symmetric (e.g., one-way streets as opposed to regular streets).

Edge A link joining two vertices in a graph.

Euler circuit A circuit that traverses each edge of a graph exactly once.

Eulerizing Adding new edges to a graph so as to make a graph that possesses an Euler circuit.

Graph A mathematical structure in which points (called vertices) are used to represent things of interest, and in which links (called edges) are used to connect vertices, denoting that the connected vertices have a certain relationship.

Management science A discipline in which mathematical methods are applied to management problems in pursuit of optimal solutions that cannot readily be obtained by common sense.

Operations research Another name for management science.

Optimal solution When a problem has various solutions that can be ranked in preference order (perhaps according to some numerical measure of "goodness"), the optimal solution is the best-ranking solution.

Path A connected sequence of edges in a graph.

Simple circuit A circuit in which every vertex has a valence equal to two.

Valence (of a vertex) The number of edges touching that vertex.

Vertex A point in a graph where one or more edges end.

SUGGESTED READINGS

Beltrami, Edward J.: *Models for Public Systems Analysis,* Academic Press, New York, 1977. Section 5.4 deals with material similar to this chapter. The rest of the book gives a nice selection of applications of mathematics to plant location, manpower scheduling, providing emergency services, and other public service areas. The mathematics is somewhat more advanced than in this chapter.

Malkevitch, Joseph, and Walter Meyer: *Graphs, Models, and Finite Mathematics,* Prentice-Hall, Englewood Cliffs, N.J., 1974. An introductory text, which includes much the same material as in this chapter, but with a little more detail. An algorithm for finding Euler circuits is given.

The following three references discuss the shortest-path and matching problems in depth; they are suitable for advanced students and faculty.

Bogart, Kenneth P.: *Introductory Combinatorics,* Wiley, New York, 1986.

Roberts, Fred S.: *Applied Combinatorics,* Prentice-Hall, Englewood Cliffs, N.J., 1984.

Tucker, Alan: *Applied Combinatorics,* 2d ed., Wiley, New York, 1984.

EXERCISES

1. In the graph below, the vertices represent cities and the edges represent roads connecting them. What are the valences of the vertices in this graph? What might the valence of city E be showing about the geography?

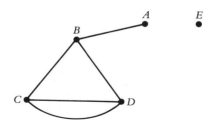

2. What are the valences of the vertices in this graph?

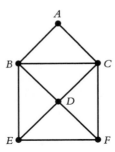

3. Which of these graphs are connected?

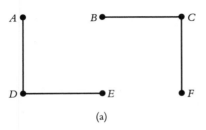

(a)

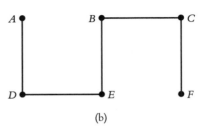

(b)

4. In the graphs below, the vertices represent cities and the edges represent roads connecting them. In which graphs could a person located in city A choose any other city and then find a sequence of roads to get from A to that other city?

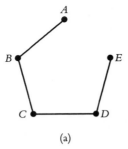

(a)

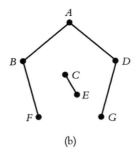

(b)

5. Draw a graph to represent five cities A, B, C, D, and E having one road connecting each pair of cities in the following list (if a pair of cities appears twice, there are two roads): (A, B), (A, C), (B, C), (B, C), (C, E), (D, E). What is the valence of each vertex in your graph? What is a real-world consequence of whether or not the graph is connected?

6. Draw a graph to represent six cities A, B, C, D, E, and F having one road connecting each pair of cities in the following list (if a pair of cities appears twice, there are two roads): *(A, B), (B, C), (D, E), (D, F)*. What is the valence of each vertex in your graph? What is a real-world consequence of whether or not the graph is connected?

7. A postal worker is supposed to deliver mail on all streets represented by edges in the graph below by traversing each edge exactly once. The first day the worker traverses the numbered edges in the order shown in (a), but the supervisor is not satisfied — why? The second day the worker follows the path indicated in (b), and the worker is unhappy — why? Is the original job description realistic — why?

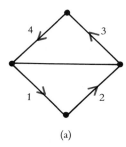

(a)

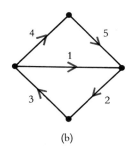

(b)

8. Examine the paths represented by the numbered sequences of edges in the figures below. Determine whether each path is a circuit. If it is a circuit, determine if it is an Euler circuit.

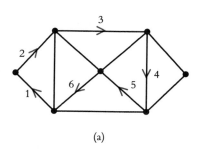

(a)

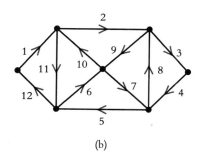

(b)

9. Which graphs in the figures below have Euler circuits? In the ones that do, find the Euler circuits by numbering the edges in the order the Euler circuit uses them. For the ones that don't, explain why no Euler circuit is possible.

(a)

(b)

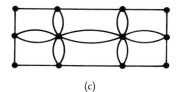

(c)

10. Each graph below represents the sidewalks to be cleaned in a fancy garden (one pass over a sidewalk will clean it). Can the cleaning be done using an Euler circuit? If so, show the circuit by numbering the edges in the order the Euler circuit uses them. If not, explain why no Euler circuit is possible.

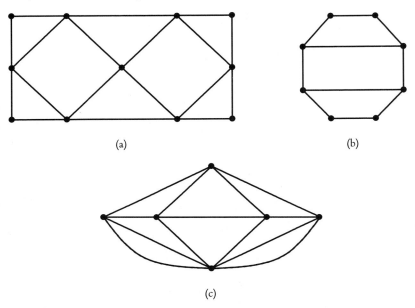

(a)   (b)

(c)

11. In the graph below, we see a territory for a parking-control officer that has no Euler circuit. Which sidewalk (edge) could be dropped in order to enable us to find an Euler circuit?

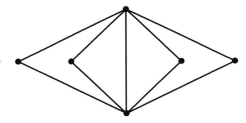

12. In the graph below, we see a territory for a parking-control officer that has no Euler circuit. Which sidewalk (edge) could be dropped in order to find an Euler circuit?

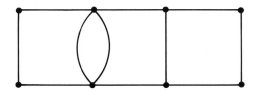

13. A college campus has a central square with sidewalks arranged like those in the graph below. Show how all the sidewalks can be traversed in one circuit; your circuit will have to repeat some edges.

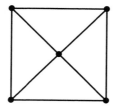

14. In the graph of the figure below, add one or more edges to produce a graph that has an Euler circuit.

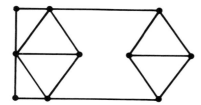

15. a. For the street network below, draw the graph that would be useful for finding an efficient route for checking parking meters. (Hint: notice that not every sidewalk has a meter; see Figure 1.12 in the text.)

b. For the same street network, draw the graph that would be useful for routing a garbage truck. Assume that all streets are two-way and that passing once down a street suffices to collect from both sides.

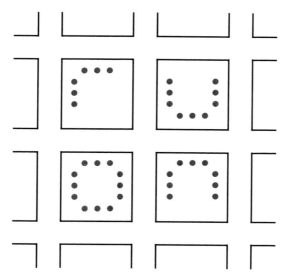

16. a. For the street network below, draw the graph that would be useful for finding an efficient route for checking parking meters. (Hint: notice that not every sidewalk has a meter; see Figure 1.12 in the text.)

b. For the same street network, draw the graph that would be useful for routing a garbage truck. Assume that all streets are two-way and that passing once down a street suffices to collect from both sides.

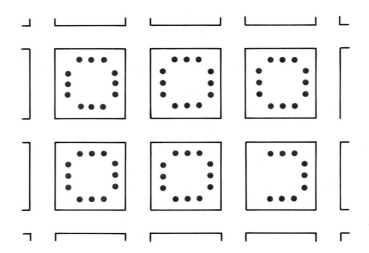

17. Squeeze the circuit shown in graph (a) onto graph (b). Show your answers by numbered arrows on the edges.

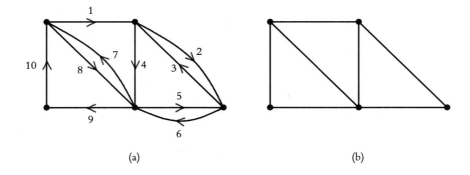

(a)   (b)

18. Squeeze the circuit shown in graph (a) onto graph (b). Show your answers by numbered arrows on the edges.

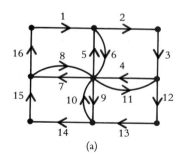 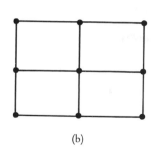

(a)   (b)

19. Find an Euler circuit on the eulerized graph (b) of the following figure. Use it to find a circuit on the original graph (a) that covers all edges and only reuses edges five times.

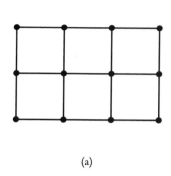 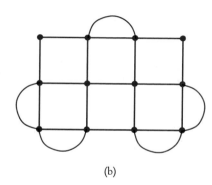

(a)   (b)

20. Find an Euler circuit on the eulerized graph (b) of the following figure. Use it to find a circuit on the original graph (a) that covers all edges and only reuses edges four times.

 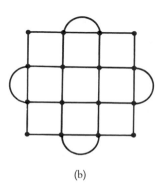

(a)   (b)

21. Can you find an eulerization with 7 added edges for a 2- by 5-block rectangular street network? Can you do better than 7?

22. Can you find an eulerization with 6 added edges for a 3- by 5-block rectangular street network? Can you do better than 6?

23. Can you find an eulerization with 9 added edges for a 3- by 6-block rectangular street network? Can you do better than 9?

24. Can you find an eulerization with 10 added edges for a 4- by 6-block rectangular street network? Can you do better than 10?

●**25.** Find a circuit in the graph below that covers every edge and has as few reuses as possible. See optional section "Proving Euler's Theorem" for hints.

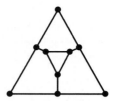

●**26.** Find a circuit in the graph below that covers every edge and has as few reuses as possible. See optional section "Proving Euler's Theorem" for hints.

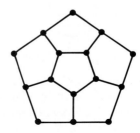

27. In the figure below, all blocks are 1000 by 1000 feet, except for the middle column of blocks, which are 1000 by 4000 feet. Can you find an eulerization in which the total length of all duplicated edges is 8000?

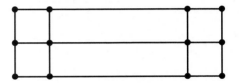

● Optional exercise.

28. In the figure below, all blocks are 1000 by 1000 feet, except for the middle column of blocks, which are 1000 by 4000 feet. Can you find an eulerization in which the total length of all duplicated edges is 8000?

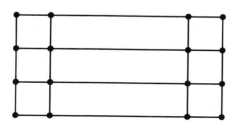

29. Eulerize these rectangular street networks using the same patterns that would be used by the "edge walkers" described in the text.

   a. A 5 × 5 rectangle

   b. A 5 × 4 rectangle

30. Eulerize these rectangular street networks using the same patterns that would be used by the "edge walkers" described in the text.

   a. A 6 × 5 rectangle

   b. A 6 × 6 rectangle

● **31.** Find good eulerizations for these graphs: try to use as few duplicated edges as you can. See the optional section "Finding Good Eulerizations" for hints.

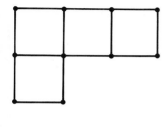  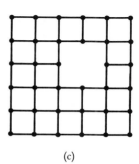

     (a)   (b)   (c)

● Optional exercise.

●**32.** Find good eulerizations for these graphs: try to use as few duplicated edges as you can. See the optional section "Finding Good Eulerizations" for hints.

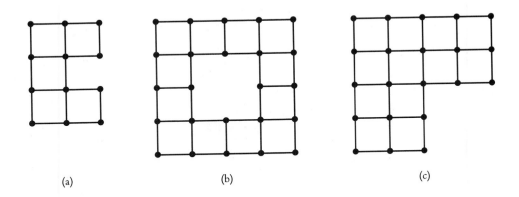

(a)   (b)   (c)

33. A graph G represents a street network to be traveled by a postal worker who must traverse every street twice, once for each side of the street. In graph G, the edges represent sidewalks. Does such a graph always have an Euler circuit? Explain your answer.

34. Suppose for a certain graph that it is possible to disconnect it by removing one edge. Explain why such a graph (before the edge is removed) must have at least one odd valence. (Hint: show that it cannot have an Euler circuit.)

35. If an arbitrary graph G has n vertices with odd valences, then show that n must be an even number (Hint: Since every edge meets two vertices, is the sum of the valences odd or even?)

36. If an arbitrary graph G has n vertices with odd valences, then show that any eulerization of G must have at least $n/2$ added edges. (Hints: Use result from Exercise 35. What is the smallest number of edges that can extend from one vertex to another?)

● Optional exercise.

·2·

Visiting Vertices

In the last chapter, we saw that it is relatively easy to determine if there is a circuit traversing the edges of a graph exactly once — for example, a route for street sweepers that covers the streets in a section of a city exactly once. However, the situation changes radically if we make an apparently innocuous change in the problem: When is it possible to find a route along distinct edges of a graph that visits each *vertex* once and only once in a simple circuit? For example, the wiggly line in Figure 2.1a shows a circuit we can take to tour that graph, visiting each vertex once and only once. This tour can be written *ABDGIHFECA*. Note that another way of writing the same circuit would be *EFHIGDBACE*. A different circuit visiting each vertex once and only once would be *CDBIGFEHAC* (Figure 2.1b). Do not be confused because *C* is written twice when we write down this list of vertices. We can think of the circuit as starting at any of its vertices, but we do start and end at the same vertex.

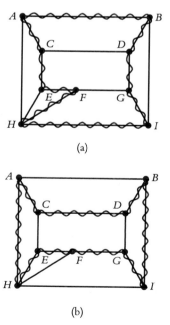

(a)

(b)

Figure 2.1 Wiggled edges illustrate Hamiltonian circuits.

HAMILTONIAN CIRCUITS

A tour, like the ones marked by wiggly edges in Figure 2.1, that starts at a vertex of a graph and visits each vertex once and only once, returning to where it started, is called a **Hamiltonian circuit.** The concept is named for the Irish mathematician William Rowan Hamilton (1805–1865), who was one of the first to study it. [We now know that the concept was discovered somewhat earlier by Thomas Kirkman (1806–1895), a British minister with a penchant for mathematics.] This problem is typical of the many new problems mathematicians create as a consequence of their professional training. In the situation here, motivated by our success in solving the problem of traversing all the edges of a graph, we will investigate visiting all the vertices of a graph.

The concepts of Euler and Hamiltonian circuits are similar in that both forbid reuse: the Euler circuit of edges, the Hamiltonian circuit of vertices. However, it is far more difficult to determine which connected graphs admit a Hamiltonian circuit than to determine which connected graphs have Euler circuits. As we saw in Chapter 1, looking at the valences of vertices tells us if a connected graph has an Euler circuit, but we have no such simple method for telling whether or not a graph has a Hamiltonian circuit. Some special classes of graphs are known to have Hamiltonian circuits, and some special classes of graphs are known to lack them. For example, the graphs in the collection of graphs in Figure 2.2 cannot have a Hamiltonian circuit if $m \neq n$, since if a Hamiltonian circuit existed it would have to alternately include vertices on the left and right of the figure. Unfortunately, it is unlikely that a method will ever be found to easily determine whether or not an arbitrarily chosen graph has a Hamiltonian circuit.

The Hamiltonian circuit problem and the Euler circuit problem are both examples of

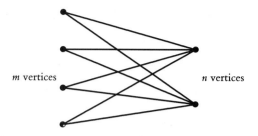

Figure 2.2 A collection of graphs that can have no Hamiltonian circuit when the number of vertices m is not equal to the number of vertices n. The case $m = 4$ and $n = 2$ is shown.

graph-theory problems. Although we posed the Hamiltonian circuit problem merely as a variant of another graph-theory problem with many applications (i.e., the Euler circuit problem), the Hamiltonian circuit problem itself has many applications. This is not unusual in mathematics. Often the mathematics used to solve some particular real-world problem leads to new mathematics that suggests applications to other real-world situations.

Suppose inspections or deliveries need to be made at each vertex (rather than along each edge) of a graph. An "efficient" tour of the graph would be a route that started and ended at the same vertex and passed through all the vertices without reuse, or repetition; that is, the route would be a Hamiltonian circuit. Such routes would be useful for inspecting traffic signals, or for delivering mail to drop-off boxes, which hold heavy loads of mail so urban postal carriers do not have to carry it long distances. There are many similar examples, but rather than pursue problems involving Hamiltonian circuits in general graphs, we will study instead a more important class of related problems.

EXAMPLE: Vacation Planning. Let's imagine that you are a college student studying in Chicago. During spring break you and a group of friends have decided to take a car trip to visit other friends in Minneapolis, Cleveland, and St. Louis. There are many choices as to the order of visiting the cities and returning to Chicago, but you want to design a route that minimizes the distance you have to travel. Presumably, you also want a route that cuts costs, and you know that minimizing distance will minimize the cost of gasoline for the trip. (Similar problems with different complications would arise for railroad or airplane trips.)

Imagine now that the local automobile club has provided you with the intercity driving distances between Chicago, Minneapolis, Cleveland, and St. Louis. We can construct a graph model with this information, representing each city by a vertex and the legs of the journey between the cities by edges joining the vertices. To complete the model, we add a number called a **weight** to each graph edge, as in Figure 2.3. In this example, the weights represent the distances between the cities, each of which corresponds to one of the endpoints of the edges in the graph. (In other examples the weight might represent a cost, time, or profit.) We want to find a minimal-cost tour

that starts and ends in Chicago and visits each other city once. Using our earlier terminology, what we wish to find is a **minimum-cost Hamiltonian circuit** — a Hamiltonian circuit with the lowest possible sum of the weights of its edges.

How can we determine which Hamiltonian circuit has minimum cost? There is a conceptually easy **algorithm,** or mechanical step-by-step process, for solving this problem:

1. Generate all Hamiltonian tours (starting from Chicago).

2. Add up the distances on the edges of each tour.

3. Choose the tour of minimum distance.

Steps 2 and 3 of the algorithm are straightforward. Thus, we need worry only about step 1, generating the Hamiltonian circuits in a systematic way. To find the Hamiltonian tours, we will use the **method of trees,** as follows. Starting from Chicago, we can choose any of the three cities to visit after leaving Chicago. The first stage of the enumeration tree is shown in Figure 2.4. If Minneapolis is chosen as the first city to visit, then there are two possible cities to visit next, namely, Cleveland and St. Louis. The possible branchings of the tree at this stage are shown in Figure 2.5. In this second stage, however, for each choice of first city we visited, there are two choices from this city to the second city visited. This would lead to the diagram in Figure 2.6.

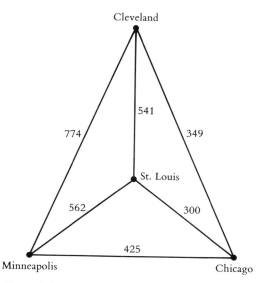

Figure 2.3 Road mileages between four cities.

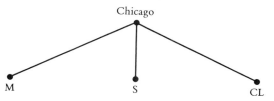

Figure 2.4 First stage in finding vacation-planning routes.

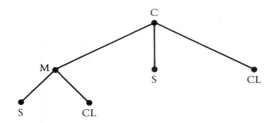

Figure 2.5 Part of second stage in finding vacation-planning routes.

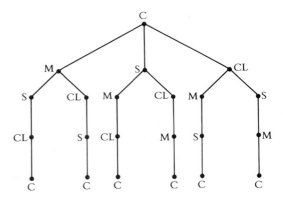

Figure 2.7 Completed tree enumeration of routes for vacation-planning problem.

Having chosen the order of the first two cities to visit, and knowing that no revisits (reuses) can occur in a Hamiltonian circuit, there is only one choice left for the next city. From this city we return to Chicago. The complete tree diagram showing the third and fourth stages for these routes is given in Figure 2.7. Notice, however, that because we can traverse a circular tour in either of two directions, the paths enumerated in the tree diagram of Figure 2.7 do *not* correspond to different Hamiltonian circuits. For example, the leftmost path (C–M–S–CL–C) and the rightmost path (C–CL–S–M–C) represent the same Hamiltonian circuit. For this problem, therefore, we find three distinct Hamiltonian circuits among the six paths in the tree diagram. These three distinct Hamiltonian circuits are shown in Figure 2.8.

Note that in generating the Hamiltonian circuits we disregard the distances involved. We are concerned only with the different patterns of carrying out the visits. But which

route is the optimal route? Adding up the distances on the edges to get each tour's length shows that the optimal tour is Chicago, Minneapolis, St. Louis, Cleveland, Chicago. The length of this tour is 1877 miles. (See Figure 2.8.)

The method of trees is not always as easy to use as our example suggests. Instead of doing our analysis for four cities, consider the general case of *n* cities. The graph model similar to that in Figure 2.3 would consist of a weighted graph with *n* vertices, with every pair of vertices joined by an edge. Such a graph is called **complete** because all possible edges are included in the graph. A complete graph with five vertices is illustrated in Figure 2.9.

Fundamental Principle of Counting

How many Hamiltonian circuits are in a complete graph of *n* vertices? We can solve this problem by using the same type of analysis that emerged in the enumeration tree. The method uses the **fundamental principle of counting,** a procedure for counting outcomes in multistage processes. Using this method we

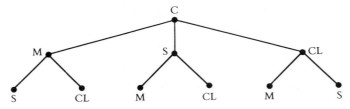

Figure 2.6 Complete second stage in finding vacation-planning routes.

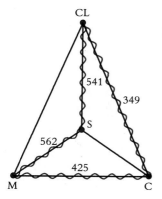

Tour length 1877

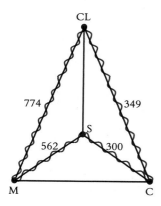

Tour length 1985

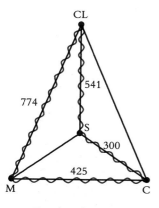

Tour length 2040

Figure 2.8 The three distinct Hamiltonian circuits for the vacation-planning problem of Figure 2.3.

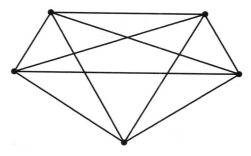

Figure 2.9 The complete graph with five vertices. Every pair of vertices is joined by an edge.

can count how many patterns occur in a situation by looking at the number of ways the component parts can occur. For example, if Jack has 9 shirts and 4 pair of trousers, he can wear 36 shirt-pants outfits. Each shirt can be worn with any of the pants. (This can be verified by drawing a tree diagram, but such a diagram is cumbersome for big numbers.)

In general, if there are a ways of choosing one thing, b ways of choosing a second after the first is chosen, . . . , and z ways of choosing the last item after the earlier choices, then the total number of choice patterns is $a \times b \times c \times \cdots \times z$.

EXAMPLES: Counting. Here are some other examples of how to use the fundamental principle of counting:

1. In a restaurant there are 4 kinds of soup, 12 entrees, 6 desserts, and 3 drinks. How many different meals can a patron chose from? The four choices can be made in 4, 12, 6, and 3 ways, respectively. Hence, applying the fundamental principle of counting, there are $4 \times 12 \times 6 \times 3 = 864$ possible meals.

2. In a state lottery a contestant gets to pick a four-digit number that contains no zero followed by an uppercase or lowercase letter. How many such patterns are there? Each of the four digits can be chosen in 9 ways (that is, 1,

2, . . . , 9), and the letter can be chosen in 52 ways (that is, A, B, . . . , Z, a, b, . . . , z). Hence there are $9 \times 9 \times 9 \times 9 \times 52 = 341,172$ possible patterns.

3. A corporation is planning a musical logo consisting of four different notes from the scale C, D, E, F, G, A, and B. How many logos are there to chose from? The first note can be chosen in 7 ways, but since reuse is not allowed, the next note can be chosen in only 6 ways. The remaining two notes can be chosen in 5 and 4 ways, respectively. Using the fundamental principle of counting, $7 \times 6 \times 5 \times 4 = 840$ possible musical logos.

Returning to the problem of enumerating Hamiltonian circuits for the complete graph with n vertices, the city visited first after the home city can be chosen in $n - 1$ ways, the next city in $n - 2$ ways, and so on, until only one choice remains. Using the fundamental principle of counting, there are $(n - 1)! = (n - 1)(n - 2) \cdots \times 3 \times 2 \times 1$ routes. [The exclamation mark in "$(n - 1)!$" is read "factorial" and is a shorthand notation for the product $(n - 1)(n - 2) \cdots \times 3 \times 2 \times 1$. For example, $5! = 5 \times 4 \times 3 \times 2 \times 1 = 120$.]

Pairs of routes correspond to the same Hamiltonian circuit because one route can be obtained from the other by traversing the cities in reverse order. Thus, although there are $(n - 1)!$ possible routes, there are only half as many, or $(n - 1)!/2$, Hamiltonian circuits. Now, if we have only a few cities to visit, $(n - 1)!/2$ Hamiltonian circuits can be listed and examined in a reasonable amount of time. To analyze a 6 city problem it would require generation of $(6 - 1)!/2 = 5!/2 = 120/2 = 60$ tours. But for, say, 25 cities, $24!/2$ is approximately 3×10^{23}. Even if these tours could be generated at the rate of 1 million a second, it would take 10 billion years to generate them all. Since large vacation-planning problems would take so long to solve using the

method we have described, this method, despite its conceptual ease, is sometimes referred to as the **brute force method.**

TRAVELING SALESMAN PROBLEM (TSP)

If the only benefit were saving money and time in vacation planning, the difficulty of finding a minimum-cost Hamiltonian circuit in a complete graph with n vertices for large values of n would not be of great concern. However, the problem we are discussing is one of the most common problems in operations research. It is usually called the **traveling salesman problem (TSP)** because of its early formulation: determine the trip of minimum cost that a salesperson can make to visit the cities in a sales territory, starting and ending the trip in the same city.

There are many situations that require solving a TSP:

1. A lobster fisherman has set out traps at various locations and wishes to pick up his catch.

2. The telephone company wishes to pick up the coins from its pay telephone booths.

3. The electric (or gas) company needs to design a route for its meter readers.

4. In drilling holes in a series of plates, the drill press operator (perhaps a robot!) must drill the holes in a predetermined order.

The meaning of cost can vary from one formulation of TSP to another. We may measure cost as distance, airplane ticket prices, time, or any other factor that is to be optimized.

In many situations, the TSP arises as a subproblem of a more complicated problem. For example, a supermarket chain may have a very large number of stores to be served from a single large warehouse. If there are fewer trucks than stores, the stores must be grouped into clusters so that one truck serves each cluster. If we then solve the TSP for every truck, we can minimize total costs for the supermarket chain. Similar vehicle-routing problems for dial-a-ride services and for delivering children to their schools or camps often involve solving the TSP as a subproblem.

STRATEGIES FOR SOLVING THE TRAVELING SALESMAN PROBLEM

Because the traveling salesman problem arises so often in situations whose complete graphs would be very large, we must find a better method than the brute force method we have described. We need to look at our original problem in Figure 2.3 and try to find an alternative algorithm for solving it. Recall that our goal is to find the minimum-cost Hamiltonian circuit.

Nearest-Neighbor Algorithm

Let's try a new approach: starting from Chicago, first visit the nearest city, then visit the nearest city that has not already been visited. We return to the start city when no other choice is available. This approach is called the **nearest-neighbor algorithm.** Applying it to the TSP in Figure 2.3 quickly leads to the tour of Chicago, St. Louis, Cleveland, Minneapolis, and Chicago, with a length of 2040 miles. The nearest-neighbor algorithm is an example of a **greedy algorithm,** because at each stage a best (greedy) choice, based on an appropriate criterion, is made. Unfortunately, as we saw earlier, this is not the optimal tour. Making the best choice at each stage may not yield the best

"global" solution. However, even for a large TSP, one can always find a nearest-neighbor route quickly.

Figure 2.10 gives another example to illustrate the ease of applying the nearest-neighbor algorithm, this time to a weighted complete graph with five vertices. Starting at vertex A, we get the tour $ADECBA$ (cost 2800) (Figure 2.10a). Note that the nearest-neighbor algorithm starting at vertex B yields the tour $BCA-DEB$ (cost 3050) (Figure 2.10b).

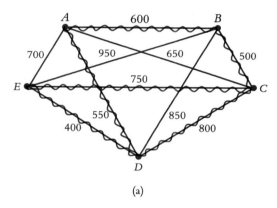

(a)

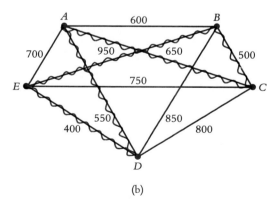

(b)

Figure 2.10 (a) A weighted complete graph with five vertices that illustrates the use of the nearest-neighbor algorithm (starting at A) and the sorted-edges algorithm. (b) TSP tour generated by the nearest-neighbor algorithm (starting at B).

Sorted-Edges Algorithm

Perhaps some other easy method would yield optimal solutions. We might start by sorting or arranging the edges of the complete graph in order of increasing cost (or equivalently, arrange the intercity distances in order of increasing distance). Then we can select at each stage that edge of least cost that (1) never requires that three edges meet at a vertex (since a Hamiltonian circuit uses up exactly two edges at each vertex), and that (2) never closes up a circular tour that doesn't include all the vertices. This algorithm will be called the **sorted-edges algorithm.**

Applying the sorted-edges algorithm to the TSP in Figure 2.3 yields the tour Chicago, St. Louis, Minneapolis, Cleveland, and Chicago, since the edges chosen would be 300, 349, 562, 774. Again, this solution is not optimal because its length is 1985. Note that this algorithm, like the nearest neighbor, is greedy.

Although the edges selected by applying the sorted-edges method to the example in Figure 2.3 are connected to each other at every stage, this does not always happen. For example, if we apply the sorted-edges algorithm to the graph in Figure 2.10a we build up the tour first with edge ED (400) and then edge BC (500), which do not touch. The edges that are then selected are AD, AB, and EC, giving the circuit EDABCE, which is the same as the nearest-neighbor circuit starting at vertex A.

Although many "quick and dirty" methods for solving the TSP have been suggested and although some methods give an optimal solution in some cases, none of these methods *guarantees* an optimal solution. Surprisingly, most experts believe that no efficient method that guarantees an optimal solution will ever be found (see Spotlight 2.1).

Recently, mathematical researchers have adopted a somewhat different strategy for dealing with TSP problems. If finding a fast

SPOTLIGHT 2.1 NP-Complete Problems

Stephen Cooke, a computer scientist at the University of Toronto, showed in 1971 that certain hard, frustrating problems are equivalently difficult. This class of problems, now referred to as **NP-complete problems,** have the following characteristic: if a "fast" algorithm for solving *one* of these problems could be found, then a fast method would exist for *all* these problems.

In this context, "fast" means that as the size n of the problem grows (the number of cities gives the problem size in the traveling salesman problem), the amount of time needed to solve the problem grows no more rapidly than a polynomial function in n. (A polynomial function has the form $a_k n^k + a_{k-1} n^{k-1} + \ldots + a_1 n + a_0$.) On the other hand, if it could be shown that any problem in the class of NP-complete problems required an exponentially growing (3^n is an example of an exponential function) amount of time to solve it as the problem size increased, then all problems in the NP-complete class would share this characteristic. If some NP-complete problems had fast solutions, it seems likely that at least one such fast solution would have been found by now. It has been known for some time that the traveling salesman problem is an NP-complete problem: for this reason it is generally thought that no "fast" algorithm for an optimal solution for the TSP will ever be found.

algorithm to generate optimal solutions for large problems is unlikely, perhaps one can show that the "quick and dirty" methods, usually called **heuristic algorithms,** come close enough to giving optimal solutions. For example, suppose one could prove that the nearest-neighbor heuristic was never off by more than 25% in the worst case or by more than 15% in the average case. For a medium-size TSP, one would then have to choose whether to spend a lot of time (or money) to find an optimal solu-

tion or to use instead a heuristic algorithm to obtain a fairly good solution. Researchers at AT&T Bell Laboratories have developed many remarkably good heuristic algorithms (see Spotlight 2.2). The best-known guarantee for a heuristic algorithm for a TSP is that it yields a cost that is no worse than 1.5 times the optimal cost. Interestingly, this heuristic algorithm involves solving a Chinese postman problem, for which a "fast" algorithm is known to exist, as part of the algorithm.

SPOTLIGHT 2.2 Solving the Elusive Traveling Salesman Problem

More than 20 years ago, Shen Lin, then a Bell Labs researcher, set out to solve the traveling salesman problem. Although the TSP defies any quick mathematical solution, Shen Lin has discovered a method that works fairly quickly on many practical problems. Today, he applies this knowledge to designing private telephone networks for AT&T's corporate customers (see Spotlight 2.5). Here, Shen Lin recalls his earlier work on the traveling salesman problem:

I started trying to solve the traveling salesman problem back in 1965 and ended in 1972, collaborating with a colleague, Brian Kernahan, in an algorithm that up to this day is considered to be the most efficient practical method of solving this problem.

In 1965, we knew very little about which kind of problems were hard and which were easy. So naively, I went in and thought I could solve this problem. Of course, I couldn't. I published my first paper on computer approximations by using the so-called iterative techniques [repeating some operation over and over] to solve them.

Today, I essentially use a *heuristic method.* The difference between this and the so-called classical method is that exact-solution methods usually take too long when you are trying to solve a hard problem like the traveling salesman problem. Also, they are formulated for problems that are very well defined. But in real life, we frequently find that no problem is so well defined, so we need some method of solving problems that is more flexible. We want something that can get at approximate solutions, which may turn out to be optimum solutions. The method must be powerful enough to solve a variety of problems. We cannot guarantee that the solution is the absolute and best one, because to find the best solution would take billions of years of computation, even by the world's fastest computer.

Within a few seconds, however, I could give you the solution to the traveling salesman problem, say, for a few hundred [geographical] points. And I'm sure no human being could look at the map, connect those points, and achieve the same result. I can't *prove* that it's the best solution that could ever be found, but it's usually quite close — within 1 to 2% of the optimum.

MINIMUM-COST SPANNING TREES

The traveling salesman problem is but one of many graph-theory optimization problems that have grown out of real-world problems in both government and industry. Here is another.

EXAMPLE: Pictaphone Service. Imagine that Pictaphone service is to be set up on an experimental basis between five cities. The graph in Figure 2.11 shows the possible links that might be included in the Pictaphone network, with each edge showing the cost in millions of dollars to create that particular link. To send a Pictaphone message between two cities, a direct communication link is not necessary because it is possible to send a message *indirectly* via another city. Thus, in Figure 2.11, sending a message from A to C could be achieved by sending the message from A to B, from B to E, and from E to C, provided the links AB, BE, and EC are part of the network. We will assume that the cost of relaying a message, compared with the link cost, is so small that we can neglect this amount. The question that concerns us, therefore, is to provide service between any pair of cities in a way that minimizes the total cost of the links.

Our first guess at a solution is to put in the cheapest possible links between cities first, until all cities could send messages to any other city. In our example, if the cheapest links are added until all cities are joined, we obtain the connections shown in Figure 2.12a.

The links were added in the order ED, AD, AE, AB, DC. However, because this graph contains the circuit $ADEA$ (wiggly edges in Figure 2.12b), it has redundant edges: we can still send messages between any pair of cities using relays after omitting the most expensive edge in the circuit—AE. After deleting an edge of a circuit, a message can still be relayed among the cities of the circuit by sending signals the long way around. After AE is deleted, messages from A to E can be sent via D (Figure 2.12c). This suggests a modified algorithm for our problem, **Kruskal's algorithm:** add the links in order of cheapest cost so that no circuits form and so that every vertex belongs to some link added (Figure 2.12d). As in the sorted-edges method for the TSP, the edges

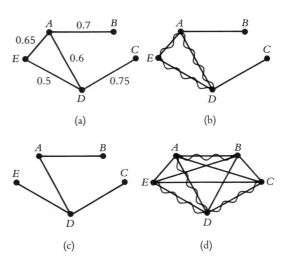

Figure 2.12 (a) Cities are linked in order of increasing cost until all cities are connected up. (b) Circuit in (a) highlighted. (c) Most expensive link in circuit in (a) deleted. (d) Highlighted edges show, as a subgraph of the original graph, those links connecting the cities with minimum cost obtained using Kruskal's algorithm.

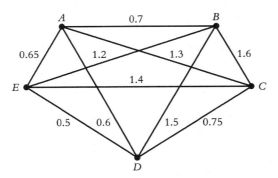

Figure 2.11 Costs of installing Pictaphone service between five cities in millions of dollars

SPOTLIGHT 2.3 Some Reminiscences on Shortest Spanning Trees

Dr. Kruskal has written the following thoughts on the minimum-cost spanning algorithm, which he refers to as shortest spanning trees.

In 1951, I entered the graduate mathematics program at Princeton. Among those active in combinatorial mathematics at Princeton then were Al Tucker, who soon became chairman of the department, Harold Kuhn, and Roger Lyndon. For a while, there was a joint Princeton–Bell Laboratories combinatorics seminar, which met alternately at two locations. My first introduction to Bell Laboratories, where I subsequently have spent several decades, came in this connection.

One day, someone in the combinatorics community handed me a blurry, carbon-copy typescript on flimsy lightweight paper about shortest spanning trees. It was in German, and appeared to be the foreign-language summary of a paper written in some other language. I had the typescript only for a limited period, and then passed it on to someone else. The typescript contained no date, and at the time I had no idea of whether it had been published or where. To this day, I have no idea where the typescript had come from, and why it was circulating. While writing my own paper stimulated by this material, I must have been told that the typescript was (part of?) a translation of a 1926 paper by Otakar Boruvka, for I give an exact citation in my paper. However, I have no memory of who told me this, nor do I remember ever seeing the full paper.

By the way, I regret that the phrase "minimum-spanning tree" has become dominant. What is actually meant is the minimum *length* spanning tree (MST), which is more concisely rendered by *shortest* spanning tree (SST). That is the phrase I used in 1954, and continue to use. If you wish to stick with MST rather than SST, however, at least do not make the common error of misconstruing MST as minimal, rather than minimum, spanning tree. The thing being minimized, length, has a unique minimum under very general conditions. [In this book, the term "minimum-cost spanning tree" is used rather than MST, as is common in the technical mathematics literature.]

The chief point of Boruvka's paper, I believe, was the uniqueness of the SST of a graph if the edges of the graph all have distinct lengths. His proof was based on a construction of the SST that I had difficulty in fully grasping. Although the construction has a certain spare elegance, I decided that it was difficult to follow because it was unnecessarily complicated. That led me to devise some much simpler methods, which are described in my 1956 paper. In 1985 Graham and Hell wrote a full history of the SST and closely related ideas. Well over 100 papers dating back to 1909 are cited, though the earliest paper specifically on the SST is Boruvka's work in 1926.

It is interesting to note that I was very uncertain about whether the material in my simple paper, which occupies only two and a half printed pages and requires few equations, was worth writing up for publication. I was a graduate student in my second year, and did not have a good sense of what was publishable. I asked a friend for advice (I don't remember whom), and fortunately received encouragement. It was my second published paper.

that are added need not be connected to each other until the end.

A subgraph formed in this way will be a **tree;** that is, it will consist of one piece and contain no circuits. It will also include all the vertices of the original graph. A subgraph that is a tree and that contains all the vertices of the original graph is called a **spanning tree** of the original graph.

Finding a **minimum-cost spanning tree,** that is, a spanning tree whose edge weights sum to a minimum value, solves the Pictaphone problem. In Figure 2.13a we have a graph model showing the costs of putting in roads to connect new houses in a suburban land development project. Applying the algorithm we have developed, adding the edges in the order of increasing cost but avoiding the creation of a circuit, yields as a minimum-cost spanning tree the tree indicated by wiggled edges in Figure 2.13b. This tree is the cheapest one that makes it possible to drive between any pair of homes, though the driving distance between some of the homes will be relatively large, since only roads corresponding to wiggled edges will be built. Remember that the weights on the edges of the graph in Figure 2.13a represent the costs of building roads, *not*

the driving distance between the houses. Note that the graph in Figure 2.13a is not a complete graph: edges that correspond to roads that would be economically prohibitive to build have not been shown in the graph model. Also, in Figure 2.13b, the two edges of weight 5 do not become part of the minimum-cost spanning tree because they would create circuits with edges already chosen.

How do we know that the spanning tree found by the algorithm we have developed achieves the minimum possible cost? Although this sounds very plausible, our experience with the traveling salesman problem should suggest caution. Remember that for the TSP, the sorted-edges algorithm, also a greedy algorithm, did not necessarily give an optimal solution! On what basis should we have more faith in our present algorithm?

Kruskal's Algorithm

It turns out that the algorithm described here was first suggested by Joseph Kruskal (AT&T Bell Laboratories) in the mid-1950s to solve a problem in pure mathematics proposed by a Czechoslovakian mathematician. In mathematics it is a surprising but not uncommon

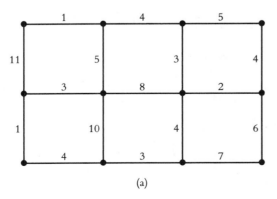

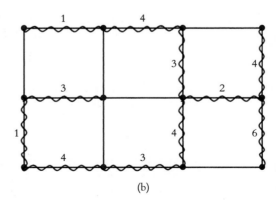

(a)   (b)

Figure 2.13 (a) A graph showing costs for constructing roads between houses. (b) Wiggled edges show a minimum-cost spanning tree for the graph in (a).

SPOTLIGHT 2.4 AT&T Manager Explains How Long-Distance Calls Run Smoothly

Although long-distance calls are now routine, it takes great expertise and careful planning for a company like AT&T to handle its vast amounts of telephone traffic. Rich Wetmore was district manager of AT&T's Communications Network Operations Center in Bedminster, New Jersey, in 1988. Here are his responses to questions about how AT&T handles its huge volumes of long-distance traffic and how it tracks its operations to keep things running smoothly.

How do you make sure that a customer doesn't run into a delayed signal when attempting a long-distance call?

We monitor the performance of our AT&T network by displaying data collected from all over the country on a special wallboard. The wallboard is configured to tell us if a customer's call is not going through because the network doesn't have enough capacity to handle it.

That's when we step in and take control to correct the problem. The typical control we use is to reroute the call. Instead of sending the customer's call directly to its destination, we'll route it via a third city — to someplace else in the country that has the capacity to complete the call.

It would seem that routing via another city would take longer. Is the customer aware of this process?

Routing a call via a third city is entirely transparent [imperceptible] to the customer. I'm an expert about the network, and even when I make a phone call, I have no idea how that individual call was routed. It's transparent both in terms of how far away the other person sounds and in how quickly the telephone call gets set up. With the signaling network we use, it takes milliseconds for switching systems to "talk" to each other to set up a call. So the fact that you are involved in a third switch in some distant city is something you would never know.

You want to be sure to keep costs down while supplying enough service to customers. So how do you balance company benefits with customer benefits?

In terms of making the network efficient, we want to do two things. First, we want our customers to be happy with our service and for all their calls to go through, which means we must build enough capacity in the network to allow that to happen. Second, we want to be efficient for our stockholders and not spend more money than we need to for the network to be at the optimum size.

There are basically two costs in terms of building the network. There is the cost of switching systems and the cost of the circuits that connect the switching systems. Basically, you can use operations-research techniques and mathematics to determine cost trade-offs. It may make sense to build direct routing between two switching systems and use a lot of circuits, or maybe to involve three switching systems, with fewer circuits between the main two, and so on.

finding that ideas used to solve problems with no apparent application often turn out to have many real-world uses. Kruskal's solution to the problem of finding a minimum-cost spanning tree in a graph with weights is a good example of this phenomenon (see Spotlight 2.3, p. 41). Kruskal showed that the greedy algorithm described does yield the minimum answer, and his work led to applications of these and related ideas in designing minimum-cost computer networks, phone connections, and road and railway systems. For additional discussion of operations research in the communications industry (see Spotlight 2.4, p. 43, and Spotlight 2.5).

Although we have mentioned many routing problems in graphs, we have not discussed one of the most obvious: finding the path between two specified distinct vertices with the sum of the weights of the edges in the path as small as possible. (Here there is no need to cover all vertices or to cover all edges.) We have seen that the weights on the edges have many possible interpretations, including time, distance, and cost. Here are some of the many possible applications:

1. Design routes to be used by a fire engine to get to a fire as quickly as possible.

2. Design delivery routes that minimize gasoline use.

3. Design routes to bring soldiers to the front as quickly as possible.

The need to find shortest paths seems natural. Next we shall investigate a situation where finding a longest path is the right tool.

CRITICAL-PATH ANALYSIS

One of the delights of mathematics is its ability to confirm the obvious in certain situations while showing that our intuition is wrong in

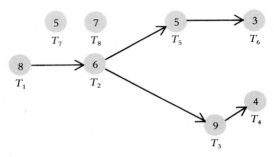

Figure 2.14 Typical order-requirement digraph.

other circumstances. Consider the common problem of scheduling a series of interrelated subtasks that are part of some major undertaking, such as building a skyscraper, installing a major computer system, or landing a man on the moon.

If the tasks cannot be performed in arbitrary sequence or order, we can specify the order in an **order-requirement digraph.** A typical example of an order-requirement digraph is shown in Figure 2.14. There is a vertex in this digraph for each task. If one task must be done immediately before another, we draw a directed edge, or arrow, from the prerequisite task to the subsequent task. The numbers within the circles representing vertices are the times it takes to complete the tasks. In Figure 2.14 there is no arrow from T_1 to T_5 because task T_2 intervenes. Also, T_1, T_7, and T_8 have no tasks that must precede them. Hence, if there are at least three processors (i.e., people or machines) available, tasks T_1, T_7, and T_8 can be worked on simultaneously at the start of the job.

EXAMPLE: Turning a Plane Around. Consider an airplane that carries both freight and passengers. The plane must have its passengers and freight unloaded and new passengers and cargo loaded before it can take off again. Also, the cabin must be cleaned before departure can occur. Thus, the job of "turning

SPOTLIGHT 2.5 Made-to-Order Phone Networks Benefit Large-Scale Users

Shen Lin was executive consultant and head of AT&T Bell Laboratories Network Configuration and Planning Department, Murray Hill, New Jersey. A scientist-turned-salesman for AT&T, he devised a method for solving the traveling salesman problem in 1972 (see Spotlight 2.2). More recently, he developed the "interactive network optimization system" (INOS) for AT&T. Here are his comments on the network:

Before our customers will sign a contract to buy a multimillion dollar network, they usually want someone to tell them about the methodology used for optimizing their networks. So I am frequently asked to present the network design results to our customers, not only to explain how it serves their needs, but also to give them an idea of how the design was worked out.

Here is how it works: Perhaps hundreds of locations in a large organization need to "talk" to each other. The people may need to talk to one another, or their terminals may need to communicate with computers.

For example, an individual may pick up his phone, make a call, and get charged for a regular long-distance call. If the calling volume is rather high, a company can use the Wide Area Telecommunications Service, commonly called WATS, and reduce the cost by 20% or 30%. Another option is to connect locations via private lines. So the problem is how to configure all communications needs into an optimal network that will minimize overall costs.

Let's consider the first step that must be taken: because not every location can be connected to every other location, a set of locations that are close together will come into what is called a "switch," or concentrator. Their calls will come into the switch and be completed thereafter.

So one of the first questions that has to be answered is, Where do you put the switch? This problem is a mathematically difficult problem in itself. Before we know where to put the concentrators, we have to estimate the cost of calls reaching their destination via the switch, and since we don't know where the switches will be located, we must use many different iterations to get at our optimal answer.

In general, a large corporation may generate about 5 million calls per month, and they may be calling from several hundred—or even 2,000—locations, to tens of thousands of destinations. The job of optimization is to group all these together in such a way that the total bill is minimized.

INOS does all the investigation and configuration. It's very computation-intensive. To do a customer network, we run something over 300 to 500 million arithmetical instructions, and that's after months of work collecting the traffic data.

Right now, INOS is moving in the direction of configuring data networks. For example, my terminal has to talk to a computer in Massachusetts at this moment, so I have a data line. People may use a dial-up or data line, and in some cases many, many terminals may be connected to use a common line called a multi drop line. All of this can amount to a great deal of savings to the customer simply by solving the mathematical problem of optimum configuration.

SPOTLIGHT 2.6 Every Moment Counts in Rigorous Airline Scheduling

When people think of airline scheduling, the first thing that comes to mind is how quickly a particular plane can safely reach its destination. But using ground time efficiently is just as important to an airline's timetable as the time spent in flight. Bill Rodenhizer, who was the manager of control operations for Eastern Airlines in Boston in 1988, is considered to be an expert on airplane turnaround time, the process by which an airplane is prepared for almost immediate takeoff once it has landed. He tells us how this well-orchestrated effort works:

> Scheduling, to the airline, is just about the whole ballgame. Everything is scheduled right to the minute. The whole fleet operates on a strict schedule. Each of the departments responsible for turning around an aircraft has an allotted period of time in which to perform its function. Manpower is geared to the amount of ground time scheduled for that aircraft. This would be adjusted during off-weather or bad-weather days or during heavy air-traffic delays.
>
> Most of our aircraft in Boston are scheduled for a 42- to 65-minute ground time. Boston is the end of the line, so it is a "terminating and originating station." In

the plane around" requires the completion of five tasks:

Task A	Unload passengers	13 minutes
Task B	Unload cargo	25 minutes
Task C	Clean cabin	15 minutes
Task D	Load new cargo	22 minutes
Task E	Load new passengers	27 minutes

The order-requirement digraph for the problem of turning an airplane around is shown in Figure 2.15. Because we want to find the earliest completion time, it might seem that finding the shortest path through the digraph (i.e., path BD with length $25 + 22 = 47$) would solve the problem. But this approach shows the danger of ignoring the relationship between the mathematical model (the digraph) and the original problem.

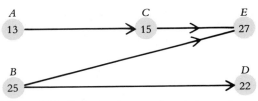

Figure 2.15 Order-requirement digraph for turning an airplane around after landing.

The time required to complete all the tasks must be at least as long as the time necessary to do the tasks on any particular path. This is true since on a particular path, each task must be completed before the next task on that path can be started. Thus, the earliest completion time actually corresponds to the length of the *longest* path. In the airplane example, this earliest completion time is 55 ($13 + 15 + 27$) min-

plain talk, that means almost every aircraft that comes in must be fully unloaded, refueled, serviced, and dispatched within roughly an hour's time.

This is how the process works: in the larger aircraft, it takes passengers roughly 20 minutes to load and 20 minutes to unload. During this period, we will have completely cleaned the aircraft and unloaded the cargo, and the caterers will have taken care of the food. The ramp service may take 20 to 30 minutes to unload the baggage, mail, and cargo from underneath the plane, and it will take the same amount of time to load it up again. We double-crew those aircraft with heavier weights so that the work load will fit the time it takes passengers to load and unload upstairs.

While this has been going on, the fueler has fueled the aircraft. As to repairs, most major maintenance is done during the midnight shift, when all but 20 of Eastern's several hundred aircraft are inactive.

We all work under a very strict time frame. There are four functional departments. If any of the four cannot fit its work into its time frame, then it advises us at the control center, and we adjust the departure time or whatever, so that the other departments can coordinate their activities accordingly.

utes, corresponding to the path *ACE*. We call *ACE* the **critical path** because the times of the tasks on this path determine the earliest completion time.

Suppose it were desirable to speed the turn-around of the plane to below 55 minutes. One way to do this might be to build a second jet-way to help unload passengers. Then, for example, we could unload passengers (task *A*) in 7 minutes instead of 13. However, reducing task *A* to 7 minutes does not reduce the completion time by 6 minutes because in the new digraph (Figure 2.16) *ACE* is no longer the critical (i.e., longest) path. The longest path is now *BE*, which has a length of 52 minutes. Thus, shortening task *A* by 6 minutes only results in a 3-minute saving in completion time. This may mean that building a new jet-way is uneconomical. Note also that shorten-

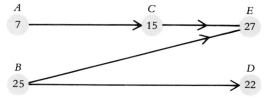

Figure 2.16 Order-requirement digraph for turning an airplane around with reduced times due to construction of new jetway.

ing the time to complete tasks that are not on the original critical path *ACE* will not shorten the completion time at all. Speeding tasks on the critical path will shorten completion time of the job only up to the point where a new critical path is created.

Not all order-requirement digraphs are as simple as the one in Figure 2.15. The order-re-

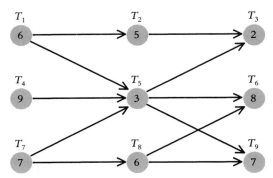

Figure 2.17 An order-requirement digraph with 12 paths to examine how to find the length of the longest path.

quirement digraph in Figure 2.17 has 12 paths, which can be found by exhaustive search. Typical examples of such paths are $T_1T_2T_3$, $T_1T_5T_9$, $T_4T_5T_9$, and $T_7T_5T_3$. (Although we have not discussed them here, fast algorithms for finding longest and shortest paths in graphs are known.) The critical path is $T_7T_8T_6$ (length 21), and the earliest completion time for all nine tasks is time 21.

These examples are typical of many scheduling problems that occur in practice (see Spotlight 2.6, p. 46). Perhaps the most dramatic use of critical-path analysis is in the construction trades. No major new building project is now carried out without first performing a critical-path analysis to ensure that the proper personnel and materials are available at the right times in order to have the project finished as quickly as possible. Many such problems are too large and complicated to be solved without the aid of computers.

The critical-path method was popularized and came into wider use as a consequence of the Apollo project. As we saw in the introduction, this project, which aimed at landing a man on the moon within 10 years of 1960, was one of the most sophisticated projects in planning and scheduling ever attempted. The dramatic success of the project can be attributed partly to the use of critical-path ideas and the related program evaluation and review technique (PERT), which helped keep the project on schedule.

REVIEW VOCABULARY

Algorithm   A step-by-step description of how to solve a problem.

Brute force method   The method that solves the traveling salesman problem (TSP) by enumerating all the Hamiltonian circuits and then selecting the one with minimum cost.

Complete graph   A graph in which every pair of vertices is joined by an edge.

Critical path   The longest path in an order-requirement digraph. The length of this path gives the earliest completion time for all the tasks making up the job consisting of the tasks in the digraph.

Fundamental principle of counting   A method for counting outcomes of multistage processes.

Greedy algorithm   An approach for solving an optimization problem, where at each stage of the algorithm the best (or cheapest) action is taken. Unfortunately, greedy algorithms do not always lead to optimal solutions.

Hamiltonian circuit   A circuit using distinct edges of a graph that starts and ends at a particular vertex of the graph and visits each vertex once and only once. A Hamiltonian circuit can be thought of as starting at any one of its vertices.

Heuristic algorithm   A method of solving an optimization problem that is "fast" but does not guarantee an optimal answer to the problem.

Kruskal's algorithm   An algorithm developed by Joseph Kruskal (AT&T Bell Laboratories) that solves the minimum-cost spanning tree problem by selecting edges in order of increasing cost but so that no edge forms a circuit with edges chosen earlier. It can be proved that this algorithm always produces an optimal solution.

Method of trees   A visual method of carrying out the fundamental principle of counting.

Minimum-cost Hamiltonian circuit   A Hamiltonian circuit in a graph with weights on the edges, for which the sum of the weights of the edges of the Hamiltonian circuit is as small as possible.

Minimum-cost spanning tree   A spanning tree of a weighted connected graph having minimum cost. The cost of a tree is the sum of the weights on the edges of the tree.

Nearest-neighbor algorithm   An algorithm for attempting to solve the TSP that begins at a "home" vertex and visits next that vertex not already visited that can be reached most cheaply. When all other vertices have been visited, the tour returns to home. This method may not give an optimal answer.

NP-complete problems   A collection of problems, which includes the TSP, that appear to be very hard to solve quickly for an optimal solution.

Order-requirement digraph   A directed graph that shows which tasks precede other tasks among the collection of tasks making up a job.

Sorted-edges algorithm   An algorithm for attempting to solve the TSP where the edges added to the circuit being built up are selected in order of increasing cost, but no edge is added that would prevent a Hamiltonian circuit's being formed. These edges must all be connected at the end, but not necessarily at earlier stages. The tour obtained may not have lowest possible cost.

Spanning tree   A subgraph of a connected graph that is a tree and includes all the vertices of the original graph.

Traveling salesman problem (TSP)   The problem of finding a minimum-cost Hamiltonian circuit in a complete graph where each edge has been assigned a cost (or weight).

Tree   A connected graph with no circuits.

Weight   A number assigned to an edge of a graph that can be thought of as a cost, distance, or time associated with that edge.

SUGGESTED READINGS

Burr, Stefan: *The Mathematics of Networks,* American Mathematical Society, Providence, R.I., 1982. A collection of articles dealing with applications of networks.

Lawler, Eugene, J. Lenstra, Rinnoy Kan, and D. Shmoys (eds.): *The Traveling Salesman Problem,* Prentice-Hall, Englewood Cliffs, N.J., 1985. This book includes survey and technical articles on all aspects of the TSP.

Lucas, William, Fred Roberts, and Robert Thrall (eds.): *Discrete and Systems Models,* vol. 3: *Modules in Applied Mathematics,* Springer-Verlag, New York, 1983. Chapter 6, "A Model for Municipal Street Sweeping Operations," by A. Tucker and L. Bodin, describes street sweeping and related models in detail. Other models described in this book detail many recent applications of mathematics.

Roberts, Fred: *Applied Combinatorics,* Prentice-Hall, Englewood Cliffs, N.J., 1984. The chapters in this book on graphs and related network optimization problems are excellent.

——: *Graph Theory and Its Applications to Problems of Society,* SIAM, Philadelphia, 1978. A very readable account of how graph theory is finding a wide variety of applications.

EXERCISES

1. For each of the graphs below and on the following page, write down a Hamiltonian circuit starting at X_1.

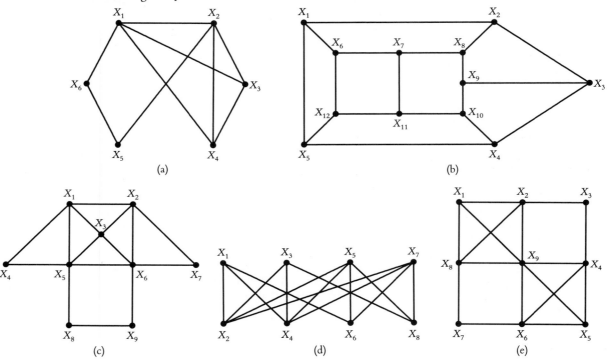

(a)   (b)

(c)   (d)   (e)

2. For each of the graphs below, add wiggly edges to indicate a Hamiltonian circuit.

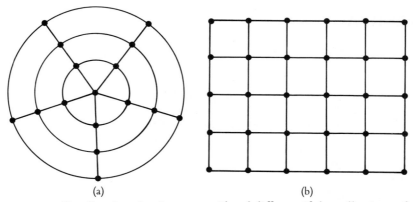

(a)   (b)

3. Suppose two Hamiltonian circuits are considered different if the collections of edges that they use are different. How many other Hamiltonian circuits can you find in the graph in Figure 2.1 different from the two discussed?

4. Explain why the tour *ACEDCBA* is not a Hamiltonian circuit for the following graph. Does this graph have a Hamiltonian circuit?

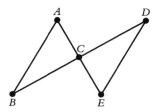

5. Each of the graphs below has no Hamiltonian circuit. Is it possible to add a single new edge to these graphs and obtain a new graph that has a Hamiltonian circuit?

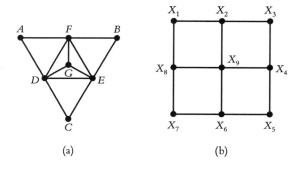

(a)   (b)

6. If the edge X_1X_2 is erased from each of the graphs in Exercise 1, does the resulting graph still have a Hamiltonian circuit?

7. Do the following graphs have Hamiltonian circuits? If not, can you demonstrate why not?

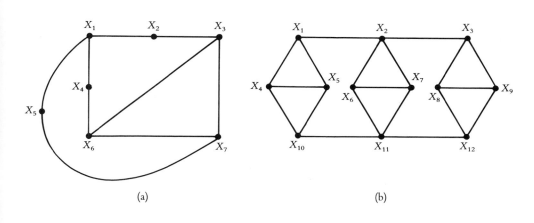

(a)   (b)

8. For each of the graphs below, determine if there is a Hamiltonian circuit.

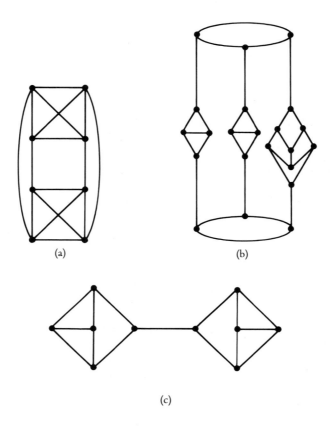

(a)   (b)

(c)

9. a. The graph below is known as a four spokes and three concentric circles graph. What conditions on m and n guarantee that an m spokes and n concentric circles graph has a Hamiltonian circuit? (Assume $m \geq 2$, $n \geq 1$.)

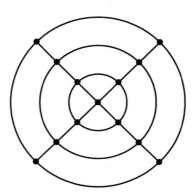

b. The graph below is known as a 3 × 4 grid graph. What conditions on *m* and *n* guarantee that an *m* × *n* grid graph has a Hamiltonian circuit?

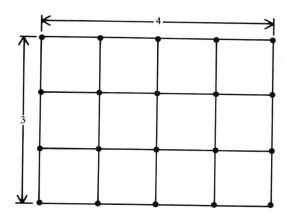

Can you invent a real-world situation in which finding a Hamiltonian circuit in an *m* × *n* grid graph would represent a solution to the problem? If an *m* × *n* grid graph has no Hamiltonian circuit, can you find a tour that repeats a minimum number of vertices and starts and ends at the same vertex?

10. The *n*-dimensional cube is obtained from two copies of an (*n* − 1)-dimensional cube by joining corresponding vertices. (The process is illustrated for the 3-cube and the 4-cube in the following figure.)

Find formulas for the number of vertices and the number of edges of an *n*-cube. Can you show that every *n*-cube has a Hamiltonian circuit? (Hint: show that if you know how to find a Hamiltonian circuit on an (*n* − 1)-cube, then you can use two copies of this to build a Hamiltonian circuit on an *n*-cube.)

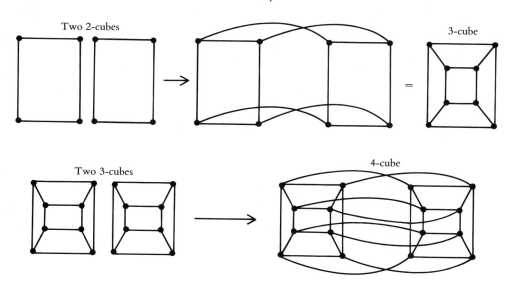

11. To practice your understanding of the concepts of Euler circuits and Hamiltonian circuits, determine for each graph below if there is an Euler circuit and/or a Hamiltonian circuit. If so, write it down.

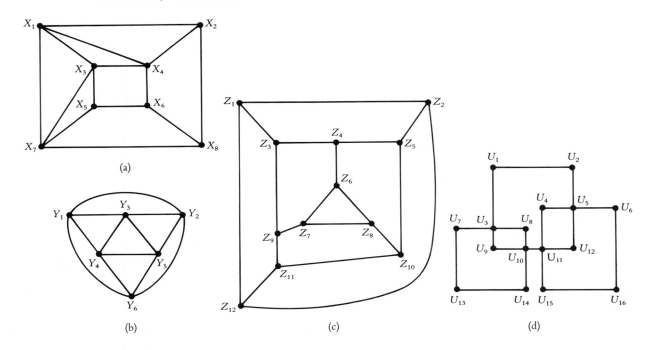

(a)

(b)

(c)

(d)

12. The following figure represents a town where there is a sewer located at each corner (where two or more streets meet). After every thunderstorm, the department of public works wishes to have a truck start at its headquarters (at vertex *H*) and make an inspection of sewer drains to be sure that leaves are not clogging them. Can a route start and end at *H* that visits each corner exactly once? (Assume that all the streets are two-way streets.) Does this problem involve finding an Euler circuit or a Hamiltonian circuit?

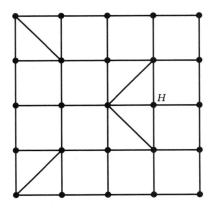

Assume that at equally spaced intervals along the blocks in this graph there are storm sewers that must be inspected after each thunderstorm to see if they are clogged. Is this a Hamiltonian circuit problem, an Euler circuit problem, or a Chinese postman problem? Can you find an optimal tour to do this inspection?

13. Give examples of real-world situations which can be modeled using a graph and for which finding a Hamiltonian circuit in the graph would be of interest.

14. In planning a camping vacation, Sally can fly to Duluth and then back home, choosing from four different airlines for each direction of her journey. On arriving at Duluth, she can choose from nine different outfitters. How many choices does she have for arrangements?

15. In designing a security system for its accounts, a bank asks each customer to choose a five-digit decimal number, all the digits to be distinct and nonzero. How many choices can the customer make?

16. a. For going outside on a cold winter day, Jill can choose from three winter coats, three wool scarfs, four pairs of boots, and three ski hats. How many outfits might her friends see her in?

b. If Jill insists on always wearing her red ski hat, how many outfits might her friends see her in?

17. a. In New York State one type of license plate has three letters followed by a three-digit number. Suppose the digits can be chosen from 0, 1, . . . , 9 except that all three digits being zero is not allowed and any letter from A to Z (repeats allowed) can be chosen. How many plates are possible?

b. Investigate what schemes for license plates are used in your state and determine how many different plates are possible.

18. Draw complete graphs with four, five, and six vertices. How many edges do these graphs have? Can you generalize to n vertices? How many TSP tours would these graphs have? (Tours yielding the same Hamiltonian circuit are considered the same).

19. Calculate the values of 5!, 6!, 7!, 8!, 9!, and 10!. Then find the number of TSP tours in the complete graph with ten vertices.

20. The table below shows the mileage between four cities: Springfield, Ill. (S), Urbana, Ill. (U), Effingham, Ill. (E), and Indianapolis, Ind. (I).

	E	I	S	U
E	–	147	92	79
I	147	–	190	119
S	92	190	–	88
U	79	119	88	–

a. Represent this information by drawing a weighted complete graph on four vertices.

b. Use the weighted graph in part **a** to find the cost of the three distinct Hamiltonian circuits in the graph. (List them starting at U.)

c. Which circuit gives the minimum cost?

d. Would there by any difference in parts **b** and **c** if the start vertex were at I?

e. If one applies the nearest-neighbor method starting at U, what circuit would be obtained? Does the answer change if one applies the nearest-neighbor algorithm starting at S? At E? At I?

f. If one applies the sorted-edges method, what circuit would be obtained? Does one get the optimal answer?

21. After a party at her house, Francine (F) has agreed to drive home Mary (M), Rachel (R), and Constance (C). If the times (in minutes) to drive between her friends' homes are shown below, what route gets Francine back home the quickest?

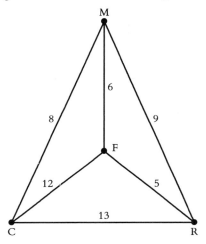

22. A fisherwoman wishes to visit three areas *A, B,* and *C* where she has set nets starting from the location where she moors her boat (*M*). If the times (in minutes) between the locales are given in the figure below, what route to visit the three sites and return to her mooring place would be optimal?

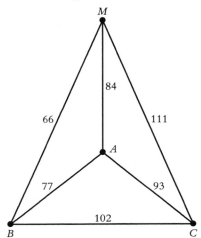

23. For each of the graphs with weights below, apply the nearest-neighbor method (starting at vertex A) and the sorted-edges method to find (it is hoped) a cheap tour.

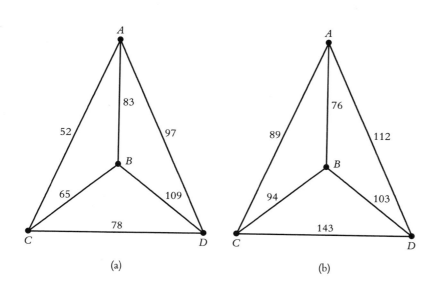

(a)   (b)

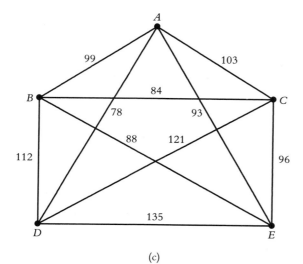

(c)

24. a. For each of the complete graphs that follow, find the costs of the nearest-neighbor tour starting at A and of the tour generated using the sorted-edges algorithm.

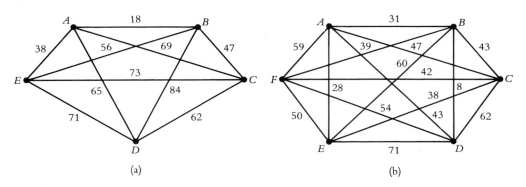

(a)   (b)

b. How many Hamiltonian circuits would have to be examined to find a shortest route for part **a** by the brute force method?

c. Can you invent an algorithm different from the sorted-edges and nearest-neighbor algorithms that is easy to apply for finding TSP solutions? [See Lawler et al., Chapter 2 suggested readings.]

25. An airport limo must take its six passengers to different downtown hotels from the airport. Is this a traveling salesman problem or an Euler circuit problem?

26. a. Solve the six-city TSP shown in the accompanying diagram using the nearest-neighbor algorithm starting at vertex A; starting at vertex B.

b. Apply the sorted-edges method.

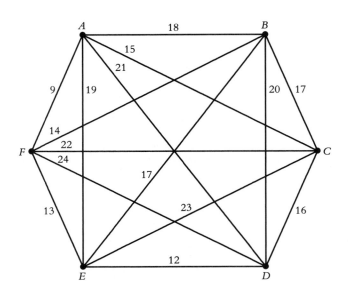

27. Construct an example of a complete graph on five vertices, with distinct weights on the edges for which the nearest-neighbor algorithm starting at a particular vertex and the sorted-edges algorithm yield different solutions for the traveling salesman problem. Can you find a five-vertex complete graph with weights on the edges in which the optimal solution, the nearest-neighbor solution, and the sorted-edges algorithm solution are all different?

28. If the brute force method of solving a 20-city TSP is employed, use a calculator to determine how many Hamiltonian circuits must be examined. How long would it take to determine the minimum-cost tour if the cost of tours could be computed at the rate of 1 billion per second? (Convert your answer to years by seeing how many years are equivalent to a billion seconds!).

29. Which of the graphs below are trees?

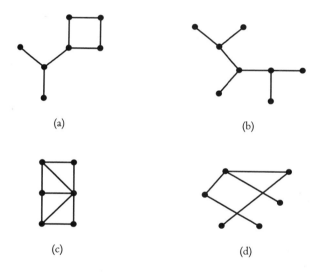

(a)   (b)

(c)   (d)

30. For each of the diagrams below explain why the wiggled edges are not:

    a. a spanning tree

    b. a Hamiltonian circuit

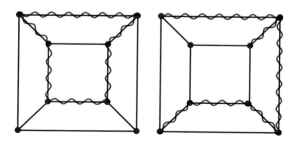

31. Find all the spanning trees in the graphs below.

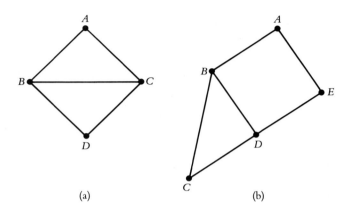

(a)   (b)

32. Use Kruskal's algorithm to find a minimum-cost spanning tree for graphs (a), (b), (c), and (d).

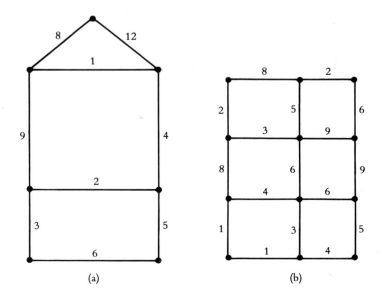

(a)   (b)

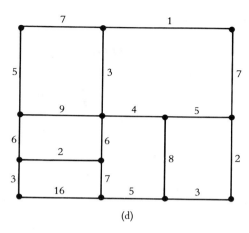

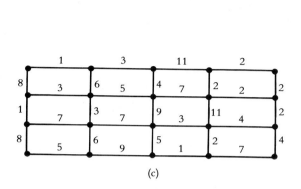

(c)

(d)

33. A large company wishes to install a pneumatic tube system that would enable small items to be sent between any of 10 locales, possibly by relay. If the nonprohibitive costs (in $100) are shown in the graph model below, between which sites should the tube be installed to minimize the total cost?

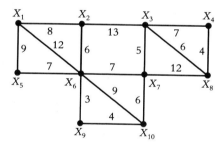

34. Give examples of real-world situations which can be modeled using a weighted graph and for which finding a minimum-cost spanning tree for the graph would be of interest.

35. Suppose that Pictaphone service is not possible between two towns where there is a hill of more than 800 feet along a straight line that runs between them. In constructing a model for solving a relay network problem, how would you handle the question of putting a link between two cities with a 1200-foot hill between them?

36. Find the cost of providing a relay network between the six cities with the largest population in your home state, using the road distances between the cities as costs. Does it follow that the same solution would be obtained if air distances are used instead?

37. Can Kruskal's algorithm be modified to find a maximum-weight spanning tree? Can you think of an application for finding a maximum-weight spanning tree?

38. Let G be a graph with weights assigned to each edge. Consider the following algorithm:

   a. Pick any vertex V of G.

   b. Select that edge E with a vertex at V that has a minimum weight. Let the other endpoint of E be W.

   c. Contract the edge VW so that edge VW disappears and vertices V and W coincide (see the following figures).

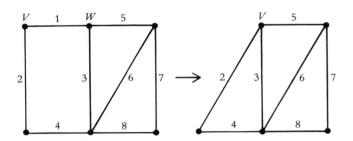

If in the new graph two or more edges join a pair of vertices, delete all but the cheapest. Continue to call the new vertex V.

   d. Repeat steps **b** and **c** until a single point is obtained. The edges selected in the course of this algorithm (called Prim's algorithm) form a minimum-cost spanning tree. Apply this algorithm to the following graphs:

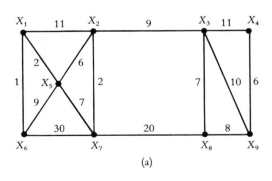

(a)

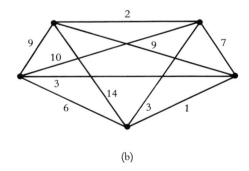

(b)

39. Determine whether each of the following items is true or false for a minimum-cost spanning tree T for a weighted connected graph G:

   a. T contains a cheapest edge in the graph.

   b. T cannot contain a most expensive edge in the graph.

   c. T contains one fewer edge than there are vertices in G.

   d. There is some vertex in T to which all others are joined by edges.

40. In the following graphs, the number in the circle for each vertex is the cost of installing equipment at the vertex if *relaying* must be done at the vertex, while the number on an edge indicates the cost of providing service between the endpoints of the edge.

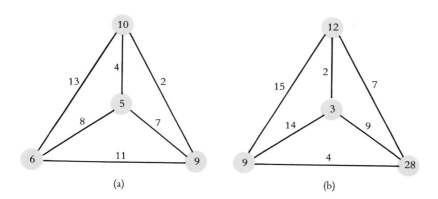

(a)   (b)

In each case, find the minimum cost (allowing relays) for sending messages between any pair of vertices, taking vertex relay costs into account. Would your answer be different if vertex relay costs are neglected? (Warning: Kruskal's algorithm cannot be used to answer the first question. This problem illustrates the value of having an algorithm over relying on "brute force.")

41. Two spanning trees of a (weighted) graph are considered different if they use different edges. Show that the graph below has different minimum-cost spanning trees, though all these different trees have the same cost.

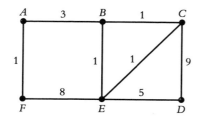

42. Suppose G is a graph such that all the weights on its edges are different numbers. Show that there is a unique minimum-cost spanning tree.

43. Find a minimum-cost spanning tree for the complete graphs in Exercise 24.

44. Suppose that a letter requires postage of p (positive integer) and that stamps of various denominations are available, say, $d_1, \ldots, d_v$ (positive integers). We are interested in finding the minimum number of stamps to choose with the available denominations to obtain the postage p exactly.

a. Give an example to show that unless other conditions are put on p and $d_1, \ldots, d_v$, it might happen that no selection of stamps will achieve the desired postage.

b. Show that even if the desired postage can be obtained for some choice of stamps, it does not follow that a minimum number of stamps can be achieved by using a greedy algorithm, that is, by selecting at each stage the largest denomination possible.

c. Show that for at least one choice of denominations, a greedy algorithm will produce the fewest stamps for any given postage!

45. Find the earliest completion time and critical paths for the order-requirement digraphs below.

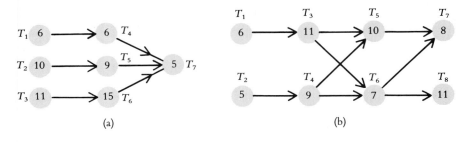

(a)   (b)

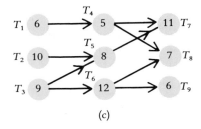

(c)

46. Construct an example of an order-requirement digraph with three different critical paths.

47. In the order-requirement digraph below, determine which tasks, if shortened, would reduce the earliest completion time and which would not. Then find the earliest completion time if task T_5 is reduced to time length 7. What is the new critical path?

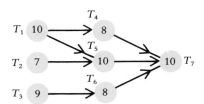

48. To build a new addition on a house, the following tasks must be completed. Construct reasonable time estimates for these tasks and a reasonable order-requirement digraph. What is the fastest time in which these tasks can be completed?

- **a.** Lay foundation.
- **b.** Erect sidewalls.
- **c.** Erect roof.
- **d.** Install plumbing.
- **e.** Install electric wiring.
- **f.** Lay tile flooring.
- **g.** Obtain building permits.
- **h.** Put in door that adjoins new room to existing house.
- **i.** Install track lighting on ceiling.

49. At a large toy store, scooters arrive unassembled in boxes. To assemble a scooter the following tasks must be performed:

TASK 1. Remove parts from the box.

TASK 2. Attach wheels to the footboard.

TASK 3. Attach vertical housing.

TASK 4. Attach handlebars to vertical housing.

TASK 5. Put on reflector tape.

TASK 6. Attach bell to handlebars.

TASK 7. Attach decals.

TASK 8. Attach kickstand.

TASK 9. Attach safety instructions to handlebars.

Give reasonable time estimates for these tasks and construct a reasonable order-requirement digraph. What is the earliest time by which these tasks can be completed.

50. For the order-requirement digraph below, find the critical path and the task in the critical path whose time, when reduced the *least,* creates a new critical path.

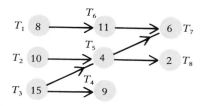

51. Construct an order-requirement digraph with six tasks that has two critical paths of length 24.

52. Use the fundamental principle of counting to predict that there are 12 paths in Figure 2.17. (Hint: All paths have three edges. Vertex T_2 has one edge coming in and one edge going out, vertex T_5 has three edges coming in and three edges going out, while vertex T_8 has one edge coming in and two edges going out.)

·3·

Planning and Scheduling

In a society as complex as ours, everyday problems such as providing services efficiently and on time require accurate planning of both people and machines. Take the example of a hospital in a major city. Around-the-clock scheduling of nurses and doctors must be provided to guarantee that people with particular expertise are available during each shift. The operating rooms must be scheduled in a manner flexible enough to deal with sudden emergencies. Equipment used for x-ray, CAT, or NMR scans must be scheduled for maximal efficiency.

Although many scheduling problems are often solved on an ad hoc basis, we can also use mathematical ideas to gain insight into the complications that arise in scheduling. The ideas we develop in this chapter have practical value in a relatively narrow range of applications, but they throw light on many characteristics of more realistic and hence more complex scheduling problems.

SCHEDULING TASKS

Assume that a certain number of identical **processors** (machines, humans, or robots) work on a series of tasks that make up a job. Associated with each task is a specified amount of time required to complete the task. For simplicity, we assume that any of the processors can work on any of the tasks. Our problem, known as the **machine-scheduling** problem, is to decide how the tasks should be scheduled so that the completion time for the tasks collectively is as early as possible.

Even with these simplifying assumptions, complications in scheduling will arise. Some tasks may be more important than others and perhaps should be scheduled first. When "ties" occur, they must be resolved by special rules. As an example, suppose we are scheduling patients to be seen in a hospital emergency room staffed by one doctor. If two patients arrive simultaneously, one with a bleeding foot, the other with a bleeding arm, which patient should be processed first? Suppose the doctor treats the arm patient first, and while treatment is going on, a person in cardiac arrest arrives. Scheduling rules must establish appropriate priorities for cases such as these.

Another common complication arises with jobs consisting of several tasks that cannot be done in an arbitrary order. For example, if the job of putting up a new house is treated as a scheduling problem, the task of laying the foundation must precede the task of putting up the walls, which in turn must be completed before work on the roof can begin.

Assumptions and Goals

To simplify our analysis, we need to make clear and explicit assumptions:

1. If a processor starts work on a task, the work on that task will continue without interruption until the task is completed.

2. No processor stays voluntarily idle. In other words, if there is a processor free and a task available to be worked on, then that processor will immediately begin work on that task.

3. The requirements for ordering the tasks are given by an order-requirement digraph. (A typical example is shown in Figure 3.1, with task times circled within each vertex. The ordering of the tasks imposed by the order-requirement digraph represents constraints of physical reality. For example, you cannot fly a plane till it has taken fuel on board.)

4. The tasks are arranged in a priority list that is independent of the order requirements. (The priority list is an ordering of the tasks according to some criterion of "importance,"

which may in no way reflect physical reality. For example, imagine a construction job with several tasks. Task A may have to be done before task B, but when task B is done a monetary payment will be made. Thus, B may be given a higher priority than A. The priority list represents, from some point of view, an ordering of the tasks. Another such point of view is to order the tasks in a manner that will help the algorithm being used construct schedules with early completion times. Usually, different points of view for giving priority to tasks are not consistent. Mathematical analysis may sometimes assist in clarifying trade-offs implicit in these different points of view.)

When considering a scheduling problem, there are various goals one might wish to achieve. Among these are

1. minimizing the completion time of the job

2. minimizing the total time that processors are idle

3. finding the minimum number of processors necessary to finish the job by a specified time

For the moment, we will concentrate on goal 1, finishing all the tasks at the earliest possible time. Note, however, that optimizing with respect to one criterion or goal may not optimize with respect to another.

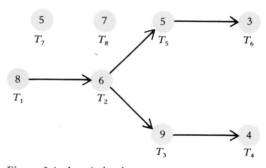

Figure 3.1 A typical order-requirement digraph.

List-Processing Algorithm

The scheduling problem we have described sounds more complicated than the traveling salesman problem (TSP). Indeed, like the TSP, it is known to be NP-complete. This means that it is unlikely that anyone will ever find a computationally fast algorithm that can find an optimal solution. Thus, we will be content to seek a solution method that is computationally fast and gives only approximately optimal answers.

The algorithm we will use to schedule tasks is the **list-processing algorithm.** In describing it, we will call a task **ready** at a particular time if all its predecessors as indicated in the order-requirement digraph have been completed at that time. For example, in Figure 3.1 at time 0 the ready tasks are T_1, T_7, and T_8, while task T_2 cannot be ready until 8 time units after T_1 is started. The algorithm works as follows: at a given time, assign to the lowest-numbered free processor the first task on the priority list that is ready at that time and that hasn't already been assigned to a processor.

In applying this algorithm, we will need to develop skill at coordinating the use of the information in the order-requirement digraph and the priority list. It will be helpful to cross out the tasks in the priority list as they are assigned to a processor to keep track of which tasks remain to be scheduled. Let's apply this algorithm to one possible priority list, T_8, T_7, T_6, . . . , T_1, using two processors and the order-requirement digraph in Figure 3.1. The result is the schedule shown in Figure 3.2,

where idle processor time is indicated by color shading. How does the list-processing algorithm generate this schedule?

Because T_8 (task 8) is first on the priority list and ready at time 0, it is assigned to the lowest-numbered free processor, processor 1. Task 7, next on the priority list, is also ready at time 0 and thus is assigned to processor 2. The first processor to become free is processor 2 at time 5. Recall that by assumption 1, once a processor starts work on a task, its work cannot be interrupted until the task is complete. Task T_6, the next unassigned task on the list, is not ready at time 5, as can be seen by consulting Figure 3.1. In fact, at time 5, the only ready task on the list is T_1, so that task is assigned to processor 2. At time 7, processor 1 becomes free, but no task becomes ready until time 13. Thus, processor 1 stays idle from time 7 to time 13. At this time, because T_2 is the first ready task on the list not already scheduled, it is assigned to processor 1. Processor 2, however, stays idle because no other ready task is available at this time.

As the priority list is scanned from left to right to assign a processor at a particular time, we pass over tasks that are not ready to find ones that are ready. If no task can be assigned in this manner, we keep one or more processors idle till such time that, reading the priority list from the left, there is a ready task not already assigned. After a task is assigned to a processor, we resume scanning the priority list, starting over at the far left, for unassigned tasks. The remainder of the scheduling shown in Figure 3.2 is completed in this manner.

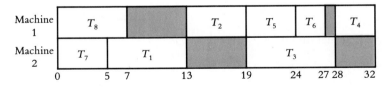

Figure 3.2 The schedule produced by applying the list-processing algorithm to the order-requirement digraph in Figure 3.1 using the list T_8, T_7, . . . , T_1.

When Is a Schedule Optimal?

The schedule in Figure 3.2 has a lot of idle time, so it may not be optimal. Indeed, if we apply the list-processing algorithm for two processors to the priority list $T_1, \ldots, T_8$, using the digraph in Figure 3.1, the resulting schedule (Figure 3.3) is optimal because the path T_1, T_2, T_3, T_4, with length 27, is the critical path in the order-requirement digraph. As we saw in Chapter 2, the earliest completion time for the job made up of all the tasks is the length of the longest path in the order-requirement digraph.

There is another way of relating optimal completion time for a scheduling problem to the completion time that is yielded by the list-processing algorithm. Suppose that we add all the task times given in the order-requirement digraph and divide by the number of processors. The completion time using the list-processing algorithm must be at least as large as this number. For example, the task times for the order-requirement digraph in Figure 3.1 sum to 47. Thus, if these tasks are scheduled on two processors, the completion time is at least $47/2 = 23.5$ (in fact, 24, since the list-processing algorithm applied to integer task times yields an integer answer), while for three processors the completion time is at least $47/3$ (in fact, 16). This method works since, even if there is no idle time for processors, the area of the rectangular diagram that represents the schedule is greater than or equal to the sum of the areas of rectangles that represent the tasks in the schedule diagram. (We think of each task that takes time t as a rectangle of size $1 \times t$ and the scheduling diagram with m processors as a rectangle whose height is m units and whose length is the completion time for all the tasks.) Sometimes the estimate for the completion time given by the list-processing algorithm from the length of the critical path gives a more useful value than the approach based on adding task times, and sometimes the opposite

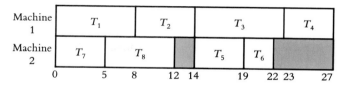

Figure 3.3 The schedule produced by applying the list-processing algorithm to the order-requirement digraph in Figure 3.1 using the list $T_1, T_2, \ldots, T_8$.

is true. For the order-requirement digraph in Figure 3.1, except for a schedule involving one processor, the critical-path estimate is superior. For some scheduling problems, both these estimates may be poor.

The number of priority lists that can be constructed if there are n tasks is $n!$, as can be computed using the fundamental principle of counting. As the priority list is changed, the list-processing algorithm we are using will schedule the tasks, subject to the constraints of the order-requirement digraph, in different ways. Different lists may yield schedules with different completion times or different schedules with the same completion time. Different priority lists may yield identical scheduling of tasks (and hence identical completion times). In general, determining which list yields an optimal completion time can be very time-consuming.

Strange Happenings

The list-processing algorithm involves four factors that affect the final schedule. The answer we get depends on

1. the times of the tasks

2. the number of processors

3. the order-requirement digraph

4. the ordering of the tasks on the list

To see the interplay of these four factors, consider another scheduling problem, this one as-

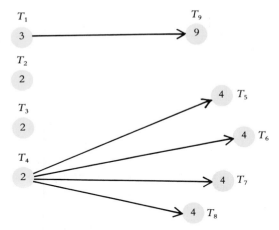

Figure 3.4 An order-requirement digraph designed to help illustrate some paradoxical behavior produced by the list-processing algorithm.

sociated with the order-requirement digraph shown in Figure 3.4 (the circled numbers are task time lengths).

The schedule generated by the list-processing algorithm applied to the list $T_1, T_2, \ldots, T_9$, using three processors, is given in Figure 3.5.

Treating the list $T_1, \ldots, T_9$ as fixed, how might we make the completion time earlier? Our alternatives are to pursue one or more of these strategies:

1. Reduce task times.

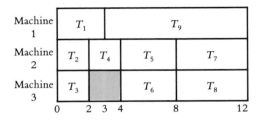

Figure 3.5 The schedule produced by applying the list-processing algorithm to the order-requirement digraph in Figure 3.4 using the list $T_1, T_2, \ldots, T_9$ with three processors.

2. Use more processors.

3. "Loosen" the constraints of the order-requirement digraph.

Let's consider each alternative in turn, changing one feature of the original problem at a time, and see what happens to the resulting schedule. If we adopt strategy 1, reducing the time of each task by one unit, it seems intuitively clear that the completion time would go down. Figure 3.6 shows the new order-requirement digraph, and Figure 3.7 shows the schedule produced for this problem, using the list-processing algorithm with three processors applied to the list $T_1, \ldots, T_9$. The completion time is now 13, longer than the completion time of 12 for the case (Figure 3.5) with the *longer* task times. Here is something unexpected! Let's explore further and see what happens.

Next we consider strategy 2, increasing the number of machines. Surely this should speed matters up. When we apply the list-processing algorithm to the original graph in Figure 3.4, using the list $T_1, \ldots, T_9$ and four machines,

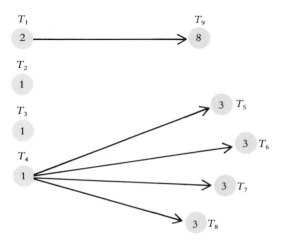

Figure 3.6 The order-requirement digraph obtained from the one in Figure 3.4 by reducing by one unit each of the task times shown there.

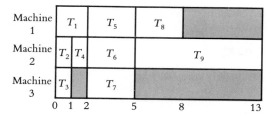

Figure 3.7 The schedule produced by applying the list-processing algorithm to the order-requirement digraph in Figure 3.6 using the list $T_1, T_2, \ldots, T_9$ with three processors.

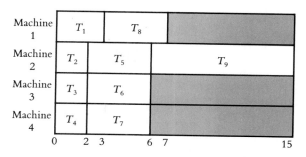

Figure 3.8 The schedule produced by applying the list-processing algorithm to the order-requirement digraph in Figure 3.4 using the list $T_1, T_2, \ldots, T_9$ with four processors.

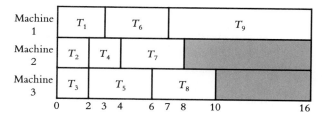

Figure 3.9 The schedule produced by applying the list-processing algorithm to the order-requirement digraph in Figure 3.4, modified by erasing all its directed edges, using the list $T_1, T_2, \ldots, T_9$ with three processors.

we get the schedule shown in Figure 3.8. The completion time is now 15! Here is a shock, an even later completion time than for the previous alteration.

Finally, we consider strategy 3, trying to shorten completion time by erasing *all* constraints (edges with arrows) in the order-requirement digraph shown in Figure 3.4. By increasing flexibility of the ordering of the tasks, we might guess we could finish our tasks more quickly. Figure 3.9 shows the schedule using the list $T_1, \ldots, T_9$; now it takes 16 units! This is the worst of our three strategies to reduce completion time.

The failures we have seen here appear paradoxical at first glance, but they are typical of what can happen when a situation is too complex to analyze with naive intuition. Sometimes our common sense leads us astray. The value of using mathematics rather than intuition or trial and error to study scheduling and other problems is that it points up flaws that can occur in unguarded intuitive reasoning.

The paradoxical behavior we see here is a consequence of the rules we set up for generating schedules. The list-processing algorithm has many nice features, including the fact that it is easy to understand and fast to implement. However, the results of the model in some cases can appear strange. Since we have been explicit about our assumptions, we could go back and make changes in these assumptions in

hopes of eliminating the strange behavior. But the price we may pay is more time spent in constructing schedules and perhaps even new types of strange behavior.

CRITICAL-PATH SCHEDULES

In our discussion so far, we have acted as though the priority list used in applying the list-processing algorithm was given to us in advance based on external considerations. We might, however, consider the question of whether there is a systematic method of *choosing* a priority list that yields optimal or nearly optimal schedules. We will show how to construct a specific priority list based on this prin-

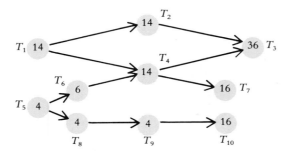

Figure 3.10 An order-requirement digraph used to illustrate the critical-path scheduling method.

ciple, to which the list-processing algorithm can then be applied.

Recall from our discussion of critical-path analysis in Chapter 2 that no matter how a schedule is constructed, the finish time cannot be earlier than the length of the longest path in the order-requirement digraph. This suggests that we should try to schedule first those tasks that occur early in long paths, because they can bottleneck the flow of tasks to be completed.

To illustrate this method, consider the order-requirement digraph in Figure 3.10. Suppose we wish to schedule these tasks on two processors. Initially, there are two critical paths of length 64: T_1, T_2, T_3 and T_1, T_4, T_3. Thus, we place T_1 first on the priority list. With T_1 "gone," there is a new critical path of length 60, which starts with T_5, so T_5 is placed second on the priority list. We continue in this fashion, placing next on the priority list whichever task is first on some current longest path. (In the case of two longest paths tying,

we place next on the priority list the lowest-numbered task heading a longest path.) Using this approach, we obtain the priority list T_1, T_5, T_6, T_2, T_4, T_3, T_8, T_9, T_7, T_{10}. The list-processing algorithm is then applied using this priority list and the order-requirement digraph in Figure 3.10. We obtain the schedule in Figure 3.11.

This example shows that **critical-path scheduling,** as this method is called, can sometimes yield optimal solutions. Unfortunately, this algorithm does not always perform well. For example, the critical-path method employing four processors applied to the order-requirement digraph shown in Figure 3.12 yields the list T_1, T_8, T_9, T_{10}, T_{11}, T_5, T_6, T_7, T_{12}, T_2, T_3, T_4 and then the schedule in Figure 3.13. (Note that T_5, T_6, T_7 are thought of as heading paths of length 10.) In fact, there can be no *worse* schedule than this one. An optimal schedule is shown in Figure 3.14.

Many of the results we have examined so far are negative because we are dealing with a general class of problems that defy our using com-

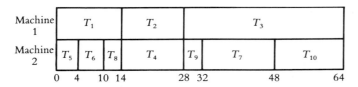

Figure 3.11 The optimal schedule produced by applying the critical-path scheduling method to the order-requirement digraph in Figure 3.10. The list used was T_1, T_5, T_6, T_2, T_4, T_3, T_8, T_9, T_7, T_{10}.

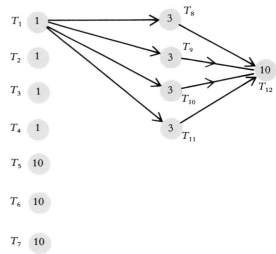

Figure 3.12 An order-requirement digraph used to illustrate how poorly the critical-path scheduling method can sometimes behave.

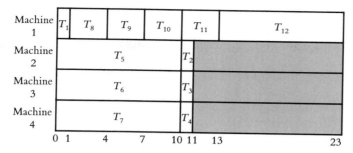

Figure 3.13 The schedule produced by applying the critical-path scheduling method to the order-requirement digraph in Figure 3.12 using four processors. The list used was T_1, T_8, T_9, T_{10}, T_{11}, T_5, T_6, T_7, T_{12}, T_2, T_3, T_4.

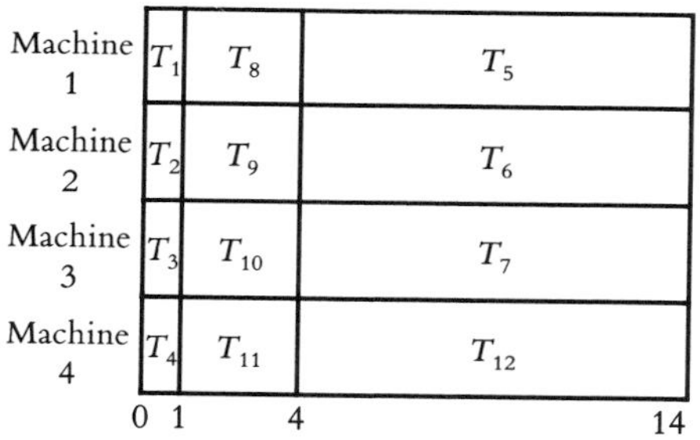

Figure 3.14 An optimal schedule for the order-requirement digraph in Figure 3.12 using four processors.

putationally efficient algorithms to find an optimal schedule. But we can close on a more positive note. Consider an arbitrary order-requirement digraph but assume all the tasks take equal time. It turns out that we can always construct an optimal schedule using two processors in this situation. Ironically, we can choose among many algorithms to produce these optimal schedules. The algorithms are easy to understand (though not easy to prove optimal) and have all been discovered since 1969! Many people think that mathematics is a subject that is no longer alive, and that all its ideas and methods were discovered hundreds of years ago. They assume mathematics consists of nothing more than the arithmetic, algebra, and trigonometry taught in high school. This stereotype is easily broken down, as we have just seen, and it is well to consider

that scheduling problems account for only a narrow portion of mathematics in general. In fact, more new mathematics has been discovered and published in the last 30 years than during any previous 30-year period.

INDEPENDENT TASKS

Mathematicians suspect that no computationally efficient algorithm for solving general scheduling problems optimally will ever be found. Due to our limited success in designing algorithms for finding optimal schedules for general order-requirement digraphs, we will consider a special class of scheduling problems for which the order-requirement digraph has no edges with arrows. In this case we say that the tasks are **independent** of each other since they can be performed in any order. (No edges with arrows in the order-requirement digraph indicates that no tasks need to precede others; that is, the tasks can be done in any order.) In this section we will consider the problem of scheduling independent tasks.

There are two approaches we can consider. To study **average-case analysis** we might ask if the average (mean) of the completion times arrived at by using the list-processing algorithm with all the possible different lists is close to the optimal possible completion time.

To study **worst-case analysis** we can ask how far from optimal a schedule obtained using the list-processing algorithm with one particular priority list can be. Average-case analysis is amenable to mathematical solution but requires methods of great sophistication. For independent tasks, the worse-case analysis can be answered using a surprisingly simple argument developed by Ronald Graham of AT&T Bell Laboratories (see Spotlight 3.1, pp. 76–77). The idea is that if the tasks are independent, no processor can be idle at a given time and then busy on a task as a later time. We will return to Graham's worst-case analysis after exploring the problem of independent tasks in more detail.

Geometrically, we can think of the independent tasks as rectangles of height 1 whose lengths are equal to the time length of the task. Finding an optimal schedule amounts to packing the task rectangles into a longer rectangle whose height equals the number of machines. For example, Figure 3.15 shows two different ways to schedule tasks of length 10, 4, 5, 9, 7, 7

on two machines. (For convenience, the rectangles in the case of independent tasks are labeled with their task times rather than their task numbers.) Scheduling basically means efficiently packing the task rectangles into the machine rectangle. Finding the optimal answer among all possible ways to pack these rectangles is like looking for a needle in a haystack. The list-processing algorithm produces a packing, but it may not be a good one.

What Graham's worst-case analysis showed for independent tasks is that no matter which list L one uses, if the optimal schedule requires time T, then the completion time for the schedule produced by the list-processing algorithm applied to list L with m processors is less than or equal to $(2 - 1/m)T$. For example, for two machines ($m = 2$), if an optimal schedule yields completion at time 30, then no list would ever yield a completion later than $(2 - \frac{1}{2})(30) = 45$. Although it is of great theoretical interest, Graham's result does not provide much comfort to those who are trying to find good schedules for independent tasks.

Decreasing-Time Lists

Is there some way of choosing a priority list that consistently yields relatively good schedules? The surprising answer is yes! The idea is that when long tasks appear toward the end of the list, they often seem to "stick out" on the right end, as in Figure 3.15a. This suggests that before trying to schedule a collection of tasks, the tasks should be placed in a list so that longest tasks are listed first. The list-processing algorithm applied to a list arranged in this fashion is called the **decreasing-time-list algorithm.** If we apply it to the set of tasks listed previously (10, 4, 5, 9, 7, 7), we obtain the times 10, 9, 7, 7, 5, 4 and the schedule (packing) shown in Figure 3.16. This packing is again optimal, but it is different from the optimal scheduling in Figure 3.15b.

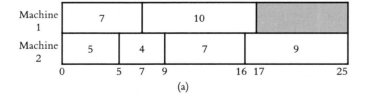

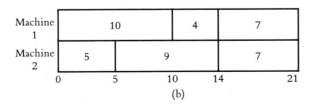

Figure 3.15 (a) A nonoptimal way to schedule independent tasks of time lengths 10, 4, 5, 9, 7, 7 using two processors. (b) An optimal way to schedule independent tasks of time lengths 10, 4, 5, 9, 7, 7 using two processors.

It is important to remember that the decreasing-time-list algorithm does not guarantee optimal solutions. This can be seen by scheduling the tasks with times 11, 10, 9, 6, 4 (Figure 3.17). However, the rearranged list 9, 4, 6, 11, 10 yields the schedule in Figure 3.18, which finishes at time 20. This solution is obviously optimal. Note that when tasks are independent, if there are m machines available, the completion time cannot be less than the sum of the task times divided by m.

The problems we have encountered in scheduling independent tasks seem to have taken us a bit far from our goal of *applying* the mathematics we have developed. Sometimes mathematicians will pursue their mathematical ideas even though they have reached a point where there appear to be no applications. Fortunately, it is very common to be able to find applications for the "abstract" extensions. This is the case in the current instance.

EXAMPLE: Photocopy Shop and Typing Pool Problems.

Imagine a photocopy shop with three photocopiers. Photocopying tasks that must be completed overnight are accepted until 5 p.m. The tasks are to be done in any manner that minimizes the finish time for all the work. Because this problem involves scheduling machines for independent tasks, the decreasing-time-list algorithm would be a good heuristic to apply.

For another example, consider a typing pool at a large corporation or college, where individual typing tasks can be assigned to any typist. In this setting, however, the assumption that the processors (typists) are identical in skill is less likely to be true. Hence, the tasks might have different times with different processors. This phenomenon, which occurs in real-world scheduling problems, violates one of the assumptions of our mathematical model.

Graham's result for the list-processing algorithm—that the finishing time is never

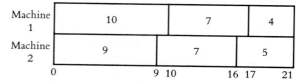

Figure 3.16 The optimal schedule resulting from applying the decreasing-time-list algorithm to a collection of independent tasks. The list used, written down in terms of task times only, is 10, 9, 7, 7, 5, 4.

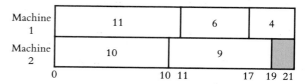

Figure 3.17 The nonoptimal schedule resulting from applying the decreasing-time-list algorithm to a collection of independent tasks. The list used, written down in terms of task times only, is 11, 10, 9, 6, 4.

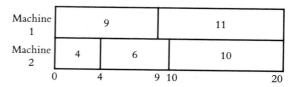

Figure 3.18 The optimal schedule resulting from applying the list-processing algorithm to a collection of independent tasks. The list used, written down in terms of task times only, is 9, 4, 6, 11, 10.

more than $(2 - 1/m)T$ (where T represents optimal completion time and m the number of processors)—offers us the small comfort of knowing that even the worst choice of priority list will not yield a completion time worse than twice the optimal time. Compared with the list-processing algorithm, the decreasing-time-list algorithm seems to improve completion times. Thus, it is not surprising that Graham was able to provide an improved *bound*, or time estimate, for this case: the decreasing-list algorithm gives a completion time no more than $[4/3 - 1/(3m)]T$, where m is the number

SPOTLIGHT 3.1 Ronald Graham on Mathematics and Mathematicians

Ronald Graham is director of the mathematics sciences research department of AT&T Bell Labs, Murray Hill, New Jersey. In addition to his own research, he supervises some of the country's most distinguished mathematicians as they investigate the puzzles of management science. Here are some of his comments:

On Scheduling

Scheduling is a very interesting area. You have some number of processors that you can think of as computers or as people, and you have a number of tasks or jobs that you would like to get done. One might think that adding more processors to the system would guarantee that you would be done sooner. In fact, just the opposite can happen. It's an example of the old adage, "Too many cooks can spoil the broth." You can add more machines or decrease the amount of time it takes for each job, or relax the constraints between the various jobs, but all of these can actually end up taking longer. To try to understand why that is and how to avoid it is the arena of mathematics.

One of the earliest applications of scheduling came in looking at the Apollo moon shot. In that case, there were three different processors, namely, the three astronauts, who had a large collection of tasks that they were supposed to do — various experiments to run on the way to and from the moon. And of course, there were various constraints. You want the astronauts to sleep a little every day. Other constraints you might not think of. For example, you have to rotate the capsule so that the same side isn't facing the sun all the time, because of heating problems. Then, within these constraints, you would next like to know: If you sequence the experiments in different orders, how much time could you save over sequencing them in orders that aren't so good?

It turns out that in fact NASA had developed very good sequencing already. All I could show them was that they couldn't improve on it very much, no matter what they might try. And of course this is one of the problems in management science where it's just impossible to enumerate the possibilities. So that's where the mathematics comes in: it enables you to say that no matter what you do, you'll never be able to "do any better than this" or "do it any faster."

You might well ask, How can you know about every one of the possibilities even though you didn't try them all? That's an interesting question, and it really goes to the root of how mathematics is used to analyze the real world. There are several well-known steps in going from a real-world problem to a model of the problem — a model in which you try to capture the essential aspects of the problem, but in a way that can be dealt with mathematically. Once you convert the problem into the world of mathematics, you can say something about all the mathematical possibilities. Then you translate it back to the real-world situation. How well it works depends crucially on how accurate the translation of the model was.

Now it's always good to check a few of the things that you've predicted are going to happen. If the translation is good and you've captured the essence of the problem, then there is a good chance that you can say something sensible about the thing you've studied. If it isn't, you can get some pretty bizarre conclusions. There are famous examples of this. One that comes to mind is the case where the people who first analyzed how bees fly were able to prove mathematically that bees *couldn't* fly. That didn't both the bees, of course, and the model was eventually modified to show that bees could fly after all.

On the Spirit of Investigation

I think in order to do the best work in any area, you have to enjoy it and be intrigued by it. You have to wonder, Why is this happening? There are many examples where something slightly out of the ordinary occurs and 99 people notice it yet go on to something else. But one in a hundred—or fewer—is fascinated by that slightly anomalous or even bizarre behavior in one mathematical area or another. They start probing it more deeply and soon find that it's really the tip of a giant iceberg. Once you start to melt it, you have a much deeper understanding about what is really going on. There are those who feel mathematics is a giant game, albeit one that is amazingly relevant to everything that is going on in the scientific world.

Often, half the battle in trying to solve a difficult problem is knowing the right question to ask. If you know what it is you are trying to look for, you can often be very far along in finding the solution. It's useful for me, and for many others who are working on a difficult problem, to look at a special case first. You can work your way up to the full problem by trying easier special cases that you still can't do. You may be bouncing the idea around a bit and not forcing it, and it's amazing how often it happens that the next day or next week something will seem obvious that was very hard to imagine even a week earlier.

On Managing the Mathematicians

One of the rules or axioms that we have here [at Bell Labs] is that each person is best equipped to know how he or she functions optimally. Some people are night people, for example, and they just don't function before 12 o'clock.

In our mathematics group, there are roughly 65 professionals, and quite a few are leading-edge, world-class researchers. People of this caliber you don't so much manage as try to keep up with what they are doing. In many cases, you act as a sounding board, because one of the most useful things you can do with a colleague who has some mathematical ideas is to act as a sympathetic listener. As the person explains the ideas or where he is stuck, he will more often than not gain a much clearer insight into what he's doing and where he's going.

Many times there is much more interactive collaboration. There is quite a lot of joint work that goes on here, not just from within the mathematics area, but among mathematicians, computer scientists, physicists, and chemists. It is a very interactive environment which, I think, works to everyone's benefit.

of processors and T is the optimal time in which the tasks can be completed. In particular, when the number of processors is 2, the schedule produced by the decreasing-time-list algorithm is never off by more than 17%! Usually, the error is much less. This result is a remarkable instance of the value of mathematical research into applied problems. Note that the optimal completion time T depends on m and that Graham's theoretical analysis is necessary precisely because there is no known algorithm to compute T easily.

BIN PACKING

Suppose you plan to build a wall system for your books, records, and stereo set. It requires 24 wooden shelves of various lengths: 6, 6, 5, 5, 5, 4, 4, 4, 4, 2, 2, 2, 2, 3, 3, 7, 7, 5, 5, 8, 8, 4, 4, and 5 feet. The lumberyard, however, sells wood only in boards of length 9 feet. If each board costs $8, what is the minimum cost to buy sufficient wood for this wall system?

Because all shelves required for the wall system are shorter than the boards sold at the lumberyard, the largest number of boards needed is 24, the precise number of shelves needed for the wall system. Buying 24 boards would of course be a waste of wood and money because several of the shelves you need could be cut from one board. For example, pieces of length 2, 2, 2, and 3 feet can be cut from one 9-foot board.

To be more efficient, we think of the boards as bins of capacity W (9 feet in this case) into which we will pack (without overlap) weights (in this case, lengths) $w_1, \ldots, w_n$, where each $w_i \leq W$. We wish to find the minimum number of bins into which the weights can be packed. In this formulation, the problem is known as the **bin-packing problem.** At first

glance, it may appear unrelated to the machine-scheduling problems we have been studying; however, there is a connection.

Let's suppose we want to schedule independent tasks so that each machine working on the tasks finishes its work by time W. Instead of fixing the number of machines and trying to find the earliest completion time, we must find the minimum number of machines that will guarantee completion by the fixed completion time (W). Despite this similarity between the machine-scheduling problem and the bin-packing problem, the discussion that follows will use the traditional terminology of bin packing.

By now, it should come as no surprise to learn that no one knows a fast algorithm that always picks the optimal (smallest) number of bins (boards). In fact, the bin-packing problem belongs to the class of NP-complete problems (see Spotlight 2.1), which means that most experts think it unlikely that any fast optimal algorithm will ever be found.

Bin-Packing Heuristics

We will think of the items to be packed, in any particular order, as constituting a list. In what follows we will use the list of 24 shelf lengths given for the wall system. We will consider various heuristic methods. Probably the easiest approach is simply to put the weights into the first bin until the next weight won't fit, and then start a new bin. (Once you open a new bin, don't use leftover space in an earlier, partially filled bin.) Continue in the same way until as many bins as necessary are used. The resulting solution is shown in Figure 3.19. This algorithm, called **next fit (NF),** has the advantage of not requiring knowledge of all the weights in advance; only the remaining space in the bin currently being packed must

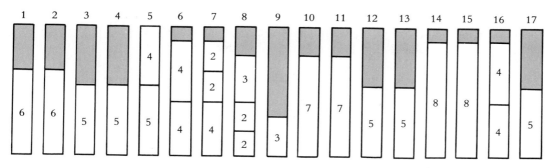

Figure 3.19 The list 6, 6, 5, 5, 5, 4, 4, 4, 4, 2, 2, 2, 2, 3, 3, 7, 7, 5, 5, 8, 8, 4, 4, 5 packed in bins using next fit.

be remembered. The disadvantage of this heuristic is that a bin packed early on may have had room for small items that come later in the list.

Our wish to avoid permanently closing a bin too early suggests a different heuristic — **first fit (FF):** put the next weight into the *first* bin already opened that has room for this weight; if no such bin exists, start a new bin. Note that a computer program to carry out first fit would have to keep track of how much room was left in *all* the previously opened bins. For the 24 wall-system shelves, first fit would generate a solution that uses only 14 bins (see Figure 3.20) instead of the 17 bins generated by next fit.

If we are keeping track of how much room remains in each unfilled bin, we can put the next item to be packed into the bin that cur-

rently has the most room available. This heuristic will be called **best fit (BF).** The solution generated by this approach looks the same as that shown in Figure 3.20. Although this heuristic also leads to 14 bins, the items are packed in a different order. For example, the first item of size 2, the tenth item in the list, is put into bin 6 in best fit, but into bin 1 in first fit.

Decreasing-Time Heuristics

One difficulty with all three of these heuristics is that large weights that appear late in the list can't be packed efficiently. Therefore, we should first sort the items to be packed in order of decreasing size, assuming that all items are known in advance. We can then pack large

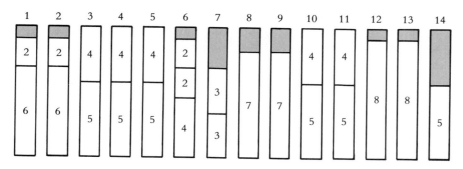

Figure 3.20 The list 6, 6, 5, 5, 5, 4, 4, 4, 4, 2, 2, 2, 2, 3, 3, 7, 7, 5, 5, 8, 8, 4, 4, 5 packed in bins using first fit. Best fit would yield a packing that would look identical.

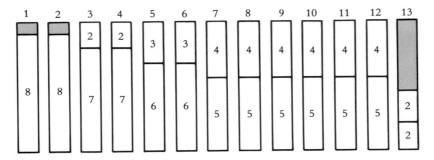

Figure 3.21 The bin packing resulting from applying first-fit decreasing to the wall-system numbers. The list involved, which uses the original list sorted in decreasing order, is 8, 8, 7, 7, 6, 6, 5, 5, 5, 5, 5, 5, 4, 4, 4, 4, 4, 4, 3, 3, 2, 2, 2, 2.

items first and pack the smaller items into leftover spaces. This approach yields three new heuristics: **next-fit decreasing (NFD), first-fit decreasing (FFD),** and **best-fit decreasing (BFD).** Here is the original list sorted: 8, 8, 7, 7, 6, 6, 5, 5, 5, 5, 5, 5, 4, 4, 4, 4, 4, 4, 3, 3, 2, 2, 2, 2. Packing using first-fit-decreasing order yields the solution in Figure 3.21. This solution uses only 13 bins.

Is there any packing that uses only 12 bins? No. In Figure 3.21, there are only 2 free units (1 unit each in bins 1 and 2) of space in the first 12 bins but 4 occupied units (two 2s) in bin 13. We could have predicted this by dividing the total length of the shelves (110) by the capacity of each bin (board): $\frac{110}{9} = 12\frac{2}{9}$. Thus, no packing could squeeze these shelves into 12 bins — there would always be at least 2 units left over for the thirteenth bin. (In Figure 3.21, there are 4 units in bin 13 because of the 2 wasted empty spaces in bins 1 and 2.) Even if this division had created a zero remainder, there would still be no guarantee that the items could be packed to fill each bin without wasted space. For example, if the bin capacity is 10 and there are weights of 6, 6, 6, 6, and 6, the total weight is 30; dividing by 10, we get 3 bins as the minimum requirement. Clearly, however, 5 bins are needed to pack the five 6s.

None of the six heuristic methods shown will necessarily find the optimal number of bins for an arbitrary problem. How, then, can we decide which heuristic to use? One approach is to see how far from the optimal solution each method might stray. Various formulas have been discovered to calculate the maximum discrepancy between what a bin-packing algorithm actually produces and the best possible result. For example, in situations where a large number of bins are to be packed, first fit (FF) can be off as much as 70%, but first-fit decreasing (FFD) and best-fit decreasing (BFD) are never off by more than 22%. Of course, FFD doesn't give an answer as quickly as FF, because extra time for sorting a large collection of weights may be considerable. Also, FFD requires knowing the whole list of weights in advance, whereas FF does not. It is important to emphasize that 22% is a worst-case figure. In many cases, FFD will perform much better. Recent results obtained by computer simulation (experiment) indicate excellent average-case performance for this algorithm.

When modeling real-world problems, we always have to look at the interface of the mathematics with the real world. Thus, first-fit decreasing usually results in fewer bins than next fit, but next fit can be used even when all the weights are not known in advance. Next fit also requires much less computer storage than first fit, because once a bin is packed, it need never be looked at again. Fine-tuning of the conditions of the actual problem often results in better practical solutions and in interesting new mathematics as well. (See Spotlight 3.2, p. 82, for a discussion of some of the tools mathematicians use to verify and even extend mathematical truths.)

CRYPTOGRAPHY

Not so many years ago secret codes existed primarily for the purpose of allowing diplomats and military personnel to exchange messages without their messages being read by people from other countries. Due to changes in telecommunications, banking, and lifestyles, secret codes are now widely used to protect computer files, electronic transfers of funds, and electronic mail.

We conclude with a variant of the bin-packing problem that is of interest to those who make and break secret codes. Given numbers (weights) $w_1, \ldots, w_n$ and a number S, find an algorithm that selects a subcollection of the weights whose sum is exactly S. This problem, also known to be NP-complete, is a special case of what is called the **knapsack problem** because S can be thought of as the capacity of a knapsack to be filled by the items selected from the list of weights. This particular knapsack problem is called the **subset sum problem.** As tame as it may sound, it is related to a revolutionary new proposal for ensuring the security of data stored in computers, such as bank records and classified military information.

The security of data has become an important new branch of **cryptography,** the study of codes and ciphers and how to break them. Traditional cryptography is based on a single key that is used to transform or encrypt a message (plaintext) into garbled form (ciphertext). The same key used to encrypt the message is used to decipher the message. However, in the new scheme, called *public-key cryptography,* there are two keys. One key, made public, is used by anyone wishing to send a secure message to X. The other key is known only to X, the receiver of the message. The keys are designed so that knowledge of the public key does not compromise the private key.

The idea behind public-key cryptography is that certain processes are very easy to carry out in one direction but very hard to reverse. For example, one can easily add $708 + 259 + 871 + 1836 + 82$ to get 3756, but it is not so easy to find a subcollection of the numbers 1086, 708, 82, 259, 589, 871, and 1836 that add to $S = 3756$. As another example, multiply 2993 by 1362 using pencil and paper or a calculator. Now, try to find two numbers whose product is 2,414,203. In case you ran out of patience, the answer is 1111×2173. If you carried out these two problems, you might guess that factoring is "harder" than multiplying.

Because knapsack problems are thought to be computationally hard to solve, schemes have been suggested for applying these ideas to public-key systems. Although recent work has shown that knapsack systems are faulty as a base for public-key systems, other public-key systems based on factoring large numbers appear to be viable. Furthermore, the idea of NP-completeness (see Spotlight 2.1), which arose as a concept in the abstract analysis of the complexity of algorithms, turned out to be related to the mundane issue of how to protect money being transferred between banks! We can see once more how the use of clever (though not necessarily impenetrable) mathematics can affect and enrich our lives.

SPOTLIGHT 3.2 Using Mathematical Tools

The tools of a carpenter include the saw, T square, level, and hammer. A mathematician also requires tools of the trade. Some of these tools are the proof techniques that enable verification of mathematical truths. Another set of tools consists of strategies to sharpen or extend the mathematical truths already known. For example, suppose that if A and B hold, then C is true. What happens if only A holds? Will C still be true? Similarly, if only B holds, will C still be true?

This type of thinking is of value because such questions will result either in more general cases where C holds or in examples showing that B alone and/or A alone can't imply C. For example, we saw that if a graph G is connected (hypothesis A) and even-valent (hypothesis B), then G has a tour of its edges using each edge only once (conclusion C). If either hypothesis is omitted, the conclusion fails to hold. The figures below illustrate this point. On the left you see an even-valent but nonconnected graph, and on the right, a connected graph with two odd-valent vertices; neither graph has an Euler circuit.

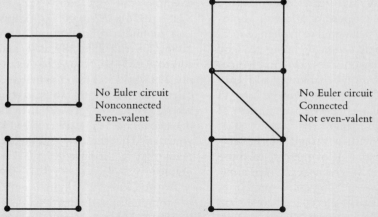

No Euler circuit
Nonconnected
Even-valent

No Euler circuit
Connected
Not even-valent

Here is another way that a mathematician might approach extending mathematical knowledge. If A and B imply C, will A and B imply both C and D, where D extends the conclusion of C? For example, not only can we prove that a connected, even-valent (hypotheses A and B) graph has an Euler circuit, but we can also show that the first edge of the Euler circuit can be chosen arbitrarily (conclusions C and D). It turns out that being able to specify the first two edges of the Euler circuit may not always be possible. Mathematicians are trained to vary the hypotheses and conclusions of results they prove in an attempt to clarify and sharpen the range of applicability of the results.

We have seen that machine scheduling and bin packing are probably computationally difficult to solve because they are NP-complete. A mathematician could then try to find the simplest version of a bin-packing problem that would still be NP-complete: What if the items to be packed can have only eight weights? What if the weights are only one and two? Asking questions like these is part of the mathematician's craft. Such questions help to extend the domain of mathematics and hence the applications of mathematics.

REVIEW VOCABULARY

Average-case analysis The study of the list-processing algorithm (more generally, any algorithm) from the point of view of how well it performs on all the types of problems it may be used on and seeing on average how well it does. (See also Worst-case analysis.)

Best fit (BF) A heuristic algorithm for bin packing in which the next weight to be packed is placed into the open bin with the largest amount of room remaining. If the weight fits in no open bin, a new bin is opened.

Best-fit decreasing (BFD) A heuristic algorithm for bin packing where the best-fit algorithm is applied to the list of weights sorted so that they appear in decreasing order.

Bin-packing problem The problem of determining the minimum number of containers of capacity W into which objects of size $w_1, \ldots, w_n$ $(w_i \leq W)$ can be packed.

Critical-path scheduling A heuristic algorithm for solving scheduling problems where the list-processing algorithm is applied to the priority list obtained by listing next in the priority list a task that heads a longest path in the order-requirement digraph. This task is then deleted from the order-requirement digraph, and the next task placed in the priority list is obtained by repeating the process.

Cryptography The study of how to make and break codes. These codes are now used primarily for data and computer security rather than for national security.

Decreasing-time-list algorithm The heuristic algorithm that applies the list-processing algorithm to the priority list obtained by listing the tasks in decreasing order of their time length.

First fit (FF) A heuristic algorithm for bin packing in which the next weight to be packed is placed in the lowest-numbered bin already opened into which it will fit. If it fits in no open bin, a new bin is opened.

First-fit decreasing (FFD) A heuristic algorithm for bin packing where the first-fit algorithm is applied to the list of weights sorted so that they appear in decreasing order.

Heuristic algorithm An algorithm that is fast to carry out but that doesn't necessarily give an optimal solution to an optimization problem.

Independent tasks Tasks are independent when there are no edges with arrows between them in the order-requirement digraph.

Knapsack problem Given a knapsack of size W and a collection of weights $w_1, \ldots w_n$, find a largest collection of weights that will fit in the knapsack.

List-processing algorithm A heuristic algorithm for assigning tasks to processors: assign the first ready task on the priority list that has not already been assigned to the lowest-numbered processor which is not working on a task.

Machine scheduling The problem of assigning tasks to processors so as to complete the tasks by the earliest time possible.

Next fit (NF) A heuristic algorithm for bin packing in which a new bin is opened if the weight to be packed next will not fit in the bin that is currently being filled; the current bin is then closed.

Next-fit decreasing (NDF) A heuristic algorithm for bin packing where the next-fit algorithm is applied to the list of weights sorted so that they appear in decreasing order.

Processor A person, machine, robot, operating room, or runway whose time must be scheduled.

Ready task A task is called ready at a particular time if its predecessors as given by the order-requirement digraph have been completed by that time.

Subset sum problem Given an integer W and integer numbers $w_1, \ldots, w_n$, find a subcollection of the numbers whose sum is W.

Worst-case analysis The study of the list-processing algorithm (more generally, any algorithm) from the point of view of how well it performs on the hardest problems it may be used on. (See also Average-case analysis).

SUGGESTED READINGS

DeMillo, R. et al. (eds.): *Applied Cryptology,* American Mathematical Society, Providence, R.I., 1983. A relatively recent survey of research work in cryptography.

French, Simon: *Sequencing and Scheduling,* Wiley, New York, 1982. A detailed account of a wide variety of scheduling models, most of them different from the ones treated in this chapter.

Graham, Ronald: "Combinatorial Scheduling Theory," in Lynn Steen (ed.), *Mathematics Today,* Springer-Verlag, New York, 1978. This essay on scheduling is one of many excellent accounts of recent developments in mathematics in this volume.

———— : "The Combinatorial Mathematics of Scheduling," *Scientific American,* 238(3):124 – 132 (March 1978). A very readable introduction to scheduling and bin packing.

Simmons, G.: "Cryptology, the Mathematics of Secure Communications," *The Mathematical Intelligencer,* 14:233 – 246 (1979). A survey of ideas and methods in cryptology, including early work on public-key cryptography.

EXERCISES

1. List as many scheduling situations as you can for the following environments:

   a. Hospital   **d.** Automobile repair garage   **g.** Your school

   b. Railroad station   **e.** Machine shop   **h.** Police station

   c. Airport   **f.** Your home   **i.** Firehouse

2. Give examples of scheduling problems for which it is *not* reasonable to assume that once a processor starts a task, it would always complete that task before it works on any other task. Give examples for which this approach would be reasonable.

3. **a.** Use the list-processing algorithm to schedule the tasks in the following order-requirement digraph on three processors, using the list $T_1, \ldots , T_{11}$. From the schedule so constructed, for each task list the start and finish time for that task.

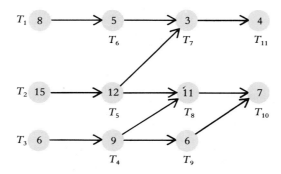

   b. Compare your answers in part **a** with scheduling these tasks on two processors with the same list.

4. For the accompanying order-requirement digraph, apply the list-processing algorithm, using three processors for lists **a** and **b**. How do the completion times obtained compare with the length of the critical path?

a. $T_1, T_2, T_3, T_4, T_5, T_6, T_7, T_8$ **b.** $T_2, T_4, T_6, T_8, T_1, T_3, T_7, T_5$

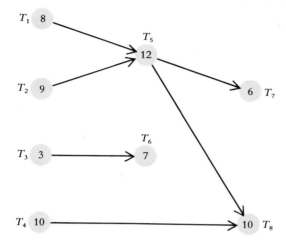

5. Discuss different criteria that might be used to construct a priority list for a scheduling problem.

6. To prepare a meal quickly involves carrying out the tasks shown (time lengths in minutes) in the following order-requirement digraph:

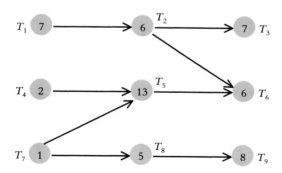

a. If Mike prepares the meal alone, how long will it take?

b. If Mike can talk Mary into helping him prepare the meal, how long will it take if the tasks are scheduled using the list $T_5, T_9, T_1, T_3, T_2, T_6, T_8, T_4, T_7$ and the list-processing algorithm?

c. If Mike can talk Mary and Jack into helping him prepare the meal, how long will it take if the tasks are scheduled using the same list as in part **b**?

d. What would be a reasonable set of criteria for choosing a priority list in this situation?

7. Consider the following order-requirement digraph:

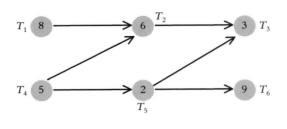

a. Find the critical path(s).

b. Schedule these tasks on one processor using the critical-scheduling method.

c. Schedule these tasks on one processor using the priority list obtained by listing the tasks in order of decreasing time.

d. Do either of these schedules have idle time? How do their completion times compare?

e. If two different schedules have the same completion time, what criteria can be used to say one schedule is superior to the other?

8. If two schedules have the same completion time, can one schedule have more idle time than the other?

9. Give examples of scheduling problems in which

a. processors available can be treated as if they are identical.

b. processors available cannot be treated as if they are identical.

10. Consider the accompanying order-requirement digraph:

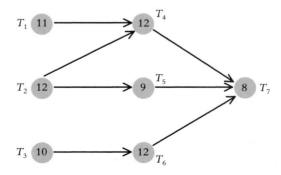

a. Find the length of the critical path.

b. Schedule these seven tasks on two processors using the list algorithm and the lists

(1) T_1, T_2, T_3, T_4, T_5, T_6, T_7

(2) T_2, T_1, T_3, T_6, T_5, T_4, T_7

c. Does either list lead to a completion time that equals the length of the critical path?

d. Show that no list can ever lead to a completion time equal to the length of the critical path (providing the schedule uses two processors).

11. For the following schedules, can you produce a list so that the list-processing algorithm produces the schedule shown when the tasks are independent? What are the task times for each task?

(a)

(b)

12. Once an optimal schedule has been found for independent tasks (e.g., see diagrams in Exercise 11), usually the scheduling of the tasks can be rearranged and the same optimal time achieved (i.e., one can, among other things, reorder the tasks done by a particular processor). Discuss criteria that might be used in implementing the rearrangement process.

13. At a large toy store, scooters arrive unassembled in boxes. To assemble a scooter the following tasks must be performed:

TASK 1. Remove parts from the box

TASK 2. Attach wheels to the footboard

TASK 3. Attach vertical housing

TASK 4. Attach handlebars to vertical housing

TASK 5. Put on reflector tape

TASK 6. Attach bell to handlebars

TASK 7. Attach decals

TASK 8. Attach kickstand

TASK 9. Attach safety instructions to handlebars

a. Give reasonable time estimates for these tasks and construct a reasonable order-requirement digraph. What is the earliest time by which these tasks can be completed?

b. Schedule this job on two processors (humans) using the decreasing-time-list algorithm.

14. Some scheduling projects have due dates for tasks (i.e., times by which a given task should be completed). Give examples of circumstances where this situation might arise.

15. Can you find a schedule (use the order-requirement digraph in Figure 3.4) with a completion time earlier than 12 as shown in the following figure, using a list other than $T_1, \ldots, T_9$? If not, why not?

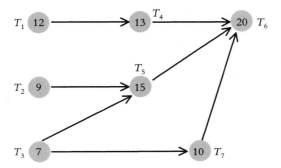

16. Given the accompanying order-requirement digraph:

a. Use the list-processing algorithm to schedule these seven tasks on two processors using these lists:

(1) The list obtained by listing the tasks in order of decreasing time

(2) T_1, T_3, T_7, T_2, T_4, T_5, T_6

(3) T_1, T_3, T_2, T_4, T_5, T_6, T_7

b. Try to determine if any of the resulting schedules are optimal.

c. Schedule the tasks using the critical-path scheduling method. Try to determine if this schedule is optimal.

17. Describe possible modifications for the list-processing algorithm that would allow for

a. a machine to be "voluntarily" idle.

b. a machine to interrupt work on a task once it has begun work on the task.

18. a. Find the completion time for independent tasks of length 8, 11, 17, 14, 16, 9, 2, 1, 18, 5, 3, 7, 6, 2, 1 on three processors, using the list-processing algorithm.

b. Find the completion time for the tasks in part **a** on three processors, using the decreasing-time-list algorithm.

c. Does either algorithm give rise to an optimal schedule?

d. Repeat for tasks of lengths 19, 19, 20, 20, 1, 1, 2, 2, 3, 3, 5, 5, 11, 11, 17, 18, 18, 17, 2, 16, 16, 2.

19. A photocopy shop must schedule independent batches of documents to be copied. The times for the different sets of documents are (in minutes): 12, 23, 32, 13, 24, 45, 23, 23, 14, 21, 34, 53, 18, 63, 47, 25, 74, 23, 43, 43, 16, 16, 76.

a. Construct a schedule using the list-processing algorithm on three machines.

b. Construct a schedule using the list-processing algorithm on four machines.

c. Repeat parts **a** and **b** but use the decreasing-time-list algorithm.

d. Suppose union regulations require that an 8-minute rest period be allowed for any photocopy task over 45 minutes. Use the decreasing-time-list algorithm, with the times above modified to take into account the union requirement, to schedule the tasks on three human-operated machines.

20. Could the following schedule have arisen from the list-processing algorithm? Could it have arisen from the application of the list-processing algorithm to a collection of independent tasks?

Machine 1	T_1	T_5	T_8
Machine 2	T_2		T_9
Machine 3	T_3	T_6	T_{10}
Machine 4	T_4	T_7	T_{11}

21. Can you give examples of scheduling problems for which it seems reasonable to assume that all the task times are the same?

22. Find a list that produces the following optimal schedule when the list-processing algorithm is applied to this list. (Assume the tasks are independent.)

What completion time and schedule are obtained when the decreasing-time-list algorithm is applied to this list?

23. Can you think of situations other than those mentioned in the text where scheduling independent tasks on processors occurs?

24. Can you think of real-world scheduling situations in which all the tasks have the same time and are independent? Can you find an algorithm for solving this problem optimally? (If there are n independent tasks of time length k, when will all the tasks be finished?)

25. a. Show that when tasks to be scheduled are independent, the critical-path method and the decreasing-time-list method are identical.

b. The (usually unknown) optimal time to complete a specific collection of independent tasks on three machines turns out to be 450 minutes.

(1) Estimate the worst possible completion time when the list-processing algorithm is used with the worst choice of priority list.

(2) Estimate the longest possible completion time using the list-processing algorithm and the decreasing-time list.

26. A radio station's policy allows advertising breaks of no longer than 2 minutes, 15 seconds. Using algorithms **a** and **b** below, determine the minimum number of breaks into which the following ads will fit (lengths given in seconds): 80, 90, 130, 50, 60, 20, 90, 30, 30, 40. Can you find the optimum solution? Do the same for these ads: 60, 50, 40, 40, 60, 90, 90, 50, 20, 30, 30, 50.

a. First fit

b. First-fit decreasing

27. It takes 4 seconds to photocopy one page. Manuscripts of lengths 10, 8, 15, 24, 22, 24, 20, 14, 19, 12, 16, 30, 15, and 16 pages are to be photocopied. How many photocopy machines would be required, using the best-fit-decreasing algorithm to guarantee that all manuscripts are photocopied in 2 minutes or less?

28. Two wooden wall systems are to be made with pieces of wood with lengths shown in the accompanying diagram. If wood is sold in 10-foot planks and can be cut with no waste, what number of boards would be purchased if one uses the first-fit-decreasing, next-fit-decreasing, and best-fit-decreasing heuristics, respectively?

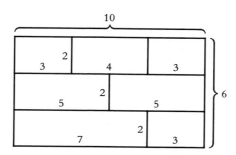

29. A typing pool gets in 30 (independent) tasks that will take the following amounts of time (in minutes) to type: 25, 18, 13, 19, 30, 32, 12, 36, 25, 17, 18, 26, 12, 15, 31, 18, 15, 18, 16, 19, 30, 12, 16, 15, 24, 16, 27, 18, 9, 14.

    a. Using these times as a priority list:

        (1) Use the list-processing algorithm to find the completion time for scheduling these tasks with four secretaries; with five secretaries.

        (2) Repeat the scheduling using the decreasing-time-list algorithm.

        (3) Can you show that any of the schedules that you get are optimal?

    b. If one needs to finish the typing in one hour:

        (1) Use the FFD heuristic to find how many typists would be needed.

        (2) Repeat for the NFD and BFD heuristics.

        (3) Can you show that any of the solutions you get are optimal?

30. Find the minimum number of bins necessary to pack items of size 8, 5, 3, 4, 3, 7, 8, 8, 6, 5, 3, 2, 1, 2, 1, 2, 1, 3, 5, 2, 4, 2, 6, 5, 3, 4, 2, 6, 7, 7, 8, 6, 5, 4, 6, 1, 4, 7, 5, 1, 2, 4 in bins of capacity **a** through **d** using the first-fit and first-fit decreasing algorithms. Can you determine if any of the packings you get are optimal?

    a. 9

    b. 10

    c. 11

    d. 12

31. Fiberglass insulation comes in 36-inch precut sections. A plumber must install insulation in a basement on piping that is interrupted often by joints. The distances between the joints on the stretches of pipe that must be insulated are 12, 15, 16, 12, 9, 11, 15, 17, 12, 14, 17, 18, 19, 21, 31, 7, 21, 9, 23, 24, 15, 16, 12, 9, 8, 27, 22, 18 inches. How many precut sections would he have to use to provide the insulation if he bases his decision on

a. next-fit?

b. next-fit decreasing?

c. best fit?

d. best-fit decreasing?

32. The files that a company has for its employees dealing with utilities occupy 100, 120, 60, 90, 110, 45, 30, 70, 60, 50, 40, 25, 65, 25, 55, 35, 45, 60, 75, 30, 120, 100, 60, 90, 85 sectors. If, after operating systems are installed, a disk can store up to 480 sectors, determine the number of disks to store the utilities if each of these heuristics is used to pack the disk with files:

a. NF

b. NFD

c. FF

d. FFD

33. Advertisements for the TV show Q are permitted to last up to a total of 8 minutes, and each group of ads can last up to 2 minutes. If the ads slated for Q last 63, 32, 11, 19, 24, 87, 64, 36, 27, 42, 63 seconds, determine if FF and FFD yield acceptable configurations for the ads.

34. A heuristic for the bin-packing problem is as follows: Place items into that currently open bin that will leave the least room left over. If an item fits in no open bin, open a new bin. This heuristic also has a "decreasing" version, where the list is first sorted in decreasing order. Using bins of capacity 10, apply this heuristic and its decreasing version to the following list: 6, 9, 5, 8, 3, 2, 1, 9, 2, 7, 2, 5, 4, 3, 7, 6, 2, 8, 3, 7, 1, 6, 4, 2, 5, 3, 7, 2, 5, 2, 3, 6, 2, 7, 1, 3, 5, 4, 2, 6.

35. How many bins are necessary to pack the numbers 1, 1, 2, 2, . . . , 20, 20 into bins of size 25? Repeat for 1, 1, 1, 2, 2, 2, . . . , 20, 20, 20. Generalize!

36. In the wall-system example in the text (see Figure 3.20), first fit and best fit required equal numbers of bins. Can you find an example where first fit and best fit yield different numbers of bins? Can you find an example where first fit, best fit, and next fit yield answers with different numbers of bins?

37. A common suggestion for heuristics for the bin-packing problem with bins of capacity W involves finding weights that sum to exactly W. Discuss the pros and cons of a heuristic of this type.

38. A record company wishes to record all the Beethoven string quartets (16 quartets, each consisting of several consecutive parts called movements) on LPs. It wishes to complete the project on as few records as possible. Is this an example of a bin-packing problem? (Defend your answer.) If the project were to record the quartets on (standard) tape cassettes or compact discs, would your answer be different?

39. Two-dimensional bin packing refers to the problem of packing rectangles of various sizes into a minimum number of $m \times n$ rectangles, with the sides of the packed rectangles parallel to those of the containing rectangle.

    a. Suggest some possible real-world applications of this problem.

    b. Devise a heuristic algorithm for this problem.

    c. Give an argument to show that the problem is at least as hard to solve as the usual bin-packing problem.

    d. If you have $1 \times m$ rectangles with total area W to be packed into a single rectangle of area $p \times q = W$, can the packing always be accomplished?

40. In what situations would packing bins of different capacities be the appropriate model for real-world situations? Suggest some possible algorithms for this type of problem.

41. Can you find an example of weights that, when packed into bins using first fit, use fewer bins than the number of bins used when the first-fit algorithm is applied with the first weight on the list removed?

42. Can you formulate "paradoxical" situations for bin packing that are analogous to those we found for scheduling processors?

·4·

Linear Programming

A corporate manager's job often involves a series of very complicated decisions. In a typical decision problem, managers have to meet objectives by careful allocation of limited resources. They also have obstacles to overcome and constraints that must be respected. All too often, there are so many alternative solutions that it is impossible to evaluate them all individually. Despite this, millions of dollars may ride on the manager's decision.

In the modern business world, diversification of products provides a company with stability in a climate of changing tastes and needs. So it is not surprising that companies would produce many products, some of which share resource needs. When resources are limited, this leads to a dilemma: how much of each product should be produced? What may not be obvious is why a company could not just specify in advance the quantity of each product it wished to produce within a given time period, calculate the amount of resources needed, and

then acquire those resources. In fact, businesses do plan what they want to produce and what resources they need.

But a business cannot always carry out its plans. It may not be able to buy just the right mix of resources it needs to produce the product mix it desires. The reasons for this inability are numerous. The world is a finite place, and at any one time there is a fixed quantity of any resource available. The amount available to a particular company is further limited by the location of the resource and the competition for, and thus the cost of, the resource. We should note here that resources include not only the raw materials used to produce a product but also labor and machinery. So resource allocation among possible products is a real problem.

When we have decisions to make in our personal lives, most of us use seat-of-the-pants methods that defy description. Traditionally, large organizations such as corporations and governments did the same — and often made bad decisions as a result. Shortly after World War II, however, an organized method of mathematical decision making called **linear programming** changed all that. Computer calculations supplemented the boss's intuition, and departments engaged in linear programming became almost as common as accounting departments. The payoff from linear programming is large, often many millions of dollars, and the method has wide applications. Estimates have been made of how frequently linear programming is used in business. Outside of routine data processing, such as payroll, linear programming is said to account for over half of all computing time used for management decisions. Some estimates range much higher, as high as 90%.

MIXTURE PROBLEMS

We will focus on a class of linear-programming problems called **mixture problems.** (Spotlight 4.1 explains how linear programming applies to other types of problems.) In mixture problems, limited resources need to be

SPOTLIGHT 4.1 Case Studies in Linear Programming

Linear programming is not limited to mixture problems. Here are two case studies that do not involve mixture problems, yet where applying linear-programming techniques produced impressive savings:

◆ The Exxon Corporation spends several million dollars per day running refineries in the United States. Because running a refinery takes a lot of energy, energy-saving measures can have a large effect. Managers at Exxon's Baton Rouge plant had over 600 energy-saving projects under consideration. They couldn't implement them all because some conflicted with others, and there were so many ways of making a selection from the 600 that it was impossible to evaluate all selections individually.

Exxon used linear programming to select an optimal configuration of about 200 projects. The savings are expected to be about $100 million over a period of years.

◆ American Edwards Laboratories uses heart valves from pigs to produce artificial heart valves for human beings. Pig heart valves come in different sizes. Shipments of pig heart valves often contain too many of some sizes and too few of others; however, each supplier tends to ship roughly the same imbalance of valve sizes on every order, so the company can expect consistently different imbalances from the different suppliers. Thus, if they order shipments from all the suppliers, the imbalances could cancel each other out in a fairly predictable way. The amount of cancellation, of course, will depend on the sizes of the individual shipments. Unfortunately, there are too many combinations of shipment sizes to consider all combinations individually.

American Edwards used linear programming to figure out which combination of shipment sizes would give the best cancellation effect. This reduced their annual cost by $1.5 million.

combined into products in a way that maximizes the profits on those products. Typical products are as diverse as juice mixtures and plywoods. We begin by looking at the typical mixture problems that might confront a juice manufacturer.

The juice manufacturer can sell many different juices and juice combinations; these are the products on which the manufacturer makes profits. Each product is either a pure juice, which would be identical to a resource for the manufacturer, or a mixture of two or more juices. There could be dozens of possible products and many varieties of fruit juice to use as resources. The manufacturer must look at the quantities and prices of unmixed juices and then determine which juice products should be produced in which quantities in order to gain the greatest, or optimum, profit. This is an enormous task, usually requiring the aid of a computer to solve. For our purposes, we take a much simpler example.

Let's examine the problem for a fruit juice manufacturer that produces and sells just two kinds of juice: appleberry and cranapple. These products are made by combining pure cranberry juice and pure apple juice in the following proportions: 3 quarts of cranberry juice and 1 quart of apple juice make 1 gallon of cranapple; 2 quarts of apple juice and 2 quarts of cranberry juice are combined to make 1 gallon of appleberry. There are 200 quarts of cranberry juice and 100 quarts of apple juice available. The manufacturer makes a profit of 3 cents on a gallon of cranapple and 4 cents on a gallon of appleberry. The manufacturer needs to determine how many gallons of cranapple and how many gallons of appleberry should be produced to obtain the highest profit without exceeding available supplies.

Let's list the features or conditions of this problem that make it a mixture problem:

◆ *Resources.* Definite resources are available in limited, known quantities for the time period in question. The resources available are the cranberry juice and the apple juice.

◆ *Products.* Definite products can be made by combining, or mixing, the resources. In this example, the products are appleberry and cranapple.

◆ *Recipes.* A recipe for each product specifies how many units of each resource are needed to make one unit of that product.

◆ *Profits.* Each product earns a known profit per unit. (It is assumed that every unit produced can be sold.)

◆ *Objective.* The objective in a mixture problem is to find how much of each product to produce in order to maximize the profit without exceeding any of the resource limitations.

Mixture problems are widespread because nearly every product produced by our economy is created by combining resources. For example, various kinds of steel alloys can be produced by combining iron with different amounts of other elements such as carbon and tungsten. How much of each alloy should be made if one has limited quantities of raw materials on hand? Or consider another example: cereal manufacturers use oats, corn, wheat, and rice to produce breakfast cereals that contain either one type of grain or a mix of grains. How many boxes of each product should be made?

MIXTURE CHARTS

One of the most important skills required to solve a mixture problem is the ability to understand its underlying structure. (This is at least as important as being able to do the subsequent algebra and arithmetic.) Understanding the underlying structure means being able to answer these questions:

1. What are the *resources?*

2. What *quantity* of each resource is available?

3. What are the *products?*

4. What are the *recipes* for creating the products from the resources?

5. What are the *unknowns?*

6. What is the *profit formula?*

We will display the answers to these questions in a diagram called a mixture chart. Then we will translate information in our mixture chart into mathematical statements that we can use to solve the mixture problem.

For the juice manufacturer, our answers to questions 1 through 6 are shown on the mixture chart in Figure 4.1. The resources (ques-

tion 1) are cranberry juice and apple juice, and the quantities available (question 2) are 200 quarts of cranberry juice and 100 quarts of apple juice, as shown in the labels at the top of the mixture chart. Questions 3 and 4, on products and recipes, are answered in the two rows of boxes in the chart.

The labels at the left side of the chart show the unknowns (question 5): how many units (gallons) of cranapple juice and how many units of appleberry to make. There is one unknown for each product, each mixture. We call our two unknowns x and y:

x = the (presently unknown) number of gallons of cranapple to make

y = the (presently unknown) number of gallons of appleberry to make

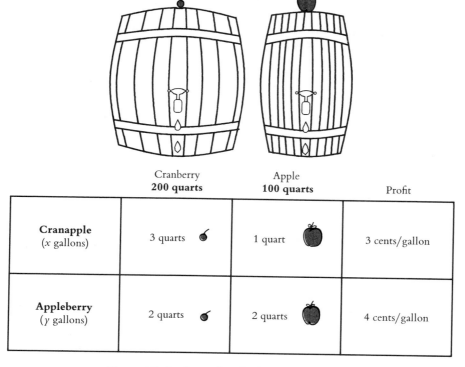

	Cranberry **200 quarts**	Apple **100 quarts**	Profit
Cranapple (x gallons)	3 quarts	1 quart	3 cents/gallon
Appleberry (y gallons)	2 quarts	2 quarts	4 cents/gallon

Figure 4.1 A mixture chart for fruit juice production.

The profit for each product (question 6) is shown in the rightmost box for that product.

Now that we have put the given data into the mixture chart, we are ready to translate it into mathematical statements. First, we note that we have one row of information for each product and one column for each resource. We also have an extra column for the profit. We will formulate a mathematical statement corresponding to each of the columns.

For each resource, we develop an inequality expressing the fact that the manufacturer cannot use more of that resource than is available. The number of quarts of cranberry juice needed for x gallons of cranapple is $3x$: 3 quarts of cranberry per gallon of cranapple times x gallons of cranapple. Similarly, $2y$ quarts of cranberry are needed for making y gallons of appleberry. So if the manufacturer makes x gallons of cranapple and y gallons of appleberry, then $3x + 2y$ quarts of cranberry juice will be used. But there are only 200 quarts of cranberry available. Maybe not all of them will be used by the manufacturer, but surely no more than the amount available can be used. Thus

$$3x + 2y \leq 200$$

Note that the numbers 3, 2, and 200 are all in the "cranberry" column. The inequality $3x + 2y \leq 200$ is called a **resource constraint** of the mixture problem; it expresses a limitation the solution must respect.

If we look at the column for the apple juice resource, we get another resource constraint, namely,

$$1x + 2y \leq 100$$

Most mixture problems also have **minimum constraints**: these are constraints that reflect the real world and tell us that the manufacturer will never make negative quantities of any mixture product. In our example we have these minimum constraints:

$$x \geq 0 \quad \text{and} \quad y \geq 0$$

Finally, we express the profit information mathematically. As with the constraints, we again deal with the presently unknown quantities x and y. Since $3x$ is the profit from making x units of cranapple and $4y$ is the profit from making y units of appleberry, we get the formula

$$\text{Profit} = 3x + 4y$$

In a mixture problem, our job is to find the values of x and y that make all the constraints true and maximize the profit. After we develop a second mixture chart, we discuss how to find these optimal values of x and y.

Here is an actual example of a mixture problem that confronted a lumber company: Plywood Ponderosa of Mexico produces various plywood products by gluing together thin sheets (veneers) of lumber. Some of these products are more profitable than others, yet it is not sensible to produce only the highest-profit product because much of the resource lumber would be left over. How much of each product — that is, which product mix — should be made in order to maximize total profit? Because Plywood Ponderosa had too many possible product mixes to consider individually, the company used linear programming to find the product mix that would give the highest overall profit. This turned out to be different from the mix they had been making. In fact, changing to the new mix increased profits by 20%. Real-world mixture problems, like Plywood Ponderosa's problem, deal with very large numbers of resources and products — too many for hand calculation. We will therefore simplify the plywood mixture problem for our example.

EXAMPLE: Making and Using a Mixture Chart. Suppose a company produces plywood by using a press to glue veneers together. The veneers come in two grades, A and B, and two kinds of plywood are made, exterior and interior. One panel of exterior plywood requires two panels of grade A veneer, two panels of grade B veneer, and 10 minutes on a press. One panel of interior plywood requires four panels of grade B veneer, no grade A veneers, and 5 minutes on the press. On a certain day there are 1000 panels of grade A veneer available and 3000 panels of grade B. There are 12 presses, each of which can press four veneer panels into a finished product. Each press can be run for 500 minutes per day, yielding a total of 6000 machine-minutes. The profit per panel is $5 for interior plywood and $6 for exterior plywood. How much of each product should be produced that day to maximize profit?

To create the mixture chart for this problem we need to identify the resources and the products. The resources are the two grades of veneer and the time available on the presses. The

products are the exterior and interior plywoods. Once the resources and products have been identified, we set up the columns and rows of the chart and enter the given data. We also assign one unknown to each product: x for exterior and y for interior. Figure 4.2 shows the mixture chart for the plywood problem.

Now we use the mixture chart to develop the mathematical statements we will need. For the constraints in the plywood problem we get these inequalities:

Grade A: $2x + 0y \leq 1000$
                   (we could omit the y term)

Grade B: $2x + 4y \leq 3000$

Time: $10x + 5y \leq 6000$

Minimums: $x \geq 0$ and $y \geq 0$

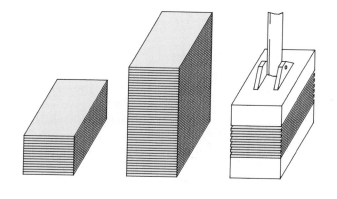

	Grade A **1000 panels**	Grade B **3000 panels**	Presses **6000 minutes**	Profit
Exterior (x panels)	2 of A	2 of B	10 minutes	$6/panel
Interior (y panels)	0 of A	4 of B	5 minutes	$5/panel

Figure 4.2 A mixture chart for plywood production.

From the column with profit data we get the profit formula

$$\text{Profit} = 6x + 5y$$

In the sections that follow, we use our mixture charts to solve these two mixture problems.

The fruit juice and plywood problems deal with different industries and different numbers, but to a mathematician they are similar because both fit into the framework of a mixture problem and can be represented by mixture charts. The task of analyzing a mixture problem (i.e., making a mixture chart for it) is mostly a matter of careful reading and logic. Some might be fooled into thinking that it is not mathematics at all. But actually, it is an extremely important skill in the practice of applied mathematics.

GRAPHING THE FEASIBLE REGION

The reader with some exposure to mathematics may sense that there must be some way to obtain the optimal x and y using a formula or algebraic procedure that applies the profit formula and the various inequalities from the mixture chart. There are, indeed, such procedures. (One of them, the simplex method, will be discussed later in this chapter.) The details of these algebraic procedures are technical and soon forgotten. Furthermore, there are computer programs available to solve linear-programming problems once the mixture chart has been constructed. We will present a pictorial view of linear programming that has been valuable to theorists and practitioners alike and that is readily understandable by nonmathematicians.

The basic notion is that a pair of numerical values, such as (5, 3) or (−1, 2) or more generally *(x, y)*, can be interpreted as a point in the plane. In a mixture problem involving two products, we are looking for a pair of values x and y that gives the most profit. This can be interpreted as a search for the best point on the plane.

We can't locate this best point directly, so we begin with another question: Which points are the reasonable candidates for this paragon of points? A candidate for best point must satisfy the constraints we developed from the mixture chart.

We recall that there were two types of constraints, those based on minimums and those based on availability of resources. Returning to our juice manufacturer, the minimum constraints simply say, for example, that you can't make −6 gallons of appleberry. To illustrate the effect of the resource constraints, suppose the fruit juice manufacturer decides to make 70 gallons of cranapple and 20 gallons of appleberry; that is, x and y are to be 70 and 20, respectively. If we substitute (70, 20) for *(x, y)* in the inequality for cranberry juice, $3x + 2y \leq 200$, we get $3(70) + 2(20)$ on the left-hand side. Evaluating this expression, we get $210 + 40$, or 250. But 250 is bigger than the 200 on the right-hand side. Thus the point (70, 20) does not represent a manufacturing decision that can be carried out; it is not feasible to do so. We are not permitted to exceed the supply of even a single resource.

Points that do not violate either type of constraint are called **feasible points.** All others are infeasible points. The collection of feasible points is called the **feasible set.** It is helpful to be able to visualize the feasible set as a portion of the plane or space. We call the visualization the **feasible region.**

What does a feasible region look like? Are the feasible points interspersed with infeasible points in a salt-and-pepper pattern? Or do the feasible points clump together in some type of connected shape? Techniques of algebra and analytic geometry allow us to use the constraint inequalities, derived from the data in the mixture chart, to determine the precise shape of the feasible region. We will do so for both the juice and the plywood manufacturers.

In the example of the juice manufacturer, we developed these constraints:

Cranberry juice:	$3x + 2y \leq 200$
Apple juice:	$1x + 2y \leq 100$
Minimums:	$x \geq 0$ and $y \geq 0$

Each of these inequalities can be broken down into two parts. Since we read the symbol $\leq$ as "less than or equal to," we have "less than" and "equal to." To graph the part of the plane that satisfies "less than or equal to," we will put together, form the union of, the portion of the plane satisfying "less than" and the portion satisfying "equal to."

Focusing first on the "equal to," we find that associated with the cranberry juice constraint, $3x + 2y \leq 200$, is the equation $3x + 2y = 200$. This is an equation for a straight line in the plane. We recognize that any equation having either an x term, a y term, or both, and some numerical constant, but no other kinds of terms, always represents a line. (The equation cannot have any squared, square root, or other kinds of algebraic combinations of x or y.) In fact, constraint inequalities always can be associated with equations for lines, hence the term *linear* programming. The *programming* does not refer to a computer but to a well-defined sequence of steps, or program of action, that solves the kinds of problems we are exploring.

EXAMPLE: Graphing a Linear Inequality. The graph of the line $3x + 2y = 200$ is shown in Figure 4.3a. There are several ways to plot a line. For many of the lines in our mixture problems, using the intercepts, the

points where the line crosses the axes, is a convenient method. We remember that when the line crosses, or intersects, the x-axis, the value of y is 0. Substituting $y = 0$ into the equation $3x + 2y = 200$ gives $3x = 200$, or $x = \frac{200}{3}$. So one point on the line $3x + 2y = 200$ is $(\frac{200}{3}, 0)$. On the y-axis, $x = 0$. Substituting $x = 0$ into our equation $3x + 2y = 200$, we get $2y = 200$, or $y = 100$. So we have a second point on our line, namely, (0, 100). In Figure 4.3a, these two points are shown and the line was drawn by connecting them. (Remember, two points determine a line.)

Every point (x, y) on the line $3x + 2y = 200$ is a pair of values for x and y that makes the equality a true statement. The line also divides the whole plane into two big halves, often

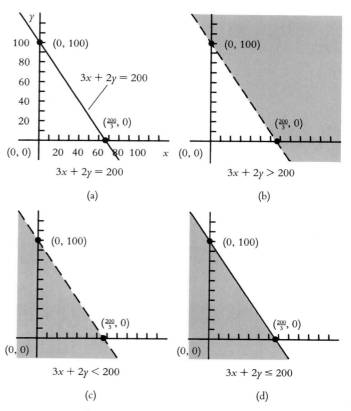

Figure 4.3 Graphing $3x + 2y \leq 200$.

called half planes. It turns out that all the points in one half plane make the inequality $3x + 2y < 200$ a true statement and all the points in the other half plane make the other inequality $3x + 2y > 200$ a true statement. To find out which half plane corresponds to the inequality $3x + 2y < 200$, we need only test one point of the plane that is not on the line $3x + 2y = 200$. Let's see how that is done.

Suppose we choose the point (50, 50). That point is not on the line because $3(50) + 2(50) = 200$ is not a true statement. In fact $3(50) + 2(50)$ evaluates to $150 + 100$, or 250, which is greater than 200. This tells us that the point (50, 50) is in the half plane corresponding to the inequality $3x + 2y > 200$. In mixture problems, many constraint inequalities correspond to lines that do not contain the point (0, 0), so that point is frequently used. Using that point we see that it lies in the half plane for which $3x + 2y < 200$. We see that points on opposite sides of the line satisfy opposite inequalities; we could use either point, or some other one, to determine which half plane corresponds to which inequality. Figure 4.3b shows the half plane corresponding to $3x + 2y > 200$, and Figure 4.3c shows the half plane corresponding to $3x + 2y < 200$.

Returning to our original constraint, $3x + 2y \leq 200$, we wish to show on a graph all the points that make it a true statement. Those are the points that make the equality $3x + 2y = 200$ true and also the points that make the inequality $3x + 2y < 200$ true. That is, all the points on the line in Figure 4.3a and all the points in the half plane in Figure 4.3c are points that make our constraint a true statement. The set of points making $3x + 2y \leq 200$ true is shown in Figure 4.3d.

EXAMPLE: Graphing a Feasible Region.
In Figure 4.4a we repeat the graph in Figure 4.3d of the cranberry resource constraint inequality $3x + 2y \leq 200$. In Figure 4.4b we show the graph of the apple constraint in-

equality $1x + 2y \leq 100$. Note that the two points used to draw that line are (0, 50) and (100, 0), the two intercepts. You can check that the correct half plane has been shaded by testing the point (0, 0). Now we know that a point in the feasible region must satisfy, or make true, every constraint inequality in the problem. Which points in the plane make both the cranberry and the apple constraint inequalities true? They are the points that are in not just one but both of the shaded regions in Figure 4.4a and 4.4b; the set we want is the intersection of the two half planes. The region satisfying both resource constraints is shown in Figure 4.4c. [In the next example we will show how coordinates of a corner point like (50, 25) can be determined.]

In the juice manufacturer mixture problem we also have two minimum constraints. The graphs for these inequalities, $x \geq 0$ and $y \geq 0$, are shown in Figure 4.4d and 4.4e, respectively. These minimum constraints tell us that the feasible region must be in the first quadrant of the plane. When we put these two together with the graph in Figure 4.4c, we get the final graph, Figure 4.4f, of the feasible region for the juice mixture problem.

We note here that the use of the two intercepts to plot a line will not work in the special cases of horizontal and vertical lines. Equations for horizontal lines have no x term; for example, $3y = 90$ is a horizontal line. It can be graphed by solving for y, giving $y = 30$, and then drawing the horizontal line all of whose y-coordinates are 30. Similarly $5x = 60$ becomes $x = 12$, whose graph is the vertical line all of whose x-coordinates are 12. The x- and y-axes are special cases of this general case: the x-axis corresponds to $y = 0$ and the y-axis to $x = 0$.

In the next section we will learn that the corner points of a feasible region are important in finding where in the feasible region we can find the maximum profit. We need to be able

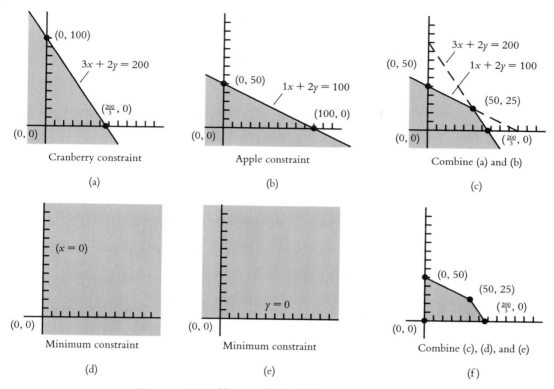

Figure 4.4 Feasible region for the fruit juice example.

to calculate the coordinates of those corner points.

Each corner point of a feasible region is the intersection of two lines; the lines, of course, each correspond to one of the constraint inequalities. It is important to keep track of which two lines are intersecting at each corner point, because in general there are intersections of lines that are not corner points of the feasible region. For example, in Figure 4.4b, the point (100, 0) is the intersection of the x-axis, that is, the line $y = 0$, and the line $1x + 2y = 100$. But (100, 0) is not a corner point of the feasible region in Figure 4.4f.

EXAMPLE: Finding Corner Points. We look at the graph in Figure 4.4f. We start at some corner and work our way clockwise around the edge of the feasible region. A convenient place to start, if it is a corner, is the origin (0, 0). Although we know the coordinates of the origin, it is useful to note that it is the intersection of two lines having equations $x = 0$ and $y = 0$ and corresponding to minimum constraints. These equations "solve" the problem of finding the coordinates. In general, we are trying to solve for values of x and y that satisfy both the linear equations; then we will have the coordinates of the intersection.

The next corner is the intersection of the lines $x = 0$ and $1x + 2y = 100$. We already found this point when we plotted the line $1x + 2y = 100$. The point is (0, 50).

Continuing clockwise, we come to the intersection of lines $1x + 2y = 100$ and $3x + 2y = 200$. This more general type of intersection can be solved by using multiplication and addition to eliminate one unknown, solving for the remaining unknown, and then substituting that value into an original equation to get the value of the eliminated unknown. First, we multiply one equation by a positive value and the other by a negative value so that when the two equations are added together, one unknown gets a coefficient of zero:

$$(-3)(1x + 2y = 100) = -3x - 6y = -300$$
$$(1)(3x + 2y = 200) = \quad 3x + 2y = 200$$

It does not matter what numbers we use to multiply as long as we get plus and minus the same coefficient on either both x terms or both y terms. Now we add the new equations together:

$$0x - 4y = -100 \quad \text{or} \quad -4y = -100$$

Solving this we get $y = 25$. Substituting this y value into $1x + 2y = 100$, we get $1x + 2(25) = 100$, which gives $x = 50$. So the point of intersection seems to be $(50, 25)$. We can check our work in the other original equation: $3(50) + 2(25) = 200$ is a true statement.

The last corner point of this feasible region comes from the intersection of $3x + 2y = 200$ and $y = 0$. As with the other intersection of a resource constraint line and an axis, we already know the coordinates: $(\frac{200}{3}, 0)$.

In some mixture problems, there is a horizontal or vertical line from a resource constraint. Thus one of the lines already tells us the value of one coordinate, and we can substitute that value into the other equation to get the value of the other coordinate. For example, to find the intersection of the vertical line $x = 5$ and the line $4x + 3y = 38$, we substitute, getting $4(5) + 3y = 38$, or $3y = 18$, which gives $y = 6$.

EXAMPLE: The Feasible Region for the Plywood Problem. We now return to the Ponderosa Plywood problem, construct its feasible region, and determine the corner points. We recall the constraint inequalities:

Grade A:	$2x \leq 1000$
Grade B:	$2x + 4y \leq 3000$
Time:	$10x + 5y \leq 6000$
Minimums:	$x \geq 0$ and $y \geq 0$

In Table 4.1 we summarize the information needed to draw the correct half plane for each resource constraint and list the graph in Figure 4.5 in which that half plane is drawn. Forming the correct feasible region from those three half planes plus the first quadrant, arising from the minimum constraints, could be tricky; remember it will be necessary to know exactly which two lines intersect to form each corner point.

One way to form the region is to start with one half plane, such as in Figure 4.5a, add to it

TABLE 4.1 Using mixture chart data to determine a feasible region

Resource	Equation	Intersections		Value at origin	Figure
		x-axis	y-axis		
Grade A	$2x = 1000$	$(500, 0)$	None	<1000	4.5a
Grade B	$2x + 4y = 3000$	$(1500, 0)$	$(0, 750)$	<3000	4.5b
Time	$10x + 5y = 6000$	$(600, 0)$	$(0, 1200)$	<6000	4.5c

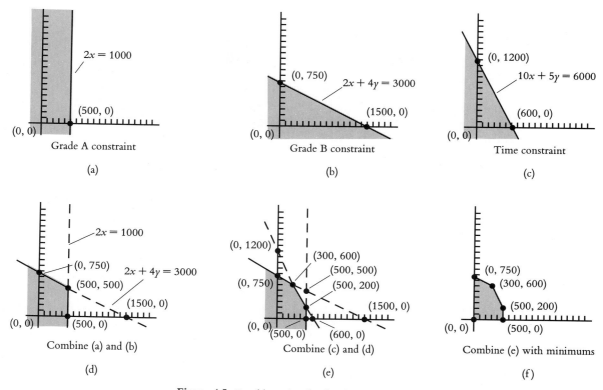

Figure 4.5 Feasible region for the plywood example.

the information from one other half plane, such as in Figure 4.5b, getting partway toward the feasible region, as we see in Figure 4.5d. Then take the new picture and add to it the information from the next half plane, here Figure 4.5c, getting Figure 4.5e. In our case, we are essentially done; all that is left is to restrict the picture to the first quadrant and we have the correct feasible region shown in Figure 4.5f. After you are familiar with constructing feasible regions, you may want to carefully draw all the relevant lines on one graph and then shade in the region.

Note that until we found the coordinates of the corner points we kept labels on all the line segments, reminding us of which lines intersect to form each corner point. The intersection of the lines $2x = 1000$ and $2x + 4y =$

3000 is not a corner point of the feasible region, so determining its coordinates is not necessary. We leave it to you to check that the coordinates of the corner points not on either axis have been properly calculated.

In general, we can say this about the shape of a feasible region:

1. For any mixture problem of the kind we are discussing, the feasible region is a polygon in the first quadrant. This is because the minimum constraints require that both x and y be nonnegative. The feasible region has one corner at the origin and one on each coordinate axis. There are usually additional corners off the axes.

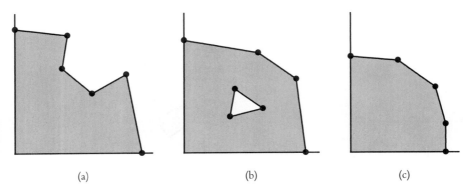

Figure 4.6 A feasible region may not have (a) dents or (b) holes. Graph (c) shows a typical feasible region.

2. The region is a polygon that has neither dents (as in Figure 4.6a) nor holes (as in Figure 4.6b). Figure 4.6c is a typical general example.

THE CORNER PRINCIPLE

To evaluate the profit that goes with a given point (x, y), use the profit formula recorded on the mixture chart. In the juice mixture problem, for example, to find the profit associated with the point (40,10), substitute $x = 40$ and $y = 10$ into the formula for profit:

$$\text{Profit} = 3x + 4y$$
$$= 3(40) + 4(10)$$
$$= 160$$

If we had some way of applying this formula to each point of the feasible region, we could then pick out the point with the highest profit. Unfortunately, the feasible region has infinitely many points, making it impossible to compute the profit for each point. Luckily, we can narrow our search because of the **corner principle:** *the highest profit value on a polygonal feasible region can always be found at a corner point.*

The corner principle is probably the most important insight in the theory of linear programming. It is the geometric nature of this principle that explains the value of creating a geometric model from the data in a mixture chart.

The corner principle allows us to use the following method to solve a mixture problem:

1. Calculate the corner points of the feasible region.

2. Evaluate the profit at each corner of the feasible region.

3. Choose the corner with the highest profit.

The corners of the feasible region for the juice mixture problem are shown in Figure 4.4f, and Table 4.2 shows the profit at each corner. The highest profit is 250 and occurs at the corner point (50, 25). Therefore, to achieve this profit, we should make 50 gallons of cranapple and 25 gallons of appleberry.

The situation here is reminiscent of Alexander the Great's approach to the problem of the Gordian knot, a legendary knot so large and tight and tangled that no one had been able to untie it. Alexander's solution was to slice the knot open with his sword. Mixture problems and other linear-programming problems are like Gordian knots because there are infinitely many feasible points—we can't calculate the profit for all of them. The corner principle functions like Alexander's sword and cuts through the problem.

TABLE 4.2 Evaluating Profit Formula at Corner Points

Corner	Profit ($= 3x + 4y$)
$(0, 0)$	$3(0) + 4(0) = 0$
$(0, 50)$	$3(0) + 4(50) = 200$
$(50, 25)$	$3(50) + 4(25) = 250$
$(\frac{200}{3}, 0)$	$3(\frac{200}{3}) + 4(0) = 200$

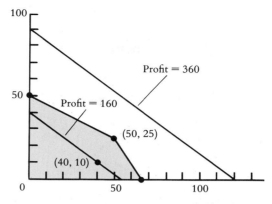

Figure 4.7 The profit line for 360 lies outside the feasible region, whereas the profit line for 160 passes through the region.

You can visualize a mathematical proof of the corner principle by imagining that each point of the plane is a tiny light bulb that is capable of lighting up. For the example in Figure 4.4f, imagine what would happen if we ask this question: Will all points with profit = 360 please light up? What geometric figure do these lit-up points form?

In algebraic terms, we can restate the profit question in this way: Will all points (x, y) with $3x + 4y = 360$ please light up? As it happens, this version of the profit question is one mathematicians learned to answer hundreds of years before linear programming was born. The points that light up make a straight line. Furthermore, it is a routine matter to determine the exact position of the line. We call this line the **profit line** for 360; it is shown in Figure 4.7. For numbers other than 360, we would get different profit lines. Unfortunately, there are no points on the profit line for 360 that are feasible, that is, that lie in the feasible region. Therefore, the profit of 360 is unachievable. *If the profit line corresponding to a certain profit doesn't touch the feasible region, then that profit can't be achieved.*

Because 360 is too big, perhaps we should ask the profit line for a more modest amount, say, 160, to light up. You can see that the new profit line of 160 in Figure 4.7 is parallel to the first profit line and closer to the origin. This is no accident: well-known elementary theorems about the slope and y-intercept of straight lines prove it must be so. All profit lines have the same coefficients for x and y, so they have the same slope. Changing the con-

stant has the effect of changing the y-intercept, but not the slope.

The most important feature of the profit line for 160 is that it has points in common with the feasible region. For example, $(40, 10)$ is on that profit line because $3(40) + 4(10) = 160$ and in addition is a feasible point. This means that it is possible to make 40 gallons of cranapple and 10 gallons of appleberry and that if we do so, we will have a profit of 160.

Can we do better than a 160 profit? As we slowly increase our desired profit from 160 toward 360, the location of the profit line that lights up shifts smoothly upward away from the origin. As long as the line continues to cross the feasible region, we are happy to see it move away from the origin, because the more it moves, the higher will be the profit represented by the line. We would like to stop the movement of the line at the last possible instant, while the line still has one or more points in common with the feasible region. It should be obvious that this will occur when the line is just touching the feasible region either at a corner point (Figure 4.8a) or along a line segment joining two corners (Figure 4.8b). This is just what the corner principle says.

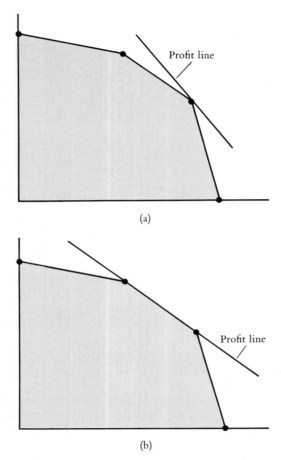

(a)

(b)

Figure 4.8 The highest profit will occur when the profit line is just touching the feasible region, either (a) at a corner point or (b) along a line segment.

Suppose now that each light bulb (point of the plane) has a color determined by the profit associated with that point. All points with the same profit (i.e., points on a profit line) have the same color. Furthermore, suppose the colors range continuously from violet to blue to green to yellow to red, just as they do in a rainbow. The cool colors represent low profits, the hot colors, higher ones: the higher the profit, the hotter the color. In effect, we are superimposing a straightened-out rainbow on the picture containing our feasible region.

Finding the highest profit point can be thought of as finding the hottest-colored point in the feasible region (Figure 4.9).

It would be convenient if all linear-programming problems yielded simple feasible regions in two-dimensional space. However, two complications arise for practical problems:

1. Sometimes, as in Figure 4.10, we have a great many corners. Naturally, the more corners there are, the more calculations we need to do to determine the coordinates of all of them and the profit at each one.

2. It will not be possible to visualize the feasible region as a part of two-dimensional space when there are more than two products. Each product is represented by an unknown, and each unknown is represented by a dimension of space. If we had 50 products, we would need 50 dimensions, and we couldn't visualize the feasible region.

Most practical linear-programming problems present us with both cases, that is, with many corners and more than two products. Thus the number of corners literally can exceed the number of grains of sand on the earth. Even with the fastest computer, computing the profit of every corner is impossible.

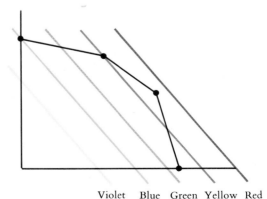

Violet   Blue   Green   Yellow   Red

Figure 4.9 If profit lines are colored according to the hues of the rainbow, the highest profit point will be the hottest-colored point in the feasible region.

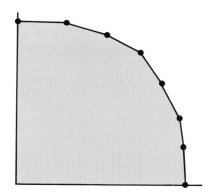

Figure 4.10 A feasible region with many corners.

THE SIMPLEX METHOD

Two main methods have been proposed to get around this difficulty. The older method is the **simplex method,** which is still the most commonly used. Discovered by the American mathematician George Dantzig (see Spotlight 4.2, p. 110), this ingenious mathematical invention makes it possible to find the best corner by evaluating only a tiny fraction of all corners. In this way, a problem that might have taken centuries to solve on a supercomputer (if each corner had to be checked) can be solved in seconds on a home computer.

The operation of the simplex method may be likened to the behavior of an ant crawling on the edges of a polyhedron (a solid with flat sides), looking for one particular target vertex (Figure 4.11). The ant cannot see where the target vertex is. As a result, if it were to wander along the edges randomly, it might take a long time to reach the target. The ant will do much better if it has a temperature clue to let it know it is getting warmer (closer to the target vertex) or colder (farther from the target vertex).

Think of the simplex method as a way of calculating these temperature hints (it turns out there are fairly simple algebraic manipulations to do this). We begin at any randomly chosen vertex. All neighboring vertices are

evaluated to see which ones are warmer and which are colder. A new vertex is chosen from among the warmer ones, and the evaluation of neighbors is repeated—this time checking neighbors of the new vertex. The process ends when we arrive at the target vertex.

Part of what the simplex method has going for it in the speed sweepstakes is that it works faster in practice than its worst-case behavior would lead us to believe. Although mathematicians have devised artificial cases for which the simplex method bogs down in unacceptable amounts of arithmetic, the examples arising from real applications are never like that. This may be the world's most impressive counterexample to Murphy's law, which says that if something can go wrong, it will.

Although the simplex method usually avoids visiting every vertex, it may require visiting many intermediate vertices as it moves from the starting vertex to the optimal one. The simplex method has to search along edges on the boundary of the polyhedron. If it happens that there are a great many small edges lying between the starting point and the optimal vertex, the simplex method must operate like a slow-moving bus that stops at every street corner.

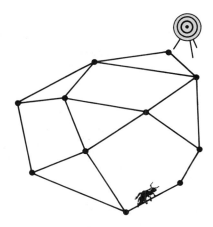

Figure 4.11 The simplex method can be compared to an ant crawling along the edges of a polyhedron, looking for one particular target vertex.

AN ALTERNATIVE TO THE SIMPLEX METHOD

In 1984, Narendra Karmarkar (see Figure 4.12), a mathematician working at Bell Laboratories, devised an alternative method for linear programming that avoids such slowdowns by making use of search routes through the interior of the feasible region (see Spotlight 4.3, p. 112). The potential applications of Karmarkar's algorithm are important to a lot of industries, including telephone communications and the airlines (see Spotlight 4.4, p. 113). Routing millions of long-distance calls, for example, means deciding how to use the long-distance landlines, repeater amplifiers, and satellite terminals to best advantage. The problem is similar to the juice company's need to find the best use of its stocks of juice to create the most profitable mix of products. We find another example in the airline business. American Airlines worked with Dr. Karmar-

SPOTLIGHT 4.2 Father of Linear Programming Recalls Its Origins

George Dantzig is professor of operations research and computer science at Stanford University. He is credited with inventing the linear-programming technique called the simplex method. For the past 40 years, the simplex method has provided solutions to linear-programming problems that have saved both industry and the military time and money. Dantzig talks about the background of his famous technique:

Initially, all the work we did had to do with military planning. During World War II, we were planning on a very extensive scale. The civilian population and the military were all performing scheduling and planning tasks, perhaps on a larger scale than at any time in history. And this was the case up until about 1950. From 1950 on, the whole emphasis shifted from military planning to practical planning for the civilian population, and industry picked it up.

The first areas of industry to use linear programming were the petroleum refineries. They used it for blending gasoline. Nowadays, they run all of the refineries in the world (except for one) using linear-programming methods. They are one of the biggest users of it, and it's been picked up by every other industry you can think of — the forestry industry, the steel industry — you could fill up a book with all the different places it's used.

The question of why linear programming wasn't invented before World War II is an interesting one. In the postwar period, various technologies just evolved that had never been there before. Computers were one example. These technologies were talked about before; you can go back in history and you'll find isolated papers on them. For example, the very famous French mathematician Joseph Fourier, who invented Fourier series, had a paper on problems similar to linear programming, as did the Belgian mathematician de la Valee Pousson. But these were isolated cases that never went anywhere.

In the immediate postwar period, everything just fermented and began to happen,

Figure 4.12 Narendra Karmarkar, a researcher at AT&T Bell Laboratories, has invented a powerful new linear-programming algorithm that solves many complex linear-programming problems faster and more efficiently than any previous method.

kar to see if his algorithm could cut fuel costs. According to Thomas Cook, director of operations research for American Airlines, "It's big dollars. We're hoping we can solve harder problems faster, and we think there's definite potential."

and one of the things that began to happen was linear programming. Mathematicians as well as economists and others who do practical planning and scheduling saw the possibilities. Things then began to happen very rapidly.

In the immediate postwar period, the whole idea of using computers to mechanize the process was an obvious one. The question was then asked, How could you formulate the process as a sort of mathematical system, and how could computers be used to make this happen?

The problems we solve nowadays have thousands of equations, sometimes a million variables. One of the things that still amazes me is to see a program run on the computer—and to see the answer come out. If we think of the number of combinations of different solutions that we're trying to choose the best of, it's akin to the stars in the heavens. Yet we solve them in a matter of moments. This, to me, is staggering. Not that we can solve them—but that we can solve them so rapidly and efficiently.

The simplex method has been used now for close to 40 years. There has been steady work going on trying to use different versions of the simplex method, nonlinear methods, and interior methods. It has been recognized that certain classes of problems can be solved much more rapidly by special algorithms than by using the simplex method. If I were to say what my field of specialty is, it is in looking at these different methods and seeing which are more promising than others.

For example, when the system is too big, we use something called the "decomposition principle" to break it into parts and solve the parts. This has been very efficient for certain classes of problems. Certain methods that have been reported in the news that are quite well known in nonlinear programming, called "interior techniques," have been experimented on very successfully with some kinds of problems, and not so successfully with others.

In my classes at Stanford, we review these different methods, and it's surprising how good some of them are. There's a lot of promise in this—there's always something new to be looked at.

In the 1980s, scientists at Bell Labs applied Karmarkar's algorithm to a problem of unprecedented complexity: deciding how to economically build telephone links between cities so that calls can get from any city to any other, possibly being relayed through intermediate cities. Figure 4.13 shows one such linking. The number of possible linkings is unimaginably large, so picking the most economical one is difficult. For any given linking that one has built, or contemplates building, there is also the problem of deciding how to economically

route calls through the network to reach their destinations.

Although difficult, these problems are definitely worth solving. Nat Levine, director of the transmission facilities planning center at Bell Labs, speculated that if one found the best

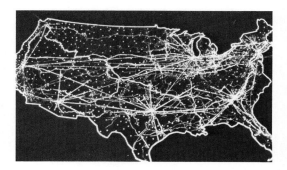

Figure 4.13 A map of the United States showing one conceivable network of major communication lines connecting major cities. Routing millions of calls over this immense network requires sophisticated linear programming techniques and high-speed computers. [Courtesy of AT&T Bell Laboratories.]

SPOTLIGHT 4.3 Linear Programming and the Cold War

In 1979, mathematics found itself on the front page of the *New York Times* as a result of a linear-programming technique invented by the Russian mathematician L. G. Khachian. This algorithm, called the ellipsoid method, seemed at first to be much better than the simplex method for large problems. Perhaps because so much is at stake in this area of applied mathematics, the story spread like wildfire, with little regard for accuracy. In the popular press it appeared as though the Russians had achieved a breakthrough that might imperil our national security. In particular, it was suggested that the method might be applied to break cryptographic codes that had previously been considered secure.

As it turned out, there was no real reason for alarm. The ellipsoid method doesn't apply to the variant of linear programming that occurs in cryptography. Furthermore, the kinds of linear-programming problems for which it does work better than the simplex method are those which are far larger than any that will ever occur in practice. Best of all, following a tradition of international communication in the mathematics community that dates back thousands of years, the Russian mathematicians made no attempt to keep the ellipsoid algorithm a secret.

In fact, the impact of the ellipsoid algorithm on American mathematics was undoubtedly positive. Although it does not seem to be a practical algorithm, it did break a certain theoretical barrier, thus creating an incentive for other researchers. Narendra Karmarkar, an American mathematician working at Bell Laboratories, broke through this barrier in another way, finding what seems to be a truly practical method for linear programming.

SPOTLIGHT 4.4 Finding Fast Algorithms Means Better Airline Service

Linear-programming techniques have a direct impact on the efficiency and profitability of major airlines. Thomas Cook, director of operations research at American Airlines, shared his ideas on why optimal solutions are essential to his business:

Finding an optimal solution means finding the best solution. Let's say you are trying to minimize a cost function of some kind. For example, we may want to minimize the excess costs related to scheduling crews, hotels, and other costs that are not associated with flight time. So we try to minimize that excess cost, subject to a lot of constraints, such as the amount of time a pilot can fly, how much rest time is needed, and so forth.

An optimal solution, then, is either a minimum-cost solution or a maximizing solution. For example, we might want to maximize the profit associated with assigning aircrafts to the schedule; so we assign large aircraft to high-need segments and small aircraft to low-load segments. Whether it's a minimum or maximum solution depends on what function we are trying to optimize.

Finding fast solutions to linear-programming problems is also essential. If we can get an algorithm that's 50 to 100 times faster, we could do a lot of things that we can't do today. For example, some applications could be real-time applications, as opposed to batch applications. So instead of running a job overnight and getting an answer the next morning, we could actually key in the data or access the data base, generate the matrix, and come up with a solution that could be implemented a few minutes after keying in the data.

A good example of this kind of application is what we call a major weather disruption. If we get a major weather disruption at one of the hubs, such as Dallas or Chicago, then a lot of flights may get canceled, which means we have a lot of crews and airplanes in the wrong places. What we need is a way to put that whole operation back together again so that the crews and airplanes are in the right places. That way, we minimize the cost of the disruption and minimize the passenger inconvenience.

We're working on that problem today, but in order to solve it in an optimal fashion, we need something as fast as Narendra Karmarkar's algorithm. In the absence of that, we'll have to come up with some heuristic ways of solving it that won't be optimal.

The simplex method, which was developed some 40 years ago by George Dantzig, has been very useful at American Airlines and, indeed, at a lot of large businesses. The difference between his solution and Karmarkar's is that if we can get an algorithm that comes up with basically the same optimal answer 50 to 100 times faster, then we can apply that technology to new problems, and even to problems that we wouldn't have tried using the simplex method. I think that's the primary reason for the excitement.

solution, "the savings could be in the hundreds of millions of dollars."

Work on these problems at Bell Labs involved a linear-programming problem with about 800,000 variables, which Karmarkar's algorithm solved in 10 hours of computer time. The scientists involved believe that the problem might have taken weeks to solve if the simplex method had been used.

Karmarkar's algorithm has been incorporated into a software product. The extensive data collected from worldwide use of the software supports the results at Bell Labs. It appears that for some kinds of linear programming problems, Karmarkar's algorithm is a big improvement over the simplex method.

REVIEW VOCABULARY

Corner principle The principle that states that there is a corner point of the feasible region that yields the optimal solution.

Feasible point A possible solution (but not necessarily the best) to a linear-programming problem.

Feasible region A representation of the feasible set as a portion of n-dimensional space.

Feasible set The set of possible solutions to a linear-programming problem.

Linear programming A set of organized methods of decision making in which the mathematical formulation of the problem involves only linear equations and inequalities. Mixture problems are usually solved by some type of linear programming.

Minimum constraint An inequality in a mix-

ture problem that reflects the fact that negative quantities of products cannot be produced.

Mixture problem A problem in which a variety of resources are available in limited quantities and in which these resources can be combined in various ways to make different products. It is usually desired to find the way of combining the resources that produces the most profit.

Profit line The set of all feasible points that yields the same profit (for two-dimensional linear-programming problems).

Resource constraint An inequality in a mixture problem that reflects the fact that no more of a resource can be used than what is available.

Simplex method One of a number of solution methods for linear-programming problems.

SUGGESTED READINGS

Anderson, David R., Dennis J. Sweeney, and Thomas A. Williams: *An Introduction to Management Science: Quantitative Approaches to Decision Making,* West, St. Paul, Minn., 1985. A business management text containing seven chapters covering many applications and aspects of linear programming.

Gass, Saul I.: *An Illustrated Guide to Linear Programming,* McGraw-Hill, New York, 1970. An engagingly written beginner's approach, which introduces the algebra gently and emphasizes the formulation of problems more than algebraic technique.

Glicksman, Abraham: *Linear Programming and the Theory of Games,* Wiley, New York, 1963. The best elementary approach to a full understanding of the simplex method, including numerous numerical examples.

Malkevitch, Joseph, and Walter Meyer: *Graphs, Models, and Finite Mathematics,* Prentice-Hall, Englewood Cliffs, N.J., 1974. A little more emphasis on "how to" than is given in this book.

Meyer, Walter: *Concepts of Mathematical Modeling,* McGraw-Hill, New York, 1984. Includes a guide to using simplex method computer programs.

EXERCISES

1. Graph each of these lines:

   a. $y = 7$

   b. $x = 2$

   c. $2x + 3y = 12$

2. Graph each of these lines:

   a. $5x + 4y = 20$

   b. $y = 3$

   c. $x = 4$

In Exercises 3 to 8, graph both lines on the same graph, showing by a dot any intersection of the lines. Use algebra to determine the x- and y-coordinates of any point of intersection.

3. $x = 5$ and $2x + 4y = 16$

4. $3x + 5y = 22$ and $y = 2$

5. $x = 7$ and $y = 3$

6. $x + 2y = 10$ and $5x + y = 14$

7. $3x + 4y = 18$ and $3x + 4y = 12$

8. $2x + 2y = 14$ and $3x + 4y = 24$

In Exercises 9 to 14, graph the half plane corresponding to the inequality.

9. $y \leq 9.5$

10. $x \geq 4$

11. $5x + 3y \leq 15$

12. $3x + 2y \leq 18$

13. $4x + 5y \leq 30$

14. $7x + 2y \leq 42$

In Exercises 15 to 18, graph the feasible region corresponding to the set of inequalities. Label each line segment bounding the region with the appropriate inequality and give the coordinates of every corner point.

15. $x \geq 0$; $y \geq 0$; $3x + y \leq 9$; $x + 2y \leq 8$

16. $x \geq 0$; $y \geq 0$; $2x + y \leq 4$; $3x + 3y \leq 9$

17. $x \geq 0$; $y \geq 0$; $x + 2y \leq 10$; $5x + y \leq 14$; $4x + 5y \leq 30$

18. $x \geq 0$; $y \geq 0$; $2x + y \leq 8$; $x + y \leq 5$; $x + 2y \leq 8$

In Exercises 19 to 22, for each of the following points, determine whether it is a point of the feasible region given: **a.** $(5, 5)$; **b.** $(2, 1)$; **c.** $(1, 3)$; **d.** $(-1, 4)$; **e.** $(2, 4)$.

19. The feasible region of Exercise 15

20. The feasible region of Exercise 16

21. The feasible region of Exercise 17

22. The feasible region of Exercise 18

23. For each of the following points, all feasible for the juice mixture example from the text whose feasible region is graphed in Figure 4.4f, compute the profit using the formula Profit $= 3x + 4y$: (20, 25); (0, 50); and (30, 10).

24. For each of the following points, all feasible for the plywood example from the text whose feasible region is graphed in Figure 4.5f, compute the profit using the formula Profit $= 6x + 5y$: (200, 250); (0, 350); and (350, 100).

In Exercises 25 to 28, for each description, write an appropriate resource-constraint inequality. The unknown to use for each product is given in parentheses.

25. One cake *(x)* requires 4 cups of flour, and one pie *(y)* requires 2 cups. There are 28 cups of flour available.

26. Mowing one lawn *(x)* takes 2 hours of gardening time, and weeding one garden *(y)* takes 1 hour. There are 40 hours of gardening time available.

27. Manufacturing one package of hot dogs *(x)* requires 6 ounces of beef, and manufacturing one package of bologna *(y)* requires 4 ounces of beef. There are 240 ounces of beef available.

28. It takes 30 feet of 12-inch board to make one bookcase *(x)*; it takes 72 feet of 12-inch board to make one table *(y)*. There are 420 feet of 12-inch board available.

29. In the juice mixture problem, x represents gallons of cranapple, y, gallons of appleberry; the feasible region is presented in Figure 4.4f.

    a. How many gallons of each juice mixture are represented by each corner point?

    b. Which corner point represents the maximum profit for each of these profit formulas:

$$\text{Profit} = 3x + 4y \quad \text{(the formula in the text)}$$
$$\text{Profit} = 2x + 5y$$
$$\text{Profit} = 5x + 3y$$

30. For the plywood problem, x represents sheets of exterior plywood, y, sheets of interior; the feasible region is presented in Figure 4.5f.

    a. How many panels of each type are represented by each corner point?

    b. Which corner point represents the maximum profit for each of these profit formulas:

$$\text{Profit} = 6x + 4y \quad \text{(the formula in the text)}$$
$$\text{Profit} = 3x + 7y$$
$$\text{Profit} = 8x + 3y$$

In Exercises 31 to 40, make a mixture chart for the problem. Write the resource- and minimum-constraint inequalities. Then draw the feasible region. Use it and the profit information to determine the point that best answers the question.

31. A cereal manufacturer has on hand 800 pounds of wheat and 40 pounds of sugar. He can make a box of Hefties from $\frac{1}{2}$ pound of wheat and no sugar, and he can make a box of Sweeties from $\frac{4}{10}$ pound of wheat and $\frac{1}{10}$ pound of sugar. The profit on a box of Hefties is $0.10, while the profit on a box of Sweeties is $0.13. How much of each product should he make to earn the maximum profit without exceeding the materials on hand?

32. A paper recycling company uses scrap paper and scrap cloth to make two different grades of recycled paper. A single batch of grade A recycled paper is made from 40 pounds of scrap cloth and 180 pounds of scrap paper, whereas one batch of grade B recycled paper is made from 10 pounds of scrap cloth and 150 pounds of scrap paper. The company has 100 pounds of scrap cloth and 660 pounds of scrap paper on hand. A batch of grade A paper brings a profit of $500, whereas a batch of grade B paper brings a profit of $250. What amounts of each grade should be made?

33. A dairy sells both regular milk and chocolate milk at respective profits of 30 and 50 cents per quart. It takes 1 quart of milk to "make" 1 quart of regular milk, and it takes 0.9 quart of milk and 0.1 quart of chocolate flavoring to make 1 quart of chocolate milk. There are 720 quarts of plain milk and 30 quarts of chocolate flavoring on hand. How many quarts of each product should the dairy make?

34. To make one package of all-beef bologna, a manufacturer uses 1 pound of beef; to make one package of regular bologna, the manufacturer uses $\frac{1}{2}$ pound each of beef and pork. The profit on regular bologna is 60 cents per pack and on all-beef bologna is 80 cents per pack. If there are 400 pounds of beef and 300 pounds of pork available, what product mix should the manufacturer make to maximize profit?

35. A manufacturer of hot dogs uses three ingredients of which these amounts are in stock: beef, 500 pounds; pork, 300 pounds; and grain filler, 100 pounds. The recipe for all-beef hot dogs calls for just beef, 1 pound per package. The recipe for regular hot dogs calls for $\frac{1}{2}$ pound of pork and $\frac{1}{4}$ pound each of beef and grain filler. If the profit on regular hot dogs is 70 cents and on all-beef hot dogs is 80 cents, how many packages of each should be made?

36. A bottler uses three pure juices—pineapple, orange, and grapefruit—to make two juice mixtures, pineapple-orange and pineapple-grapefruit, sold in 1-quart bottles. The profit is 50 cents per bottle of pineapple-orange and 40 cents per bottle of pineapple-grapefruit. Each mixture is made by using equal proportions of the two juices in its name. The amounts of juice on hand are 100 quarts of pineapple juice, 70 quarts of grapefruit juice, and 40 quarts of orange juice. How many of each juice mixture should be produced?

37. A refinery mixes high-octane and low-octane fuels to produce regular and premium gasoline. The profits per gallon on the two gasolines are 30 and 40 cents, respectively. One gallon of premium gasoline is produced by mixing $\frac{1}{2}$ gallon of each of the fuels. One gallon of regular gasoline is produced by mixing $\frac{1}{4}$ gallon of high octane with $\frac{3}{4}$ gallon of low octane. If there are 500 gallons of high octane and 600 gallons of low octane available, how many gallons of each gasoline should the refinery make?

38. A car maintenance shop must decide how many oil changes and how many tune-ups can be scheduled in a typical week. The oil change requires 10 minutes of junior mechanics' time plus 5 minutes of senior mechanics' time. The tune-up requires 15 minutes of junior mechanics' time plus 25 minutes of senior mechanics' time. The maintenance shop makes a profit of $20 on an oil change and $30 on a tune-up. What mix of services should the shop schedule if the typical week has available 4000 minutes of junior mechanics' time and 2000 minutes of senior mechanics' time?

39. In a certain medical office, a routine office visit requires 10 minutes of nurses' time, 5 minutes of doctors' time, and 5 minutes of lab time. The comprehensive office visit requires 5 minutes of nurses' time, 25 minutes of doctors' time, and 10 minutes of lab time. In a typical week, there are 6250 minutes of nurses' time, 11,000 minutes of doctors' time, and 5000 minutes of lab time available. If the medical office clears $30 from a routine visit and $50 from a comprehensive, how many of each should be scheduled per week?

40. A student has decided that a prospective employer looking at a transcript will value passing a tough course twice as much as passing an easy course. (Assume that the student and the prospective employer rate the course difficulty in the same way.) The student estimates that to pass a typical tough course $75 will be needed in texts and supplies and 150 hours will be needed to study and do homework. The student estimates that the typical easy course demands $50 and 50 hours. This year the student has available 600 hours and $450. How many of each kind of course should the student take? (Hint: The profit could be viewed as 2 "toughness points" for passing a tough course and 1 "toughness point" for passing an easy course.) Would the optimal point be the same if a tough course were valued at three times as much as an easy one?

41. In the text example of the juice manufacturer, how does the profit formula change if the profit per gallon doubles for both products? Explain whether this change will affect the mixture that gives the maximum profit?

42. How many variables and resource constraints would be associated with a mixture chart for a firm that uses six resources to make four products?

43. In the text example, suppose the available amounts of apple juice and cranberry juice are doubled. Using algebra, we can show that doubling resources causes the two boundary lines of the feasible region that don't contain (0, 0) to move to new positions parallel to the original positions but farther from the origin. Can you predict what effect this will have on the best profit?

44. In our colored-light-bulb visualization (see Figure 4.9), we superimposed a straightened-out rainbow on the feasible region. Suppose it were not straightened out, as in the figure on top of page 119, and we still wanted to find the hottest-colored point. Would the corner principle still hold? (This is an example of nonlinear programming.)

Exercises 45 to 50 are designed to indicate areas in which the underlying assumptions of the linear-programming model may not always fit the reality of a specific situation. These exercises are meant to stimulate discussion and do not have "right answers." Exercise 45 gives a concrete situation that could be used in the other questions.

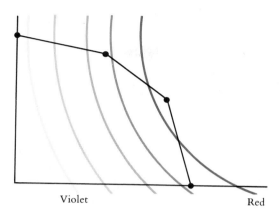

Violet   Red

▲ **45.** You are in charge of a business that produces sandwiches for snack bars. Make a list of your products (kinds of sandwiches), the resources you would need (sandwich ingredients), and the profit you might expect to get for each product. You need not specify the recipes numerically or the amounts of the resources available.

▲ **46.** Discuss the validity of the corner principle if the solution (x, y) to the mixture problem is required not only to lie in the feasible region but also to have integer coordinates. In the sandwich problem (Exercise 45), would there be a useful meaning to a fractional number of some kind of sandwich?

▲ **47.** In a mixture problem we view recipes as fixed, but in practice this is not always true. For example, bolognas having only slightly different percentages of beef and pork all taste the same to the customers. What might prompt the manufacturer to vary a recipe? What effect might the varying have on the profit function, the feasible region, and the optimal product mix?

▲ **48.** In the real world, firms often give discounts for large-volume purchases. Does this necessarily contradict the assumption of a fixed (constant) profit on each unit sold? In examining this situation, you should note that discounts for large-volume purchases might not only apply to the products sold by a manufacturer but also to the prices paid by the manufacturer for resources.

▲ **49.** We learn in economics that prices are determined by the interplay of supply and demand. For example, the price of a product may fall if a large quantity of it is available. However, in mixture problems we assume a fixed (constant) profit regardless of how much is produced. Is there a contradiction here? Could the model be adjusted to incorporate this economic fact of life?

▲ **50.** In economics, it is often useful to distinguish between a firm that has a monopoly (for example, is the only supplier of a product) and firms that supply only a small share of the market. How would the presence of a monopoly affect the relation between production and price? Would the presence of a monopoly tend to ensure the fixed-profit assumption of linear programming, or would it make it more likely that the interplay of supply and demand would have to be considered in order to have a truly realistic model?

▲ Discussion exercise.

PART

·II·

Statistics: The Science of Data

Numerical facts, or data, make up an increasing part of the information we need in order to understand our world. Business executives base decisions on data about the national economy, financial markets, and their firm's own costs, sales, and profits. Engineers gather data on the performance, quality, and reliability of their products. The medical professions watch data on costs as well as data from medical research. Advertisers fine-tune their messages in response to market research data, and politicians use polls of public opinion to shape their campaigns. Citizens and consumers are surrounded by data from all these and other sources. We must be able to understand and communicate with data just as we understand and communicate with words.

The information conveyed by data may be as vital as the fact that 7.2% of the American labor force is out of work, or as trivial as the fact that 57% of American adults think they look younger than their true age. But whether numerical facts are vital or trivial, we must understand where they come from and whether the information they convey is trustworthy. *Statistics* is the science of data — of producing data, of putting them into clear and usable form, and of interpreting them to draw conclusions about the world around us. Just as literacy enables us to use and understand words, a basic knowledge of statistics allows us to use data honestly and skillfully.

Statistics provides quantitative insights into the similarities and differences exhibited by populations, whether they are composed of people, animals, or things. City Stadium, Richmond, Virginia. [Henley and Savage.]

·5·

Collecting Data

The news media often present information in the form of numbers. Headlines announce that the unemployment rate has dropped to 7.2%. The Gallup poll claims that 45% of Americans are afraid to go out at night because of crime. Where do these numbers come from? Most people aren't personally interviewed to determine their employment status. And the Gallup poll asks only a few of us if fear of being mugged keeps us indoors at night.

The unemployment rate and other data about the labor force are estimated by the Bureau of Labor Statistics from data gathered by the Current Population Survey. The Census Bureau conducts this survey, interviewing 60,000 households each month. Considering that the country has over 90 million households, it seems remarkable that data about such a small group—less than 1 in 1000 households—could represent the unemployment rate of the whole nation. How can infor-

mation collected from a small number of people justify conclusions about a much larger group?

Even when numbers are not in the headlines, they often affect our lives. When a popular antiarthritis drug is recalled because of doubts about its safety or when the Food and Drug Administration approves a new medication for use in treating AIDS, these decisions are based on clinical trials using relatively few patients. Public confidence about the safety and effectiveness of new drugs relies on statistical conclusions, or *inferences,* drawn from clinical data.

Any use of numbers as evidence depends on the proper collection of data. The straightforward way to calculate the nation's unemployment rate or to gauge public opinion would be to interview each and every member of the population. However, the cost of such a large-scale operation would be enormous, and the effort required is obviously impractical. Instead, we must rely on information about only a part of the population — in statistical language, a sample.

We often draw conclusions about a whole on the basis of a sample. Everyone has sipped a spoonful of soup and judged the entire bowl on the basis of that taste. But a bowl of soup is homogeneous; the taste of a single spoonful clearly represents the whole. Other kinds of observations, such as fear of crime or reaction to a drug, are not so easily gauged. They vary from person to person or from region to region.

Dealing with this kind of variability is the central task of statistics. The first step is to collect data that represent a large group of individuals. To do this, we must state carefully which group we want information about. Statisticians call this group the **population.**

For example, if we want to measure unemployment, we must first define the population we want to describe. Which age groups will we include? Will we include illegal aliens or people in prisons? The Bureau of Labor Statistics must answer these questions in order to collect monthly unemployment information (see Spotlight 5.1). In fact, the Current Popu-

SPOTLIGHT 5.1 Statistician Heads Agency That Measures U.S. Work Force

Dr. Janet Norwood, U.S. commissioner of labor statistics.

Dr. Janet L. Norwood's position as commissioner of labor statistics makes her one of the nation's most influential statisticians. As head of the Bureau of Labor Statistics, she supervises the collection and interpretation of data on employment, unemployment, prices, wages, and productivity. These data have a large impact on the U.S. economy. For example, price levels collected by the Bureau are used to adjust many federal and private payments for the effect of inflation, including social security payments and union wage scales.

Data on unemployment and prices represent politically sensitive issues. For this reason, the agency collecting and interpreting the data must remain independent and objective. To safeguard the independence of the Bureau of Labor Statistics, the commissioner is appointed by the President and confirmed by the Senate for a fixed term of four years. "That is tremendously important in protecting the basic integrity of the Bureau and the office," Norwood says. "There have been times in the past when commissioners have been in open disagreement with the Secretary of Labor or, in some cases, with the President. We have guarded our professionalism with great care."

Norwood has served as commissioner of labor statistics since 1979 and was reappointed for a third term in 1987. Her tasks require statistical skill, administrative ability, and a facility for working with both Congress and the President.

lation Survey defines its population as all U.S. residents (whether citizens or not) 16 years of age and over who are civilians and are not institutionalized. The civilian unemployment rate published in newspapers and in other reports always refers to this specific population.

The clinical testing of a new drug for treating ulcers would also have a particular population in mind. This population could consist of all individuals who have ulcers, or it might be limited to patients with specific types of ulcers or to those in a specific age group. The detailed plan (called the *protocol*) of the clinical trial must specify exactly the eligible population.

SAMPLING

A population need not consist of people; it may also consist of animals or objects. Let's say we want to test a population of newly produced fuses to see if they blow under excessive current, as they should. In this case, gathering information about every member of the population would destroy the entire population. In other cases, such as taking inventory of all items in a large warehouse, boredom and fatigue could prevent an accurate accounting. In such situations, as well as in population surveys and clinical trials, information is gathered about only a few items. The part of the population used to draw conclusions about the whole is called a **sample.**

How can we choose a sample that is truly representative of the population? The easiest way to select a sample is to choose individuals close at hand. If we are interested in finding out how many people have jobs, for example, we might stand at a busy intersection and ask people passing by if they are employed. A sample selected by taking the members of the population that are easiest to reach is called a *convenience sample.*

Convenience samples often produce unrepresentative data. When gathering employment data from passersby, we will tend to choose well-dressed, middle-income subjects and we will tend to avoid poorly dressed, unfriendly, or tough-looking individuals. It is also likely that people who are currently employed would be at work, not shopping or sightseeing in the middle of the day. In short, our intersection interviews would not accurately reflect the nation's rate of unemployment.

If we sample individuals on the basis of friendliness, appearance, or income level — whether consciously or unconsciously — we will leave some parts of the population underrepresented. Our street-corner survey, for example, would probably overrepresent middle-class and affluent people and underrepresent blue-collar workers and the poor. Such a systematic difference between the results obtained by sampling and the truth about the whole population is called **bias.** To collect accurate data, we must take specific steps to eliminate bias. In particular, statisticians go to great lengths to eliminate the role of personal choice in the selection of the sample to be measured.

EXAMPLE: Call-in Polls. Television makes heavy use of call-in polls, in which viewers are invited to register their opinions by telephone. Some television stations poll the public daily, asking a question on the 6 o'clock news and reporting the responses on the 11 o'clock news. Viewers are urged to call special 900-prefix telephone numbers with their responses. Dialing one number indicates a yes reply; dialing the other, no. The system works so that talking is not necessary. Completing the telephone call registers a viewer's answer.

It is not difficult to see the sources of bias in a call-in poll. Households without telephones are automatically excluded. (Although about 93% of U.S. households have telephones, about one-quarter of those in Alaska and one-fifth of those in Mississippi do not.) Because dialing the 900-prefix number incurs a small charge, people in lower-income households may be reluctant to pay a fee in order to telephone the station.

The most serious source of bias in call-in surveys is **voluntary response** — the respondents select themselves. Only those who go to the time, trouble, and expense of calling are counted. Any sample chosen by voluntary response draws people with strong feelings, most often negative feelings. So when the newscaster asks the audience if they are afraid to go out at night because of crime, people angry about crime are more likely to call in than those who are not. Voluntary response is a common and serious source of bias in polls.

It is also possible that the call-in poll could be manipulated. If crime is a local issue with political impact, one political party could arrange for its workers to spend the evening dialing in. Or, because talking isn't needed for registering a response, a computer could be programmed to dial the 900-prefix number repeatedly.

Personal choice is a common cause of bias in sampling. Whether it is expressed in the way an interviewer or study team selects its subjects or in the respondents' voluntary response, personal choice can distort the results. To reduce bias, it is vital to reduce the influence of personal choice. The way statisticians eliminate personal choice is to select the sample by *chance.*

RANDOM SAMPLING

Imagine that we have a glass box containing thousands of beads. The beads are all identical except that most of them are light and some of them are dark. The beads form a population. We can say that the beads represent the American labor force and that dark beads represent those who are unemployed (see Figure 5.1).

Our task is to estimate what percentage of the population of beads is dark without examining each bead individually. Suppose that we thoroughly mix the beads in the box and then draw out a sample, using a scoop with 50 recesses in it. Each bead then has the same chance as every other bead of being selected, and each group of 50 beads also has the same chance as any other group of 50 to be selected. This is a simple random sample of size 50.

> A **simple random sample** of size n is a sample drawn in such a way that every possible sample of n members of the population has the same chance to be the sample actually chosen.

Because every bead has an equal chance to be chosen, the simple random sample has eliminated bias in selecting the sample. In a survey of employment, a simple random sample gives everyone — rich or poor, male or female, employed or unemployed — an equal chance to be selected.

Figure 5.1 A random-sampling demonstrator.

The first time we draw a sample from our box of beads, we get 12 dark beads. Because 12 out of 50 is the fraction $\frac{12}{50}$, or 0.24, 24% of this sample are dark. The results of simple random sampling are free of bias, so we can use the sample percentage to estimate the truth about the entire population. If 24% of the beads in the sample are dark, we estimate that 24% of the beads in the box are dark. Here is an important statistical technique: *to estimate a characteristic of a population, take a simple random sample and use the sample characteristic as the estimate.*

In principle, we now know how to estimate the nation's rate of unemployment. Put the names of all workers in a hat, mix the names well, and draw out 60,000 names. Because this is a simple random sample, the percent unemployed among those 60,000 persons is an unbiased estimate of the percent unemployed in the entire labor force.

In practice, however, when a population is too large to fit into a hat, other methods of sampling are needed. For example, each member of the population can be tagged with a numerical label, and then a sample can be chosen from the set of numerical labels in a way that gives every label an equal chance of being selected. If the population is relatively small, tagging and drawing samples can be done easily with a **table of random digits** such as Table 5.1. For larger populations, a computer can be programmed to generate random numbers and then to print out the sample chosen.

A table of random digits is a string of the digits 0, 1, 2, 3, 4, 5, 6, 7, 8, and 9 chosen in such a way that each entry is equally likely to be any of the 10 possibilities and each entry is independent of all the other entries. You can think of writing each of the 10 digits on a tag, putting the tags in a hat, mixing thoroughly, and drawing one. That's the first entry in the table. Put back the tag you drew, mix again, and draw another. That's the second entry. Continue drawing and mixing for hours and

you will have a table of random digits like Table 5.1. The digits in Table 5.1 appear in groups of five to make the table easier to read and the rows are numbered so we can refer to them, but the groups and row numbers are just for convenience. The entire table is one long string of randomly chosen digits.

There are two steps in using the random-digit table to choose a simple random sample.

STEP 1. Give each member of the population a numerical label of the *same length*. Up to 100 items can be labeled with two digits, up to 1000 items can be labeled with three digits, and so on.

STEP 2. To choose a simple random sample, go through Table 5.1 looking at successive groups of digits of the length used as labels. This works because, for example, any two-digit group in the table is equally likely to be any of the 100 possible labels 00, 01, . . . , 99. Any group of digits that was not used as a label or that duplicates a label already in the sample is simply ignored.

Here is an example that illustrates the technique.

EXAMPLE: Sampling Autos. An auto manufacturer wants to select 5 of the last 50 cars produced on an assembly line for a very detailed quality inspection. To avoid bias, a simple random sample will be chosen.

STEP 1. Give each car a numerical label. Because two digits are needed to label 50 cars, all labels will have two digits. Let's begin with 00. Then the labels are as shown in Figure 5.2.

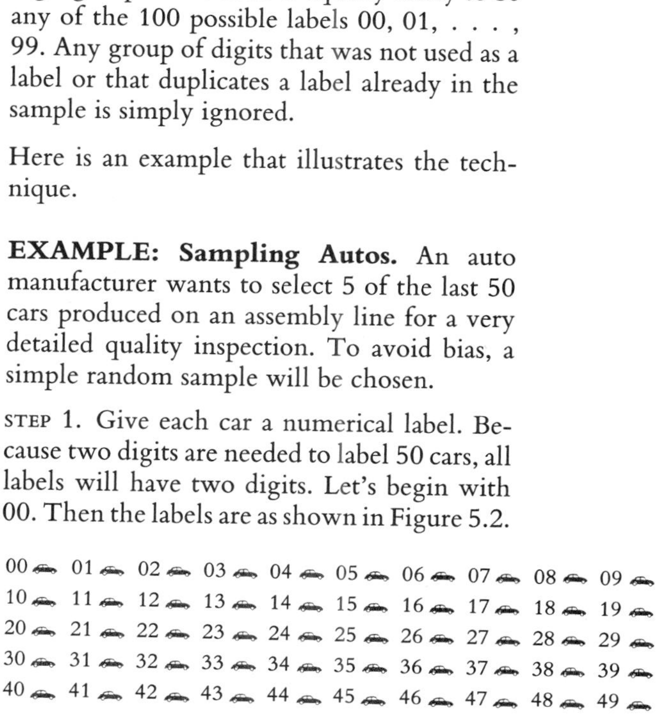

Figure 5.2 The first step in random sampling: assigning labels to 50 cars.

TABLE 5.1 **Random digits**

101	03918	86495	47372	21870	28522	99445	38783	83307
102	10041	35095	66357	64569	08993	20429	28569	63809
103	43537	58268	80237	17407	89680	04655	24678	61932
104	64301	47201	31905	60410	80101	33382	95255	10353
105	43857	42186	77011	93839	28380	49296	63311	49713
106	91823	39794	47046	78563	89328	39478	04123	19287
107	34017	87878	35674	39212	98246	29735	09924	27893
108	49105	00755	39242	50472	39581	44036	54518	46865
109	72479	02741	75732	99808	02382	77201	44932	88978
110	84281	45650	28016	77753	39495	41847	19634	82681
111	61589	35486	59500	20060	89769	54870	75586	07853
112	25318	01995	87789	41212	74907	90734	31946	24921
113	40113	37395	51406	98099	43023	70195	07013	72306
114	58420	43526	15539	24845	15582	16780	95286	69021
115	18075	45894	09875	42869	20618	07699	80671	54287
116	52754	73124	93276	71521	59618	44966	37502	15570
117	05255	53579	08239	99174	75548	95776	42314	13093
118	76032	35569	28738	38092	74669	00749	17832	64855
119	97050	31553	32350	51491	53659	89336	36912	05292
120	29030	43074	84602	95131	22769	44680	68492	33987
121	28124	29686	63745	12313	15745	11570	20953	17149
122	97469	41277	90524	36459	22178	63785	20466	67130
123	91754	40784	38916	12949	76104	20556	34001	59133
124	84599	29798	57707	57392	91757	76994	43827	69089
125	06490	42228	94940	10668	62072	58983	10263	08832
126	30666	02218	89355	76117	75167	69005	42479	79865
127	87228	15736	08506	29759	74257	85594	75154	48664
128	45133	49229	32502	99698	68202	44704	39191	73740
129	55713	98670	57794	64795	27102	83420	26630	95009
130	20390	38266	30138	61250	07527	02014	43972	49370
131	13400	68249	32459	41627	56194	93075	50520	96784
132	08900	87788	73717	19287	69954	45917	80026	55598
133	86757	47905	16890	99047	78249	73739	97076	00525
134	19862	54700	18777	22218	25414	13151	54954	80615
135	96282	11576	59837	27429	60015	40338	39435	94021
136	17463	26715	71680	04853	55725	87792	99907	67156
137	44880	55285	95472	57551	24602	98311	63293	58110
138	61911	78152	96341	31473	58398	61602	38143	93833
139	07769	22819	58373	88466	71341	32772	93643	92855
140	73063	63623	29388	89507	78553	62792	89343	27401
141	24187	60720	74055	36902	22047	09091	79368	35408
142	06875	53335	91274	87824	04137	77579	54266	38762
143	23393	37710	46457	03553	58275	11138	18521	59667
144	00980	73632	88008	10060	48563	31874	90785	78923
145	46611	39359	98036	25351	88031	72020	13837	03121
146	56644	79453	49072	30594	73185	81691	29225	70495
147	98350	36891	04873	71321	29929	37145	95906	41005
148	17444	61728	86112	76261	92519	61569	65672	95772
149	45785	21301	89563	23018	60423	50801	70564	45398
150	54369	08513	36838	19805	67827	74938	66946	01206

STEP 2. Look at successive two-digit groups in Table 5.1. Starting at line 140 (any line will do), we read

73063   63623   29388

The cars labeled 06, 36, 23, 29, and 38 are chosen. The initial 73 is ignored because it is not used as a label, and the second 36 is ignored as a duplicate.

Clearly, the Bureau of Labor Statistics doesn't use beads in a box or names in a hat to determine who will be interviewed for the Current Population Survey. In fact, the Bureau doesn't even use simple random samples. National opinion polls and the Bureau of Labor Statistics must use more complex versions of random sampling. Chance still determines the sample, but the process of selecting from the entire nation is done in stages instead of all at once. The Census Bureau, under contract with the Bureau of Labor Statistics, does the job. The entire country is first divided into primary sampling units, or PSUs (see Figure 5.3). Each PSU is a group of neighboring counties. The Census Bureau selects a random sample of PSUs. Within each PSU selected, smaller areas of about 500 inhabitants, called *census enumeration districts,* are chosen at random. Finally, the Census Bureau selects, also at random, individual households within each district.

Such a *multistage random sample* offers several practical advantages over a simple random sample. For one, drawing the sample does not require listing every household in the nation. Moreover, the households to be interviewed are clustered together in relatively few locations so that the travel costs for the interviewers are reduced. The price paid for practicality, however, is complexity in actually choosing the sample and in interpreting the results (see Spotlight 5.2, p. 130). Because simple random sampling is the essential principle behind all random sampling and because it is

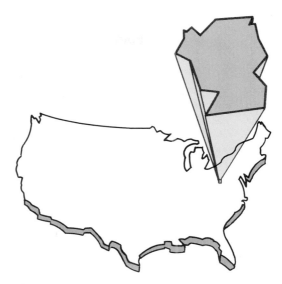

Figure 5.3 A primary sampling unit (PSU) for the Current Population Survey.

also the main building block for more complex designs, we will focus our study on simple random sampling.

SAMPLING VARIABILITY

Our first sample from the box of beads contained 12 dark beads. If we draw another sample, we will probably not get 12 dark beads again. In fact, when we remix the beads and take a second sample, we find that only 8 out of 50, or 16%, are dark, a figure quite different from the 24% we got earlier.

Suppose that the Gallup poll conducted its weekly survey of public opinion twice, selecting two random samples, sending out interviewers, and asking the same question of the two samples. Two random samples, each selecting 1500 of the more than 185 million U.S. residents age 18 and over, will certainly contain different people. And two distinct groups of people will have somewhat different opinions.

SPOTLIGHT 5.2 Sampling and Surveys: Taking the National Pulse

Dr. Janet L. Norwood, commissioner of labor statistics, shares a number of observations on her agency and on the task of sampling the nation at large through the Current Population Survey.

How We Go About the Business of Sampling

The household survey is a basic labor-force survey that provides us with an enormous amount of data. Specifically, it provides information on the demographic characteristics of people and about their labor-force status: their employment, the length of time they are employed, whether they are job seekers, whether they are job losers, whether they are employed part-time or full-time, and so on. That survey, which we call the Current Population Survey, is conducted for us on contract by the Bureau of the Census. Interviewers go out to a sample of 60,000 households spread throughout the United States in more than 400 areas of the country. The reason that the sample is so large — and it is a very large sample for a household survey — is that so much data comes out of it. We do not set out simply to collect data on how many people are employed or unemployed. We also must have information on men versus women, on blacks, on whites, on Hispanics, and on trends in the youth population. We need to have some information by region and by state. And as you get down to smaller and smaller groups, it is necessary to have a sample size that is sufficient to provide adequate data.

On the Impact of Our Agency

Our price data are probably among our best-known indicators. This is because the consumer price index (CPI) — the best measure we have in this country of inflation —

Gallup's announcement that 45% of Americans hesitate to go out at night because they are afraid of crime refers to the 1500 specific individuals in a specific sample. A second sample would no doubt produce a different result. If the Current Population Survey were conducted twice this month, the two samples would differ, producing two different officially announced unemployment rates. Random sampling eliminates bias, but it does not eliminate *variability*. The variation from sample to sample in repeated random samples is an inevitable consequence of variability in the population.

So how can we trust the result of a random sample, knowing that a second sample will give a different result? How can we base economic and political decisions on the unemployment rate, knowing that the rate would vary if the Bureau of Labor Statistics took a second sample? In fact, random sampling *can* be trusted. To see why, we need to look more closely at sampling variability.

There are different kinds of variability. The answers obtained by sending an interviewer to a busy intersection vary in a haphazard and unanalyzable way. However, repeated random samples vary in a regular manner because a

is used in a number of government programs as an escalator to keep income up with inflation. In fact, if you take dependents into account, we estimate that more than half the population of this country has income in some way affected by the CPI. And in 1985, income tax brackets will have begun to be adjusted by the CPI for purposes of calculating income tax.

Dealing with Uncertainty

Let's face it. Everyone wants to live in a world of certainty. Even the statistician wants to live in a world of certainty. Our job is to explain to people that there is no such thing as absolute certainty. Every word we use is looked at to be sure that we understand it and that someone looking at it isn't going to read more into it than we intended.

What we *can* do is to provide a set of statistics that have some error surrounding them, something that we often call sampling error or variance, which will tell what the basic tolerances are. If the unemployment rate, for example, moves by one-tenth of a point, we in the Bureau of Labor Statistics say that it was about the same. If it goes up or down two-tenths of a point, we will say that there has been a change, because the sampling error surrounding that number is such that a two-tenths change is outside of that limit.

If somebody wants to know something about BLS data, they pick up a phone and they call me or they call someone on my staff who is an expert in the area. We have made it our business to provide information as rapidly as we can and as openly as we can. I happen to believe that a statistical agency can only be a good one if it is completely open.

specific chance mechanism is used in random selection. The long-run results are not haphazard. We see such regular variation in the results of repeated coin tosses or of successive spins of a roulette wheel. Tossing a balanced coin 1500 times is much like choosing a simple random sample of 1500 from a large population, if we imagine that opinion in this population is evenly divided so that heads represents yes and tails represents no.

Let's look more closely at the outcomes of coin tossing. It is unlikely that a person tossing a coin 1500 times will obtain exactly 750 heads (a 50% yes response). The first sample might produce 732 heads (49% in favor), a second, 781 heads (52% in favor), and so on. But coin tossing is subject to the laws of *probability,* which dictate how frequently each outcome will occur in the long run (see Chapter 7). In particular, 1500 tosses of a balanced coin will almost never (only 1 chance in 1000) produce less than 46% or more than 54% heads.

Therefore, 1500 tosses of a balanced coin can be trusted to give a result close to 50% heads. Similarly, a Gallup poll of 1500 Americans can be trusted to give a result close to the result obtained by polling all 185 million adult Americans. Probability deals with phenomena

TABLE 5.2 Bead-sampling results

Percent of dark beads	Number of samples	Percent of dark beads	Number of samples
8	1	22	14
10	1	24	14
12	5	26	8
14	8	28	3
16	11	30	3
18	19	32	0
20	12	34	1

that are variable but that nonetheless show a regular pattern in the long run. Tossing a coin and choosing a random sample are such phenomena.

In the case of choosing a simple random sample of 50 beads from a box, we can easily take repeated samples and observe the pattern of outcomes.

EXAMPLE: A Bead-Sampling Experiment. We took 100 simple random samples of size 50 from the box of beads. Our first sample had 12 dark beads out of 50, or 24%. The second sample had 16%, the third 26%, and so on. Table 5.2 displays the results from our 100

samples. We can display the outcomes of all 100 samples in a **histogram** (Figure 5.4). The height of each bar in the histogram shows how often the outcome marked at the base of the bar occurred. For example, the height of the bar marked 16% is 11 because 11 of our samples had 16% dark beads. More detail about histograms appears in the next chapter.

The histogram for our 100 samples in Figure 5.4 shows a lot of scatter, but most of the samples cluster around the 20% mark. In fact, 20% is the actual proportion of dark beads in the box from which these samples were drawn. Only 12 of the samples we took contained exactly 20%, but 70 of the 100 samples had between 16% and 24% dark beads.

The results of random sampling usually do come quite close to the truth about a population. And as we'll discover in Chapter 7, the laws of probability make possible exact statements about the accuracy of simple random sampling. We will see, for example, that the outcome of a sample of size 1500 is much more likely than that of a sample of size 50 to come close to the truth.

The deliberate use of chance in collecting data is one of the fundamental ideas of statistics. It is important as a way of removing bias.

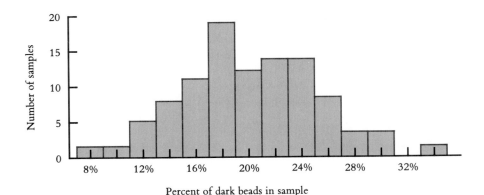

Figure 5.4 Histogram showing sampling variability of the outcomes of a bead-sampling demonstration.

More important, the proper use of random samples enables us to draw confident conclusions from data because the mathematics of probability provides support for these conclusions.

As we saw earlier, simple random sampling is the building block of all sampling designs. Thus, even though the Current Population Survey and the Gallup poll use more complex random-sampling procedures, the results of their multistage random samples are still described by the laws of probability. Consequently, these samples usually give results close to the truth about the population. A careful statement of a sample result expresses "usually" in terms of probability and "close to the truth" in terms of a *margin of error.*

For example, the Gallup poll takes a sample of about 1500 people each week. To describe the accuracy of the poll results, we ask: What would happen if Gallup took many samples of 1500 people, asking the same question each time? This is similar in principle to our many samples from the same box of beads. The results would also be similar: there is a regular pattern of outcomes that allows us to state how often the sample result will come close to the truth about the population. Specifically, the probability is 0.95 that a sample percentage from a Gallup poll is within ±3% of the true population value.

Probability describes the regular behavior that appears when random sampling is repeated many times. The probability 0.95 means that in the long run 95% of Gallup's many sample results fall within a margin of error of ±3 percentage points about the truth for the entire population. If the poll reveals that 45% of the sample fear to go out at night, we can be quite confident that between 42% and 48% of the entire population share that fear. The variability in sampling has not vanished, but probability allows us to describe the variability by announcing a margin of error that most samples will meet. Gallup and other polling organizations report these facts in their press releases, but news editors often cut that part of the story.

What about variability in the unemployment rate? Here are the facts in this case. The Census Bureau prefers to announce the margin of error that 90% of Current Population Survey samples will meet: it is about ±0.2%, or two-tenths of one percent. Thus, when the unemployment rate drops from 8.4% to 8.3%, sampling variability may well account for the change. But a change from 8.4% to 8.0% very probably reflects a real drop in the percent of the American labor force who are without a job.

The small margin of error in the unemployment rate is due to the larger sample size. An error of ±3% is acceptable in polling public opinion, but an unemployment rate of 8% ± 3% would be of little value. Because data on employment and unemployment are important for economic planning, the federal government is willing to undertake the expense of a large monthly sample. The margin of error attained with any desired probability (such as 0.90 or 0.95) depends on the size of the sample and on the exact sampling design. The mathematics of probability describes these relationships exactly.

EXPERIMENTATION

We have seen that sample surveys gather information on part of the population in order to draw conclusions about the whole. When the goal is to describe a population, as the Current Population Survey's aim is to describe employment and unemployment in the United States, statistical sampling is the right tool to use.

Suppose, however, that we want to study the response to a stimulus, to see how one variable affects another when we change existing conditions. Will a new mathematics curriculum improve the scores of sixth graders on a standard test of mathematics achievement? Will taking small amounts of aspirin daily reduce the risk of a heart attack? Does smoking increase the risk of lung cancer? Observational studies, such as sample surveys, are ineffective tools for answering these questions. Instead, we prefer to carry out experiments.

An **experiment** differs from observation in that the experimenter intervenes actively by imposing a *treatment* on the subjects. A treatment can be any condition that the experimenter is interested in, such as a new math curriculum or an aspirin tablet every day. In sampling, on the other hand, we observe or measure the state of the subjects without trying to change that state by a treatment.

Experiments are the preferred method for examining the effect of one variable on another. By imposing the specific treatment of interest and controlling other influences, we can pin down cause and effect. A sample survey, in contrast, may show that two variables are related, but it cannot demonstrate that one causes the other. Statistics has something to say about how to arrange experiments, just as it suggests methods for sampling.

EXAMPLE: An Uncontrolled Experiment. The Bigfoot Mountain school system, concerned about the poor mathematics preparation of American children, adopts an ambitious new mathematics curriculum. After three years of the new curriculum, students completing sixth grade have an average achievement score 10% higher than they had before the treatment. Bigfoot Mountain pronounces the curriculum a success, and other systems adopt it.

This experiment had a very simple design. A group of subjects (the students) were exposed to a treatment (the new curriculum), and the outcome (achievement-test scores) was observed. Here is the design:

$$\text{Treatment} \longrightarrow \text{Observation}$$

$$\text{New curriculum} \longrightarrow \text{Improved test scores}$$

Most laboratory experiments use a design like that in the example: apply a treatment and measure the response. In the controlled environment of the laboratory, simple designs are often adequate. But field experiments and experiments with human subjects are exposed to more variable conditions and deal with more variable subjects. They require control of outside factors that can influence the outcome. With greater variability comes a greater need for statistical design.

In Bigfoot Mountain, an atmosphere of concern for education brought about a number of simultaneous changes that could have influenced the students' achievement-test scores. Elementary teachers were given additional training in mathematics. A parent group began to provide classroom tutors to give children individual help with mathematics. Public concern led parents to pay more attention to their children's progress and teachers to assign more homework.

In these circumstances, mathematics achievement would have increased without a new curriculum. In fact, the new curriculum could even be *less* effective than the old. The Bigfoot Mountain experiment cannot distinguish the effects of the changes in parents and teachers from the effects of the new curriculum. Variables, whether part of a study or not, are said to be **confounded** when their effects on the outcome cannot be distinguished from each other.

The remedy for confounding is to do a *comparative experiment* in which some children are taught from the new curriculum and others from the old. Changes in parents' attitudes and involvement, teacher retraining, and other such variables now operate equally on both groups of students, so direct comparison of the two curricula is possible. Most well-designed experiments compare two or more treatments.

Once we decide to do a comparative experiment, we need to find a way to assign the students to the two groups. If the groups differ markedly when the experiment begins, bias will result. For example, if we allow students to volunteer for the new curriculum, only adventurous children who are interested in math are likely to sign up for our experimental treatment, and these students are likely to perform well. Personal choice will bias our results in the same way that volunteers bias the results of call-in opinion polls. The solution to the problem of bias is the same for experiments and for samples: use impersonal chance to select the groups.

Let's say the Bigfoot Mountain school system decides to compare the progress of 100 students taught under the new mathematics curriculum with that of 100 students taught under the old curriculum. In this case, the students who will be taught the new curriculum are selected by taking a simple random sample of size 100 from the 200 available subjects. The remaining 100 students form the **control group;** they will continue in the old curriculum.

The selection procedure is exactly the same as it is for sampling: all 200 members of the population are tagged with numerical labels, beginning with 000 and ending with 199. Next, we consult a table of random digits, inspecting successive three-digit groups. The first 100 labels encountered select the group that will be taught from the new curriculum. Repeated labels and groups of digits not used as labels are ignored. For example, if we begin at line 125 in Table 5.1, the first few students chosen are those labeled 064, 106, 102, 022, 188. The remaining 100 students form the

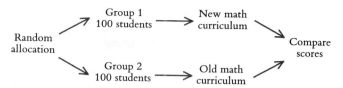

Figure 5.5 Outline of the design of a randomized comparative experiment to evaluate a new mathematics curriculum.

control group, to be taught from the old curriculum.

The result is a **randomized comparative experiment** with two groups (see Figure 5.5). The experiment is comparative because two treatments (the two math curricula) are compared; it is randomized because the subjects are assigned to the treatments by chance.

Randomized comparative experiments are used whenever environmental variables, such as behavioral change in parents or teachers, threaten to confound the results. The results of such experiments are as reliable as the results of random samples, and for the same reasons: random selection is governed by the laws of probability. Here is an example with three treatments.

EXAMPLE: Raising Turkeys. Turkeys raised commercially for food are often fed the antibiotic salinomycin to prevent infections from spreading among the birds. Salinomycin can damage the birds' internal organs, especially the pancreas. A researcher believes that adding vitamin E to the diet may prevent injury. He wants to explore the effects of three levels of vitamin E added to the diet along with the usual dose of salinomycin. There are 30 turkeys available for the study. At the end of the study, the birds will be killed and each pancreas examined under a microscope.

The researcher decides on a randomized comparative design that allocates 10 birds chosen at random to each of the three levels of

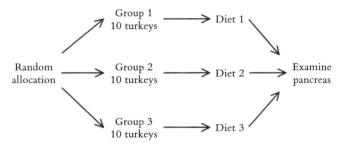

Figure 5.6 The design of a randomized comparative experiment to compare three diets for turkeys.

vitamin E. See Figure 5.6 for an outline of the design. The turkeys are labeled with tags marked 00 to 29 (01 to 30 is also acceptable, but be sure that each label has two digits). Read two-digit groups starting in line 115 of Table 5.1 until 10 turkeys are chosen to make up the first group. Those chosen have labels 18, 07, 09, 28, 20, 15, 24, 27, 21, and 05. Then continue in the table to choose 10 more birds for the second group. The 10 that remain form the third group.

Randomized comparative experiments are common tools of industrial and academic research. They are also widely used in medical research. For example, federal regulations require that the safety and effectiveness of new drugs be demonstrated by randomized comparative experiments. Let's look at a typical medical experiment.

EXAMPLE: The Physicians' Health Study. There is some evidence that taking low, regular doses of aspirin will reduce the risk of heart attacks. Some physicians also suspect that regular doses of beta carotene (which the body converts into vitamin A) will help prevent some types of cancer. The Physicians' Health Study was a large experiment designed to test these claims (see Spotlight 5.4, p. 138). The subjects of this study were 22,000 male physicians over 40 years of age. Each physician took a pill every other day over a period of several years. There were four treatments: aspirin alone, beta carotene alone, both, and neither. The subjects were randomly assigned to one of these treatments at the beginning of the experiment.

The Physicians' Health Study introduces several new ideas important to the proper design of experiments. The first is the importance of counteracting the **placebo effect.** A placebo is a fake treatment, a dummy pill that contains no active ingredient but looks and tastes like the real thing. The placebo effect is the tendency of subjects to respond favorably to any treatment, even a placebo. If subjects given aspirin, for example, are compared with subjects who receive no treatment, the first group gets the benefit of both aspirin and the placebo effect. Any beneficial effect that aspirin may have is confounded with the placebo effect. To prevent confounding, it is important that some treatment be given to *all* subjects in any medical experiment.

In the Physicians' Health Study, all subjects took pills that appeared identical, but some pills contained aspirin or beta carotene and some were placebos. The study was designed as a **double-blind experiment:** subjects did not know which treatment they were receiving, because this knowledge might influence their reaction. In addition, knowing the subjects' treatments might influence researchers who interview and examine them. Therefore, experimental workers were also kept "blind." Only the study's statistician knew which treatment each subject received.

The Physicians' Health Study (see Figure 5.7) is a more elaborate experiment than our earlier examples. Not only are four treatments compared, but two distinct experimental variables are present: aspirin or not, and beta carotene or not. A *two-factor experiment,* or two-variable experiment, allows us to study the interaction, or joint effect, of the two drugs as

SPOTLIGHT 5.3 Sir Ronald A. Fisher, 1890–1962

Sir Ronald A. Fisher.

While employed at the Rothamsted agricultural experiment station in the 1920s, British statistician and geneticist R. A. Fisher revolutionized the strategy of experimentation. Experimenters there were comparing the effects of several treatments, such as different fertilizers, on field crops. Because fertility and other variables can change as we move in any direction across the planted field, the experimenters used elaborate checkerboard planting arrangements to avoid bias. Fisher realized that random assignment of treatments to growing plots was simpler and better. He introduced randomization, described more complex random arrangements, such as blocks and Latin squares, and worked out the mathematics of the *analysis of variance* to analyze data from randomized comparative experiments.

Fisher contributed many other ideas, both mathematical and practical, to the new science of statistics. His influential books organized the field. Fisher was both opinionated and combative. From the 1930s until his death, he was engaged in sometimes vitriolic debates over the appropriate use of statistical reasoning in scientific inference.

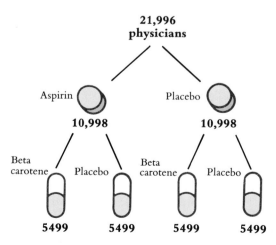

Figure 5.7 The design of the Physicians' Health Study, an experiment with two factors.

well as the separate effects of each. For example, beta carotene may reinforce (or counteract) the effect of aspirin on future heart attacks. By comparing these four groups, we can study all these possible interactions.

STATISTICAL EVIDENCE

A properly designed experiment, in the eyes of a statistician, is an experiment employing the principles of *comparison* and *randomization:* comparison of several treatments and randomization in assigning subjects to the treatments. As we saw in the math curriculum example, comparison eliminates confounding by envi-

SPOTLIGHT 5.4 Using Statistics to Study Disease

Dr. Julie Buring, associate director of the Physicians' Health Study, discusses the ways in which epidemiologists gather data on disease.

As an epidemiologist, I look at factors that are involved in the distribution and disease frequency in human populations. What is it about what we do, what we eat, what our environment is, what our occupations are, our history—our medical history, our family history—that leads one group of people to be more or less likely to develop a disease than another group of people? It is these factors that we are trying to identify.

We go at it from a couple of different angles. One is called *descriptive epidemiology,* or looking at the trends of diseases over time, trends of diseases in one population relative to another population.

Another is called *observational epidemiology,* in which we observe what people do. We take a group of people who have a disease and a group of people who don't have a disease. We look at their patterns of eating or drinking, medical history, and what their exposures may have been. The other way to go about it is to take a group of people who have been exposed to something such as smoking and a group of people who haven't and follow them up over time to see whether they develop the disease or not. Whether we do one design or another, both are observational. That means we don't interfere in the process. We just observe it.

A second type of epidemiology, of which the Physicians' Health Study is an example, is *experimental epidemiology,* sometimes called an intervention study. We take a group of people who have the treatment and a group of people who do not. The difference between the experimental and observational approach is that in experimental epidemiology, the investigators determine who will be in what treatment group, who will receive the treatment and the placebo, who will be exposed, and who won't be exposed.

From these different approaches—descriptive epidemiology and observational epidemiology—we can judge whether a particular factor causes or prevents the disease that we are looking at.

ronmental variables. The immediate appeal of randomization is the elimination of bias by creating groups equivalent in all respects save the treatment they receive.

The future health of the subjects of the Physicians' Health Study may depend on age, past medical history, emotional status, smoking habits, and many other variables known and unknown. Randomization will, on average, balance groups simultaneously in all such variables. Because the groups were exposed to exactly the same environmental variables, except for the actual content of the pills, differences among the groups can be attributed to the effect of the medication. That is the logic of randomized comparative experiments.

Let's be a bit more specific: any difference among the groups is due *either* to the medication *or* to the accident of chance in the random assignment of subjects. It could happen, for example, that men about to have a heart attack were, by chance, overrepresented in one of the groups. Once again, statistics calls on probability. Because chance was deliberately used in making assignments, the laws of probability tell us how large the differences among the four groups are likely to be if nothing but chance were operating. As in sampling, larger numbers of subjects increase our confidence in the results. The Physicians' Health Study followed 22,000 subjects in order to be quite certain that any medically important differences among the groups would be detected and that these differences could be attributed to aspirin or to beta carotene. In fact, there were significantly fewer heart attacks among the men who took aspirin than among men who took the placebo. As a result of the Physicians' Health Study, doctors often recommend that men over 50 take small amounts of aspirin regularly.

The logic of experimentation, the statistical design of experiments, and the mathematics of probability combine to give compelling evidence of cause and effect. Only experimentation can produce fully convincing evidence of causation.

By way of contrast, consider the statistical evidence linking cigarette smoking to lung cancer. This evidence is based on observation rather than experiment. The most careful studies have selected samples of smokers and nonsmokers, then followed them for many years, eventually recording the cause of death. These are called *prospective studies* because they follow the subjects forward in time. Prospective studies are comparative, but they are not experiments because the subjects themselves choose whether or not to smoke. Remember that an experiment must actually impose treatments on the subjects. A large prospective study of British doctors found that the lung cancer death rate among cigarette smokers was 20 times that of nonsmokers; another study of American men aged 40 to 79 found that the death rate from lung cancer was 11 times higher among smokers than among nonsmokers. Thus, the observed connection between smoking and lung cancer is strong.

This connection is statistically significant: it is far larger than would occur by chance. We can be confident that something other than chance links smoking to cancer. But observation of samples cannot tell us *which* factors other than chance are at work. Perhaps there is something in the genetic makeup of some people that predisposes them both to nicotine addiction and to lung cancer. In that case, a strong link would be observed even if smoking itself had no effect on the lungs.

The statistical evidence that points to cigarette smoking as a cause of lung cancer is about as strong as nonexperimental evidence can be. First, the connection has been observed in many studies in many countries. This eliminates factors peculiar to one group of people or to one specific study design. Second, specific ways in which smoking could cause cancer have been identified. Cigarette smoke contains tars that can be shown by experiment to cause tumors in animals. Third, no really plausible alternative explanation is available. For example, the genetic hypothesis cannot explain the rise in lung cancer rates among women that occurred as more women became smokers. Lung cancer, which has long been the leading cause of cancer deaths in men, is now challenging breast cancer as the most fatal cancer for women. Moreover, genetics cannot explain why lung cancer death rates drop among smokers who quit.

This evidence is convincing to most people, and almost all physicians accept it. But it is not quite as strong as the conclusive statistical evidence we get from randomized comparative experiments.

Optional MORE ELABORATE EXPERIMENTS

Many experiments are more complex than a simple comparative randomized design, just as many samples are not simple random samples. If we anticipate that male and female patients will respond differently to aspirin and beta carotene, we can first divide the pool of patients into two **blocks** (males and females) and then randomly assign patients to treatments separately within each block. The division into blocks controls one influential variable, the patient's sex, by including it in the design. If aspirin, for example, has different effects for women and men, these effects are separated in the two blocks. Other influences are still averaged out by randomization. (The Physicians' Health Study was restricted to male subjects because there were not enough female physicians over 40 to form a second block.)

Even more common than blocking is the simultaneous study of several *factors,* or experimental variables. We saw earlier that the Physicians' Health Study was a two-factor experiment. Industry provides us with another example.

EXAMPLE: An Industrial Experiment. A chemical engineer trying to determine the most efficient temperature-pressure setting for a production process would be foolish to rely on experiments taking into account only one variable at a time, because the most productive temperature will change according to the pressure, and vice versa. An effective experiment must change both temperature and pressure from treatment to treatment. There are therefore two factors. The engineer chooses three temperatures and two pressures that cover the range of interest. Combinations of these temperatures and pressures form six treatments (see Figure 5.8).

Factor 1:
Temperature (°C)

		95°	110°	120°
Factor 2: Pressure	100 psi	1	2	3
	150 psi	4	5	6

Figure 5.8 The treatments in a two-factor experiment.

The combination of several experimental variables with the use of blocks can make experiments prohibitively large and expensive. However, *combinatorics,* the mathematical study of arrangements, can sometimes provide clever arrangements of treatments that help hold down the size and cost of experiments. We can illustrate this by a comparison of motor oils.

EXAMPLE: Comparing Motor Oils. Some makes of motor oil claim that using their product improves gasoline mileage in cars. To test this claim, we want to compare the effect of 4 different oils on mileage. However, because the car model and the driver's habits greatly influence the mileage obtained, the effect of an oil may vary from car to car and from driver to driver. To obtain results of general interest, then, we must compare the oils in several different cars and with several different drivers. If we choose 4 car models and 4 drivers, we will have 16 car-driver combinations. The type of car and driving habits are so influential that we consider each of these 16 combinations a block. If we then complete test drives with 4 oils (in random order) in each block, we need 4 × 16, or 64, test drives.

However, combinatorics gives us a way to test the effects of cars, drivers, and oils that requires only 16 test drives. Call the oils *A, B, C,* and *D* and assign *one* oil to each car-driver combination in the arrangement shown in

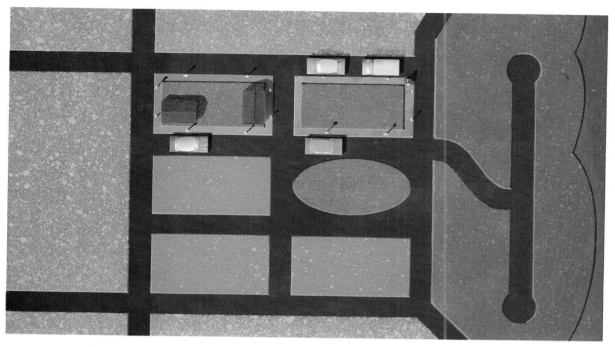

COLOR PLATE 1 A model town showing an aerial view of its street network. (See pp. 4–7.)

(a)

(b)

COLOR PLATE 2 (a) Computer-generated street network, showing an overview of part of a city. (b) A similar computer-generated street network, but showing more detail in a zoomed in view. [Sidney Bowne & Son, Consulting Engineers.] (See p. 19.)

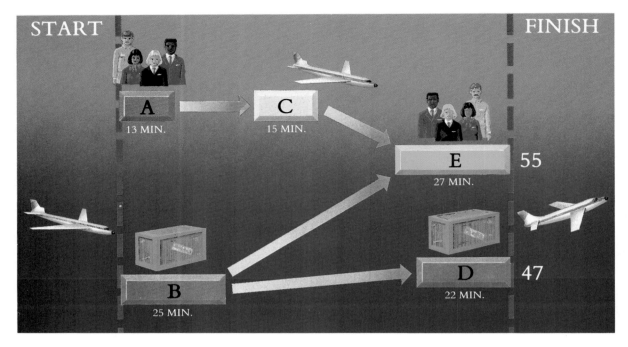

COLOR PLATE 3a Critical-path scheduling for servicing an airplane. (See pp. 42–48.)

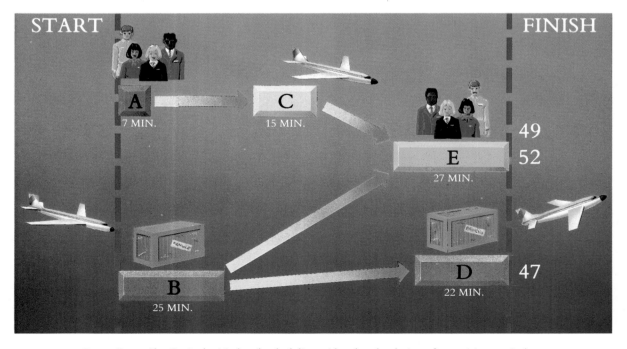

COLOR PLATE 3b Revised critical-path scheduling with reduced task times for servicing an airplane. (See pp. 42–48.)

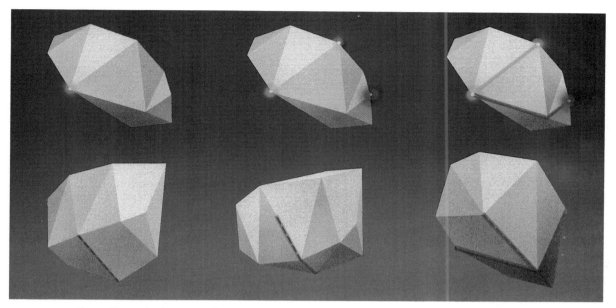

COLOR PLATE 4   Sequence of polyhedrons showing graphically how the simplex method of linear programming finds a solution. (See pp. 109–111.)

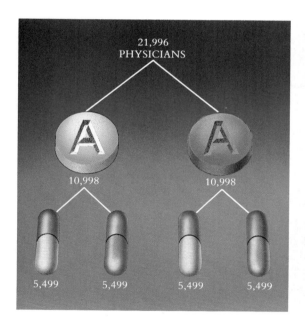

COLOR PLATE 5   Two-part tree diagram showing the Physicians' Health Study two-factor experiment involving aspirin and beta carotene. (See pp. 136–139.)

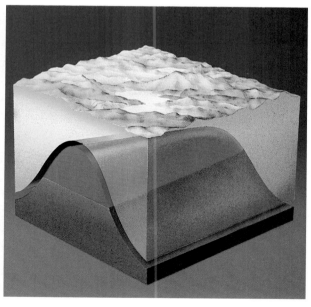

COLOR PLATE 6   A computer graphics representation of geologic seismic data plotted in three dimensions that shows a promising oil reservoir site. (See pp. 152–153.)

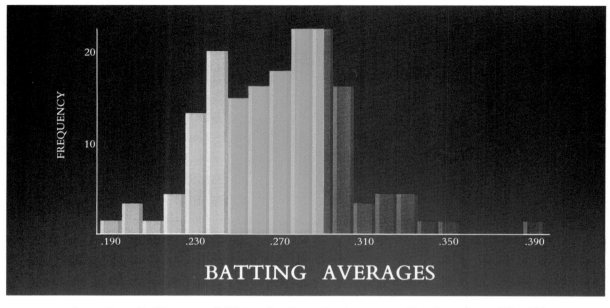

COLOR PLATE 7 Histogram of 1980 American League batting averages. Note the outlier at .390, which was George Brett's batting average that year. (See pp. 153–155.)

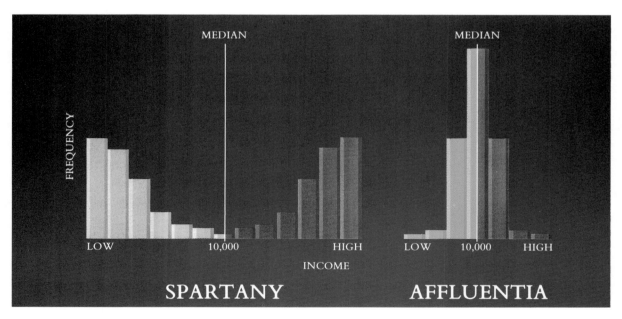

COLOR PLATE 8 Histogram of the wage spread in Spartany and Affluentia. (See pp. 159.)

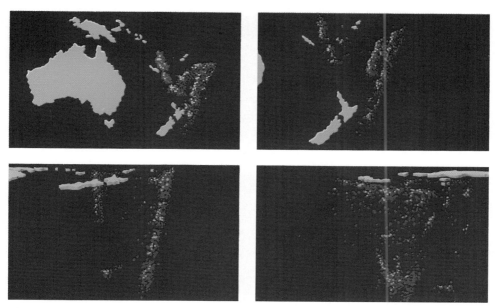

COLOR PLATE 9   A series of computer–generated graphics displays of earthquake data gathered near the Fiji islands. The top images show a traditional plotting of earthquake epicenters on a two-dimensional map. The remaining images show how the computer can manipulate the data to give a three-dimensional view of the epicenters, that is, their depth within the earth. (See pp. 166–168.)

COLOR PLATE 10   A normal distribution curve showing three standard deviations of probability 0.68, 0.95, and 0.997, respectively. This property of any normal curve is known as the 68-95-99.7 rule. The area under the curve of three standard deviations equals 99.7% of the total area. This is also shown in the nearly filled cup. (See pp. 193–196.)

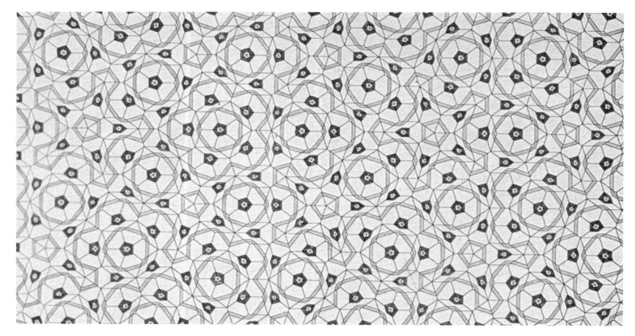

COLOR PLATE 11a A Penrose nonperiodic tiling made with two rhombus shapes. [Tiling by Roger Penrose.] (See pp. 498–501.)

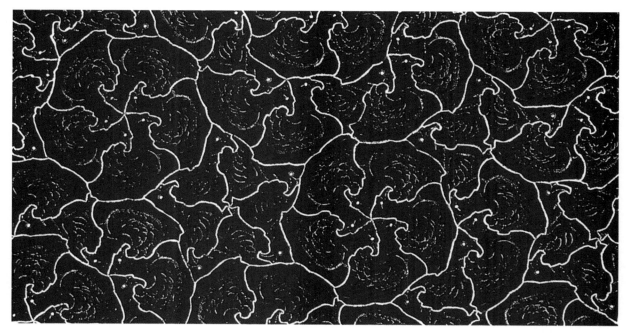

COLOR PLATE 11b A modification of a Penrose tiling by refashioning kites and darts into bird shapes. [Tiling by Roger Penrose.] (See pp. 498–501.)

COLOR PLATE 11c A Penrose tiling by kites and darts, colored with five colors. A Penrose tiling can always be colored in such a way using four colors that two tiles that share an edge have different colors. Whether a Penrose tiling can be colored in such a way using only three colors is an unsolved problem; we know, though, that if one Penrose tiling can be colored using three colors, all Penrose tilings can. [Tiling and coloring by Roger Penrose.] (See pp. 498–501.)

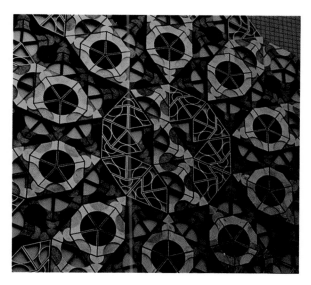

COLOR PLATE 12 A Penrose tiling with specially marked tiles, forming what is known as the cartwheel tiling. [From Roger Penrose.] (See pp. 498–501.)

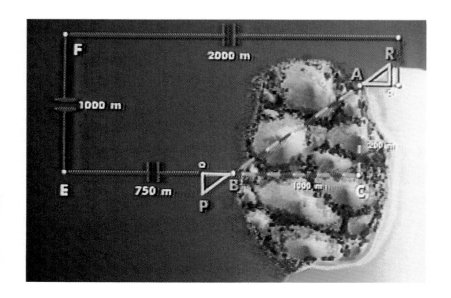

COLOR PLATE 13 The method of similar triangles was probably used 2500 years ago by Greek engineers on the island of Samos to dig a tunnel through Mount Castro to bring water to the capital city. (See pp. 416–417.)

COLOR PLATE 14a *Seven Butterflies Pattern*. This pattern is based on a repeating pattern of the Euclidean plane by M. C. Escher in which six butterflies meet at left front wing tips; increasing this number to seven results in a repeating pattern of the hyperbolic plane. A fundamental region for *Seven Butterflies Pattern*, as a colored pattern, consists of 168 butterflies; within any "ring" of butterflies of one color there are three different ways to arrange the other seven colors around the center. [Pattern designed by Douglas Dunham, University of Minnesota, Duluth.] (See pp. 452–454, 457–460.)

COLOR PLATE 14b *Six Fish Pattern*. This pattern is based on M. C. Escher's similar pattern of fish, *Circle Limit III*. The difference is that six fish meet at left fin tips in this pattern (instead of the four fish that meet at left fin tips in *Circle Limit III*). In *Six Fish Pattern*, as in *Circle Limit III*, 12 fish form a fundamental region for the colored pattern; of course one fish serves as a fundamental region for either pattern if color is disregarded. Also, as in *Circle Limit III*, each white backbone is a circular arc making acute angles with the bounding circle; thus in hyperbolic geometry, such an arc is not a hyperbolic line but is an equidistant curve from the circular arc with the same endpoints that is orthogonal (at right angles) to the bounding circle (and thus is a hyperbolic line). [Pattern designed by Douglas Dunham, University of Minnesota, Duluth.] (See pp. 452–454, 457–460.)

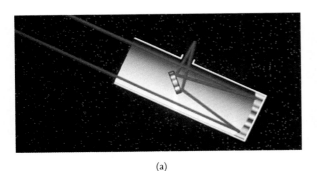
(a)

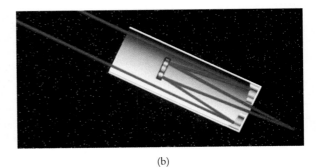
(b)

COLOR PLATE 15 Diagrams of the Newtonian and Cassegrain telescopes, both of which use parabolic mirrors to collect and focus light rays. (a) The light path in a Newtonian reflection telescope. (See pp. 436–437.) (b) The hyperbola, as well as the parabola, is applied in the Cassegrain telescope. Whereas Newton's telescope used a plane mirror to transmit the focused image to the viewer, the Cassegrain telescope uses a hyperbolic mirror for this purpose, again taking advantage of the focal properties of conic sections. (See pp. 448–450.)

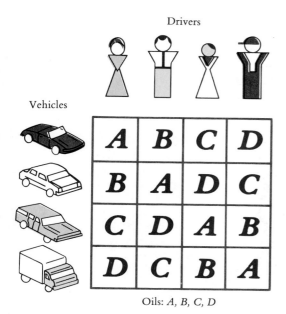

Drivers

Vehicles

Oils: *A, B, C, D*

Figure 5.9 The layout of a Latin square design to compare four motor oils in four cars with four drivers.

long before the statistical design of experiments was invented. The modern use of Latin squares to design experiments illustrates how the study of pure mathematics, originally pursued simply out of curiosity, often turns out to have practical importance.

The Latin square is a clever arrangement of comparisons. It is almost the opposite of a random assignment. However, the randomization principle of statistical design still governs the application of Latin squares. The arrangement of the labels in rows and columns that appears in Figure 5.9 is the Latin square itself. In applying this Latin square to an experiment, it is essential to assign the four drivers at random to the columns, the four cars at random to the rows, and the four motor oils at random to the labels *A, B, C, D*. Then the laws of probability will tell us whether the gas mileage figures we obtain using various oils differ by a greater amount than can be accounted for by chance. •

Figure 5.9. If we study the arrangement, we find that each oil appears exactly once in each row (for each car) and also exactly once in each column (for each driver). This setup can, in fact, show how each oil performs with each car and with each driver.

An arrangement like that in the motor oil example is called a **Latin square.** Figure 5.9 is a 4 × 4 Latin square because it arranges 4 kinds of objects (the oils) in a 4 × 4 square array so that each object appears exactly once in each row (vehicle) and in each column (driver). There are several ways of assigning the labels *A, B, C, D* so that each appears once in each row and once in each column; that is, there are several different 4 × 4 Latin squares. Exercise 35 asks you to find one that is not the same as Figure 5.9. Mathematicians studied Latin squares and cataloged many of them

STATISTICS IN PRACTICE

There is more to the wise use of statistics than a knowledge of such statistical techniques as Latin square designs. A statistician must know which techniques are applicable and appropriate. In the motor oil experiment, we knew that cars and drivers would influence gas mileage. A Latin square allowed us to compare different motor oils without getting confused by different kinds of cars and different drivers; we had three known variables. However, in a study of heart disease or cancer, we don't even know all the important influences; in such cases the Latin square does not apply.

Moreover, good data collection requires more than a well-designed sample or experiment. We must also *measure* the variables of interest. That is easy in the case of gas consumption, but measuring "intelligence" or

SPOTLIGHT 5.5 Experiments and Ethics

Dr. Charles Hennekens, director of the Physicians' Health Study, had to concern himself with the goals, design, and implementation of his large-scale study. But other questions also arise in the course of such an experiment. Dr. Hennekens was asked about the ethics of experimenting on human health:

Dr. Charles Hennekens.

Much has been made of the ethical concerns about randomized trials. There are instances where it would not be ethical to do a randomized trial. When penicillin was introduced for the treatment of pneumococcal pneumonia, which was virtually 100% fatal, the mortality rate plummeted significantly. Certainly it would have been unethical to do a randomized trial, to withhold effective treatment from people who need it.

There's a delicate balance between when to do or not to do a randomized trial. On the one hand, there must be sufficient belief in the agent's potential to justify exposing half the subjects to it. On the other hand, there must be sufficient doubt about its efficacy to justify withholding it from the other half of subjects who might be assigned to the placebos, the pills with inert ingredients. It was just these circumstances that we felt existed with regard to the aspirin and the beta carotene hypotheses.

"attitude toward abortion" is much harder. We must decide exactly what we want to measure and design a procedure for measuring it.

Even with a good design and careful procedures for measurement, other sources of error can creep in. Some of the subjects in a random sample of people may not be at home. Others may misunderstand the questions, refuse to co-

operate, or not tell the truth. In experiments with human subjects, it is often hard to apply realistic treatments. A psychologist who studies stress by exposing student volunteers to an artificial situation must ask if a few hours of laboratory stress can simulate months or years of hard living and elicit the same responses.

When we are planning a statistical study, we must also face some ethical questions. Does the

A randomized trial done on a newer therapy or drug is best done when the procedure is first introduced. It becomes very difficult, with regard to both feasibility and ethics, to do such trials after a long period of time has elapsed. Treatments begin to be so accepted by the population that it is difficult to find people willing to have the treatment withheld. Others might feel ethically that it would be difficult to withhold treatment.

One example in contemporary times regards the treatment of breast cancer. William Halsted of Johns Hopkins, the father of American surgery, invented the radical mastectomy in the early 1900s as a therapy for breast cancer. It's been used widely for more than half a century. However, after all this time, it became apparent to some investigators that less extensive procedures might accomplish the same results, that is, to keep the age-specific mortality rate from breast cancer in affected women at a low level. It's only been in recent years that randomized trials of less extensive forms of treatment for breast cancer have been done. It is important that such research be done. It is optimal to do it early so that we get a clear answer at the beginning of the development of new procedures and new drugs.

Dr. Julie Buring, associate director of the Physicians' Health Study, adds:

Sometimes, in the case where certain procedures are not tested in a trial immediately, the reason is that the procedures seem to make sense intuitively. For example, it would make sense that if you remove more of the breast tissue, it might reduce your risk of having a recurrence of the disease. Yet most clinicians, or most medical professionals, only see a small number of patients. It is very hard to get a pattern, to see that there really is no difference between those who receive a radical mastectomy and a less invasive procedure. It is only when you are able to do these trials in thousands of patients that you are able to see that there really is no difference, or just a slight difference. You must have large numbers to be able to pick up those patterns.

knowledge gained from an experiment or study justify the possible risk to the subjects? In the Physicians' Health Study, doctors gave their informed consent to take either aspirin, beta carotene, or placebos, in any combination, as prescribed by the study designers. When it became clear that men taking aspirin had fewer heart attacks, the experiment was stopped so that all the subjects could take advantage of this new knowledge. In Spotlight 5.5 the directors of the Physicians' Health Study explain why randomized comparative experiments are the mainstay of medical, agricultural, and many other kinds of research and when such clinical trials are justified. Practical and ethical problems are never far from the surface when statistics is applied to real problems.

REVIEW VOCABULARY

Bias A systematic error that tends to cause the observations to deviate in the same direction from the truth about the population whenever a sample or experiment is repeated.

Block A group of experimental subjects that is homogeneous on one or more variables, for example, a same-sex group. Division into blocks controls for the effects of that variable. Complex experimental designs often assign treatments to subjects at random separately within each of several blocks.

Confounding Two variables are confounded when their effects on the outcome of a study cannot be distinguished from one another.

Control group A group of experimental subjects who are given a standard treatment or no treatment (such as a placebo).

Double-blind experiment An experiment in which neither the experimental subjects nor the persons who interact with them know which treatment each subject received.

Experiment A study in which treatments are applied to people, animals, or things in order to observe the effect of the treatment.

Histogram A graph that displays how often various outcomes occur by means of bars. The height of each bar is the frequency of occurrence of an outcome or group of outcomes.

Latin square An arrangement of k kinds of objects in a $k \times k$ square array so that each kind appears exactly once in each row and exactly once in each column.

Placebo effect The effect of a dummy treatment (such as an inert pill in a medical experiment) on the response of subjects.

Population The entire group of people or things that we want information about.

Randomized comparative experiment An experiment that compares two or more treatments in which people, animals, or things are assigned to treatments by chance.

Sample A part of the population that is actually observed and used to draw conclusions, or inferences, about the entire population.

Simple random sample A sample chosen by chance so that every possible sample of the same size has an equal chance to be the one selected.

Table of random digits A table whose entries are the digits 0, 1, 2, 3, 4, 5, 6, 7, 8, 9 in a completely random order. That is, each entry is equally likely to be any of the 10 digits and no entry gives information about any other entry.

Voluntary-response survey A sample survey in which the sample chooses itself by responding to a general invitation to write or call with their opinions. Such a survey is usually strongly biased.

SUGGESTED READINGS

Box, George E. P., William G. Hunter, and J. Stuart Hunter: *Statistics for Experimenters,* Wiley, New York, 1978, chapters 4, 7, and 8. This more advanced text places greater emphasis on concepts and experimental design than most books at a similar level. A good source for more information about such topics as experiments with several factors and Latin squares.

Freedman, David, Robert Pisani, and Roger Purves: *Statistics,* Norton, New York, 1978, chapters 1, 2, and 19. Excellent, but rather lengthy, conceptual discussion with good examples. Slightly higher in level than *For All Practical Purposes.*

Moore, David S.: *Statistics: Concepts and Controversies,* 2d ed., Freeman, New York, 1985, chapters 1 and 2. Written for liberal arts students, this book provides more extensive discussion at about the same mathematical level as *For All Practical Purposes.*

EXERCISES

In each of Exercises 1 to 3, identify the *population* about which information is desired and the *sample* that is actually observed. If the exact population is not specified by the information given, complete the description of the population in a reasonable way.

1. Ms. Caucus is her party's candidate in the Second Congressional District of Indiana. The party wants to know what percent of registered voters would vote for Ms. Caucus if the election were held tomorrow. A polling firm contacts 800 voters, of whom 456 say they would vote for Ms. Caucus.

2. Home canners sometimes can vegetables in used mayonnaise jars to avoid buying special canning jars. *Organic Gardening* magazine wondered what percent of mayonnaise jars would break when used for canning. They obtained 100 mayonnaise jars and canned tomatoes in them. Only 3 of the jars broke.

3. A maker of electronic instruments buys megabyte RAM memory chips from another company. The company wants to know the percent of substandard chips made by the supplier, so it tests all chips received and keeps records. Last year 32,000 out of 400,000 chips received failed to meet standards.

▲ **4.** A magazine for health foods and organic healing wants to establish that large doses of vitamins will improve health. They ask readers who have regularly taken vitamins in large doses to write in, describing their experiences. Of the 2754 readers who reply, 93% report some benefit from taking vitamins.

Is the sample proportion of 93% probably higher than, lower than, or about the same as the percent of all adults who would perceive some benefit from large vitamin intake? Why? (In answering these questions, you have identified a source of bias in the sampling method.)

▲ **5.** A Mississippi television station cancels a program aimed at black audiences because a rating service shows that only 12% of the total viewers in the program's time slot watch it. The ratings are based on a telephone survey of households in the station's market.

Is the 12% rating probably higher than, lower than, or about the same as the true percent of all viewers who watch the program? Why? (In answering these questions, you have identified a source of bias in the rating service's survey.)

6. A newspaper advertisement for *USA Today: The Television Show* said:

Should handgun control be tougher? You call the shots in a special call-in poll tonight.

If yes, call 1-900-720-6181
If no, call 1-900-720-6182

Why is this opinion poll almost certainly biased?

7. Sampling from a list that contains only part of the population is a common cause of bias in sampling. In each of the following examples, explain why this source of bias may be present.

▲ Discussion exercise.

a. To assess public opinion on a proposal to reduce welfare and unemployment payments, a polling firm selects a sample by random-digit dialing (RDD). RDD uses a machine that dials residential telephone numbers at random.

b. To assess the reaction of her constituents to the same proposal, a member of Congress uses her free-mailing privilege to send a questionnaire to every registered voter in her district.

8. You are worried about the problem of false credentials being offered by candidates for employment at your firm. You decide to investigate some of the credentials at random from now on. Use line 123 of Table 5.1 to choose a simple random sample of four of the following group of candidates for investigation.

Adams	Cleveland	Jack	Miller	Russell	Weinstein
Alvarez	Drasin	Kodaira	Ogden	Sanguillen	Wren
Bartkowsky	Edwards	LaMay	Pierce	Toon	
Bishop	Frank	Marsden	Pollack	Turco	
Borchardt	Hoffer	Martinez	Riersol	Ungarn	
Chan	Hohenstein	Michel	Rubin	Vlasov	

9. Your class in ancient Ugaritic religion is poorly taught and has decided to complain to the dean. The class decides to choose three of its members at random to carry the complaint. The class list appears below. Choose a simple random sample of three by using Table 5.1, starting at line 105.

Anderson	Fuller	Kempthorne	Robertson
Aspin	Grant	Landis	Rodriguez
Bennett	Gutierrez	Laskowsky	Siegel
Bock	Green	Olds	Tompkins
Breiman	Harter	Patnaik	Vandegraff
Cochran	Henderson	Pirelli	Williams
Dixon	Hughes	Rao	
Edwards	Johnson	Rider	

10. You must allocate five tickets to a rock concert among 25 clamoring members of your club. Choose five at random to receive the tickets, using line 135 of Table 5.1 (ignore the asterisks).

Agassiz	Darwin	Herrnstein	Myrdal	Vogt*
Binet*	Epstein	Jerison*	Perez*	Went
Blumenbach	Ferri	Lombrosco	Spencer*	Wilson
Chase*	Goddard*	Moll*	Thomson	Yerkes
Cuvier*	Hall	McKim*	Toulmin	Zimmer

11. Let us illustrate sampling variability in a small sample from a small population. Ten of the 25 club members listed in Exercise 10 are female. Their names are marked with asterisks in the list.

a. Draw five at random 20 times, using a different part of Table 5.1 each time. Record the number of females in each of your samples. Make a histogram to display your results. What is the average number of females in your 20 samples?

b. Do you think the club members should suspect discrimination if none of the five tickets go to women?

● **12.** Random digits can be used to *simulate* the results of random sampling. Suppose that you are drawing simple random samples of size 25 from a large bowl of beads and that 20% of the beads in the bowl are dark. To simulate this experiment, let 25 consecutive entries in Table 5.1 stand for the 25 beads in your sample. The digits 0 and 1 stand for dark beads, while other digits stand for light beads. This is an accurate imitation of the sampling experiment because 0 and 1 make up 20% of the 10 equally likely digits. Simulate the results of 50 samples by counting the number of 0s and 1s in the first 25 entries in each of the 50 rows of Table 5.1. Note that the percent of dark beads in each sample is just four times the count; for example, 7 dark beads out of 25 is 28%. Make a histogram like Figure 5.4 to display the results of your 50 samples. Is the truth about the population (20% dark) near the center of your histogram? What are the smallest and largest percents you obtained in your 50 samples?

13. A student wishes to study the opinions of faculty at her college on the advisability of setting up a state board of higher education to oversee all colleges in the state. The college has 380 faculty members.

    a. What is the population in this situation?

    b. Explain carefully how you would choose a simple random sample of 50 faculty members.

    c. Use Table 5.1, starting at line 135, to choose *only the first five* members of this sample.

14. The number of students majoring in political science at Ivy University has increased substantially without a corresponding increase in the number of faculty. The campus newspaper plans to interview 25 of the 450 political science majors to learn student views on class size and other issues. You suggest a simple random sample. Explain carefully how you would choose this sample. Then use Table 5.1, starting at line 120, to select *only the first five* members of your sample.

▲ **15.** The advice columnist Ann Landers regularly invites her readers to respond to questions asked in her newspaper column. On one occasion, she asked, "If you had it to do over again, would you have children?" Almost 10,000 people wrote in, of whom 70% said no. Shortly afterward, a national poll asked a random sample of 1400 people the same question; 90% of this sample said yes. Which of these polls is more trustworthy, and why?

16. An opinion poll asks a sample of 1450 adults whether they jog regularly; 224 say yes. What percent of the sample jog? The margin of error for this poll is ±3%. What interval are you confident covers the percent of all adults who jog? (This margin of error will hold in 95% of all samples, that is, with probability 0.95.)

17. A sample survey asks a sample of 1324 adults whether they believe that life exists on other planets; 609 say yes. What percent of the sample believe in extraterrestrial life? The polling organization announces a margin of error (holding with probability 0.95) of ±3%. What conclusion can you draw about the percent of all adults who believe that life exists on other planets?

● Optional exercise.
▲ Discussion exercise.

18. National opinion polls such as the Gallup poll usually take weekly samples of about 1500 people. This sample size gives a margin of error of about ±3 percentage points. (That is, 95% of all such samples are accurate within ±3%.) Just before a presidential election, however, the polls often increase the size of their samples to about 4000 people. Is the margin of error now more than ±3%, less than ±3%, or still equal to ±3%? Why?

The studies in Exercises 19 to 21 may produce invalid data because of confounding of outside influences with the treatment of interest. Explain in each case how confounding could influence the outcome.

▲ **19.** A college student believes that rose hip tea has remarkable curative powers. To demonstrate this, she and several friends visit a local nursing home several times a week, talking with the residents and serving them rose hip tea. After a month, the head nurse reports that the residents visited are indeed more cheerful and alert.

▲ **20.** A language teacher believes that study of a foreign language improves command of English. He examines the records at his high school and finds that students who elect a foreign language do indeed score higher on English achievement tests.

▲ **21.** A job-training program is being reviewed. Critics claim that because the unemployment rate in the manufacturing region affected by the program was 8% when the program began and 12% four years later, the program was ineffective.

22. It has been suggested that there is a "gender gap" in political party preference in the United States, with women more likely than men to prefer Democratic candidates. A political scientist asks each of a group of men and a group of women whether they voted for the Democratic or Republican candidate in the last Congressional election. Explain carefully why this study is *not* an experiment.

23. Will reducing blood-cholesterol levels prevent heart attacks? You have available a drug that will lower blood cholesterol. You also have 3000 men aged 50 to 65 who are willing to participate in a study. Outline the design of an experiment to settle the question. Your outline should follow the model of Figure 5.5.

24. Does regular exercise reduce the risk of a heart attack? Several ways of studying this question suggest themselves.

a. A researcher takes a sample of 2000 men in their forties who recently suffered their first heart attack. He matches each with another man of the same age, occupation, and other demographic characteristics who has not had a heart attack. Both groups are questioned about their past exercise habits. Is this an experiment? Is it a prospective study? Explain your answers.

b. Another researcher finds 2000 men over 40 who exercise regularly and have not had heart attacks. She matches each with a similar man who does not exercise regularly, and she follows both groups for 10 years. Is this an experiment? Is it a prospective study? Explain your answers.

c. You have 4000 men over 40 who have not had a heart attack and who are willing to participate in a study. Outline the design of an experiment to investigate the effect of regular exercise on heart attacks.

▲ Discussion exercise.

25. Should either or both the experiments in Exercises 23 and 24c be double-blind studies? Explain your answer.

26. Ignoring all practical difficulties and moral issues, outline the design of an experiment that would settle the question of whether cigarette smoking causes lung cancer.

27. In a test of the effects of persistent pesticides, rats are to be fed a diet contaminated with DDT for 60 days after weaning. Then measurements will be made of their nerve responses to assess the effects of the DDT.

a. Explain why the experimenter should also study a control group of rats that are fed the same diet uncontaminated with DDT.

b. If 20 newly weaned male rats are available, outline the design of the experiment and use Table 5.1, starting at line 123, to carry out the randomization.

28. A college allows students to choose either classroom or self-paced instruction in a basic statistics course. To compare the effectiveness of self-paced and regular instruction, someone proposes administering the same final exam to all students in both versions of the course and comparing the average score of those who took the self-paced option with the average score of students in regular sections.

a. Explain how confounding makes the results of that study worthless.

b. Given 30 students who are willing to use either regular or self-paced instruction, outline an experimental design to compare the two methods of instruction. Then use Table 5.1, starting at line 108, to carry out the randomization.

29. Below are the names of 20 patients who have consented to participate in a trial of surgical treatments for angina. Outline an experiment to compare surgical treatment with a placebo (sham surgery), and use table 5.1, beginning at line 101, to do the required randomization. (Ignore the asterisks.)

Ashley	Cravens*	Lippmann	Strong*
Bean*	Dorfman	Mark*	Tobias
Block	Epstein	Morton*	Voison*
Burt	Huang*	Popkin	Washington
Chavez*	Kidder	Spearman	Williams

30. Unknown to the researchers in Exercise 29, the eight subjects whose names are marked by asterisks will have a fatal heart attack during the study period. We can observe how sampling variability operates in a randomized experiment by keeping track of how many of these eight subjects are assigned to the group that will receive the new surgical treatment. Carry out the random assignment of 10 subjects to the treatment group 20 times, keeping track of how many asterisks are on the names you choose each time. Then make a histogram of the count of heart attack victims assigned to the treatment. What is the average number in your 20 tries?

31. Explain clearly the advantage of using several thousand subjects, rather than just 20, in the experiment of Exercise 29.

● **32.** Explain carefully how you would randomly assign the 20 subjects named in Exercise 29 to the four treatments in the Physicians' Health Study. (Assign 5 of the 20 to each group.) Then enter Table 5.1 at line 120 to carry out the randomization.

● **33.** New varieties of corn with altered amino acid patterns may have higher nutritive value than standard corn, which is low in the amino acid lysine. An experiment is conducted to compare a new variety, called floury-2, with normal corn. Corn–soybean meal diets using each type of corn are formulated at each of three protein levels: 12%, 16%, and 20% protein. There are thus six diets in all. Ten one-day-old male chicks are assigned to each diet, and their weight gains after 21 days are recorded. The weight gain of the chicks is a measure of the nutritive value of their diet.

    a. This experiment has two factors. What are they?

    b. Outline the design of the experiment. Be sure to use randomization. (You need not actually carry out the randomization required by your design.)

● **34.** An engineer wants to study the effect of the speed of a conveyor belt carrying electronic circuit boards on the performance of a wave-soldering machine that simultaneously solders all the connections on a board as the conveyor moves the board through a wave of molten solder. The speeds to be compared are 20, 25, and 30 feet per minute. The outcome variable is the number of improperly soldered connections among the 2000 connections on a circuit board.

    a. The engineer plans to process 10 boards at each conveyor speed. Why should she assign the speeds at random to the 30 boards rather than simply process the first 10 at 20 feet per minute, the second 10 at 25 feet per minute, and so on?

    b. Outline the design of a randomized comparative experiment, beginning with boards numbered 1 through 30 in the order in which they will be soldered.

    c. Enter Table 5.1 at line 130 to carry out the randomization required. List the sequence of 30 conveyor speeds that the engineer will use when she carries out the experiment.

● **35.** Find a 3×3 Latin square arrangement of three treatments A, B, and C. Then find a 4×4 Latin square arrangement of four treatments A, B, C, and D that is different from the one illustrated in Figure 5.9.

● **36.** You wish to compare the durability of four types of wood treatment by burying the treated wood in soil for one year, then measuring decay. Because the variety of wood and the soil type also influence decay, you choose four wood species and four locations with different soil types. Lay out a 4×4 Latin square as the basis for an experimental design. Then randomly assign the wood species to the rows, the soil types to the columns, and the four types of wood treatment to the letters in the body of the Latin square.

● Optional exercise.

·6·

Describing Data

The proliferation of data is a prominent feature of modern society. Data, or numerical facts, are essential for making correct decisions in almost every area of our lives. Like the proverbial trees in the forest, however, these numbers threaten to overwhelm us unless we can take control of them through careful organization and interpretation. Many modern scientific activities, for example, seismic exploration for oil, produce hundreds of thousands of numbers each day. All these data must somehow be interpreted — a task that requires a way of presenting the numbers so that their message is immediately clear.

To use data for human purposes, we must compress, summarize, and describe them. A few numbers computed from the data — averages, percents, and the like — can be very helpful. Numbers computed from the data are also the raw material of *statistical inference,* the

science of drawing conclusions from data with the aid of the mathematics of probability.

The method of formal statistical inference is to ask specific questions in advance and then collect data to answer those specific questions. However, this neat process is not always possible. We may have to analyze a mass of data collected for other purposes, such as government records, before we know whether the data can help to answer our questions. In other cases we may not even know what questions we should ask.

Exploratory data analysis is the art of letting the data speak, of seeing patterns in data that we may not have anticipated. Exploratory analysis is informal; unlike inference, it does not seek answers to specific questions. However, exploratory data analysis is not at all a diversion from the mainstream of statistics. Even in the most carefully planned experiment, exploring the data is an essential first step. It may reveal obvious errors or an important effect that was not anticipated.

Exploratory analysis of data combines numerical summaries with graphical display. We can grasp pictures more easily than columns of numbers, so the most powerful tools of data analysis are graphs. As long as our emphasis is on description rather than inference, pictorial display of data will occupy first place.

Pictorial display of data is not a new idea. In 1861, Charles Minard, a French engineer, used an elaborate and original graph to show the impact of harsh winter conditions on Napoleon's troops during the unsuccessful campaign of 1812. Some 422,000 French soldiers entered Russia and 100,000 reached Moscow, but only 10,000 straggled back. The pink and black bands across the map of eastern Europe in Minard's graph show the route of Napoleon's Grand Army (see Figure 6.1). The width of the bands is proportional to the number of surviving troops. As the number of survivors dwindles, the river of soldiers narrows to a trickle. At the bottom of his graph, Minard showed the temperature during the winter retreat from Moscow. The falling temperatures and shrinking army march together in that famous defeat.

Visual representations of data abound in books and magazines, although not many examples are as imaginative or as striking as Minard's. The progress of computer graphics has given a new emphasis to pictorial display of data in business, medicine, and technical fields. The computing power of machines combined with the unique ability of the human eye and brain to recognize visual patterns provides powerful new tools for data analysis.

For example, to discover where oil reservoirs lie, geologists probe the structure of the

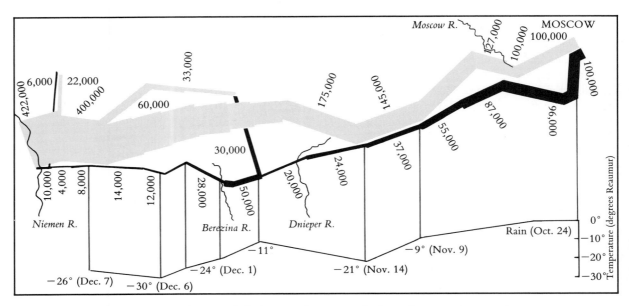

Figure 6.1 Redrawing of Charles Minard's 1861 graph of Napoleon's Russian campaign.
(In the Reaumur temperature scale, water boils at 80°R and freezes at 0°R.)

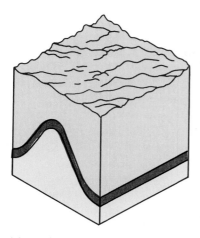

Figure 6.2   Underground structures reconstructed from seismic data.

earth with explosions. The shock waves of these explosions bounce off underlying rock strata and are reflected up to seismic sensors scattered around the exploration site. A single explosion can produce half a million separate data points, and a skilled seismic crew operating by helicopter can set off up to 20 such tests a day. This abundance of data has to be organized into meaningful and easily understood patterns.

The results of the tests are processed numerically and then presented using the tool of computer graphics (see Figure 6.2). An oil expert could look at this display and instantly interpret it as a promising site for an oil reservoir. The computer does the calculating and graphics, then a trained eye "sees" the oil in the graphically displayed data. Such analysis is an effective collaboration between the computational power of machines and the ability of humans to see and understand.

We will use both numbers and pictures to explore data. We will organize our thinking by three principles:

1. Move from describing a single variable to describing relations among several variables.

2. Look first for an overall pattern in the data and then for important deviations from this pattern.

3. Use graphical display in combination with numerical description.

The pattern of values of a numerical variable is called its **distribution.** Data analysis begins with graphical displays of the distribution of a single variable.

DISPLAYING DISTRIBUTIONS

Baseball fans have long memories for the statistics of the game (Spotlight 6.1, p. 157). For example, how many home runs does it take to lead the league? Table 6.1 gives the American League leaders and their home run totals from 1970 to 1989. We can get a quick picture of the distribution of the league-leading home run total by making a **dotplot.** First, draw a horizontal axis marked off to span the range of the data. Then mark each observation with a dot above the axis. Here is the result:

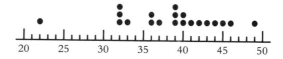

We see, for example, that 32 and 39 each led the league three times during these 20 seasons. The data cover the range from 32 to 49 without any strong pattern. A systematic pattern is often hard to see in small data sets such as this. But there is a clear **outlier,** an individual observation that falls outside the pattern of the remaining data. In 1981, it took only 22 homers to lead the league. Outliers often point to errors in recording the data or to unusual circumstances. In this case, the 1981 baseball season was interrupted by a players' strike that

TABLE 6.1 American League home run leaders, 1970–1989

Year	Player	Home runs	Year	Player	Home runs
1970	Frank Howard	44	1980	Reggie Jackson	41
1971	Bill Melton	33	1981	Four players	22
1972	Dick Allen	37	1982	Thomas and Jackson	39
1973	Reggie Jackson	32	1983	Jim Rice	39
1974	Dick Allen	32	1984	Tony Armas	43
1975	George Scott and Jackson	36	1985	Darrell Evans	40
1976	Graig Nettles	32	1986	Jesse Barfield	40
1977	Jim Rice	39	1987	Mark McGwire	49
1978	Jim Rice	46	1988	Jose Canseco	42
1979	Gorman Thomas	45	1989	Fred McGriff	36

reduced the number of games played from the usual 162 to 108.

Dotplots are quick to draw and work well for small numbers of observations. They become awkward when there are many observations or when the observations do not have well-separated values. (Whole-number values like counts of home runs are ideal for dot-plots.) When a dotplot is not satisfactory, we display a distribution by a more formal type of graph, a **histogram.** To see how histograms work, let's ask another baseball question: How well do major league batters hit?

EXAMPLE: Drawing a Histogram. The simplest measure of how well a baseball player hits is his batting average, which is simply the proportion of times at bat that the player gets a hit. However, the batting average of a player who has been at bat only a few times may be due to luck — either good or bad. To eliminate these cases, we will consider only players who have been at bat 200 or more times in a season. Figure 6.3 is a histogram of the distribution of batting averages for all 167 American League players who batted 200 or more times in the 1980 season. We can learn from this example how to draw a histogram of any distribution.

1. Divide the range of the data into classes of equal width. In this case, each class covers a 10-point range of batting averages. The first two classes are

$$.185 \leq batting\ average < .195$$

$$.195 \leq batting\ average < .205$$

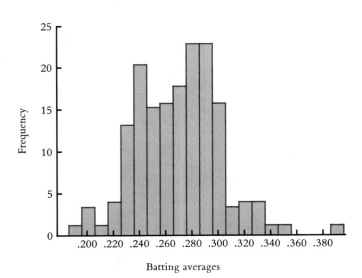

Figure 6.3 Histogram of 1980 American League batting averages.

Leave no space between classes and be careful about the endpoints of the classes; for example, .194 falls in the first class and .195 in the second.

2. Count the number of observations in each class. These counts are called **frequencies.** The frequencies for the batting average classes are shown in Table 6.2

3. Draw the histogram. The batting average scale is horizontal and the frequency scale vertical. Each bar represents a class. The base of the bar covers the class, and the bar height is the class frequency. The graph is drawn with no horizontal space between the bars (unless a class is empty, so its bar has 0 height).

There is no single right number of classes in a histogram. You must use judgment. The goal is to give a clear picture of the distribution. Avoid using too few classes, which produces a skyscraper picture with all the observations bunched together; also avoid using too many classes, which gives a pancake histogram with few observations in each class. Figure 6.3 is a successful histogram because the overall pattern of the distribution is immediately apparent. Almost all major league regulars hit between about .225 and .305. A typical player hit about .270.

Another aspect of the histogram is also immediately apparent. The single observation at .390 is an outlier. It is a fact of life that any sizable set of data is likely to contain errors that must be tracked down and corrected. Outliers are often the result of errors and must be carefully investigated. It would have been easy, for instance, to have typed .390 in place of .290 while recording the data. However, this outlier is not a mistake. George Brett of the Kansas City Royals hit .390 in 1980, the highest major league batting average since Ted Williams's average of .406 in 1941. Brett therefore deserves his isolated point on the histogram.

George Brett and his more-average colleagues illustrate an important principle in the exploration of data: always look for an *overall pattern,* and then look for *deviations* from that pattern. In short,

$$\text{Data} = \text{smooth} + \text{rough}$$

The *smooth* in any data is an overall pattern. In this histogram, the smooth is the regular distribution of batting averages, with a few in the .220s, many between .240 and .290, then falling off to a few in the .330s and above. The *rough* is the outlier recording Brett's performance. Both the smooth and the rough — the pattern and the exception — carry useful information.

In general, the smooth in a histogram is the overall shape of the distribution. A distribution is **symmetric** if the right and left halves are mirror images of each other. A **skewed** distribution has one tail that is much longer than the other. Distributions of real data will, of course, be only approximately symmetric. We consider Figure 6.3 — without George Brett — to be approximately symmetric.

TABLE 6.2 Frequencies for the histogram in Figure 6.3

Class	Frequency	Class	Frequency
.185 to .194	1	.295 to .304	16
.195 to .204	3	.305 to .314	3
.205 to .214	1	.315 to .324	4
.215 to .224	4	.325 to .334	4
.225 to .234	13	.335 to .344	1
.235 to .244	20	.345 to .354	1
.245 to .254	15	.355 to .364	0
.255 to .264	16	.365 to .374	0
.265 to .274	18	.375 to .384	0
.275 to .284	23	.385 to .394	1
.285 to .294	23		

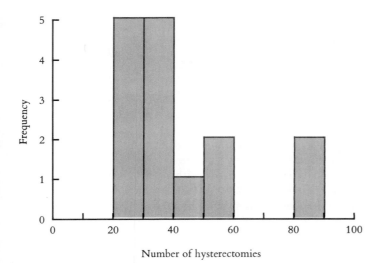

Figure 6.4 Histogram of hysterectomies performed by male Swiss doctors.

EXAMPLE: How Many Operations Do Doctors Perform?

A study of the number of surgical procedures performed by physicians obtained the following data on the number of hysterectomies performed in a year by each of a sample of 15 male doctors in Switzerland:

$$27, 50, 33, 25, 86, 25, 85, 31, 37, 44, 20, 36,$$
$$59, 34, 28$$

Figure 6.4 is a histogram of this distribution; the classes are 20 to 29, 30 to 39, and so on. The distribution has a long right tail not matched by corresponding observations to the left of the bulk of the data. That is, a few doctors perform many hysterectomies. We say that the distribution is *skewed to the right*. Notice that the direction of the skew is the direction of the long tail, not the direction in which most observations cluster.

The rough in a distribution often takes the form of *outliers* or *gaps*. The baseball strike of 1981 and George Brett's exceptional performance were responsible for outliers in the distributions of home runs and of 1980 batting averages. The next example features a gap.

EXAMPLE: Quality Control.

Spotting patterns in quality control data can increase a manufacturer's productivity and profitability. Although Japanese products are known today for high reliability, this wasn't always the case. W. Edwards Deming, an American expert in quality control, was a leader in teaching Japanese industrialists about statistical methods in the 1950s. His efforts were so successful that Japan's annual award for excellence in quality control is named after him.

One of Deming's studies concerned the size of steel rods used in a manufacturing process. The histogram in Figure 6.5 summarizes the data. It displays the diameters of 500 steel rods as measured by the manufacturer's inspectors. An overall pattern — the smooth — is apparent in the size of the rods: the distribution is approximately symmetric, centered at 1.002 centimeters and falling off rapidly both above and below. The departure from regularity — the rough — is the empty space at 0.999 centimeter.

One centimeter, or precisely 1.000, is the lower specification limit for these rods. Rods

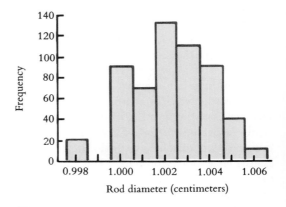

Figure 6.5 Deming's illustration of the effects of improper inspection: a histogram with a gap.

SPOTLIGHT 6.1 Baseball: The Great Statistics Game

Baseball fans have a love affair with statistics. There's absolutely no doubt about it: it's the greatest statistics game there is, probably because baseball goes back so far in this country: it's embedded in most of the population. They really understand a home run and an RBI, a batting average and an earned run average—all those basic statistics that have been with baseball throughout its history. The basics have never changed, so people know and love them.

DICK BRESCIANI
Red Sox Chief Statistician

George Brett of the Kansas City Royals hit .390 in 1980, the highest major league batting average since Ted Williams's average of .406 in 1941. [Kansas City Royals.]

There is a whole lore of baseball history involving statistics. The great thing is to compare the players of old with the players of today. Many times on talk shows or panels people will say, "Could Dave Winfield or Jim Rice or George Brett have played with a Ty Cobb or a Mickey Cochran or a Ted Williams or a Joe DiMaggio?" What you have to argue is statistics. You have to go back and examine DiMaggio's years in the big leagues. You look at what he did year by year: he was on average a .300-and-some hitter; he drove in so many runs; he did thus-and-so defensively in the outfield. The statistics are all that remain of the career of that player. So you lay them out against a Brett or a Winfield and try to compare them. That is the fun of the game.

LOU GORMAN
Red Sox General Manager

that are any smaller than this will be loose in their bearings and should therefore be rejected. The empty 0.999 class in the histogram, together with the higher-than-expected spike in the 1.000 class, provides strong evidence that the inspectors were passing rods that fell just below this limit by recording them in the 1-centimeter class. The inspectors didn't realize that just $\frac{1}{1000}$ of a centimeter can be crucial. With better training of the inspec-

tors, the missing 0.999 class filled in and the distribution became quite regular.

The strong visual impact of a graph helped pinpoint an important problem in the quality control example. Interpreting the smooth portion of a histogram can also be helpful. In this example, it might suggest that the entire process should be adjusted so that more of the distribution (i.e., more of the steel rods produced) exceed 1 centimeter in diameter.

NUMERICAL DESCRIPTION OF DISTRIBUTIONS

Dotplots and histograms display the overall shape of a distribution of values. We can describe specific aspects of this overall shape by a few carefully chosen numbers. Two important aspects of the overall shape of a distribution are its *center* (sometimes called *location*) and its *spread*. Center and spread are visible in a graph, but now we want to describe them using numbers.

Perhaps the most obvious way to describe the center of a distribution is to find the number with half the values falling below it and the other half above. This is the **median** of the observations. The median can be regarded as the typical value. We will call the median M for short. Here is the exact rule for finding the median:

1. Arrange all observations in order of size, from smallest to largest.

2. If the number n of observations is odd, the median M is the center observation in the ordered list. The location of the median is found by counting $(n + 1)/2$ observations up from the bottom of the list.

3. If the number n of observations is even, the median M is the average of the two center observations in the ordered list. The location of the median is again $(n + 1)/2$ from the bottom of the list.

Be sure to write down each individual observation in the data set, even if several observations repeat the same value. And be sure to arrange the observations in order of size before locating the median. The middle observation in the haphazard order in which the observations first come has no importance.

Another way to measure the center of a set of observations is to find their ordinary arithmetic average, which statisticians call the **mean.** The mean is usually written as $\bar{x}$, which is read as "x bar." To find the mean of a set of observations, add the values and divide by the number of observations. For an example of the calculation of the mean and the median, let's return to the Swiss doctors.

EXAMPLE: Calculating Mean and Median. Our sample data for male doctors were

27, 50, 33, 25, 86, 25, 85, 31, 37, 44, 20, 36, 59, 34, 28

The mean of this sample is

$$\bar{x} = \frac{27 + 50 + 33 + \cdots + 28}{15}$$

$$= \frac{620}{15} = 41.3$$

To find the median, first arrange the observations in order:

20, 25, 25, 27, 28, 31, 33, 34, 36, 37, 44, 50, 59, 85, 86

There are $n = 15$ observations, so

$$\frac{n + 1}{2} = \frac{16}{2} = 8$$

and the median is the eighth observation in the ordered list. Therefore $M = 34$.

A sample of 10 female Swiss doctors was studied at the same time. The numbers of hysterectomies performed by these doctors (arranged in order) were

5, 7, 10, 14, 18, 19, 25, 29, 31, 33

Here $n = 10$ is even, and the location of the median is

$$\frac{n+1}{2} = \frac{11}{2} = 5.5$$

The location 5.5 means "halfway between the fifth and sixth observations in the ordered list." So the median is $M = 18.5$.

The typical female doctor performed many fewer hysterectomies than the typical male doctor. This was one of the important conclusions of the study.

This example illustrates an important difference between the mean and the median. The mean is strongly affected by a few large observations, so the mean of a right-skewed distribution is larger than the median. The median number of hysterectomies performed by the male doctors was 34, but the few large values in the right tail of the distribution pull the mean up to 41.3. In practice, you must ask yourself whether the "typical value" (the median) or the "arithmetic average value" (the mean) is a better description of the center of the data.

The mean and median provide two different measures of the center of a distribution. But a measure of center alone can be misleading. In 1988, the median income of American families was $32,491. Half of all families earned less than $32,491, and half had incomes that were higher. But these figures do not tell the whole story.

EXAMPLE: Affluentia and Spartany. Imagine two countries with the same median family income, say, $32,491. But these countries have vastly different economic systems. In Affluentia, all families earn nearly the same income. In Spartany, on the other hand, there are extremes of wealth and poverty. Figure 6.6 displays the two income distributions. Despite the fact that median income is identical in Affluentia and Spartany, income patterns in the two countries are dramatically different. Family incomes in Spartany have much greater spread or variation than those in Affluentia. The median tells us nothing about the spread of a distribution.

The simplest useful numerical description of a distribution consists of both a measure of center and a measure of spread. One way to measure spread is to calculate the *range,* the difference between the highest and lowest observations. For example, the highest and lowest 1980 American League batting averages were .390 and .188. The range is therefore .390 to .188, or .202. But here we see a weakness in the range as a measure of spread: it is determined by the two most extreme observations. George Brett alone adds 40 points to the range of 1980 batting averages. For a more useful indication of variability, one that is less

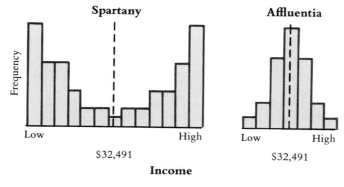

Figure 6.6 Two distributions with the same center but unequal spread.

influenced by extremes, we can limit the range to the middle 50% of observations.

We find this range by identifying points called **quartiles.** The *first quartile* is the value below which one-fourth of the observations fall. Three-fourths of the observations lie below the *third quartile.* The second quartile is the median, since two-fourths of the observations lie below the median. The exact rule for calculating the quartiles uses the rule for the median: To calculate the quartiles, first locate the median in the ordered list of observations. The first quartile Q_1 is the median of the observations below the location of the median, and the third quartile Q_3 is the median of the observations above that location.

EXAMPLE: Calculating Quartiles. The numbers of hysterectomies performed by our sample of 15 male doctors were (arranged in order)

20, 25, 25, 27, 28, 31, 33, 34, 36, 37, 44, 50,
59, 85, 86

The median is the eighth in the list, $M = 34$. The first quartile is the median of the seven observations to the left of the median, so $Q_1 = 27$. The third quartile, $Q_3 = 50$, is the median of the seven observations to the right of the median.

Suppose that another sample gave

20, 22, 26, 27, 27, 27, 28, 30, 32, 32

The median of these 10 observations is halfway between the fifth and sixth values. Because both these values are 27, $M = 27$. The first quartile is the median of the first five observations, because these are the observations below the location of the median. Be sure to note that it is the location of the median, not its numerical value, that decides which observations are included in finding the quartiles. That location is halfway between the fifth and sixth observations, even though both these observations have the same value. Check that $Q_1 = 26$ and $Q_3 = 30$.

A quick description of both the center and the spread of a set of data is provided by the **five-number summary.** This consists of the median, the two quartiles, and the two extremes (the smallest and largest individual observations). The five-number summary is always given in order from the smallest number to the largest. For example, we could calculate from the list of 167 American League batting averages that the five-number summary is .188, .244, .271, .290, .390. This summary contains much useful information. For example, a typical major league regular player hit .271, and the middle half of all such players hit between .244 and .290.

A graph known as a **boxplot** helps to make the five-number summary more vivid.

> A boxplot consists of a central box that spans the quartiles, with a line marking the median and "whiskers" extending out from the box to the extremes.

The boxes in a boxplot can be drawn either horizontally or vertically, as you prefer. In either case, it is essential that the plot be accompanied by a scale marked off in units of the variable being described. The median, quartiles, and extremes are located on this scale in order to draw the boxplot.

Boxplots are particularly helpful for comparing several distributions. Figure 6.7, for example, displays the distributions of the number of hysterectomies performed in a year by our samples of male and female Swiss doctors. The scale in this example is simply the count of hysterectomies. Figure 6.7 quickly conveys a lot of information. You can tell at a glance that female doctors in general perform far fewer hysterectomies than men. In fact, the upper extreme for females falls below the male median. You can see, too, that the female dis-

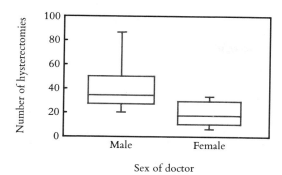

Figure 6.7 Side-by-side boxplots of the distributions of hysterectomies performed by male and female doctors.

tribution has less spread. In particular, it lacks the few very large observations that stretch out the upper whisker for the men. Figure 6.7 illustrates, once again, the effectiveness of visual displays. Before concluding that male doctors are more ready to perform hysterectomies, however, we need more information about the samples. Both males and females should be drawn from doctors with the same specialty and with practices of similar size. That was, in fact, the case for the Swiss study.

DISPLAYING RELATIONS BETWEEN TWO VARIABLES

In the examples we have looked at to this point, there was only one variable of interest, such as a hitter's batting average or the number of hysterectomies performed by a doctor. Now we will examine data for two variables, emphasizing the nature and strength of the relationship between the variables. Because it is in practice impossible to spot a relationship from columns of numbers, graphs are essential.

EXAMPLE: Natural Gas Consumption. Sue is concerned about the amount of energy she uses to heat her home. She keeps a record of the natural gas consumed over a period of nine months. Because the months are not all equally long, she divides each month's consumption by the number of days in the month to get cubic feet of gas used per day. Then from local weather records, Sue obtains the number of degree-days for each month. She divides this total by the number of days in the month, giving the average number of degree-days per day during the month. (Degree-days are a measure of demand for heating; one degree-day is accumulated for each degree that a day's average temperature falls below 65°F. For example, a day with an average temperature of 30°F gives 35 degree-days.)

Table 6.3 shows Sue's data: nine measurements of the two variables. Looking at the numbers in the table, we can see that more degree-days go with higher gas consumption. But the shape and strength of the relationship are not fully clear. Out problem is to display and interpret these data.

Data on two variables can be displayed in a **scatterplot,** a plot of all data points in two

TABLE 6.3 Sue's household consumption of natural gas compared to the need for heat

	Oct	Nov	Dec	Jan	Feb	Mar	Apr	May	June
Degree-days per day	15.6	26.8	37.8	36.4	35.5	18.6	15.3	7.9	0.0
Gas consumed per day (in cubic feet)	5.2	6.1	8.7	8.5	8.8	4.9	4.5	2.5	1.1

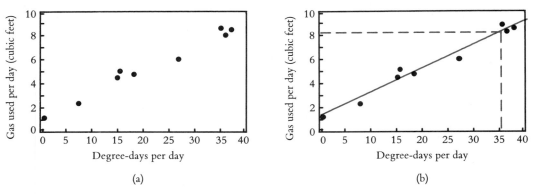

Figure 6.8 Natural gas consumption versus degree- days. (a) A scatterplot. (b) A fitted line and its use for prediction.

dimensions. Figure 6.8a shows a scatterplot of Sue's data. Always plot the explanatory or causal variable on the horizontal scale and the outcome or response variable on the vertical scale. Degree-days appear on the horizontal scale and gas consumption on the vertical scale because the weather affects gas consumption; gas consumption does not explain the weather.

The relationship between degree-days and gas consumption is very strong and takes the form of a straight line. We can represent the overall pattern of the relationship by drawing a straight line through the points of the scatterplot. Figure 6.8b shows such a line. The smooth in our "data = smooth + rough" motto is represented by the line; the rough is the scatter of points about the line. In this case, the number of degree-days explains most of the variation in gas consumption. The scattered points above and below the line reflect the effects of other factors, such as use of gas for cooking or turning down the thermostat when the family is away from home. These effects are relatively small.

How can we draw a line through the points of a scatterplot? When the plot shows a strong straight-line relationship, it is easy to fit a line on the graph by using a stretched black thread or a transparent straightedge. This gives us a line on the graph, but not an equation for the line. There is also no guarantee that the line we fit by eye is the "best" line. There are statistical techniques for finding from the data the equation of the best line (with various meanings of "best"). The most common of these techniques, called *least squares regression,* is discussed in the next section. All statistical computer software packages and some calculators will calculate the least squares line for you, so a line is often available with little work. You should therefore know how to use a fitted line even if you do not learn the details of how to get the equation from the data.

In writing the equation of a line, we use x for the explanatory variable because this is plotted on the horizontal or x-axis and y for the response variable. The equation has the form

$$y = b_0 + b_1 x$$

The number b_1 is the *slope* of the line, the amount by which y changes when x increases by one unit. The number b_0 is the *intercept,* the value of y when $x = 0$.

EXAMPLE: Interpreting Slope and Intercept. A computer program tells us that the least squares regression line computed from Sue's data is

$$y = 1.23 + 0.202x$$

The slope of this line is $b_1 = 0.202$. This means that gas consumption increases by 0.202 cubic foot per day when there is one more degree-day per day. The intercept is $b_0 = 1.23$. When there are no degree-days (i.e., when the average temperature is 65°F or above), gas consumption will be 1.23 cubic feet per day. The slope and intercept are, of course, estimates based on fitting a line to the data in Table 6.3. We do not expect every month with no degree-days to average exactly 1.23 cubic feet of gas per day. The line represents only the overall pattern of the data.

One of the most common goals in fitting a line to data is to predict the value of the response variable for a given value of the explanatory variable. A line drawn on a scatterplot can be used for making predictions with only a ruler and pencil. If the equation of the line is available, we can simply substitute the given value of the explanatory variable into the equation.

EXAMPLE: Predicting Gas Consumption. How much natural gas should Sue predict that she will consume in a month with 35 degree-days per day? Figure 6.8b illustrates the use of the line drawn on the scatterplot. First locate 35 on the horizontal axis. Draw a vertical line up to the fitted line and then a horizontal line over to the gas consumption scale. As the figure shows, we predict that slightly more than 8 cubic feet per day will be consumed.

We can give a more exact prediction using the equation of the least squares regression line. This equation is

$$y = 1.23 + 0.202x$$

In this case, x is the number of degree-days per day during a month and y is the number of cubic feet of gas consumed per day. Our predicted gas consumption for a month with $x = 35$ degree-days per day is

$$y = 1.23 + (0.202)(35)$$
$$= 8.3 \text{ cubic feet per day}$$

This prediction will almost certainly not be exactly correct for the next month that has 35 degree-days per day. But the past data points lie so close to the line that we can be confident that gas consumption in such a month will be quite close to 8.3 cubic feet per day.

Sue had a very practical reason for working through this exercise in statistics. She plans to add insulation to her house during the summer, and at the end of next winter she will want to know how much she has saved in heating costs. She cannot simply compare before-and-after gas usage, because the winters before and after will not be equally severe. Nor can she do a comparative experiment to weigh the costs of insulated and uninsulated houses during the same winter, for she has only one house. However, once Sue has next winter's degree-day data, she can use the fitted line to predict how much gas she would have used before insulating. Comparing this prediction with the actual amount used after insulation will show her savings.

Scatterplots are a good tool for exploring the relationships between two or more variables, but often the overall pattern is not as simple as the straight-line pattern in Figure 6.8. Combining scatterplots with some of the numerical descriptions mentioned earlier can help us see the pattern in data.

Consider the case of the 1970 draft lottery. To eliminate distinctions among men eligible for the draft, Congress decided to allow chance to determine who would be selected for military service. The first draft lottery was

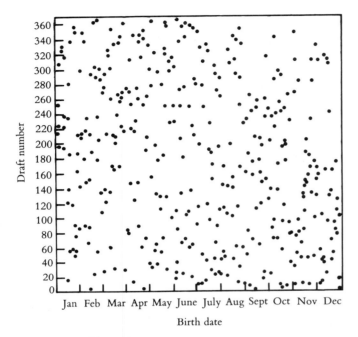

Figure 6.9 Scatterplot of draft selection number versus birth date for the 1970 draft lottery.

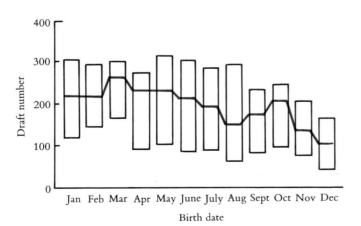

Figure 6.10 Monthly medians (line) and quartiles (boxes) for the 1970 draft lottery.

held in 1970. Birth dates for all men born between 1943 and 1952 were to be drawn at random and assigned selection numbers in the order drawn. Men whose birth dates matched selection number 1 were drafted first. They were followed by men with numbers 2, 3, 4, and so on.

This procedure sounds fair, but the random drawing was mishandled. Birth dates were placed into identical capsules, and the capsules were placed in a drum for the drawing. But the capsules were not mixed thoroughly enough. December dates, which were added last, remained on top and had a greater chance of being drawn early. Thus, men born in December were drafted and sent to Vietnam in greater numbers than those born in January.

EXAMPLE: The 1970 Draft Lottery. Figure 6.9 is a scatterplot of all 366 birth dates and the selection numbers assigned to them by the 1970 lottery. Having a low selection number, nearer the bottom of this graph, means being drafted earlier. The alleged association between birth dates late in the year and low selection numbers isn't easily seen. The scatterplot alone is not very helpful because the graph looks much like a random scatter of points. A formal analysis by probability would show that a result this unfair to men born late in the year would happen in less than 1 in 1000 truly random lotteries. The inequity is really there, but to see it, we need to be more imaginative in looking at the data.

In Figure 6.10, we combine graphical methods with some basic numerical descriptions to get a more detailed picture of the draft lottery. Because medians and quartiles give a compact summary of both the center and spread of a distribution, we compute these for each month's selection numbers. We then replace each month's points on the scatterplot with a box whose ends are at the upper and lower quartiles. These boxplots allow us to instantly compare the distribution of selection numbers from month to month. A line within each box marks the median for the month. To make the picture even clearer, we can draw a line connecting the medians. The misfortune of men born late in the year is now evident.

The media noticed this inequity, and in 1971 a new and genuinely random selection process was designed by statisticians at the National Bureau of Standards. Two drums instead of one were used in the 1971 lottery: one contained birth dates; the other, selection numbers. The capsules in both drums were thoroughly mixed, and then a randomly drawn birth date was paired with a randomly drawn selection number. When we add to our plot the medians of the reformed 1971 lottery, we can see the difference (Figure 6.11). There is a good bit of random variation in the 1971 medians, but no systematic trend as in 1970.

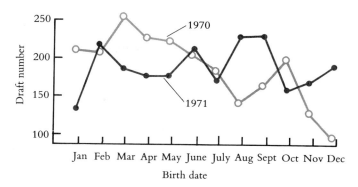

Figure 6.11 Median draft number by month for the 1970 and 1971 lotteries.

`Optional` LEAST SQUARES REGRESSION

When a scatterplot shows a clear straight-line relation between an explanatory variable x and a response variable y, we want to describe the relation by a straight line. The points will rarely lie exactly on a line, so our problem is to find the line that best fits the points. To do this, we must first say what we mean by the "best-fitting" line.

Suppose that we want to use our line to predict y for given values of x. The error in our prediction is measured in the y, or vertical, direction. So we want to make the vertical distances of the points from the line as small as possible. A line that fits the data at all well will not pass entirely above or below the plotted points, so some of the errors will be positive and some negative. Their squares, however,

will all be positive. The **least squares regression line** is the line that makes the sum of the squares of the vertical distances of the data points from the line as small as possible.

The least squares idea says what we mean by the best-fitting line. We must still learn how to find this line from the data. Given n observations on variables x and y, what is the equation of the least squares line? Here is the solution to this mathematical problem: the least squares regression line of y on x calculated from n observations on these two variables is given by $y = b_0 + b_1 x$, where

$$b_1 = \frac{n\Sigma xy - (\Sigma x)(\Sigma y)}{n\Sigma x^2 - (\Sigma x)^2}$$

$$b_0 = \bar{y} - b_1 \bar{x}$$

Here Σ is just an abbreviation for "sum of," so Σx means the sum of all the x values. The quantities $\bar{x}$ and $\bar{y}$ are the means of the x and y values. These algebraic formulas are shorthand for a series of operations that for clarity can be arranged in the form of a table, as the following example shows.

EXAMPLE: Gas Consumption Least Squares Line. The data for natural gas consumption y versus heating degree-days x are

TABLE 6.4 **The arithmetic of least squares**

x	y	x^2	xy
15.6	5.2	243.36	81.12
26.8	6.1	718.24	163.48
37.8	8.7	1428.84	328.86
36.4	8.5	1324.96	309.40
35.5	8.8	1260.25	312.40
18.6	4.9	345.96	91.14
15.3	4.5	234.09	68.85
7.9	2.5	62.41	19.75
0.0	1.1	0.0	0.0
$\Sigma x = 193.9$	$\Sigma y = 50.3$	$\Sigma x^2 = 5618.11$	$\Sigma xy = 1375.0$

given in Table 6.3. Table 6.4 lists the quantities needed to obtain the least squares regression line, beginning with the values of x and y in the first two columns.

At the bottom of each column we write its sum. These sums are the building blocks for the least squares calculation. After writing down the x and y values, the steps in the calculation are:

STEP 1. Sum the x and y columns and compute the two means. Here

$$\Sigma x = 193.9 \quad \text{and} \quad \bar{x} = \frac{193.9}{9} = 21.54$$

$$\Sigma y = 50.3 \quad \text{and} \quad \bar{y} = \frac{50.3}{9} = 5.59$$

STEP 2. Calculate x^2 and xy for each (x, y) data point, enter in the proper column, and sum.

STEP 3. Substitute into the formulas for the slope b_1 and then for the intercept b_0.

$$b_1 = \frac{n\Sigma xy - (\Sigma x)(\Sigma y)}{n\Sigma x^2 - (\Sigma x)^2}$$

$$= \frac{(9)(1375.0) - (193.9)(50.3)}{(9)(5618.11) - (193.9)^2}$$

$$= \frac{2621.83}{12965.78} = 0.202$$

$$b_0 = \bar{y} - b_1\bar{x}$$
$$= (5.59) - (0.202)(21.54) = 1.23$$

The equation of the least squares line is therefore

$$y = 1.23 + 0.202x$$

It is more common in statistical practice to use a statistical calculator or computer software to obtain b_1 and b_0 with less work. •

GRAPHICS IN MANY DIMENSIONS

Graphic analysis of the draft lottery shows how much insight we can gain using only simple graphs and some basic statistical calculations. Looking at data in several ways can yield results that we wouldn't see using a single method such as a scatterplot.

But so far, we have looked only at scatterplots for two variables. What if we want to display a third variable in the same plot? Because we have already used the horizontal and vertical directions of the graph, there is only one space dimension left, moving out of and into the page. Unfortunately, such three-dimensional scatterplots are very hard to see clearly unless color or motion (or both) are used to help us gain perspective. Computer graphics can add color and motion, allowing us to see a scatterplot in three dimensions.

Computer graphics makes it possible to see relations and detect outliers in high-dimensional data sets. From most viewing angles, an outlier would appear as part of the main group, blending invisibly with the mass of points. However, if we change the viewing angle by rotating the plot, the outlier would eventually

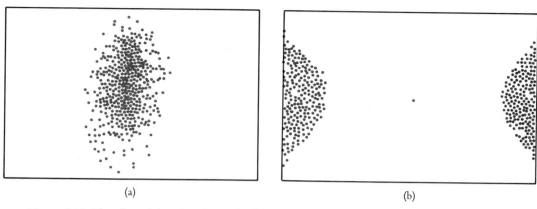

(a)   (b)

Figure 6.12 The effect of changing viewing angle in a three-dimensional scatterplot. (a) From this angle, the data form a single cloud. (b) From another angle, two clouds and an outlier are visible.

be seen apart — like the Death Star appearing from behind the moon in *Star Wars.* Imagine, for example, a manufacturing process in which three key measurements are plotted in three dimensions. Figure 6.12a shows that from most points of view, the data appear as a cloud, like gnats swarming in space. But if we rotate the image (Figure 6.12b), we find that there are two clouds, one of which was hidden behind the other from our first viewpoint. What is more, an outlier lies between the two clouds.

Remember that every point in this picture is positioned according to three measurements. If we tried to spot the outlier by looking at a long table with three columns of measurements for hundreds of points, we would quickly lose our bearings. This is a case where graphic display creates clarity out of the chaos of raw numbers.

Computer graphics allows us to move around the cloud of data points and look at it from any direction. Moreover, the computer can move around a 10-dimensional cloud of points almost as easily as a three-dimensional one. We cannot, of course, see 10 dimensions. In fact, we do not directly see three dimensions on a piece of paper or video screen. In both cases, the computer presents a *projection* of the

cloud of data onto the two-dimensional surface of the video screen. The computer can be instructed to compute and display changing projections, just as if we were moving around a many-dimensional cloud of data. Outliers and other important relations among the variables come clearly into view when they are scanned from the correct angle. To reduce the amount of time needed to discover the most meaningful viewing angles, the computer can even be programmed to search for interesting projections.

These multidimensional displays have practical applications. At Harvard University, for example, a graphics system designed by Professor Peter Huber aids geologists who are studying the pattern of earthquakes near the Fiji islands (see Spotlight 6.2, p. 169). For many years, the only view scientists had of earthquake epicenters was a standard map projection showing the two-dimensional location in latitude and longitude (see Figure 6.13). Crucial data about the depths of the epicenters were therefore not reflected in their maps.

A computer graphics system allows earthquake locations to be seen in more detail. A fully three-dimensional view of the epicenters that adds depth to latitude and longitude can be presented and manipulated. This kind of

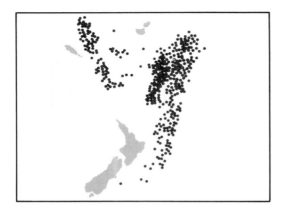

Figure 6.13 Earthquake epicenters in the Fiji islands as seen from the earth's surface.

1. When looking at a set of data, progress from the simple to the complex. Examine a single variable, perhaps by means of a histogram. Then, if the problem warrants it, look at the relationship between two variables, using a scatterplot or other methods. Then move on to the complexity of picturing and understanding the relationship between three or more variables.

2. Remember that data = smooth + rough. Organize your thinking to look for an overall pattern and then focus on deviations from that pattern.

3. Remember the power of both pictures and numbers. The human eye is the best device known for seeing both the smooth and the rough. Combine graphics with numerical calculation for greater effectiveness.

Today, the computational power and display features of computers allow us to use both graphics and calculation quickly and easily. To an ever-increasing extent, computers will be doing our arithmetic and drawing our pictures for us. But it remains for the human eye and brain to see and understand.

picture permits geologists to examine the Fiji island earthquake pattern in light of plate tectonics, the geological theory of movements of vast plates that make up the earth's crust. These plate movements give rise to such geological events as the eruption of volcanoes and the creation of mountain ranges.

Geologists explain that in the region of the Fiji islands, a plate moving in from the east has bent beneath the plate on the west so that the eastern plate is diving straight down into the earth's mantle. The collision and resulting redirection of the plate accounts for most of the earthquakes in this region. The three-dimensional view that emerges as the computer continuously changes our viewing angle (see Figure 6.14 and Color Plate 9) shows that earthquakes occur along the boundary between the plates deep inside the earth. The graphics system turns the image constantly, giving a clearer view than a still picture provides. The detailed computer images even enable geologists to spot wrinkles and other features in the surfaces of the plates.

Statistics makes it possible to look at complex relationships in an organized way. Our examples of exploring data have fundamental themes in common.

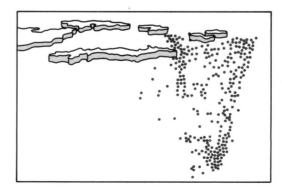

Figure 6.14 Earthquake epicenters after changing viewing angle to show depth beneath the surface.

SPOTLIGHT 6.2 Visual Statistics: Eyeballing the Data

Dr. Peter Huber, a statistician now at the Massachusetts Institute of Technology, comments on the computer's ability to display complex, multidimensional arrangements of data, giving new meaning to the term *descriptive statistics:*

Dr. Peter Huber, MIT.

For most of the twentieth century, statistics focused on mathematical rigor and on small samples. In the past 10 years or so we have seen renewed interest in descriptive statistics, driven by the computer. It would be a mistake, however, to view this new emphasis as just a reaction of statistics to high tech. What we are witnessing now is a return to the descriptive statistics of the nineteenth century, completing unfinished business left over from that era. The computer makes it possible to do things that couldn't be done before.

In my view, descriptive statistics and mathematical statistics are complementary. In descriptive statistics, you see things, but you cannot test them in a formal way. In mathematical statistics, you may test for something preconceived, but you may overlook something you hadn't built into the test. So you simply have to do both so as to complement one with the other.

You also need subject specialists to complement the data analyst. In our experience, doing data analysis usually turns out to be a close conversational collaboration between the scientist (the one with the data) and the data analyst. They sit in front of the screen, discuss what they see, suggest the next action to be taken. Data analysis requires dynamic collaboration between the statistician and the subject-matter specialist.

The first step in any data analysis is data inspection, mainly to get familiar with the data and find extraordinary features. The next step is modification; enhance the picture by lines, maybe color groups, cluster and label selected points, maybe even fit some model to it. This leads to the most important step in data analysis: *comparison.* Without comparing things, you are not able to interact with your data. To help with interpretation, you almost always have to compare things — either several data sets or a data set and a model or different models for the same data set. Interpretation is the next step. And after interpretation, one usually has to begin another round of modeling. Very often the entire cycle starts again.

REVIEW VOCABULARY

Boxplot   A graph of the five-number summary. A box spans the quartiles, with an interior line marking the median. Whiskers extend out from this box to the extreme high and low observations.

Distribution   The pattern of outcomes of a variable, listing the values that the variable takes and how often each value occurs.

Dotplot   A graphical display for small data sets in which each observation is represented by a dot above its value on a scale of the variable being measured.

Exploratory data analysis   The practice of examining data for unanticipated patterns or effects as opposed to seeking answers to specific questions.

Five-number summary   A summary of a distribution of values consisting of the median, the upper and lower quartiles, and the largest and smallest observations.

Frequency   The number of times an outcome or group of outcomes occurs in a set of data.

Histogram   A graph of the frequencies of all outcomes (often divided into classes) for a single variable. The height of each bar is the frequency of the class of outcomes covered by the base of the bar. All classes should have the same width.

Least squares regression line   A line drawn on a scatterplot that makes the sum of the squares of the vertical distances of the data points from the line as small as possible. The fitted line can be used for prediction.

Mean   The ordinary arithmetic average of a set of observations; to find the mean, divide the sum of all the observations by the number of observations summed.

Median   The midpoint of a set of data; half the observations fall below the median and half fall above.

Outlier   A data point that falls well outside the overall pattern of a set of data.

Quartiles   The first quartile of a distribution is the point with 25% of the observations falling below it; the third quartile is the point with 75% below it.

Scatterplot   A graph of the values of two variables as points in the plane; the value of the explanatory variable is plotted on the horizontal axis and the corresponding value of the response variable on the vertical axis.

Skewed distribution   A distribution in which observations on one side of the median extend notably farther from the median than do observations on the other side. In a right-skewed distribution, the larger observations extend farther to the right of the median than the smaller observations extend to the left.

Symmetric distribution   A distribution with a histogram or dotplot in which the part to the left of the median is roughly a mirror image of the part to the right of the median.

SUGGESTED READINGS

Chambers, John M., William S. Cleveland, Beat Kleiner, and Paul A. Tukey: *Graphical Methods for Data Analysis,* Wadsworth, Belmont, Calif., 1983. A detailed survey of modern graphical methods, many requiring a computer for effective use.

Cleveland, William S.: *The Elements of Graphing Data,* Wadsworth, Belmont, Calif., 1985. A careful study of the most effective elementary ways to present data graphically, with much sound advice on improving simple graphs.

Moore, David S., and George P. McCabe: *Introduction to the Practice of Statistics,* Freeman, New York, 1989. The first three chapters of this text provide a more extensive treatment of displaying and describing data. The material of this chapter is covered in more detail, and much new material on both technique and interpretation is presented.

Tufte, Edward R.: *The Visual Display of Quantitative Information,* Graphics Press, Cheshire, Conn., 1983. Beautifully printed, with classic examples such as Minard's work and suggestions for both statisticians and graphic artists.

EXERCISES

1. The data below are the number of days on which hail was observed at Evansville, Indiana, in each of 11 consecutive years.

$$4, 5, 3, 2, 4, 2, 0, 5, 0, 1, 1$$

Draw a dotplot of these data. Are there any outliers or other unusual features?

2. In an experiment on the effect of a drug on reaction time, a subject is asked to depress a button whenever a light flashes. Her reaction times for 11 trials are (in milliseconds)

$$96, 101, 102, 138, 93, 99, 107, 93, 95, 100, 100$$

Make a dotplot of these observations. Are there any outliers or other unusual features?

3. The following histogram displays data on the hour at which the first flash of lightning was observed each day during a study in Colorado. Describe this distribution: Is it roughly symmetric or distinctly skewed? Where is the center? Are there any outliers or gaps?

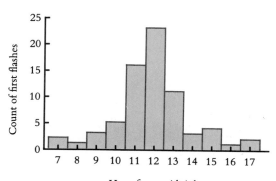

Hour from midnight

4. Members of a health maintenance organization (HMO) can make an unlimited number of visits to its member physicians for a fixed annual fee. The following histogram displays the distribution of the number of visits made by members of one HMO. (The vertical scale gives the proportion of the members in each class rather than the frequency.) Describe this distribution: Is it roughly symmetric or distinctly skewed? Are there any outliers or gaps?

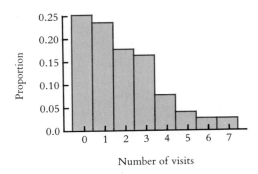

Number of visits

5. In 1798 the English scientist Henry Cavendish measured the density of the earth in a careful experiment with a torsion balance. Here are his 23 repeated measurements of the same quantity (the density of the earth relative to that of water) made with the same instrument:

5.36	5.57	5.44	5.27	5.63	5.75
5.29	5.53	5.34	5.39	5.34	5.68
5.58	5.62	5.79	5.42	5.46	5.85
5.65	5.29	5.10	5.47	5.30	

[Source: S. M. Stigler, "Do Robust Estimators Work with Real Data?" *Annals of Statistics,* 5:1055–1078 (1977).]

a. Make a histogram of these data.

b. Describe the distribution: Is it approximately symmetric or distinctly skewed? Are there gaps or outliers?

6. A fisheries researcher compiled the following data on lengths of six-year-old white female crappies (in millimeters).

217	230	220	221	225	223
219	217	225	228	234	222
231	222	220	222	222	223
225	214	221	233	227	234
223	225	253	220	213	224
235	283	210	218	235	231

a. The data range from 210 to 283 millimeters. Group them into five classes of width 15 millimeters, starting with $210 \leq$ length < 225 as the leftmost class. Draw a well-labeled frequency histogram of the grouped data.

b. Describe the distribution: Is it roughly symmetric or clearly skewed? Are there gaps or outliers?

7. Return to the data on annual number of days with hail given in Exercise 1.

a. Find the mean and the median number of hail days.

b. Find the first and third quartiles of the number of hail days.

8. Find the five-number summary of the reaction times given in Exercise 2.

9. Here are the percents of the popular vote won by the successful candidate in each of the presidential elections from 1948 to 1988.

Year	1948	1952	1956	1960	1964	1968	1972	1976	1980	1984	1988
%	49.6	55.1	57.4	49.7	61.1	43.4	60.7	50.1	50.7	58.8	53.9

a. Make a dotplot of these percents. Are there any outliers?

b. What is the median percent of the vote won by the successful candidate in presidential elections?

c. Call an election a landslide if the winner's percent falls at or above the third quartile. Find the third quartile. Which elections were landslides?

10. Find the mean of the reaction times in Exercise 2. Explain from the nature of the distribution why the mean is larger than the median.

11. A random sample of 12 Wisconsin farms showed the following soybean yields (bushels per acre).

$$41, 28, 34, 36, 26, 44, 39, 32, 40, 35, 36, 33$$

a. Make a dotplot of these data. Are there any outliers or other unusual features?

b. Find the mean and median yields.

c. Find the first and third quartiles of the yields.

12. Return to the data on lengths of fish given in Exercise 6.

a. Find the five-number summary of this distribution. What range of lengths contains the middle 50% of the distribution?

b. From the shape of the distribution, do you expect the mean to be larger than the median, smaller than the median, or about the same as the median? Find the mean and verify your expectation.

13. Find the median and quartiles of Cavendish's measurements of the density of the earth in Exercise 5. Then give a five-number summary. How is the symmetry of the distribution reflected in the five-number summary?

14. The mean of the 23 measurements in Exercise 5 is Cavendish's best estimate of the density of the earth. Find this mean.

15. The following table lists the amount spent per student by the public schools of each state.

    a. Make a histogram of these data.

    b. Describe the overall shape of the distribution: Is it roughly symmetric, clearly skewed to the right, or clearly skewed to the left? Are there any outliers?

TABLE 6.5 Average dollars spent per pupil in public elementary and secondary schools, 1988

State	Dollars	State	Dollars	State	Dollars
Alabama	2752	Kentucky	3355	North Dakota	3353
Alaska	7038	Louisiana	3211	Ohio	4019
Arizona	3265	Maine	4276	Oklahoma	3051
Arkansas	2410	Maryland	4871	Oregon	4574
California	3994	Massachusetts	5396	Pennsylvania	5063
Colorado	4359	Michigan	4122	Rhode Island	5456
Connecticut	6141	Minnesota	4513	South Carolina	3075
Delaware	4994	Mississippi	2760	South Dakota	3159
D.C.	5643	Missouri	3566	Tennessee	3189
Florida	4389	Montana	4061	Texas	3462
Georgia	2939	Nebraska	3641	Utah	2658
Hawaii	3894	Nevada	3829	Vermont	4949
Idaho	2814	New Hampshire	3990	Virginia	4145
Illinois	4217	New Jersey	6910	Washington	4083
Indiana	3616	New Mexico	3880	West Virginia	3895
Iowa	3846	New York	6864	Wisconsin	4991
Kansas	4262	North Carolina	3911	Wyoming	6885

Source: *Statistical Abstract of the United States,* 1989.

16. The following table gives the survival times (in days) of 72 guinea pigs after they were infected by tubercle bacilli in a medical study. Make a histogram of these data. Is the survival-time distribution approximately symmetric or strongly skewed? Compute the five-number summary for this distribution. How is the shape of the distribution reflected in the distances of the two quartiles and the two extremes from the median?

Guinea pig survival times

43	45	53	56	56	57	58	66	67	73
74	79	80	80	81	81	81	82	83	83
84	88	89	91	91	92	92	97	99	99
100	100	101	102	102	102	103	104	107	108
109	113	114	118	121	123	126	128	137	138
139	144	145	147	156	162	174	178	179	184
191	198	211	214	243	249	329	380	403	511
522	598								

Source: T. Bjerkedal, "Aquisition of Resistance in Guinea Pigs Infected with Different Doses of Virulent Tubercle Bacilli," *American Journal of Hygiene,* 72:130–148 (1960).

17. Find the mean and the median per student expenditures on public education from Table 6.5. Explain from the overall shape of the distribution the relationship between the two measures of location.

▲ **18.** Choose a set of interesting data from the *Statistical Abstract* or an almanac (for example, populations of the states or per capita incomes of nations). Make a histogram of the data and describe the pattern and any outliers.

19. A common criterion for detecting suspected outliers in a set of data is as follows:

 a. Find the quartiles Q_1 and Q_3 and the *interquartile range IQR* $= Q_3 - Q_1$. The interquartile range is the spread of the central half of the data.

 b. Call an observation an outlier if it falls more than $1.5 \times IQR$ above the third quartile or below the first quartile.

Find the quartiles for the education-spending data in Table 6.5. Find the interquartile range *IQR.* Are there any outliers according to the $1.5 \times IQR$ criterion?

▲ **20.** A study of the size of jury awards in civil cases (such as injury, product liability, and medical malpractice) showed that the median award in Cook County, Illinois, was about $8000. But the mean award was about $69,000. Explain how this great difference between two measures of location can occur.

21. In June 1989, two measures of the center of the distribution of prices for new homes were $129,900 and $159,100. Which of these numbers is the mean and which is the median? Explain your answer.

▲ **22.** Colleges announce an "average" Scholastic Aptitude Test score for their entering freshmen. Usually the college would like this average to be as high as possible. A *New York Times* article noted that "private colleges that buy lots of top students with merit scholarships prefer the mean, while open-enrollment public institutions like medians." Use what you know about the behavior of means and medians to explain these preferences.

▲ Discussion exercise.

23. Give an example of a small set of data whose mean is larger than the upper quartile.

24. Find the five-number summary for the home run data in Table 6.1 and make a boxplot of this distribution. What range contains the middle half of league-leading home run totals?

25. Make a boxplot of the public-school-expenditures data in Table 6.5.

26. Joe DiMaggio played center field for the Yankees for 13 years. He was succeeded by Mickey Mantle, who played for 18 years. Here are the number of home runs hit each year by DiMaggio:

> 29, 46, 32, 30, 31, 30, 21, 25, 20, 39, 14, 32, 12

and by Mantle:

> 13, 23, 21, 27, 37, 52, 34, 42, 31, 40, 54, 30, 15, 35, 19, 23, 22, 18

Compute the five-number summary for each player and make side-by-side boxplots of the home run distributions. What does your comparison show about the relative effectiveness of DiMaggio and Mantle as home run hitters?

▲ **27.** Using Table 6.5, consider the states east of the Mississippi. Omit the District of Columbia and divide the other states into two groups, the northeast and the southeast. Following boundaries between Census Bureau regions, the northeast states are Connecticut, Illinois, Indiana, Maine, Massachusetts, Michigan, Minnesota, New Hampshire, New Jersey, New York, Ohio, Pennsylvania, Rhode Island, Vermont, and Wisconsin. The southeast states are Alabama, Delaware, Florida, Georgia, Kentucky, Louisiana, Maryland, Mississippi, North Carolina, South Carolina, Tennessee, Virginia, and West Virginia.

   Make side-by-side boxplots of public school expenditures for both groups and comment on the comparison of the two distributions.

28. The table at the top of page 177 gives data on the lean body mass (kilograms) and resting metabolic rate for 12 women and 7 men who are subjects in a study of obesity. The researchers suspect that lean body mass (i.e., the subject's weight leaving out all fat) is an important influence on metabolic rate.

   a. Make a scatterplot of the data for the female subjects. Which is the explanatory variable?

   b. Does metabolic rate increase or decrease as lean body mass increases? What is the overall shape of the relationship? Are there any outliers?

29. Compare the distribution of lean body mass among the male subjects in Exercise 28 to the distribution for female subjects by making side-by-side boxplots. Describe what the plots show.

30. When observations on two variables fall into several categories, you can display more information in a scatterplot by plotting each category with a different symbol or a different color. Add the data for male subjects to your scatterplot in Exercise 28, using a different color or symbol than you used for females. Does the type of relationship you found for females hold for men also? How do the male subjects as a group differ from the female subjects as a group?

▲ Discussion exercise.

Obesity-study data

Subject	Sex	Mass	Rate	Subject	Sex	Mass	Rate
1	M	62.0	1792	11	F	40.3	1189
2	M	62.9	1666	12	F	33.1	913
3	F	36.1	995	13	M	51.9	1460
4	F	54.6	1425	14	F	42.4	1124
5	F	48.5	1396	15	F	34.5	1052
6	F	42.0	1418	16	F	51.1	1347
7	M	47.4	1362	17	F	41.2	1204
8	F	50.6	1502	18	M	51.9	1867
9	F	42.0	1256	19	M	46.9	1439
10	M	48.7	1614				

31. The table below lists the retail prices of several consumer items in Washington, D.C., and in Japan. Make a scatterplot with the U.S. price as the explanatory variable. Describe the overall pattern of the relationship between U.S. and Japanese consumer prices. Are there any items with prices that fall clearly outside the overall pattern? If so, describe how these items differ from the overall pattern.

Retail prices in Washington, D.C., and Japan

Item	Washington	Japan
First-run movie	$5.00	$13.00
Pound of rice	0.50	1.89
Big Mac hamburger	1.78	3.08
Electricity (kWh)	0.06	0.33
Gallon of gasoline	0.95	3.50
Cantaloupe	0.75	13.00
Dozen eggs	1.09	1.70
Subway	0.80	1.00
Quart of milk	0.55	1.48

Source: *USA Today*, January 9, 1989.

32. We saw that the least squares regression line for the home-heating data of Table 6.3 is $y = 1.23 + 0.202x$. Use this line to predict the daily gas consumption for this home in a month averaging 20 degree-days per day and in a month averaging 40 degree-days per day.

33. Concrete road pavement gains strength over time as it cures. Highway engineers use regression lines to predict the strength after 28 days (when curing is complete) from measurements made after 7 days. Let x be strength (in pounds per square inch) after 7 days and y the strength after 28 days. One set of data gave the least squares regression line to be

$$y = 1389 + 0.96x$$

A test of some new pavement after 7 days shows that its strength is 3300 pounds per square inch. Predict the strength of this pavement after 28 days.

34. The table below shows the true number of calories in 10 common foods and the average number of calories estimated for these same foods in a sample of 3368 people.

    a. Make a scatterplot of these data, with true calories on the horizontal axis. Is there a general straight-line pattern? Which foods are outliers from the pattern?

    b. Fit a line by eye to the data (ignoring the outliers). If a food product contains 200 calories, what would you guess the general public's estimated calorie level for that food to be?

    c. The least squares regression line, computed without dropping the outliers, is

$$y = 58.6 + 1.30x$$

Here y = estimated calories and x = true calories. Draw this line on your scatterplot. [Hint: use the equation to find y for two values of x, then plot the two (x, y) points and draw the line through them.] Use the least squares line to predict y when $x = 200$.

The difference between your results in **b** and **c** reflects the influence of the outliers. It is often difficult to decide whether to include outliers in making a prediction.

True and estimated calories in 10 common foods

Product	True calories	Estimated calories
8 oz whole milk	159	196
5 oz spaghetti with tomato sauce	163	394
5 oz macaroni and cheese	269	350
1 slice wheat bread	61	117
1 slice white bread	76	136
2-oz candy bar	260	364
Saltine cracker	12	74
Medium-size apple	80	107
Medium-size potato	88	160
Cream-filled snack cake	160	419

Source: *USA Today,* October 1983.

35. The following scatterplot of the average SAT math score (y) versus the average SAT verbal score (x) for high school seniors in each of the 50 states shows a strong linear pattern. The least squares regression line for scores is

$$y = 27 + 1.03x$$

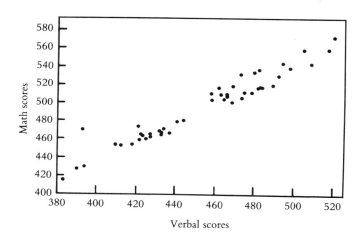

a. Do math SAT scores tend to be higher or lower than verbal SAT scores?

b. The average verbal SAT score in New York was 422. Use the regression line to predict New York's average math score. New York's actual math score was 466. What is the error (observed score minus predicted score)?

c. The only outlier among the states is Hawaii, where $x = 393$ and $y = 471$. Is Hawaii's math score higher or lower than would be predicted from its verbal score? Can you suggest why Hawaii might be an outlier?

36. An old study in Iowa produced the following data on corn yield (bushels per acre) from 1910 to 1919 and value per acre in 1920 for farmland in 10 counties. (Both corn yields and land prices have changed since 1920!)

County	1	2	3	4	5	6	7	8	9	10
Yield	40	36	34	41	39	42	40	31	36	30
Value ($)	87	133	174	285	263	274	235	104	141	115

a. Make a scatterplot of these data. There is an overall straight-line pattern with one outlier. Circle the outlier on your plot.

b. Corn yield should help to predict the value of farmland. When the outlier is omitted, the least squares regression line of land value y on yield x is

$$y = -371.3 + 15.4x$$

Draw this line on your scatterplot. Use the equation to predict the value of land that yields 35 bushels per acre. Then use your graph to do this prediction as in Figure 6.8b and compare the two results.

37. A study of sewage treatment measures the oxygen demand of decomposing solid wastes. If y is the logarithm of the oxygen demand (milligrams per minute) and x is the total solids (milligrams per liter of waste), measurements on 20 occasions give the following data:

x	7.2	7.8	7.1	6.4	6.4	5.1	5.9	5.3	5.0	5.0
y	1.56	0.9	0.75	0.72	0.31	0.36	0.11	0.11	−0.22	−0.15

x	4.8	4.4	4.3	3.7	3.9	3.6	4.4	3.3	2.9	2.8
y	0.0	0.0	−0.09	−0.22	−0.4	−0.15	−0.22	−0.4	−0.52	−0.05

a. Make a scatterplot of these data. Is the relationship roughly linear? Are there outliers?

b. Draw a fitted line on your scatterplot by eye. Use your line to predict the log of the oxygen demand y when $x = 4$ milligrams per liter of solids.

▲ **38.** The table on p. 181 gives the price and weight of 70 automobiles of the 1986 model year. The models marked with an asterisk are made by foreign companies. Make a scatterplot of weight (y) versus price (x), with domestic and foreign distinguished either by different colors or by different symbols (such as **x** and **o**). Discuss carefully what the plot shows about the relationship between the three variables: weight, price, and domestic versus foreign make.

▲ Discussion exercise.

Price (dollars) and weight (pounds) of 1986 cars (four-door sedan-type models only)

Model	Price	Weight	Model	Price	Weight
Chevrolet Nova	7,435	2250	*Volkswagon Quantum	13,595	2640
Chevrolet Spectrum	6,928	1920	*Volvo	14,370	2920
*Honda Civic	7,798	2035	*Audi 5000	18,065	2915
*Isuzu I-Mark	7,249	1920	Buick Century	10,228	2775
*Mazda 323	7,495	2240	Cadillac Seville	26,756	3425
*Mitsubishi Tredia	7,849	2400	Chevrolet Celebrity	8,931	2800
*Nissan Sentra	7,169	2090	Chrysler Fifth Avenue	14,910	3575
Plymouth Colt	6,862	2185	Chrysler Le Baron	10,127	2585
*Renault Alliance	6,199	2010	Chrysler New Yorker	13,409	2720
*Subaru	7,391	2295	Dodge 600	9,370	2600
*Toyota Corolla	7,348	2270	Dodge Aries	7,301	2535
*Volkswagon Jetta	8,545	2310	Dodge Diplomat	10,086	3575
AMC Eagle	10,719	3480	Ford Taurus	9,645	2870
*Audi 4000	14,230	2360	Lincoln Continental	24,556	3800
*BMW 325	20,055	2550	Mercury Sable	10,700	2870
Buick Skyhawk	8,073	2500	Olds Cutlass Ciera	10,354	2745
Buick Skylark	9,620	2645	Olds Cutlass Supreme	10,872	3355
Cadillac Cimarron	13,128	2620	Plymouth Caravelle	9,241	2720
Chevrolet Cavalier	6,888	2555	Plymouth Gran Fury	10,086	3575
Ford Tempo	7,508	2600	Plymouth Reliant	7,301	2535
*Honda Accord	9,679	2590	Pontiac 6000	9,729	2535
*Mazda 626	8,995	2600	Pontiac Bonneville	10,249	3350
*Mercedes-Benz 190	25,080	2745	*Saab 9000	22,145	3005
Mercury Topaz	8,235	2600	*Volvo 740	18,240	2980
*Mitsubishi Galant	13,219	2840	*Volvo 760	22,960	3065
*Nissan Maxima	14,459	3150	Buick Electra	15,588	3300
*Nissan Stanza	10,069	2440	Buick Le Sabre	12,511	3195
Oldsmobile Calais	9,478	2645	Cadillac De Ville	19,990	3345
Oldsmobile Firenza	8,035	2535	Chevrolet Caprice	10,243	3630
*Peugeot 505	12,651	3075	Ford LTD Crown Victoria	12,562	3885
Pontiac Grand Am	8,749	2725	Lincoln Town Car	20,764	4060
Pontiac Sunbird	7,495	2435	Mercury Grand Marquis	13,504	3930
*Saab 900	12,985	2845	Olds Delta 88 Royale	12,760	3195
*Toyota Camry	9,678	2690	Olds 98 Regency	15,989	3265
*Toyota Cressida	16,630	3335	Pontiac Parisienne	11,169	3630

* Manufactured by foreign companies.

39. Manatees are large, gentle sea creatures that live along the Florida coast. Many manatees are killed or injured by powerboats. The table below gives data on powerboat registrations (in thousands) and the number of manatees killed by boats in Florida in the years 1977 to 1987.

a. Which is the explanatory variable? Make a scatterplot of these data. Describe the overall pattern and any outliers or other important deviations.

b. Fit a line to the scatterplot by eye and draw the line on your graph. If in a future year powerboat registrations increase to 700,000, predict the number of manatees that will be killed by boats.

Manatees killed by boats, 1977–1987

Year	Boats (thousands)	Manatees killed	Year	Boats (thousands)	Manatees killed
1977	447	13	1983	526	15
1978	460	21	1984	559	34
1979	481	24	1985	585	33
1980	498	16	1986	614	33
1981	513	24	1987	645	39
1982	512	20			

40. Suppose that in some far future year 2 million powerboats are registered in Florida. Extend the fitted line you found in Exercise 39 and use it to predict manatees killed. Explain why this prediction is very unreliable. (Using a fitted line to predict the response to an x value outside the range of the data used to fit the line is called *extrapolation*. Extrapolation often produces unreliable predictions.)

● **41.** Calculate the least squares regression line of estimated calories y on true calories x from the data in exercise 34. Verify that your result agrees with that given in Exercise 34c.

● **42.** Calculate the least squares regression line of land value y on corn yield x for the data in Exercise 36, omitting county 1. Verify that your result agrees with that given in Exercise 36.

● **43.** Find the equation of the least squares regression line of y on x for the sewage treatment data of Exercise 37. Use your line to do the prediction asked for in Exercise 37**b**.

● **44.** Find the equation of the least squares regression line for the manatee data in Exercise 39. Use the equation to predict manatee deaths in a future year when 700,000 powerboats are registered in Florida.

● Optional exercise.

·7·

Probability: The Mathematics of Chance

Have you ever wondered how gambling, which is a recreation or an addiction for individuals, can be a business for the casino? A business requires predictable revenue from the service it offers, even when the service is a game of chance. Individual gamblers may win or lose; they can never say whether a day at Lake Tahoe or Atlantic City will turn a profit or a loss. But the casino itself does not gamble. Casinos are consistently profitable, and lotteries are now an important source of revenue for many state governments.

It is a remarkable fact that the aggregate result of many thousands of chance outcomes can be known with near certainty. The casino need not load the dice, mark the cards, or alter the roulette wheel. It knows that in the long run, each dollar bet will yield its 5 cents or so of revenue. It is therefore good business to con-

Figure 7.1 The more dollars bet, the more money a casino is guaranteed to take in. [Las Vegas News Bureau.]

centrate on free floor shows or inexpensive bus fares to increase the flow of dollars bet. The flow of profit will follow.

Gambling houses are not alone in profiting from the fact that a chance outcome repeated many times is firmly predictable. For example, although a life insurance company does not know *which* of its policyholders will die next year, it can predict quite accurately *how many* will die. It sets its premiums by this knowledge, just as the casino sets its jackpots.

A phenomenon is called **random** if individual outcomes are uncertain but the long-term pattern of many individual outcomes is predictable. Many phenomena, both natural and of human design, are random. The life spans of insurance buyers and the hair color of children are examples of natural randomness. Indeed, quantum mechanics asserts that at the subatomic level the natural world is inherently random. Probability theory, the mathematical description of randomness, is therefore essential to much of modern science.

Games of chance are examples of randomness deliberately produced by human effort. Casino dice are carefully machined, and their drilled holes—called pips—are filled with material equal in density to the plastic body.

This guarantees that the six side has the same weight as the opposite side, which has only one pip. Thus, each side is equally likely to land upward. All the odds and payoffs of dice games rest on this carefully planned randomness.

Statisticians and casino managers have the same vested interest in planned randomness, although statisticians use tables of random digits rather than dice and cards. The reasoning of statistical inference rests on planned randomness and on the mathematics of probability, the same mathematics that guarantees the profits of casinos and insurance companies. The mathematics of chance is the topic of this chapter.

WHAT IS PROBABILITY?

The mathematics of chance, the mathematical description of randomness, is called the *theory of probability*. Probability describes the predictable long-run patterns of random outcomes.

Toss a coin in the air. Will it land heads or tails? Sometimes it lands heads and sometimes tails. We cannot say what the next outcome will be. Perhaps you would argue that the coin has an equal chance of falling heads or tails on the next toss. You might explain your personal opinion by saying that if forced to bet on the next toss, you would accept even odds on the two outcomes. Probability as the expression of personal opinion is a reasonable idea. But we have in mind a different meaning, one based on *observation* of random phenomena.

Suppose that we toss a coin not once but 10,000 times. John Kerrich, an English mathematician, actually did this while interned by the Germans during World War II. Figure 7.2 shows Kerrich's results. His first 10 tosses gave 4 heads, a proportion of 0.4. The proportion of heads increased to 0.5 after 20 tosses and to 0.57 after 30 tosses. Figure 7.2 shows how this

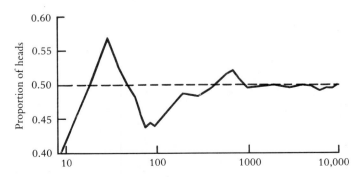

Figure 7.2 Percent of heads versus number of tosses in Kerrich's coin-tossing experiment. [David Freedman et al., *Statistics,* Norton, New York, 1978.]

proportion changes as more and more tosses are made. In a small number of tosses, the proportion of heads fluctuates — it is still essentially unpredictable. But many tosses produce a smoothing effect. A proportion of 0.507, or 50.7%, heads resulted after Kerrich threw the coin 5000 times. And in all 10,000 trials, he scored 5067 heads, again 50.7% of the total. After many trials, the proportion of heads settles down to a fixed number. This number is the probability of a head. The **probability** of an outcome is the proportion of trials in which the outcome occurs in a very long run of trials.

The idea of probability involves using what would happen in many trials to describe the uncertain outcome of a single trial. This "definition" of probability would not satisfy either a mathematician or a philosopher, but it fixes in mind the idea we will deal with. Strictly speaking, Kerrich's 0.507 only estimates the probability that his coin will come up heads. The proportion after 100,000 tosses would be closer to the true probability. In fact, we can't even say that Kerrich's coin is weighted in favor of heads. There is still enough unpredictability in the results of 10,000 tosses that 70 excess heads could easily occur even if heads and tails were equally probable.

PROBABILITY MODELS

Gamblers have known for a long time that the fall of coins, cards, or dice stabilizes into definite patterns in the long run. France gave birth to the mathematics of probability when gamblers in the seventeenth century turned to mathematicians for advice (see Spotlight 7.1, p. 186). The idea of probability rests on the observed fact that the average result of many thousands of chance outcomes can be known with near certainty. But a definition of probability as "long-run proportion" is not precise enough. Who can say what the "long run" is?

Instead, we give a mathematical description of how probabilities behave, based on our understanding of long-run proportions. This mathematical description of probability has the advantage that it applies equally well to probability thought of as personal assessment of chance. The same mathematics describes two different informal concepts of probability.

Our first task in assigning probabilities to outcomes is to list all the possible outcomes. This set of outcomes is called the **sample space** S. The set S may be very simple or very complex. When we toss a coin once, there are only two outcomes, heads and tails. So the sample space is $S = \{H, T\}$. If we draw a random sample of 1500 U.S. residents age 18 and over, as the Gallup poll does, the sample space contains all possible choices of 1500 of the over 185 million adults in the country. This S is extremely large — to count its members would require a number over 8000 digits long! Each member of S is a possible Gallup poll sample, which explains the term *sample space*.

EXAMPLE: Rolling Dice. Suppose that we roll a single die and observe the number of pips on the up face. The possible outcomes are shown in Figure 7.3. We take the sample space, the set of these outcomes, to be

$$S = \{1, 2, 3, 4, 5, 6\}$$

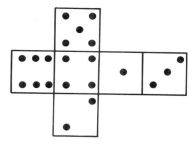

Figure 7.3 The possible outcomes for rolling one die.

SPOTLIGHT 7.1 The Mathematical Bernoullis

Few families have made more contributions to mathematics than the Bernoullis of Basel, Switzerland. No fewer than seven Bernoullis, over three generations spanning the years between 1680 and 1800, were distinguished mathematicians. Five of them helped build the new mathematics of probability.

Jakob Bernoulli

Johann Bernoulli

Jakob (1654–1705) and Johann (1667–1748) were sons of a prosperous Swiss merchant, but they studied mathematics against the will of their practical father. Both were among the finest mathematicians of their times, but it was Jakob who concentrated on probability. Several seventeenth-century mathematicians had started the study of games of chance, concentrating on counting outcomes to find chances. Jakob Bernoulli was the first to see clearly the idea of a long-run proportion as a way of measuring chance. He proved that if in a very large population (say, size N), K members have a property A, then the proportion of members of a sample of size n having property A must approach K/N (the probability of A) as the sample size n increases. This *law of large numbers* helped to connect probability to the study of sequences of chance outcomes observed in human affairs.

Johann's son Daniel (1700–1782) and Jakob and Johann's nephew Nicholas (1687–1759) also studied probability. Nicholas saw that the pattern of births of male and female children could be described by probability. Despite his own rebellion against his father's strictures, Johann tried to make his son Daniel a merchant or a doctor. Daniel, undeterred, became yet another Bernoulli mathematician. He studied mainly the mathematics of flowing fluids (later applied to designing ships and aircraft) and elastic bodies. In the field of probability, he worked to fairly price games of chance and gave evidence for the effectiveness of inoculation against smallpox.

The Bernoulli family in mathematics, like their contemporaries the Bachs in music, are an unusual example of talent in one field appearing in successive generations. Their work helped probability to grow from its birthplace in the gambling hall to a respectable tool with worldwide applications.

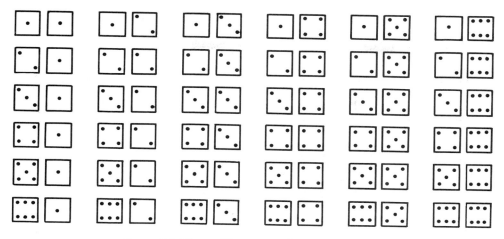

Figure 7.4 The possible outcomes for rolling two dice.

What if we roll *two* dice, as many games of chance require? Figure 7.4 shows the possible combinations of up faces on the two dice. The sample space consists of these 36 outcomes.

In craps and other games, all that matters is the *sum* of the pips on the up faces. Let's change the random outcomes we are interested in: roll two dice and count the pips on the up faces. Now there are only 11 possible outcomes, from a sum of 2 for rolling a double one through 3, 4, 5, and on up to 12 for rolling a double six. The sample space is now

$$S = \{2, 3, 4, 5, 6, 7, 8, 9, 10, 11, 12\}$$

As this example shows, it is important in choosing a sample space to be clear which random outcomes you will look for. Tossing two dice and recording both up faces is not the same as tossing two dice and recording only the sum of the two up faces.

The next step in the mathematical description of chance is to assign probabilities to the outcomes that make up the sample space. There are many ways to assign probabilities, so it is convenient to start with some general rules

that any assignment of probabilities to outcomes must obey. Let $P(s)$ stand for the probability of any outcome s in the sample space S. Here are the fundamental laws for assigning probabilities to outcomes:

LAW 1. Every probability $P(s)$ is a number between 0 and 1.

LAW 2. The sum of the probabilities $P(s)$ over all outcomes s in S is exactly 1.

These laws are based on our understanding of probability as long-run proportion. The first law reflects the fact that any proportion is a number between 0 and 1. If an outcome *never* occurs, it has probability 0; if it *always* occurs, the probability is 1; if it *sometimes* occurs, the proportion of trials producing the outcome is a number between 0 and 1. The second law says that some outcome must always occur, so the sum of all their probabilities is 1. Laws 1 and 2 describe any legitimate assignment of probabilities to outcomes. An assignment that does not satisfy these laws does not make sense. A **probability model** consists of a sample space S and an assignment of probabilities that satisfies laws 1 and 2.

EXAMPLE: Tossing One Die. Let's return to tossing a single die to see what a probability model looks like. We have already seen that the sample space (see Figure 7.3) is

$$S = \{1, 2, 3, 4, 5, 6\}$$

If the die is a carefully made casino model, each face should be equally likely to land up. Because the probabilities of the six outcomes must sum to 1 by law 2 (and none can be negative, by law 1), we are forced to assign the following probabilities:

$$P(1) = P(2) = P(3) = P(4) = P(5) = P(6)$$
$$= 1/6 = 0.166$$

Dice and other chance devices that are carefully made so as to have equally likely outcomes are often called *fair*.

It is important to understand what the definition of a probability model does and does not do for us. It does state which assignments of probabilities to outcomes make sense. But it does not say which assignments are correct, in the sense of accurately describing a real die; this can only be determined by actual trial. Professional dice are well described by the equal-probabilities model of the example. But cheap dice with hollowed-out pips fall unequally. The six face is lightest and is located opposite the one face, which is heaviest. Thus, a cheap die might be described by a probability model in which lighter faces are more likely, such as

$$P(1) = 0.159 \qquad P(4) = 0.166$$
$$P(2) = 0.163 \qquad P(5) = 0.171$$
$$P(3) = 0.166 \qquad P(6) = 0.175$$

This assignment also satisfies laws 1 and 2. It is legitimate, even if the die is not.

Assigning probabilities to individual outcomes is not enough. We also want to assign probabilities to **events,** which are collections of outcomes. For example, what is the probability of rolling an odd number in one toss of a fair die? The odd outcomes are 1, 3, and 5. The proportion of tosses on which one of these numbers comes up must be the sum of the proportions on which each alone comes up. So thinking of probabilities as long-run proportions leads us to find the probability of any event by summing the probabilities of the outcomes that make up the event. In this case,

$$P(\text{outcome is odd}) = P(1) + P(3) + P(5)$$
$$= 3/6 = 0.5$$

Probability models for random phenomena with only finitely many possible outcomes have a simple form: assign probabilities to outcomes in a way that satisfies laws 1 and 2, then find probabilities of events by adding up the probabilities of the outcomes that make up the event.

EXAMPLE: Household Size. A household is a group of people living together, regardless of their relationship to each other. Sample surveys such as the Current Population Survey select a random sample of households. Here is the probability model for the number of people living in a randomly chosen American household:

Household size	1	2	3	4	5	6	7
Probability	.236	.320	.181	.156	.069	.024	.014

These probabilities are the proportions for all households in the country and so give the probabilities that a single household chosen at random will have each size. (The very few households with more than 7 members are placed in the 7 group.) Check that laws 1 and 2 are satisfied. The probability that a randomly chosen household has more than two members is

$$P(\text{size} > 2) = P(3) + P(4) + P(5) \\ + P(6) + P(7) \\ = .181 + .156 + .069 \\ + .024 + .014 \\ = .444$$

A simple random sample gives all possible samples an equal chance to be chosen. Dealing from a well-shuffled deck gives all possible card hands an equal chance to be the hand you are dealt. When randomness is the product of human design, it is often the case that the outcomes in the sample space are all equally likely. Laws 1 and 2 force the assignment of probabilities in this case:

> If a random phenomenon has k possible outcomes, each equally likely, then each individual outcome has probability $1/k$.

The probability of any event in the equally likely case is found as usual by adding the individual probabilities of the outcomes making up the event. Because each of these probabilities is the same $1/k$, we have this new rule:

> When all outcomes have equal probabilities, the probability of any event A is

$$P(A) = \frac{\text{number of outcomes in } A}{\text{number of outcomes in } S}$$

EXAMPLE: Rolling Two Dice. Roll two fair dice and record the pips on each of the two up faces. The sample space consists of the 36 outcomes pictured in Figure 7.4. Because of the balance of the dice, these outcomes are all equally likely. So each has probability 1/36.

What is the probability of rolling a 5? The event "roll a 5" contains the four outcomes

and the probability is therefore 4/36, or 0.111. What about the probability of rolling a score that is a multiple of 3? The event in question contains scores 3, 6, 9, and 12. A look at Figure 7.4 will count 12 outcomes for which the sum of the pips is one of these numbers. The probability is therefore 12/36, or 0.333.

Be certain that you understand that the method of finding probabilities by counting outcomes applies *only* when all outcomes are equally likely. The S shown in Figure 7.4 does have equally likely outcomes. But if we choose to use the sample space for rolling two dice *and counting the pips,* we get

$$S = \{2, 3, 4, 5, 6, 7, 8, 9, 10, 11, 12\}$$

These outcomes do *not* have equal probabilities; in particular, the probability of a 5 is *not* 1/11.

When outcomes are equally likely, finding probabilities leads to the study of counting methods, called **combinatorics.** Combinatorics is an important area of mathematics in its own right. Although we will not study combinatorics, the following example illustrates a useful multiplication method of counting.

EXAMPLE: Code Words. A computer system assigns log-in identification codes to users by choosing three letters at random. All three-letter codes are therefore equally probable. What is the probability that the code assigned to you has no x in it?

We must count the number of code words. There are 26 letters that can occur in each position in the word. Any of the 26 letters in the first position can be combined with any of the 26 letters in the second position to give 26×26 choices. (This is true because the order of the letters matters, so *ab* and *ba* are different choices.) Any of the 26 letters can then follow in the third position. The number of different codes is

$$26 \times 26 \times 26 = 17{,}576$$

Codes without an *x* are made up of the other 25 letters. There are

$$25 \times 25 \times 25 = 15{,}625$$

such codes. The probability that your code has no *x* is therefore

$$P(\text{no } x) = \frac{\text{number of codes with no } x}{\text{number of codes}}$$
$$= \frac{15{,}625}{17{,}576} = 0.889$$

Suppose that the computer is programmed to avoid repeated letters in the identification codes. Any of the 26 letters can still appear in the first position. But only the 25 remaining letters are allowed in the second position, so there are 26×25 choices for the first two letters in the code. Any of these choices leaves 24 letters for the third position. The number of different codes without repeated letters is

$$26 \times 25 \times 24 = 15{,}600$$

Codes with no *x* are allowed one fewer choice in each position; there are

$$25 \times 24 \times 23 = 13{,}800$$

such codes. The probability that your code has no *x* is then

$$P(\text{no } x) = \frac{\text{number of codes with no } x}{\text{number of codes}}$$
$$= \frac{13{,}800}{15{,}600} = 0.885$$

Eliminating repeats slightly decreases your chance of avoiding an *x*.

EXPECTED VALUES

Suppose you are offered this choice of bets, each costing the same: bet A pays $10 if you win and you have probability 1/2 of winning, while bet B pays $10,000 and offers probability 1/10 of winning. You would very likely choose B even though A offers a better chance to win, because B pays much more if you win. It would be silly to decide which bet to make just on the basis of the probability of winning. How much you can win is also important. When a random phenomenon has numerical outcomes, we are concerned with the amount as well as with the probability of the outcomes. In particular, we might ask what will be the average payoff of our two bets in many plays. Recall that the probabilities are the long-run proportions of plays on which each outcome occurs. Bet A produces $10 half the time in the long run and nothing half the time. So the average payoff should be

$$(\$10)(\tfrac{1}{2}) + (\$0)(\tfrac{1}{2}) = \$5$$

Bet B, on the other hand, pays out $10,000 on 1/10 of all bets in the long run. Bet B's average payoff is

$$(\$10{,}000)(\tfrac{1}{10}) + (\$0)(\tfrac{9}{10}) = \$1000$$

If you can place many bets, you should certainly choose B. Here is a general definition of the kind of "average outcome" we used to compare the two bets.

Suppose that the possible outcomes $s_1, s_2, \ldots , s_m$ in a sample space S are numbers and that p_j is the probability of outcome s_j. The **expected value** E of the random outcome is

$$E = s_1 p_1 + s_2 p_2 + \ldots + s_m p_m$$

EXAMPLE: Expected Household Size.
The probability model for the number of people living in a randomly chosen household is

Household size	1	2	3	4	5	6	7
Probability	.236	.320	.181	.156	.069	.024	.014

The expected value is therefore

$$E = (1)(.236) + (2)(.320) + (3)(.181)$$
$$+ (4)(.156) + (5)(.069) + (6)(.024)$$
$$+ (7)(.014)$$
$$= 2.63$$

In this case, the expected value is the average size of all American households.

The expected value is an average outcome in two senses. The definition of E says that it is the average of the possible outcomes, not weighted equally but weighted by their probabilities. More likely outcomes get more weight in the average. An important fact of probability, the **law of large numbers,** says that E is the average outcome in another sense as well. The law of large numbers states that if the random phenomenon is repeated a large number of times, the mean of the actually observed outcomes will get closer and closer to the expected value. The proportion of trials on which each outcome occurs will similarly get closer and closer to the probability of that outcome. These facts can actually be proved mathematically for any assignment of probabilities that satisfies laws 1 and 2. The law of large numbers brings the study of basic probability back to a natural completion. We first observed that some phenomena are random in the sense of showing long-run regularity. Then we used the idea of long-run proportions to motivate the basic laws of probability. Those laws are mathematical idealizations that can be used without interpreting probability as proportion in many trials. Now the law of

large numbers tells us that in many trials the proportion of trials on which an outcome occurs will always approach its probability.

SAMPLING DISTRIBUTIONS

Sampling, in a way, is a lot like gambling. Both rely on the deliberate use of chance. To see how probability applies to sampling, let's look again at the population of light and dark beads described in Chapter 5.

We have a large container filled with beads of the same size and shape. How can we estimate the percentage of dark beads in the box? We plunged a scoop with 50 bead-sized hollows into the container and drew out 50 beads. This is a simple random sample of size 50. Of the 50 beads in the sample, 12, or 24% of the sample, were dark. From this we estimated that 24% of the population is made up of dark beads. Our sampling strategy illustrates a basic form of statistical inference: take a simple random sample and use the sample result to estimate the unknown truth about a population. The Gallup poll and the Current Population Survey carry out more elaborate versions of this method.

When we drop the scoop into the box once more, drawing a second sample, perhaps only 8 of the 50 beads, or 16%, will be dark. A third sample might yield 7 dark beads, or 14%. This chance variation is called **sampling variability.** We saw in Chapter 5 how sampling variability affects the results of sample surveys. Now we can use the laws of probability to describe it more precisely.

The sample space in the bead-sampling experiment is made up of all the possible samples of 50 beads drawn from the beads in the container. This sample space is very large. Random sampling guarantees that each possible sample is equally likely to be drawn. So we could find the probability of getting 16% dark beads by counting the possible samples that have 16% dark beads and dividing by the number of outcomes in the entire sample space. That's hopelessly tedious; we need a shortcut method that will find the probability without counting.

To get a clue about the value of the probabilities, let's return to actually observing the outcomes of many samples. Figure 7.5a records the results of 200 samples. For example, 21 of the 200 samples had exactly 16% (8 of 50) dark beads. So we estimate the probability of getting 16% dark beads to be about 21/200, or 0.105. The figure is a new kind of histogram in which the heights of the bars are the proportions of outcomes in each class rather than the counts of outcomes.

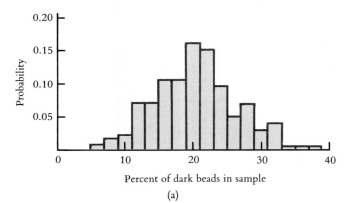

(a)

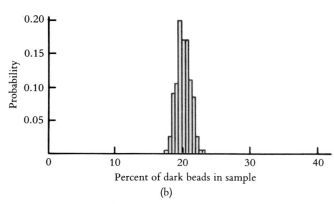

(b)

Figure 7.5 Sampling distributions of the percent of dark beads in samples of two sizes from the same population. (a) Sample size 50. (b) Sample size 1500.

Look at the overall pattern of the histogram of estimated probabilities in Figure 7.5a. The percentages of dark beads range from 6% to 38%. The histogram is fairly symmetric. The center is close to the 20% bar (representing 10 dark beads out of 50). This is also the outcome that occurred most often. There are no extreme outliers.

Statisticians call a number that is computed from a sample a **statistic.** The percentage of dark beads in our sample is a statistic. The histogram displays the sampling variability of this statistic. In fact, the histogram assigns probabilities to each value of the statistic. These probabilities make up the **sampling distribution** of the statistic. Of course, the probabilities are only approximate because they are based on only 200 trials. But the overall pattern of the sampling distribution is clear.

Let's try a second experiment in the hope of confirming the pattern. Remember that public opinion polls often rely on a sample of about 1500 people. We will take 200 simple random samples of 1500 beads each. To do this, we must imagine a very large container of beads. In fact, we instruct a computer to simulate the experiment. Figure 7.5b displays the outcomes, using the same scale as Figure 7.5a.

The histogram in Figure 7.5b for samples of size 1500 is taller and narrower than that in Figure 7.5a for samples of size 50. It is narrower because the sample variability is much smaller in the larger sample. All 200 outcomes fall between 17% and 23% dark beads. And if we add up the estimated probabilities near the middle, we find there is a probability 0.93 that a sample will fall between 18.25% and 21.75%. We can rely much more confidently on a single sample of size 1500 than on a single sample of size 50. In fact, all these samples were drawn from a population with 20% dark beads. Samples of size 1500 almost always give estimates quite close to this true value.

Our sampling experiment has both produced an approximate assignment of probabilities (without counting) and taught us a bit about how the sampling distribution behaves when we increase the size of the sample. The following two sections will explore these two topics in more detail.

NORMAL DISTRIBUTIONS

Although they differ in variability, the histograms in Figure 7.5 have similar shapes in other respects. Both are symmetric, with centers close to 20%. The tails fall off smoothly on either side, with no outliers. Suppose that we represent the shape of each histogram by drawing a smooth curve through the tops of the bars. If we do this carefully — using the actual probabilities of the outcomes rather than estimates from only 200 samples — the two curves we obtain will be quite close to two members of the family of *normal curves.* The two normal curves appear in Figure 7.6.

Normal curves introduce a new way of describing probabilities. The assignment of probabilities to the values of a statistic can be described by a histogram like those of Figure 7.5. A histogram makes probability visible. The height of any bar represents the probability of the outcomes spanned by the base of that bar. Because all bars are of equal width, their areas also represent probability. Normal curves can be thought of as approximations to a histogram of probabilities; they are easier to work with than histograms because many bars are replaced by a single smooth curve. Normal curves have the property that the total area under the curve is exactly 1, corresponding to the fact that all outcomes together have probability 1. The probability of any interval of outcomes is the area under the normal curve above that interval.

Our basic method of assigning probabilities to events was to first give a probability to each individual outcome. Probability as area under a

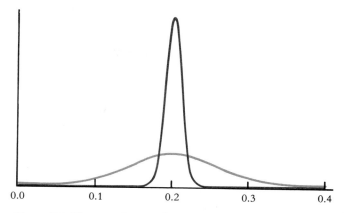

Figure 7.6 The normal curves that approximate the sampling distributions of the percent of dark beads for samples of sizes 50 (light curve) and 1500 (dark curve).

curve is the second important method of assigning probability and is easier when there are many individual outcomes falling close together. Curves of different shapes describe different assignments of probability. We will emphasize the normal curves, because they describe probability in a number of important situations. An assignment of probabilities to outcomes by a normal curve is a **normal probability distribution.** Figures 7.5 and 7.6 demonstrate that the sampling distribution of a sample proportion from a simple random sample is close to a normal distribution. This is not just a matter of artistic judgment. It is a mathematical fact, first proved by Abraham De Moivre in 1718. Other common statistics, such as the mean of a large sample, also have sampling distributions that are approximately normal.

There is a close connection between describing an assignment of probability to numerical outcomes and describing a set of data. Histograms are used for both tasks. Similarly, smooth curves such as the normal curves can replace histograms for describing large sets of data. Many sets of data are approximately described by normal distributions. The normal distributions therefore deserve more detailed study.

Normal curves can be specified exactly by an equation, but we will be content with pictures like Figure 7.6. All normal curves are symmetric and bell-shaped, with tails that fall off rapidly. The center of the symmetric curve is the center of the distribution in several senses. It is the expected value for the assignment of probabilities. If we consider a normal distribution as an idealized distribution of data, then the center is both the median and the mean of the data. As we saw earlier, the median is the midpoint of the outcomes, while the mean is their arithmetic average. Both lie at the center because of the symmetry of the curve. Thus, the mean and median of any normal distribution are the same.

This is not true of all distributions. Figure 7.7, for example, shows a right-skewed distribution. The right tail of the curve is much longer than the left. The prices of new houses are an example of a skewed distribution—there are many moderately priced houses and a few extravagantly priced mansions out in the right tail. Those mansions pull the mean—or average price—up, so it is greater than the median. For example, the mean price of a new house in 1989 was $159,100, but the median price was only $129,900.

As we saw in Chapter 6, even the most cursory description of data on a single variable should include a measure of spread in addition to a measure of center or location. What about the spread of a normal curve? Normal curves have the special property that their spread is completely measured by a single number, the **standard deviation.** The standard deviation, like the mean, can be calculated from a set of data. But for normal distributions, the standard deviation (again like the mean) can be found directly from the curve. Because the standard deviation is most useful for normal distributions, we will work with normal

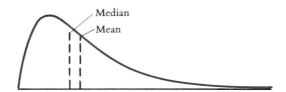

Figure 7.7 The mean of a skewed distribution is located farther toward the long tail than the median.

curves rather than start from individual observations.

> To find the standard deviation of a normal distribution, run a pencil along the normal curve from the center (the mean) outward. At first the curve opens downward (falls ever more steeply as you go out); farther from the mean it opens upward (falls ever less steeply). The two points where the curvature changes are located one standard deviation on either side of the mean.

With a little practice you can locate the change-of-curvature points quite accurately. For example, Figure 7.8 shows the distribution of heights of American women ages 18 to 24. The shape of the curve is normal, with the mean (and median) height at 64 inches. The two change-of-curvature points are at 61.5 and 66.5 inches. The standard deviation of the distribution is the distance of either of these points from the mean, or 2.5 inches.

In Chapter 6, we used the quartiles to indicate the spread of a distribution. Because the standard deviation completely describes the spread of any normal distribution, it fixes both quartiles as follows:

> The first quartile of any normal distribution is located 0.67 standard deviation below the mean; the third quartile is 0.67 standard deviation above the mean.

EXAMPLE: Locating the Quartiles. The distribution of heights of young women, shown in Figure 7.8, is approximately normal with mean 64 inches and standard deviation

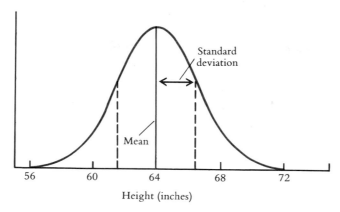

Figure 7.8 Locating the mean and standard deviation on a normal curve.

2.5 inches. The quartiles lie 0.67 standard deviation, or

$$(0.67)(2.5) = 1.7 \text{ inches}$$

on either side of the mean. The first quartile is $64 - 1.7$, or 62.3 inches. The third quartile is $64 + 1.7$, or 65.7 inches. Figure 7.9 marks the quartiles on the normal curve. They contain between them the middle 50% of women's heights.

The mean and standard deviation of normal curves have a special property: the shape of a normal distribution is completely specified once the mean and the standard deviation are given. This information is not sufficient to determine the exact shape of most distributions of data, but it is enough when the distribution is normal. Changing the mean of a normal curve does not change its shape; it only slides the curve to a new location. Changing the standard deviation does change its shape. A normal curve with a smaller standard deviation is taller and narrower (has less spread) than one with a larger standard deviation. You can see this by comparing the two normal curves for our bead-sampling experiments in

Figure 7.6. Both normal curves have the same mean, but the curve for samples of size 1500 has the smaller standard deviation.

Another consequence of the fact that the mean and standard deviation completely specify a normal distribution is that all normal distributions are the same when we record observations in terms of how many standard deviations they lie from the mean. In particular, the probability that an observation falls within one, two, or three standard deviations

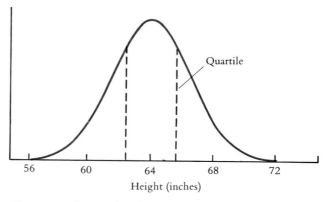

Figure 7.9 The quartiles of a normal distribution are located 0.67 standard deviation on either side of the mean.

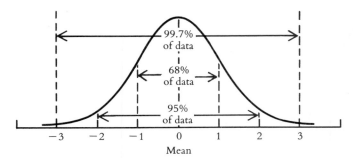

Figure 7.10 The 68-95-99.7 rule for normal distributions.

of the mean is the same for all normal distributions. The probability of an outcome falling within one standard deviation on either side of the mean is 0.68. If we go out two standard deviations from the mean, we find a probability of 0.95. Finally, the probability of falling within three standard deviations of the mean is almost 1, or 0.997 to be exact. These facts can be derived mathematically from the equation of a normal curve. They are not true for distributions with other shapes.

Figure 7.10 illustrates these facts expressed in terms of percents. Together, we call them the **68-95-99.7 rule** for normal distributions.

• 68% of the observations in any normal distribution fall within one standard deviation of the mean.

• 95% of the observations fall within two standard deviations of the mean.

• 99.7% of the observations fall within three standard deviations of the mean.

By using the three numbers in the 68-95-99.7 rule, we can quickly derive helpful information about any normal distribution. More detailed information can be gleaned from tables of areas under the normal curves, but the 68-95-99.7 rule is adequate for our purposes.

EXAMPLE: Heights of Young Women. The heights of American women between the ages of 18 and 24 are roughly normally distributed, with mean 64 inches and standard deviation 2.5 inches. One standard deviation below the mean is $64 - 2.5$, or 61.5 inches. Similarly, one standard deviation above the mean is $64 + 2.5$, or 66.5 inches. The 68 part of the 68-95-99.7 rule says that about 68% of women are between 61.5 and 66.5 inches tall. Because two standard deviations are 5 inches, we know that 95% of young women are between $64 - 5$ and $64 + 5$, that is, between 59 and 69 inches tall. Almost all women have heights within three standard deviations of the mean, or between 56.5 and 71.5 inches. Very few women are 6 feet tall or over.

EXAMPLE: SAT Scores. The distribution of scores on tests such as the Scholastic Aptitude Test (SAT) is close to normal. SAT scores are adjusted so that for a reference population of students the mean score is 500 and the standard deviation is 100. (The mean and standard deviation of the scores for students taking the test in any one year will differ from those of the reference population.) This information allows us to answer many questions about SAT scores.

• *How high must a student score to fall in the top 25%?* The third quartile is $(0.67)(100) = 67$ points above the mean. So scores above 567 are in the top 25%.

• *What percent of scores fall between 200 and 800?* Note that 200 and 800 are three standard deviations on either side of the mean. The 99.7 part of the 68-95-99.7 rule says that 99.7% of all scores lie in this range. (In fact, 200 and 800 are the lowest and highest scores that are reported on the SAT. The few scores higher than 800 are reported as 800.)

• *What percent of scores are above 700?* First note that 700 is two standard deviations above the mean. By the 95 part of the 68-95-99.7 rule, 95% of all scores fall between 300 and 700 and 5% fall below 300 or above 700. Because normal curves are symmetric, half of this 5% are above 700. So a score above 700 places a student in the top 2.5% of the reference population.

THE CENTRAL LIMIT THEOREM

The significance of normal distributions is explained by a key fact in probability theory known as the **central limit theorem.** This theorem says that the distribution of any random phenomenon tends to be normal if we average it over a large number of independent repetitions. The central limit theorem allows us to analyze and predict the results of chance phenomena if we average over many observations.

We have already seen the central limit theorem at work in our bead-sampling experiment. A single bead drawn at random is either dark or light. Only two outcomes are possible, and there is no normal curve in sight. However, the percent of dark beads when 50 beads are drawn at random roughly follows a normal distribution. You can think of the percent of dark beads as an average of "dark" or "light" over the 50 beads drawn. When 1500 beads are drawn, the percent of dark beads represents an average over a larger number of beads and is even closer to a normal curve.

Our sampling experiment showed that samples of 1500 beads have much less spread than samples of 50. Spread can be described by the standard deviation of the normal distribution of outcomes. The central limit theorem makes this explicit. Here is a more exact statement.

A sample mean or sample proportion from n trials on the same random phenomenon has a distribution that is approximately normal when n is large. The mean of the normal distribution is the expected value for a single trial. The standard deviation of the normal distribution is the standard deviation for a single trial divided by $\sqrt{n}$.

Because we have not discussed how to find the standard deviation for distributions that are not normal, we will not concern ourselves with finding the standard deviation for a single trial. The important fact is that the standard deviation of a mean or proportion decreases with the square root of the number of observations, $\sqrt{n}$.

To cut the standard deviation of a sample mean or proportion in half, the sample size must be multiplied by 4, not just by 2. For example, an average over 100 observations ($\sqrt{100} = 10$) has a standard deviation 1/10 as large as the standard deviation of a single observation. An average over 25 observations ($\sqrt{25} = 5$) from the same population has standard deviation 1/5 as large as the standard deviation for an individual. Because 1/10 is half of 1/5, increasing the sample size from 25 to 100 cuts the standard deviation of the normal distribution of the sample mean in half. The same $\sqrt{n}$ effect applies to sample proportions. For example, a sample of 1500 beads is 30 times larger than a sample of 50 beads. So the standard deviation of the percentage of dark beads is $\sqrt{30}$, or 5.5, times smaller for samples of 1500 than for samples of 50. The two normal curves in Figure 7.6 display exactly this difference in standard deviations.

The central limit theorem can help to answer our opening question: How can gambling be a business for a casino? Let's look at just one of the many bets that a casino offers.

EXAMPLE: Red or Black in Roulette. An American roulette wheel has 38 slots, of which 18 are black, 18 are red, and 2 are green (see Figure 7.11). When the wheel is spun, the

Figure 7.11 A gambler may win or lose at roulette, but in the long run the casino always wins.

ball is equally likely to come to rest in any of the slots. Gamblers can place a number of different bets in roulette. One of the simplest wagers chooses red or black. A bet of $1 on red will pay off an additional dollar if the ball lands in a red slot. Otherwise, the player loses. (When gamblers bet on red or black, the two green slots belong to the house.)

If we decide to bet on red, there are only two possible outcomes: win or lose. We win if the ball stops in one of the 18 red slots; we lose otherwise. Because casino roulette wheels are carefully balanced so that all slots are equally likely, the probabilities are

$$P(\text{win } \$1) = 18/38$$
$$P(\text{lose } \$1) = 20/38$$

The expected value of a single bet on red is found in the usual way:

$$E = (1)(\tfrac{18}{38}) + (-1)(\tfrac{20}{38})$$
$$= -\tfrac{2}{38} = -0.053$$

The law of large numbers says that in the long run gamblers will lose (and the casino will win) an average of 5.3 cents per bet.

Just as when only one bead is selected from the container of beads, there is no normal curve in sight when only one bet is made on red in roulette. But the central limit theorem ensures that the average outcome of many bets follows a distribution that is close to normal. Suppose that we place 50 bets in an evening's play, again betting $1 each time. The mean outcome is the average winnings, the overall gain (or loss) divided by 50. If we win 30 and lose 20 times, the overall gain is $10, an average winnings of $0.20 per bet. If we continue to gamble night after night, placing 50 bets each night, our average winnings per bet will vary from night to night. But a histogram of these values will follow a normal distribution. Figure 7.12 shows the results of many trials of 50 bets each. The normal curve superimposed on the histogram is the distribution given by the central limit theorem in this case.

We know that the mean of the normal distribution in Figure 7.12 is the expected value of a single bet, −0.053. What is the standard deviation? The full spread of outcomes observed was −0.47 to 0.37, or 0.84 in all. By the 99.7 part of the 68-95-99.7 rule, the outcomes should span about three standard deviations on each side of the mean. The standard deviation is therefore about one-sixth of 0.84, or 0.14 (14 cents). Check this by locating the change-of-curvature points of the normal curve in the figure. From this combination of calculation and experiment we conclude that the average winnings in 50 bets follow approximately the normal distribution with mean −0.053 and standard deviation 0.14.

What will be the experience of a habitual gambler who places 50 bets per night? Almost all average nightly winnings will fall within three standard deviations of the mean, that is, between

$$-0.053 + (3)(0.14) = 0.367$$

and

$$-0.053 - (3)(0.14) = -0.473$$

The total winnings after 50 bets will therefore fall between

$$(0.367)(50) = 18.35$$

and

$$(-0.473)(50) = -23.65$$

The gambler may win as much as $18.35 or lose as much as $23.65. Gambling is exciting because the outcome, even after an evening of bets, is uncertain. It is possible to walk away a winner. It's all a matter of luck.

The casino, however, is in a different position. It doesn't want excitement, just a steady income. The house bets with all its customers — perhaps 100,000 individual bets on black or red in a week. The distribution of average customer winnings on 100,000 bets is very close to normal, and the mean is still the expected value of one bet, -0.053, a loss of 5.3 cents per dollar bet.

The central limit theorem says in addition that the standard deviation of the distribution of average winnings decreases with the square root of the number of bets over which we are averaging. Now, 100,000 is 2000 times as much as 50. So the standard deviation of the casino's distribution (average winnings over 100,000 bets) is

$$\sqrt{2000} = 44.72$$

times as small as the standard deviation of the gambler's distribution (average over 50 bets). The gambler has standard deviation 0.14; the casino therefore has standard deviation

$$\frac{0.14}{44.72} = 0.003$$

There you have it. The individual gambler will experience wide variation in winnings; he

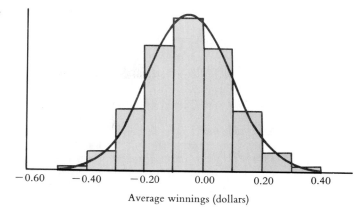

Figure 7.12 The distribution of winnings in repeated bets on red or black in roulette.

or she gets excitement. The casino experiences very little variation; it has a business. Here is what the spread in the casino's average winnings per bet looks like after 100,000 bets:

$$
\begin{aligned}
\text{Spread} &= \text{mean} \pm 3 \text{ standard deviations} \\
&= -0.053 \pm (3)(0.003) \\
&= -0.053 \pm 0.009 \\
&= -0.044 \text{ to } -0.062
\end{aligned}
$$

Because the casino covers so many bets, the standard deviation of the average winnings per bet becomes very narrow. And because the mean is negative, almost all outcomes will be negative. Thus, the gamblers' losses and the casino's winnings are almost certain to average between 4.4 and 6.2 cents for every dollar bet.

The gamblers who collectively placed those 100,000 bets will lose money. We are now in a position to estimate the probable range of their losses:

$$(-0.044)(100,000) = -4400$$
$$(-0.062)(100,000) = -6200$$

The gamblers are almost certain to lose — and the casino is almost certain to take in —

SPOTLIGHT 7.2 State Lotteries

Both public and private lotteries were common in the early years of the United States. After disappearing for a century or so, government-run gambling reappeared in New Hampshire and New York in the mid-1960s and has spread to almost all states outside the South. State lotteries sold over $15 billion worth of tickets in 1988 and continue to expand rapidly. Their growth is fed by constant advertising and by new games designed to hold the interest of the public. The most popular game in most states is lotto, in which players choose (for example) 6 numbers out of 54 in the hope of matching the randomly chosen winning numbers.

State lotteries differ from casino gambling in several respects. For one thing, the lotteries offer much poorer returns to their customers. Las Vegas and Atlantic City pay out between about 85% and 95% of the dollars bet, depending on the game. State lotteries typically pay out about half the dollars bet; about 15% goes for advertising and expenses, and the remaining 35% flows into the state's treasury. The states also pay lotto winners over time, usually 20 years. Because money earns interest over time, a jackpot advertised as $10 million actually costs the state only about $4.8 million.

Finally, gamblers can't rely on the central limit theorem to predict their long-run average lotto winnings as it predicts their long-run average winnings in roulette. Lotto offers one very large but very improbable jackpot, a few small prizes, and nothing to the remaining millions of tickets. The expected value of a $1 ticket is about 50 cents, but the variation in outcomes is so large that a gambler's average winnings over even thousands of tickets remain highly unpredictable. The central limit theorem remains true for any game of chance, of course. But when the variation in a single play is extremely large, no humanly possible number of plays, even aided by that $\sqrt{n}$, can reduce the variation in the average outcome enough for a useful prediction. The only compensation almost all lotto players will receive is the pleasure of imagining themselves rich.

between $4400 and $6200 on those 100,000 bets. What's more, we have seen from the central limit theorem that the more bets that are made, the narrower is the range of possible outcomes. That is how a casino can make a business out of gambling. The more money that is bet, the more accurately the casino can predict its profits.

REVIEW VOCABULARY

Central limit theorem The average of many independent random outcomes is approximately normally distributed. When we average n independent repetitions of the same random phenomenon, the resulting distribution of outcomes has mean equal to the expected value of a single trial and standard deviation proportional to $1/\sqrt{n}$.

Combinatorics The branch of mathematics that counts arrangements of objects.

Event Any collection of possible outcomes of a random phenomenon. An event is a subset of the sample space.

Expected value The average outcome of a random phenomenon with numerical values, found by multiplying each possible outcome by its probability and then summing all the products.

Law of large numbers As a random phenomenon is repeated many times, the mean of the observed outcomes approaches the expected value of the probability model.

Normal distributions A family of probability models that assign probabilities to events as areas under a curve. The normal curves are symmetric and bell-shaped. A particular curve is completely described by giving its center (the mean) and a measure of spread (the standard deviation).

Probability A number between 0 and 1 that gives the long-run proportion of repetitions of a random phenomenon on which an event will occur.

Probability model A sample space S together with an assignment of probabilities $P(s)$ to all outcomes s in S. The probabilities $P(s)$ must satisfy two laws:

1. For every outcome s, $0 \leq P(s) \leq 1$.

2. The sum of the probabilities $P(s)$ over all outcomes s is exactly 1.

Random phenomenon A phenomenon is random if it is uncertain what the next outcome will be but each outcome nonetheless tends to occur in a fixed proportion of a very long sequence of repetitions. These long-run proportions are the probabilities of the outcomes.

Sample space A list of all possible outcomes of a random phenomenon.

Sampling distribution An assignment of probabilities to the possible values of a statistic. This distribution describes the sampling variability of the statistic.

Sampling variability The random variability in the value of a statistic (such as a sample mean or proportion) when random samples are drawn repeatedly from the same population.

68-95-99.7 rule In any normal distribution, 68% of the observations lie within one standard deviation on either side of the mean, 95% lie within two standard deviations of the mean, and 99.7% lie within three standard deviations of the mean.

Standard deviation A measure of spread that is particularly appropriate for normal distributions. The standard deviation of a normal curve is the distance from the mean to the change-of-curvature points on either side.

Statistic A number computed from a sample. In random sampling, the value of a statistic will vary in repeated sampling.

SUGGESTED READINGS

Mosteller, Frederick, Robert E. K. Rourke, and George B. Thomas: *Probability with Statistical Applications,* Addison-Wesley, Reading, Mass., 1970. A rich treatment of basic probability that requires only high school algebra but is somewhat sophisticated.

Olkin, Ingram, Leon J. Gleser, and Cyrus Derman: *Probability Models and Applications,* Macmillan, New York, 1980. This book is distinguished by an emphasis on the use of probability to describe real phenomena and by outstanding examples of modeling. In level it falls between Mosteller et al. and Snell.

Snell, J. Laurie: *Introduction to Probability,* Random House, New York, 1988. A calculus-based text aimed at undergraduate mathematics majors that is recommended here because of its excellent examples and historical remarks and because Snell makes good use of BASIC programs that are included in the text.

EXERCISES

You can estimate an unknown probability by actually observing many repetitions of the random phenomenon in question. Exercises 1 to 5 produce rough estimates based on a small number of repetitions. You can see the random behavior in more detail by making a graph like Figure 7.2 rather than just reporting the final proportion of outcomes.

1. Toss a thumbtack on a hard surface 100 times. How many times did it land with the point up? What is the approximate probability of its landing point up?

2. Hold a penny upright on its edge under your forefinger on a hard surface, then snap it with your other forefinger so that it spins for some time before falling. Based on 50 spins, what is the probability of heads?

3. Open your local telephone directory to any page and note whether the last digit of each of the first 100 telephone numbers on the page is odd or even. How many of the digits were odd? What is the approximate probability that the last digit of a telephone number is odd?

4. The table of random digits (Table 5.1) was produced by a random mechanism that gives each digit a probability 0.1 of being a 0. What proportion of the first 200 digits in the table are 0s? This proportion is an estimate of the true probability, which in this case is known to be 0.1.

5. Pick up a book and open to any page. Count the words in the first complete paragraph on that page and note how many of them begin with a vowel. (If the paragraph contains fewer than 100 words, include the next paragraph as well.) What do you estimate to be the probability that a word chosen at random from this book begins with a vowel?

In each of Exercises 6 to 10, describe a reasonable sample space for the random phenomena mentioned. In some cases, more than one choice is possible.

6. Toss a coin 10 times.

    a. Count the number of heads observed.

    b. Compute the percentage of heads among the outcomes.

    c. Record whether or not at least five heads occurred.

7. A female lab rat is about to give birth. You count the number of offspring in the litter. (We don't know how large rat litters can be, but you can set a reasonable upper limit if you want.)

8. A couple plans to have three children.

    a. Record the sex (M or F) of each child in order of birth.

    b. Record the number of girls.

9. You choose a student at random and record the number of dollars in bills (ignore change) that he or she is carrying. (We don't know the largest amount that a student could reasonably carry, so you will have to make a choice in stating the sample space.)

10. Subjects in a clinical trial are assigned at random to either the new treatment group or the control group. For the next subject, you record treatment or control, male or female, and smoker or nonsmoker.

11. Which of the following are legitimate probability models for tossing three (possibly unfair) coins? Explain your answer in each case.

Outcome	Model A	Model B	Model C	Model D
H, H, H	0.125	0	0.125	0.250
H, H, T	0.125	0.375	0.250	0.125
H, T, H	0.125	0	−0.125	0.250
H, T, T	0.125	0.125	0.125	0.125
T, H, H	0.125	0	0.125	0.250
T, H, T	0.125	0.375	0.125	0.125
T, T, H	0.125	0	0.250	0.250
T, T, T	0.125	0.125	−0.125	0.125

12. M&M candies come in several colors mixed together in a bag. Which of the following are legitimate probability models for drawing a single M&M and recording its color?

Color	Brown	Red	Yellow	Green	Orange	Tan
Model A	.3	.2	.2	.2	.1	.1
Model B	.2	.2	.2	.1	.1	.1
Model C	.3	.2	.2	.2	.1	0

13. Here is the distribution of the blood type of a randomly chosen black American. If this is to be a legitimate assignment of probabilities, what must be the probability of type AB blood?

Blood type	O	A	B	AB
Probability	.49	.27	.20	

14. Here is the distribution of marital status for American women aged 25 to 29 years. If this is to be a legitimate probability model, what must be the probability that a woman in this age group is married?

Outcome	Single	Married	Widowed	Divorced
Probability	.288		.003	.076

15. A bridge deck contains 52 cards, four of each of the 13 face values ace, king, queen, jack, ten, nine, . . . , two. You deal a single card from such a deck and record the face value of the card dealt. Give an assignment of probabilities (Model A) to these outcomes that should be correct if the deck is thoroughly shuffled. Give a second assignment of probabilities (Model B) that is legitimate (i.e., obeys laws 1 and 2) but differs from your first choice. Then give a third assignment of probabilities (Model C) that is *not* legitimate, and explain what is wrong with this choice.

16. Exactly one of Brown, Chavez, and Williams will be promoted to partner in the law firm that employs them all. Brown thinks that she has probability 0.25 of winning the promotion and that Williams has probability 0.2. What probability does Brown assign to the outcome that Chavez is the one promoted?

17. Suppose that A and B are events that have no outcomes in common and thus cannot occur simultaneously. For example, in tossing three coins we could have $A = \{$first coin gives H$\}$ and $B = \{$first coin gives T$\}$. Starting from the definition of a probability model and the fact that the probability of any event is the sum of the probabilities of the outcomes making up the event, explain why

$$P(A \text{ or } B \text{ occurs}) = P(A) + P(B)$$

must always be true for two such events.

18. If a fair die is rolled once, is it reasonable to assign probability 1/6 to each of the six faces? If we accept this probability model, what is the probability of rolling a number less than 3?

19. In the rolling dice example on p. 189 we gave a probability model for rolling two fair dice and recording the two up faces that assigned equal probability to each of the 36 possible outcomes in Figure 7.4. Starting from this model, give a probability model for rolling two fair dice and recording the sum of the faces showing. Then use this model to answer the following questions.

    a. What is the probability of rolling a 7 or an 11?

    b. What is the probability of rolling a number 7 or greater?

20. Toss three coins and record heads or tails for each. Exercise 11 shows the eight members of the sample space. If the coins are fair, these outcomes are equally likely. Starting from this probability model, find the probability model for tossing three fair coins and counting the number of heads. What is the probability of at least two heads?

21. A computer assigns three-letter log-in identification codes at random, as in the code words example on p. 189. If we take the vowels to be *a, e, i, o, u,* and *y,* what is the probability that your code contains no vowels if repeated letters are allowed? If no repeats are allowed?

22. The computer of Exercise 21 is programmed to assign log-in codes of the form consonant-vowel-consonant. The consonants and vowels are both chosen at random and the consonants can repeat. What is the probability that your code does not contain an *x*?

23. Suppose that the computer of Exercise 21 assigns three-character log-in codes that may contain the digits 0 to 9 as well as letters, with repeats allowed. What is now the probability that your code contains no *x*? What is the probability that your code contains no digits?

24. Automobile license plate numbers in Indiana consist of seven characters. The first three describe the county in which the car is licensed, while the last four are digits assigned at random. You are hoping for a plate on which these four digits are identical (like 7777). What is your probability of receiving such a plate?

25. A monkey at a keyboard presses three keys and hits the letters *a, g,* and *s* in random order. How many possible three-letter "words" can the monkey type using only these letters? Which of these are meaningful English words? What is the probability that the word the monkey typed is meaningful?

26. You are about to visit a new neighbor. You know that the family has four children, but you do not know their age or sex. Write down all possible arrangements of girls and boys in order from youngest to oldest, such as BBGG (the two youngest are boys, the two oldest girls). The laws of genetics say that all of these arrangements are equally likely.

   a. What is the probability that the oldest child is a girl?

   b. What is the probability that the family has at least three boys?

   c. What is the probability that the family has at least three children of the same sex?

27. What is the expected number of pips observed in rolling a single fair die?

28. In Exercise 19, you found a probability model for rolling two fair dice and counting the pips on the two up faces. What is the expected number of pips obtained?

29. In Exercise 20 you found the probability model for tossing three fair coins and counting the heads observed. Compute the expected number of heads.

30. Teachers in the Lost Valley Central School District are allowed up to seven days of paid sick leave each year. Here is the distribution of the number of days of sick leave taken by the teachers last year. What is the expected number of days of sick leave that a teacher will take in a year?

Days taken	0	1	2	3	4	5	6	7
Percent of teachers	15	15	10	10	10	12	8	20

31. A study selected a sample of fifth-grade pupils and recorded how many years of school they eventually completed. Based on this study we can give the following probability model for the years of school that will be completed by a randomly chosen fifth grader:

Years	4	5	6	7	8	9	10	11	12
Probability	.010	.007	.007	.013	.032	.068	.070	.041	.752

    a. Verify that this is a legitimate probability model.

    b. What outcomes make up the event "the student completed at least one year of high school"? (High school begins with the ninth grade.) What is the probability of this event?

    c. What is the expected number of years of school completed?

32. In an experiment on the behavior of young children, each subject is placed in an area with five toys. The response of interest is the number of toys that the child plays with. Past experiments with many subjects have shown that the probability model for the number of toys played with is as follows:

Number of toys	0	1	2	3	4	5
Probability	.03	.16	.30	.23	.17	.11

    a. What is the probability that a child will play with more than one toy during the experiment?

    b. What is the expected number of toys a child will play with?

33. An American roulette wheel has 38 slots numbered 0, 00, and 1 to 36. The ball is equally likely to come to rest in any of these slots when the wheel is spun. The slot numbers are laid out on a board on which gamblers place their bets. One column of numbers on the board contains multiples of 3, that is, 3, 6, 9, . . . , 36. A gambler places a $1 column bet that pays out $3 if any of these numbers comes up.

    a. What is the probability of winning?

    b. What is the expected value of a play, taking into account the $1 cost of each play?

34. Another common casino game is keno, in which 20 numbers between 1 and 80 are chosen at random and gamblers attempt to guess some of the numbers in advance. As in roulette, a bewildering variety of keno bets are available. Here are some of the simpler keno bets. Give the expected winnings for each.

    a. A $1 bet on "Mark 1 number" pays $3 if the single number you mark is one of the 20 chosen; otherwise you lose your dollar.

b. A $1 bet on "Mark 2 numbers" pays $12 if both your numbers are among the 20 chosen. The probability of this is about 0.06. Is Mark 2 a more or a less favorable bet than Mark 1?

● **35.** Return to Exercises 10 and 11 of Chapter 5. Working as a team with other students, draw 100 simple random samples of size 5 from this population. Compute the sample proportion $\hat{p}$ of females in each sample. What probability model, based on your experiment, describes the sampling distribution of $\hat{p}$? Make a histogram of this distribution. Then find the expected number of females in a sample.

36. The following table contains the results of 100 repetitions of the drawing of a simple random sample of size 200 from a large lot of bearings, 10% of which do not conform to the specifications. The numbers in the table are the percents of nonconforming bearings in each sample of 200.

8.5	11.5	9	13.5	7.5	8.5	9	6.5	8	9
10	7.5	9	8	10.5	8.5	9	9.5	8	11.5
10	9	9	8.5	9.5	6.5	13.5	11	11.5	13
8.5	6.5	8	7	12	11	8	10.5	12	10.5
15	12	8.5	7	8	8	8.5	12	10.5	8
8.5	11.5	9	11.5	11	12	11.5	11.5	10	9.5
10	9	10	12.5	8	12	12	12	7.5	11
11	8	14	7.5	11	4.5	9.5	8	9.5	9.5
12.5	12	10	7.5	10.5	12.5	12	9.5	9.5	10
14	9	8.5	8.5	12.5	8.5	8.5	9	9.5	9

a. Make a histogram of these outcomes and describe the shape of the distribution. Is the center close to 10%?

b. Give an estimated sampling distribution for the sample proportion in this situation by recording each outcome and the proportion of trials on which it occurred. Then find the expected outcome from your distribution.

37. The distribution of heights of adult American men is approximately normal with mean 69 inches and standard deviation 2.5 inches. Draw a normal curve on which this mean and standard deviation are correctly located. (Hint: Draw the curve first, then mark the horizontal axis.)

38. Using the normal distribution described in Exercise 37, answer the following questions about the heights of adult American men.

a. What percent of men are taller than 74 inches?

b. Between what heights do the middle 95% of American men fall?

c. What percent of men are shorter than 66.5 inches?

● Optional exercise.

39. What are the quartiles of the distribution of heights of American men in Exercise 37?

40. The figure that follows is a probability distribution that is not symmetric. The mean and median do not coincide. Which of the points marked is the mean of the distribution, and which is the median?

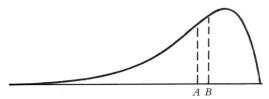

41. The concentration of the active ingredient in capsules of a prescription painkiller varies according to a normal distribution with mean 10% and standard deviation 0.2%.

a. What is the median concentration? Explain your answer.

b. What range of concentrations covers the middle 95% of all capsules?

c. What range covers the middle half of all capsules?

42. Answer the following questions for the painkiller in Exercise 41.

a. What percent of all capsules have a concentration of active ingredient higher than 10.4%?

b. What percent have a concentration higher than 10.6%?

43. Scores on the Wechsler Adult Intelligence Scale (a standard IQ test) for the 20 to 34 age group are approximately normally distributed with mean 110 and standard deviation 25.

a. About what percent of people in this age group have scores above 110?

b. About what percent have scores above 160?

44. The army reports that the distribution of head circumference among soldiers is approximately normal with mean 22.8 inches and standard deviation 1.1 inches.

a. What percent of soldiers have head circumference greater than 23.9 inches?

b. What percent of soldiers have head circumference between 21.7 inches and 23.9 inches?

45. The length of human pregnancies from conception to birth varies according to a distribution that is approximately normal with mean 266 days and standard deviation 16 days.

a. Between what values do the lengths of the middle 95% of all pregnancies fall?

b. How short are the shortest 2.5% of all pregnancies?

46. The *deciles* of a distribution are the points having 10% (lower decile) and 90% (upper decile) of the observations falling below them. The lower and upper deciles contain between them the central 80% of the data. The lower and upper deciles of any normal distribution are located 1.28 standard deviations on either side of the mean. What score is needed to place you in the top 10% of the distribution of SAT scores in the reference population (normal with mean 500 and standard deviation 100)?

47. Based on the information in Exercises 45 and 46, how short are the shortest 10% of human pregnancies?

48. A student repeats a chemistry laboratory measurement three times and uses the mean $\bar{x}$ in her lab report. When many students do this, the standard deviation of their $\bar{x}$'s is 10. How many times must the measurement be repeated to reduce the standard deviation of $\bar{x}$ to 5?

49. A student organization is planning to ask a sample of 50 students if they have noticed AIDS education brochures on campus. The sample percentage who say yes will be reported. Their statistical advisor says that the standard deviation of this percentage will be about 7%. What would the standard deviation be if the sample contained 100 students rather than 50?

50. If the laboratory measurements in Exercise 48 were repeated six times rather than three times, what would be the standard deviation of the mean result $\bar{x}$?

51. How large a sample is required in the setting of Exercise 49 to reduce the standard deviation of the percentage who say yes from 7% to 3.5%?

· 8 ·

Statistical Inference

Inference is the process of reaching conclusions from evidence. Evidence can come in many forms. In a murder trial, evidence might be presented by the testimony of a witness, by a record of telephone conversations, or by a weapon. Evidence can also be more subtle. For example, if we walk into an office filled with papers, journals, library books, class notes, and a computer terminal, we might infer that the office belongs to a college professor. In the case of statistical inference, the evidence is provided by data. Informal statistical evidence is often based on graphical presentation of data. Formal evidence, the topic of this chapter, requires a statement of probabilities.

Informal evidence is sometimes compelling. The gap in the histogram in Figure 6.5, for example, demands an investigation of the inspection process. But in many cases it is difficult to reach a firm conclusion from informal evidence. We saw from Figure 6.10 that the 1970 draft lottery appeared to favor men born early in the year. However, that inequity was rather small, so small that it is not clearly visible in the scatterplot of Figure 6.9. We might well ask whether the 1970 outcome was simply due to chance rather than to systematic bias in the lottery. After all, any lottery will show some deviation from perfect uniformity due to the play of chance.

The purpose of formal statistical inference is to verify appearance by calculation. Statistical inference can be compared to an engineer's calculation of the load on a beam—we are more confident after the mathematics is done than we are if the engineer merely says that the beam looks large enough. In the case of the 1970 draft lottery, calculation shows that in a truly random lottery a trend as strong as the one actually observed has probability less than

1 in 1000. This calculation of probability shows that the observed trend is strong evidence that the 1970 lottery was not random.

One of the most intriguing aspects of statistical inference is the fact that *chance* — which we usually associate with uncertainty — is the ally rather than the enemy of confident conclusions. At first glance, the opposite seems to be true. Suppose, for example, that the Gallup poll decided to take its weekly public opinion survey twice, separately and simultaneously selecting two random samples, sending out interviewers, and asking the same questions. Two random samples, each selecting 1500 of 185 million U.S. residents aged 18 and over, will contain different people. And these different people will hold somewhat different opinions. Thus Gallup's recent finding that 45% of Americans are afraid to go out at night for fear of crime really refers only to the 1500 people in one particular sample. If Gallup had taken a simultaneous second survey, no doubt the results would be different. Random sampling may eliminate *bias,* but it can't eliminate *variability.*

How can we trust the results of a random sample, knowing that a second sample may yield a different result? For that matter, how can we trust the results of a randomized experiment? As we saw in Chapter 5, the Physicians' Health Study tested the effects of aspirin and beta carotene on reducing heart disease and cancer. We know, however, that different people react differently to drugs. What is more, the people in the four treatment groups were assigned by chance. A second trial with other participants would distribute drug reactions and persons with a high risk of cancer and heart disease differently among the treatments. We must ask whether the conclusions drawn from this particular experiment are convincing or merely the result of chance.

This chapter will address the issue of confidence in statistical conclusions. Formal statistical inference enables us to quantify our confidence in the results of random samples and randomized experiments and thus to verify our impressions by calculation.

CONFIDENCE INTERVALS

We will use a simplified version of the Gallup crime survey to introduce an important type of statistical inference. Like most national sample surveys, the Gallup poll uses a complex multistage sampling design. Suppose that we instead drew a *simple random sample* of 1500 adults and discovered that 45% were afraid to go out at night because of crime. We will call a **sample proportion** that refers to the 1500 people in this particular sample $\hat{p}$ (read as "p hat"). In this case, $\hat{p} = 45\%$. What we really want to know is the *population proportion,* the percent (call it p) of all adult Americans who stay home at night for fear of crime. To discuss statistical inference intelligently, it is essential to keep straight which numbers describe the sample and which describe the population. A number such as p that describes a population is called a **parameter;** a number such as $\hat{p}$ that describes a sample is called a **statistic.**

It is easy to remember that **p**arameters belong to **p**opulations and **s**tatistics belong to **s**amples because the first letters agree. In an inference problem, parameters are usually unknown. We do not know, for example, the true proportion p of all adults who stay home at night for fear of crime. We use the statistic $\hat{p}$, which we know because we actually interviewed the sample, to estimate the unknown p. *Our goal is not simply to estimate p, but to say how accurate our estimate is.* How close to the unknown p will the estimate $\hat{p}$ usually fall?

To answer this question, we turn to the *sampling distribution* of $\hat{p}$. This is the distribution of values taken by the sample proportion as it varies from sample to sample in a large number of repeated samples. If the sample is relatively large, such as our poll's sample of 1500 people,

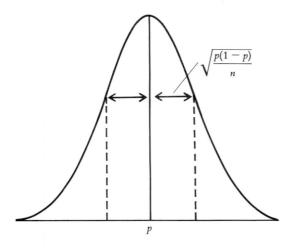

Figure 8.1 The sampling distribution of the sample proportion $\hat{p}$.

the sampling distribution will be very close to a normal curve like that in Figure 8.1. This figure illustrates several important facts about the sampling distribution. First, the mean of the curve is the true proportion p of people afraid to go out at night. This fact says that $\hat{p}$ has no bias or systematic error as an estimator of the unknown p. In repeated sampling our result will sometimes be high and sometimes low, but the long-run average result will be correct.

Being correct on the average is not enough; a good estimator must also be highly repeatable in the sense of giving nearly the same answer in repeated samples. Repeatability is described by the spread of the sampling distribution, as measured by its standard deviation. Here are the facts for the sample proportion $\hat{p}$:

The standard deviation of the sampling distribution of $\hat{p}$ depends on the population proportion p and the sample size n. It is

$$\text{Standard deviation of } \hat{p} = \sqrt{\frac{p(100 - p)}{n}}$$

(Throughout this chapter, p and $\hat{p}$ are measured in percent.)

If we repeated the sampling many times, sending out waves of interviewers across the nation, each time we would get a value of the sample proportion somewhere along the curve in Figure 8.1.

EXAMPLE: Sampling Distribution for the Crime Survey. Suppose that in fact 40% of all adults fear to go out at night because of crime. That is, $p = 40\%$. Take a simple random sample of size $n = 1500$ people. In repeated samples, the sample percent $\hat{p}$ will vary according to a normal distribution with

$$\text{Mean} = p = 40\%$$

$$\text{Standard deviation} = \sqrt{\frac{p(100 - p)}{n}}$$

$$= \sqrt{\frac{(40)(60)}{1500}}$$

$$= \sqrt{1.6} = 1.26\%$$

If, instead, the truth about the population is $p = 50\%$, the mean of the sampling distribution moves to 50% as well. The standard deviation changes to

$$\sqrt{\frac{p(100 - p)}{n}} = \sqrt{\frac{(50)(50)}{1500}}$$

$$= \sqrt{1.67} = 1.29\%$$

Notice that the standard deviation does not change very much when p changes. That is, when we take a sample of the same size from different populations, the center of the sampling distribution of $\hat{p}$ moves to the true p for each population, but the spread stays about the same. The size of the sample is the major influence on the spread. Suppose that we took a sample of only $n = 400$ instead of 1500 people from a population for which $p = 40\%$. The mean of the distribution of $\hat{p}$ is 40% — this fact is not affected by the sample size — but the standard deviation increases to

$$\sqrt{\frac{p(100-p)}{n}} = \sqrt{\frac{(40)(60)}{400}}$$
$$= \sqrt{6} = 2.45\%$$

Our poll of 1500 people, in fact, found that $\hat{p} = 45\%$. This is our best guess for the population percent p. How close to the true p is our guess likely to be? Well, $\hat{p}$ varies normally. The 95 part of the 68-95-99.7 rule says that $\hat{p}$ falls within two standard deviations of the true p (the mean of the sampling distribution) in 95% of all samples. So our guess based on this one sample is likely to be within two standard deviations, that is,

$$2\sqrt{\frac{p(100-p)}{1500}}$$

of the true p.

The catch is that the standard deviation depends on the unknown p. Fortunately, as the example demonstrates, the standard deviation changes only slowly as p changes, as long as p is not very close to either 0% or 100%. Because $\hat{p}$ is close to p, we simply substitute $\hat{p} = 45\%$ for the unknown p in the formula for the standard deviation. The standard deviation of a statistic is called its **standard error** when any unknown parameters are replaced by estimates based on the data.

EXAMPLE: Standard Error for the Crime Survey. We need to estimate the standard deviation of our observed sample proportion. The sample size was $n = 1500$, and for p we use the estimate $\hat{p} = 45\%$, based on our survey. The standard error is

$$\sqrt{\frac{(45)(55)}{1500}} = \sqrt{1.65} = 1.28\%$$

Here at last is our conclusion: in 95% of all samples, the sample proportion $\hat{p}$ will fall within 2×1.28, or about 2.6%, of the unknown population proportion p. We took one

sample and got $\hat{p} = 45\%$. So we conclude that the p lies in the interval

$$45\% \pm 2.6\%$$

or between 42.4% and 47.6%. We say that we are *95% confident* in this conclusion because we got the interval by calculating how close to p the sample proportion will lie in 95% of all samples. Our interval is a 95% *confidence interval* for estimating the unknown population proportion.

In mathematical terms, the probability is 0.95 that the sample proportion will fall within $\pm 2.6\%$ of the unknown true fraction of people in the total population afraid to go out at night because of crime. Figure 8.2 makes the idea clearer. The normal curve at the top of

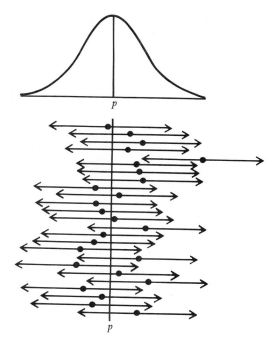

Figure 8.2 The behavior of confidence intervals in repeated sampling.

the figure is the sampling distribution of $\hat{p}$. As we take many samples, the actual values of $\hat{p}$ vary according to this distribution. The values of $\hat{p}$ observed in 25 samples appear as dots below the curve, together with the confidence intervals that extend out 2.6% on either side of the observed $\hat{p}$. The true population proportion p is marked by the vertical line. Although the intervals vary from sample to sample, all but one of these samples gave a confidence interval that covers the true p. To say that these are 95% confidence intervals is just to say that the interval covers the true p in 95% of all samples and misses in only 5%. Be sure you understand that this 95% and 5% refer to what would happen if we continued to take samples forever. In a small number of samples, the number of confidence intervals that fail to cover the true p may be a bit more or less than 5% of the samples. In Figure 8.2, for example, 1 out of 25, or 4%, of the confidence intervals fail to contain p.

A 95% **confidence interval** is an interval obtained from the sample data by a method that in 95% of all samples will produce an interval containing the true population parameter.

You can see in Figure 8.2 that a confidence interval from one particular sample can either hit or miss the unknown true parameter. We don't know whether our sample is one of the 95% that hit or one of the 5% that miss. To say that our interval 45% ± 2.6% is a 95% confidence interval means, "We got this interval by a method that catches the true parameter 95% of the time."

We have now accomplished two things: we have seen what "95% confidence" means, and we have actually found a 95% confidence interval for estimating a population proportion. Here, as a summary, is the recipe for this interval.

If a simple random sample of size n is drawn, then a 95% confidence interval for the population proportion p is

$$\hat{p} \pm 2\sqrt{\frac{\hat{p}(100 - \hat{p})}{n}}$$

Remember that both p and $\hat{p}$ are measured in percent. This recipe is only approximately correct, but is quite accurate when the sample size n is large.

EXAMPLE: Germination of Seeds. A simple random sample of 100 seeds from a new lot is tested for germination; 87 of the 100 germinate. The sample proportion that germinate is

$$\hat{p} = \frac{87}{100} = 0.87 = 87\%$$

The 95% confidence interval for estimating the proportion p of all seeds in the lot that will germinate is

$$\hat{p} \pm 2\sqrt{\frac{\hat{p}(100 - \hat{p})}{n}} = 87 \pm 2\sqrt{\frac{(87)(13)}{100}}$$
$$= 87 \pm 2\sqrt{11.31}$$
$$= 87\% \pm 6.7\%$$

We are 95% confident that between 80.3% and 93.7% of the entire lot of seeds will germinate.

Any confidence interval has two essential pieces: the interval itself and the confidence level. The interval usually has the form

Statistic ± margin of error

The statistic (such as $\hat{p}$) estimates the unknown parameter, and the margin of error indicates how accurate this estimate is. In the germination example, the margin of error is ±6.7%.

The confidence level states how confident we are that our interval contains the true parameter. Although 95% confidence is common, you can hold out for higher confidence, such as 99%, or be satisfied with lower confidence, such as 90%. Our 95% confidence interval was based on the middle 95% of a normal distribution. A 99% confidence interval requires the middle 99% of the distribution and so is wider (has a larger margin of error). Similarly, a 90% confidence interval is shorter than a 95% interval obtained from the same data. So there is a trade-off between how closely we can pin down the parameter (the margin of error) and how confident we can be in the result.

Understanding confidence intervals helps us read newspapers and listen to TV news broadcasts (see Spotlight 8.1, p. 216). Sample survey results are common in the news, often with a margin of error attached. The margin of error, together with the basic result of the survey, forms a confidence interval. A news report of our crime survey would say, "The survey found that 45% of all Americans are afraid to go out at night because of crime. The margin of error in the survey is plus or minus 2.6 percentage points." Although reputable sample surveys such as Gallup, Harris, or the Current Population Survey always announce a margin of error to help us interpret their results, editors often cut this information from their stories. It is also common for news reports to give the margin of error without the confidence level; we need to know both figures because higher confidence requires a larger margin of error. Almost all public opinion polls announce 95% confidence intervals. So if a story about an opinion poll gives a margin of error without a confidence level, you can usually assume 95%.

The Bureau of Labor Statistics, on the other hand, chooses to announce its unemployment-survey results at a 90% level of confidence. Basing its conclusions on the monthly Current Population Survey of 60,000 households, the Bureau states that the unemployment rate they announce is within ±0.2% (two-tenths of 1 percent) of the figure they would get if they counted everyone. So when the headlines announce a 7.9% unemploy-ment rate, the Bureau is saying — with 90% confidence — that between 7.7% and 8.1% of the labor force is out of work.

Opinion polls often have margins of error of about ±3%. The much smaller margin of error for the announced unemployment rate is due to the much larger sample interviewed by the Current Population Survey. Larger samples give smaller margins of error at the same confidence level. However, the square root of *n* that appears in the calculations shows that in order to reduce our margin of error by half, we need a sample size four times bigger. To obtain a very small margin of error, the Current Population Survey goes to the trouble to interview a sample of 60,000 people, compared with the Gallup poll's usual 1500. The Gallup poll can afford to be 3% off. The unemployment rate must be more exact because so many economic and political decisions depend upon it.

ESTIMATING A POPULATION MEAN

The statistician's tool kit contains many different confidence intervals, matching the many different population parameters that we may wish to estimate. We have met the confidence interval for estimating a population proportion *p*. Now we want to estimate a population mean. We have regularly used the **sample mean** $\bar{x}$ of a sample of observations to describe the center of a set of data. Now we will use the sample mean $\bar{x}$ to estimate the unknown mean of the entire population from which the sample is drawn. To distinguish the mean of a population from the sample mean $\bar{x}$ we use a new symbol. The mean of a population is denoted by μ, the Greek letter "mu." The sample mean $\bar{x}$ is a statistic, while the population mean μ is a parameter. Fortunately, the new confidence interval for estimating μ is quite similar to the familiar confidence inter-

SPOTLIGHT 8.1 How the Poll Was Taken

In June of 1989, the *New York Times* conducted a national opinion poll on women's issues. In response to one of the questions, 41% of the women interviewed agreed that "all things considered, there are more advantages in being a man in America today." Among men, 30% agreed. The poll estimated that 37% of all adults share this opinion. We know that these are sample results that are subject to a margin of error when used to draw conclusions about the population as a whole. Here is the *Times*'s statement about the conduct of the poll (from the August 21, 1989, edition).

How the Poll Was Taken

The New York Times Poll on women's issues is based on telephone interviews conducted June 20 through 25 with 1,497 adults around the United States, excluding Alaska and Hawaii.

The sample of telephone exchanges called was selected by a computer from a complete list of exchanges in the country. The exchanges were chosen so as to assure that each region of the country was represented in proportion to its population. For each exchange, the telephone numbers were formed by random digits, thus permitting access to both listed and unlisted numbers. The numbers were then screened to limit calls to residences.

Women were sampled at a higher rate than men so that there would be enough women interviewed to provide statistically reliable comparisons among various subgroups of women. The results of the interviews with 1,025 women and 472 men were then weighted to their correct proportions in the population.

Results were also weighted to take account of household size and number of residential telephone lines and to adjust for variations in the sample relating to region, race, age, and education.

---

val for estimating p, because both intervals are based on a normal sampling distribution.

EXAMPLE: Scholastic Aptitude Test Scores. How well would high school seniors do on the mathematics part of the Scholastic Aptitude Test (SAT)? Although about a million students take the SAT each year, these students are planning to attend college and are not representative of the entire population of high school seniors. We therefore select a simple random sample of 500 seniors and administer the mathematics SAT to them. Their average score is $\bar{x} = 451$. What can we say about the mean score μ for the entire popula-

tion if we want to be 95% confident in our conclusion?

We once again use a statistic — the mean $\bar{x}$ of our sample — to estimate an unknown parameter — the mean score μ for the entire population. To give a confidence interval, we need to know the sampling distribution of $\bar{x}$. The central limit theorem tells us that this distribution is close to normal. The mean of the sampling distribution is the same as the mean μ of the population from which we drew our sample. That is, the sample mean has no bias or systematic error as an estimator of the unknown μ. To find the standard deviation,

A group of 978 of these respondents were interviewed a second time from July 25 through 30, after the Supreme Court's decision allowing states more freedom to restrict abortion. Respondents in the second survey amounted to 79 percent of the 1,236 randomly selected people who were to be asked to participate, but 258 declined or were not reached despite several attempts.

In theory, in 19 cases out of 20 the results based on either of such samples will differ by no more than three percentage points in either direction from what would have been obtained by seeking out all American adults.

The percentages reported are the particular results most likely to match what would be obtained by seeking out all adult Americans. Other possible percentages are progressively less likely the more they differ from the reported results.

The potential sampling error for smaller subgroups is larger. For example, for men it is plus or minus five percentage points in both the first and second surveys. For women it is plus or minus three percentage points in the first survey and plus or minus four percentage points in the second survey. For women aged 18 to 29 in the first survey, it is plus or minus six percentage points.

In addition to sampling error, the practical difficulties of conducting any survey of public opinion may introduce other sources of error into the poll.

The methods and margins of error described by the *Times* are typical of national opinion polls. Do note the mention of "practical difficulties" and "other sources of error" at the end of the description. The poll was no doubt unable to contact some members of the sample, even in repeated calls. Some of the respondents may have given a socially acceptable answer rather than express their true opinion. Households without telephones could not be included in the survey. These practical difficulties do lead to additional errors, and these errors are *not* included in the announced margin of error.

we need to know something about the population. SAT scores for any large population follow a distribution that is close to normal. Moreover, the tests are arranged so that the standard deviation for the population used to develop the tests is 100. We will use σ, the Greek letter "sigma," for the standard deviation of a population. For SAT scores, the standard deviation σ may vary a bit among different populations of students. We will simplify our work by assuming that we know that $\sigma = 100$ for the population we are interested in.

Recall from Chapter 7 that the sampling distribution of the sample mean $\bar{x}$ has a standard deviation that decreases with the square root $\sqrt{n}$ of the sample size. Because individual SAT scores have a distribution that is close to normal, we don't even have to call on the central limit theorem. It's a fact that if the distribution of the population is normal, then the sampling distribution of $\bar{x}$ is also normal, no matter how small the sample is.

Suppose that a population is described by a normal distribution with mean μ and standard deviation σ. Draw a simple random sample of size n from this population. Then the sampling distribution of the sample mean $\bar{x}$ is normal with mean μ and standard deviation $\sigma/\sqrt{n}$.

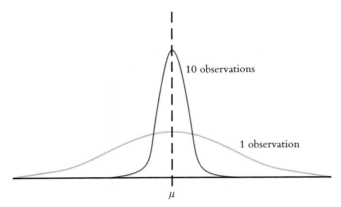

Figure 8.3 The sampling distribution of the sample mean $\bar{x}$ compared with the distribution of a single observation.

Figure 8.3 shows the relation between the distribution of a single observation drawn from the population and the distribution of the mean of several (in this case 10) observations. The mean of several observations is less variable than individual observations. Now we have in hand the facts we need to give a confidence interval for the mean SAT mathematics score for all high school seniors based on our sample of 500 students from this population.

EXAMPLE: Estimating the Mean SAT Score. The normal sampling distribution of $\bar{x}$ has mean equal to the unknown population mean μ. The standard deviation of the sampling distribution is

$$\frac{\sigma}{\sqrt{n}} = \frac{100}{\sqrt{500}} = 4.47$$

By the 95 part of the 68-95-99.7 rule, $\bar{x}$ will fall within two standard deviations of μ in 95% of all samples. Two standard deviations is 2×4.47, or about 9 points. We observed $\bar{x} = 451$

in our sample. So we are 95% confident that the population mean μ lies in the interval

$$451 \pm 9$$

or between 442 and 460.

As in the case of the confidence interval for a population proportion p, we can give a recipe that summarizes our development. The confidence interval again has the form

$$\text{Statistic} \pm \text{margin of error}$$

where the statistic is now the sample mean $\bar{x}$.

Suppose that a population is described by a normal distribution with unknown mean μ and known standard deviation σ. Draw a simple random sample of size n from this population and calculate the sample mean $\bar{x}$. A 95% confidence interval for the population mean μ is

$$\bar{x} \pm 2\frac{\sigma}{\sqrt{n}}$$

Because of the central limit theorem, this recipe is also approximately correct when the population is not described by a normal distribution, if a large sample is drawn. Often in practice the standard deviation σ of the population is not known in advance. Then we must estimate σ from the sample data. We will not discuss how to do this. Here is another example of estimating a population mean.

EXAMPLE: Estimating Dust in Coal Mines. Because the mean of several observations is less variable than a single observation, it is good practice to take the average of several observations when accuracy is important. The amount of dust in the atmosphere of coal mines is measured by exposing a filter in the mine and then weighing the dust collected by the filter. The weighing is not perfectly precise; in fact, repeated weighings of the same filter will vary according to a normal distribution. The values that would be obtained in many weighings form the population we are interested in. The mean μ of this population is the true weight (i.e., there is no bias in the weighing). The population standard deviation describes the precision of the weighing; it is known to be $\sigma = 0.08$ milligram (mg). Each filter is weighed three times and the mean weight is reported.

For one filter the three weights are

$$123.1 \text{ mg} \qquad 122.5 \text{ mg} \qquad 123.7 \text{ mg}$$

What is the 95% confidence interval for the true weight μ? First compute the sample mean

$$\bar{x} = \frac{123.1 + 122.5 + 123.7}{3}$$

$$= \frac{369.3}{3} = 123.1 \text{ mg}$$

Then the 95% confidence interval is

$$\bar{x} \pm 2\frac{\sigma}{\sqrt{n}} = 123.1 \pm (2)\left(\frac{0.08}{\sqrt{3}}\right)$$

$$= 123.1 \pm (2)(0.046)$$

$$= 123.1 \pm 0.09$$

or between 123.01 and 123.19 mg.

STATISTICAL PROCESS CONTROL

Statistical methods are widely used to gather social and economic information and in research on a wide variety of subjects. Most of our examples to this point, such as the Current Population Survey and the Physicians' Health Study, have illustrated these two types of application of statistics. Statistics is also heavily involved in the drive to improve the quality of manufactured products. Along with new technology such as robots and new management emphases such as cooperating with workers and suppliers, statistical ideas are an important part of any manufacturer's efforts to compete in the worldwide marketplace. In this section we will look at one simple but important statistical tool for monitoring and improving quality, the control chart.

At its Oklahoma City plant, AT&T manufactures the computerized electronic switches that interconnect our telephones. These switches are largely composed of complex electronic elements called circuit packs. AT&T needs efficient methods to check newly manufactured circuit packs for defects. The best method is to prevent defects by monitoring the manufacturing process to catch problems early rather than to wait to inspect the product and fix defective circuit packs later (see Spotlight 8.2, p. 220).

One important step in manufacturing a circuit pack is the soldering of the 2000 electrical connections that attach components to the printed wiring board. All 2000 connections are soldered at once as a conveyor carries the circuit pack through a wave of hot liquid solder. This wave-soldering operation is delicate. If the speed of the conveyor, the temperature of the solder, or other variables are not quite right, bad connections will appear both in the circuit pack and in our telephone conversations. AT&T therefore monitors the performance of the wave-soldering machine constantly and takes immediate action if something goes wrong.

To accomplish this goal, workers take a sample of five newly soldered circuit boards every hour and inspect them carefully for defects. Once a board is checked, the worker calculates a number that expresses the quality of soldering for that pack. A score of 100 represents the standard of quality that AT&T believes the process should attain. Lower scores

SPOTLIGHT 8.2 Check the Process before the Product

Process control engineer Connie Moore of AT&T.

At its Oklahoma City plant, AT&T uses statistical process control to ensure smooth production of circuit packs.

Connie Moore, a process control engineer at AT&T, gives her views on ensuring product quality.

> If we looked at 100% of the product it would take more time, more people, and would not give us any better information about the process. There was a time when industry thought that a quality control department's function was to inspect quality at the end of the line. Now we know that the only reasonable philosophy is to build it right the first time.
>
> I've been told that unless you measure how you're doing as you go along, you'll never know if you're done or if you succeeded. That's why I think that statistics and people are such an important combination. Statistics is the tool that tells us how we're doing as we go along and people are the force that drives us until we've succeeded.

represent poorer quality, while higher scores mean that the quality is higher than the target. The sample mean of the five quality scores is plotted on a **process control chart.** This point is a sample estimate of the quality of that hour's production.

There will always be some chance variation in the mean quality scores over time. Any industrial process will produce some variability. Constant fiddling with the process in response to small variations is unnecessary and wasteful. The purpose of the control chart is to help us distinguish the usual natural variation in the process from the added variation that indicates that the process has been disturbed. When unusual variation is spotted, we look for a specific cause. The disturbance may be caused by a new operator who hasn't been properly trained or by a malfunction in the machine.

Just plotting the mean quality scores against time can be helpful. We can see if there is a trend up or down, for example. Or we can see whether the last point plotted falls outside the pattern of the earlier points and so suggests that something has gone wrong. Figure 8.4 is a plot against time for the mean quality scores for 20 hourly samples. The graph appears to show that the level of quality dropped shortly after sample number 10. The horizontal center line at the target value 100 helps us see the trend. Once again, we would like to confirm this appearance by calculation. Adding the result of a simple calculation turns the plot against time into a control chart.

When the wave-soldering machine is performing its task properly, the quality index for individual circuit packs will vary according to a normal distribution. Let's imagine that we know from long experience that the mean of this distribution should be 100 and the standard deviation 4. We are plotting the mean $\bar{x}$ of five observations. What range of variation do we expect to see in the values of $\bar{x}$? We know that the distribution of $\bar{x}$ in many samples is itself normal, with mean 100 and standard deviation

$$\frac{\sigma}{\sqrt{n}} = \frac{4}{\sqrt{5}} = 1.79$$

By the 95 part of the 68-95-99.7 rule, 95% of all values will fall within two standard deviations of the mean, that is, between

$$100 - (2)(1.79) = 96.42$$

and

$$100 + (2)(1.79) = 103.58$$

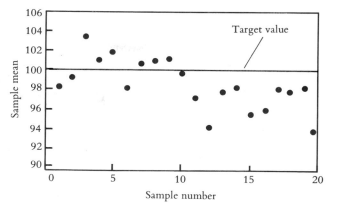

Figure 8.4 Plot of sample mean quality scores against time.

As we draw many samples, only 5% of the values of $\bar{x}$ will fall outside this range if the process is operating undisturbed. In particular, only 2.5% (half of 5%) of all samples will give an $\bar{x}$ less than 96.42. A mean quality score this low is good evidence that something has gone wrong with the wave-soldering process.

Figure 8.5 is the control chart for the observations from Figure 8.4. The dashed line is the control limit 96.42 that indicates when action should be taken. The control chart shows convincingly that the quality of the process has deteriorated. Not only are the means for samples 12, 15, 16, and 20 below the control limit,

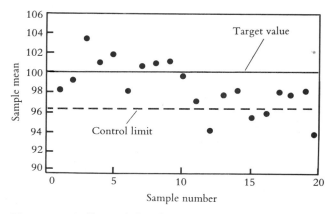

Figure 8.5 An $\bar{x}$ control chart for soldering quality.

but the last 11 means are all below the center line. In the long run, only half these means should be below the center line if the process mean is really 100. It appears that the process quality shifted downward at about sample 9 or 10. In practice, the first out-of-control point at sample 12 would trigger an investigation to find and correct the cause of this trend. Here is a summary of the steps in constructing a control chart such as Figure 8.5.

> Suppose that a process follows a normal distribution with mean μ and standard deviation σ when operating undisturbed. To monitor the process, take samples of size n at regular intervals. An $\bar{x}$ *control chart* for the process is a plot of the sample means against time with a solid *center line* at μ and dashed *control limits* at $\mu - 2\sigma/\sqrt{n}$ and $\mu + 2\sigma/\sqrt{n}$.

There are many variations on the control chart idea. Although most control charts have both upper and lower control limits lying at equal distances above and below the center line, only the lower limit is drawn in Figure 8.5. This is because only decreases in the quality index concern us in this example. Our control limits are placed two standard deviations out from the center line. It is more common in industry to place the control limits three standard deviations out in order to minimize the number of false alarms. Such limits contain 99.7% of all values of $\bar{x}$ if the process has not been disturbed. It is also common to keep control charts for statistics other than the sample mean $\bar{x}$.

More important than these details are the statistical ideas that are the basis for control charts. First, our goal is to distinguish expected from unexpected variation. Second, we use the normal sampling distribution of $\bar{x}$ and the 68-95-99.7 rule to specify the range of expected variation. Third, we combine this formal inference with a graph of the data that can be used in the factory by people with little statistical training.

Why sample only five circuit boards each hour? The purpose of statistical process control is not to check the function of the circuit packs; they will be rigorously tested when completed. The goal is rather to monitor the soldering process and correct any malfunctions quickly. It is not practical to check every circuit pack at every stage of manufacture. Instead, statistical sampling techniques give a quick and economical way to keep the process running smoothly. Control charts based on samples not only help to keep the quality of the final product at a high level but also keep down costs by catching malfunctions quickly, allowing a faulty process to be corrected immediately. This eliminates the need to repair or scrap parts at the end of the assembly line.

EXAMPLE: Control Chart for a Machining Operation. An important operation in producing cast aluminum aircraft frame parts is the machining of the raw casting. When the machining process is operating in control, a critical dimension of the parts varies according to a normal distribution with mean $\mu = 1.50$ centimeters (cm) and standard deviation $\sigma = 0.20$ cm. A sample of four parts is measured each hour in order to keep an $\bar{x}$ control chart. Here are the data for the past 20 hours.

Hour	1	2	3	4	5
$\bar{x}$	1.60	1.54	1.31	1.45	1.40

Hour	6	7	8	9	10
$\bar{x}$	1.61	1.47	1.45	1.45	1.52

Hour	11	12	13	14	15
$\bar{x}$	1.49	1.66	1.60	1.57	1.73

Hour	16	17	18	19	20
$\bar{x}$	1.68	1.58	1.60	1.57	1.73

The control chart appears in Figure 8.6. The center line is at $\mu = 1.50$. The control limits are

$$\mu \pm 2\frac{\sigma}{\sqrt{n}} = 1.50 \pm (2)\left(\frac{0.20}{\sqrt{4}}\right)$$
$$= 1.50 \pm 0.20$$
$$= 1.3 \text{ and } 1.7$$

The points for hours 15 and 20 are outside the control limits. Moreover, the last nine points all lie above the center line. This is very unlikely to occur if the mean remains at 1.5, so it is additional evidence that some outside cause has disturbed the process. A common criterion is to look for a disturbing cause when eight straight points fall on the same side of the center line. This criterion would lead to action at hour 19. In this case, the point out of control at hour 15 would already have called for action.

THE PERILS OF DATA ANALYSIS

Statistical designs for collecting data may, like the Physicians' Health Study, involve experimentation. Or they may use sampling procedures such as those used in the Current Population Survey and also for process control. In both cases, we rely on randomization and the mathematics of probability to compute sampling distributions. From a sampling distribution we can obtain results that have known levels of confidence.

However, formal statistical inference, as reflected in levels of confidence, is secondary to well-designed data collection and to insight into the behavior of data. Inference is often unable to correct basic flaws in the data, such as the use of voluntary response samples. Moreover, the effects of *hidden variables* can make even an apparently clear inference misleading. We saw in Chapter 5 that a well-designed ex-

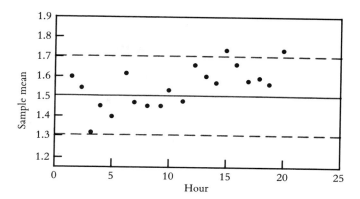

Figure 8.6 An $\bar{x}$ control chart for a machining operation.

periment can control for the effects of hidden variables. However, if an experiment is not possible, we may need to do some statistical detective work. Let's look at an example. Although the example is imaginary, it is based on a study of admissions to graduate programs at the University of California, Berkeley.

EXAMPLE: Admissions Discrimination?
Metro University has several limited-enrollment courses that admit only some of the students who apply. There are complaints about sexual discrimination in the admissions process. These complaints seem to be based on clear numerical evidence. Of the 80 men who applied to limited-enrollment courses at Metro, 35 were admitted. The percent of male applicants admitted was

$$\frac{35}{80} = 0.44 = 44\%$$

On the other hand, only 20 of the 60 women who applied were accepted. The success rate among women was

$$\frac{20}{60} = 0.33 = 33\%$$

Almost half the men, but only one-third of the women, were admitted. A probability calculation shows that this difference is much larger than could reasonably be expected to occur simply by chance. That is, formal inference backs up the appearance that a systematically higher percent of men are being admitted. Are men being favored over women?

The data in the example can be displayed in a **two-way table** (see Table 8.1). When variables simply place subjects into categories, such as male-female or admit-deny, we cannot draw a scatterplot to display the relationship between them. Instead we display the counts and percents in a two-way table and see the relationship by comparing percents. Comparing the percents of successful male and female applicants (44% and 33%) shows that men are more likely to be admitted. Metro is concerned. A closer look at admissions data is called for.

TABLE 8.1 Metro admissions

	Male		Female	
	Count	Percent	Count	Percent
Admit	35	44	20	33
Deny	45	56	40	67
Total	80	100	60	100

Investigation discloses that men and women tend to apply for different courses. The organic chemistry course is very rigorous. Of the 40 women who applied to it, only 10 got in, or one-fourth. Of the 20 men who also applied, 5 were admitted—again, one-fourth. These figures do not seem to point to discrimination.

The only other limited course offered at Metro is the history and sociology of the TV sitcom. Compared with organic chemistry, this course is something of a soft option; 60 men applied, with 30 admitted, whereas 10 out of 20 female applicants were accepted.

Once again, we find no apparent discrimination because 50% of both groups were admitted. We can display the data in a pair of two-way tables, Tables 8.2 and 8.3.

TABLE 8.2 Organic chemistry

	Male		Female	
	Count	Percent	Count	Percent
Admit	5	25	10	25
Deny	15	75	30	75

TABLE 8.3 TV sitcom

	Male		Female	
	Count	Percent	Count	Percent
Admit	30	50	10	50
Deny	30	50	10	50

When added together, these tables give exactly the single two-way table for overall admissions, Table 8.1. The numbers are remarkable. Each program appears to make no distinction between men and women. Yet when we consider the totals only, we see that 44% of the men who applied were admitted, compared with 33% of the women. This happened because two-thirds of the women applied to the course that is harder to get into, whereas three-quarters of the men signed up for the easier sitcom class. The hidden variable explains the apparent inequity.

This example provides a warning about statistical evidence, especially when an experiment was not done. Without walking through our hypothetical example, we would not suspect that data showing equality in each of several cases can appear as evidence of inequality when the cases are lumped together. First appearances can be deceiving, even in statistics. Statistical evidence not based on experiments can often show that an effect is present — more men than women get into limited courses at Metro — but does not show *why* the effect is present. At Metro, a hidden variable was responsible. In another case, an investigation might reveal discrimination. At a time when statistical evidence of all kinds is increasingly being used to formulate social policy and resolve legal disputes, it is crucial that we select, analyze, and interpret our data with great care.

Even if carefully collected data are properly analyzed, correct conclusions are not absolutely guaranteed. There is always some chance, however small, that random selection will lead to a false conclusion. The strength of statistical inference is that the chance of a false conclusion is known and can be controlled by setting the confidence level as high as we think necessary.

Statistics does not produce proof. But in a world where proof is always wanting and most evidence is uncertain, statistical evidence is often the best evidence available.

REVIEW VOCABULARY

Confidence interval An interval computed from a sample by a method that has a known probability of producing an interval containing the unknown parameter. This probability is called the *confidence level*. Confidence intervals usually have the form

Statistic ± margin of error

Parameter A number that describes the population. In statistical inference, the goal is often to estimate an unknown parameter or make a decision about its value.

Process control chart A graph showing the value of a statistic for successive samples (for example, one sample each hour or one sample each shift). The graph also contains a *center line* at the target value for the process parameter and *control limits* that the statistic will rarely fall outside of unless the process drifts away from the target. The purpose of a control chart is to monitor a process over time and signal when some unusual source of variation interferes with the process.

Sample mean The mean (arithmetic average) $\bar{x}$ of the observations in a sample. The sample mean from a simple random sample is used to estimate the unknown mean μ of the population from which the sample was drawn.

Sample proportion The proportion $\hat{p}$ of the members of a sample having some characteristic (such as agreeing with an opinion poll question). The sample proportion from a simple random sample is used to estimate the corresponding proportion p in the population from which the sample was drawn.

Standard error The standard deviation of a statistic with any unknown parameters appearing in the standard deviation formula replaced by estimates based on the sample data.

Statistic A number that describes a sample. A statistic can be calculated from the sample data alone and does not involve any unknown parameters of the population.

Two-way table A table showing the frequencies (counts) or percentages of outcomes that are classified according to two variables (such as applicants classified by both sex and admission decision).

SUGGESTED READINGS

Freedman, David, Robert Pisani, and Roger Purves: *Statistics,* Norton, New York, 1978. Chapters 21 and 23 discuss sampling distributions and confidence intervals for proportions and means.

Moore, David S., and George P. McCabe: *Introduction to the Practice of Statistics,* Freeman, New York, 1989. Chapter 6 of this text discusses sampling distributions and control charts. Chapter 7 is devoted to the reasoning of formal inference, including confidence intervals. Later chapters present confidence intervals for use in many specific situations.

EXERCISES

Identify each of the boldface numbers in Exercises 1 to 3 as either a parameter or a statistic.

1. A random sample of female college students has a mean height of **64.5** inches, which is greater than the **63**-inch mean height of all adult American women.

2. A sample of students of high academic ability under 13 years of age was given the SAT mathematics examination, which is usually taken by high school seniors. The mean score for the females in the sample was **386,** whereas the mean score of the males was **416.**

3. About **4%** of all U.S. households are without telephones, but another **20%** have unlisted numbers.

4. Exercises 1 to 3 in Chapter 5 describe three samples. Find the sample proportion $\hat{p}$ in each case, first as a decimal fraction and then as a percent.

Each of the statistics described in Exercises 5 to 8 has a normal sampling distribution (at least approximately). Give the mean and standard deviation of the distribution in each case.

5. About 35% of residential telephones in the San Francisco area have unlisted numbers. A telephone sales organization uses random-digit dialing to dial a random sample of 200 residential telephone numbers. The percent of these that are unlisted is the statistic of interest.

6. A shipment of machined parts has a critical dimension that is normally distributed with mean 12 centimeters and standard deviation 0.01 centimeter. The acceptance sampling team measures a random sample of 25 of these parts; the sample mean $\bar{x}$ of the critical dimension for these parts is the statistic of interest.

7. The Acculturation Rating Scale for Mexican Americans (ARSMA) is a psychological test that evaluates the degree to which Mexican Americans have adapted to Anglo/English culture. The scores in a large population are normally distributed with mean 3.0

and standard deviation 0.8. A researcher gives the test to a random sample of 12 Mexican Americans. Their average score is the statistic of interest.

8. In a midwestern state, 84% of the households have Christmas trees at holiday time. A sample survey asks a random sample of 400 households, "Did you have a Christmas tree this year?" The percent who say yes is the statistic of interest.

9. The standard deviation of a sample proportion $\hat{p}$ varies with the true value of the population proportion p. Fortunately, it does not vary greatly unless p is near 0% or 100%. Suppose that the size of the sample is $n = 1500$. Evaluate the standard deviation of $\hat{p}$ for $p = 30\%$, 40%, 50%, 60%, and 70%.

10. The report of a sample survey of 1500 adults says, "With 95% confidence, between 27% and 33% of all American adults believe that drugs are the most serious problem facing our nation's public schools." Explain to someone who knows no statistics what the phrase "95% confidence" means in this report.

11. A Gallup poll of a random sample of 1540 adults asked, "Do you happen to jog?" Fifteen percent answered yes. The news item stated that these results have a 3% margin of error. Explain carefully, to someone who knows no statistics, what is meant by a "3% margin of error."

12. Suppose that the poll in Exercise 11 had used a simple random sample of size 1540, of whom 15% answered yes. Give a 95% confidence interval for the percent of all adults who would have answered yes if asked.

13. The Forest Service is considering additional restrictions on the number of vehicles allowed to enter Yellowstone National Park. To assess public reaction, the Service asks a simple random sample of 150 visitors if they favor the proposal. Of these, 89 say yes. Give a 95% confidence interval for the proportion of all visitors to Yellowstone who favor the restrictions. Are you 95% confident that more than half are in favor? Explain your answer.

14. A simple random sample of students at Upper Wabash Tech is asked whether they favor limiting enrollment in crowded majors as a way of keeping the quality of instruction high. The student government suspects that the plan will be unpopular among freshmen, who have not yet been admitted to a major. Here are the responses for freshmen and seniors:

	Favor	Oppose
Freshman	40	160
Seniors	80	20

a. Give a 95% confidence interval for the percent of all freshman who support the plan.

b. Give a 95% confidence interval for the percent of all seniors who support the plan.

15. A Gallup poll of 1514 adults taken between July 30 and August 2 of 1983 asked, "Do you approve of the way Ronald Reagan is handling his job as President?" Of these, 41% said yes.

    a. If the poll had used a simple random sample, what would have been the margin of error in a 95% confidence interval?

    b. The actual margin of error for a Gallup poll of this same size is ±3%. Why does this not agree with your result in part **a**?

16. In August 1983, *Organic Gardening* magazine reported the results of a test to see whether mayonnaise jars would break when used for home canning. Here is their conclusion:

> The mayonnaise jars didn't do badly—only 3 out of 100 broke. Statistically this means you'd expect between 0% and 6.4% to break.

Verify this statistical statement by giving a 95% confidence interval.

Exercises 17 to 20 are based on the following situation. A news report says that a national opinion poll of 1500 randomly selected adults found that 43% thought they would be worse off during the next year. The news report went on to say that the margin of error in the poll result is ±3 percentage points with 95% confidence.

17. Which of the following sources of error are included in the poll's margin of error?

    a. The poll dialed telephone numbers at random and so missed all people without phones.

    b. The poll could not contact some people whose numbers were chosen.

    c. There is chance variation in the random selection of telephone numbers.

18. Would a 90% confidence interval based on the poll results have a margin of error less than, equal to, or greater than ±3 percentage points? Explain your answer.

19. If the poll had interviewed 1000 persons rather than 1500 (and still found 43% believing they would be worse off), would the margin of error for 95% confidence be less than, equal to, or greater than ±3 percentage points? Explain your answer.

20. Suppose that the poll had obtained the outcome 43% by a similar random sampling method from all adults in New York State (population 18 million) instead of from all adults in the United States (population 250 million). Would the margin of error for 95% confidence in New York be less than, equal to, or greater than ±3 percentage points? Explain your answer.

21. In the text we used the sampling distribution of $\hat{p}$ and the 68-95-99.7 rule to give a 95% confidence interval for a population proportion p.

    a. Explain carefully why

$$\hat{p} \pm \sqrt{\frac{\hat{p}(100 - \hat{p})}{n}}$$

is a 68% confidence interval for p.

    b. Give the recipe for a 99.7% confidence interval for p.

22. Use the result of Exercise 21**b** and the data in Exercise 11 to give a 99.7% confidence interval for the percent of all adults who jog. Compare the width of the 99.7% interval with that of the 95% confidence interval from Exercise 12. What is the reason for the difference in widths?

23. Use the result of Exercise 21**a** and the data in Exercise 13 to give a 68% confidence interval for the percent of visitors to Yellowstone who support restricting the number of vehicles allowed in the park. Compare the width of the 68% interval with that of the 95% interval from Exercise 13 and explain the difference in plain language.

24. Electrical pin connectors for use in computers are gold-plated for better conductivity. The specified plating thickness is 0.001 inch. Due to variations in the plating process, the actual plating thickness on different pins has a normal distribution with mean 0.001 inch and standard deviation 0.0001 inch.

    a. What range of plating thickness contains 95% of all pins?

    b. Quality control samples of four pins are taken regularly during production. The plating thickness is measured and the sample mean of the four measurements is recorded on a control chart. What range of plating thickness contains 95% of the recorded sample means?

25. Scores on the American College Testing (ACT) college admissions examination for the reference population used to develop the test vary normally with mean $\mu = 18$ and standard deviation $\sigma = 6$. The range of reported scores is 1 to 36.

    a. What range of scores contains the middle 95% of all students in the reference population?

    b. If the ACT scores of 25 randomly selected students are averaged, what range contains the middle 95% of the averages $\bar{x}$?

26. Errors in careful measurements often have a distribution that is close to normal. Experience shows that the error in a certain surveying method varies when a measurement is repeated according to a normal distribution with mean 0 (that is, the procedure does not systematically overestimate or underestimate the true distance) and standard deviation 0.03 meter. Each measurement is repeated three times, and the mean $\bar{x}$ of the three measurements is used as the final value.

    a. What is the distribution of the mean error $\bar{x}$ when this surveying method is used to measure many distances?

    b. Between what values do 95% of the errors fall?

27. A laboratory scale is known to have a standard deviation of $\sigma = 0.001$ gram in repeated weighings. Suppose that scale readings in repeated weighings are normally distributed, with mean equal to the true weight of the specimen. Three weighings of a specimen give (in grams)

$$3.412 \quad 3.414 \quad 3.415$$

Give a 95% confidence interval for the true weight of the specimen.

28. An instrument in a chemistry laboratory measures the concentration of trace substances in specimens. When the instrument makes repeated measurements on the same specimen, the readings are known to vary normally with standard deviation

$\sigma = 0.03$. It is customary to make three readings and use the sample mean as the final result. For a particular specimen, the readings are

$$53.12 \qquad 53.08 \qquad 53.17$$

Give a 95% confidence interval for the mean of the distribution of readings for this specimen. (This mean is the true concentration if the instrument has no bias.)

29. The Family Adaptability and Cohesion Evaluation Scales (FACES) is a psychological test that measures two different aspects of family behavior. One of these is *cohesion,* which is the degree to which family members are emotionally connected to each other. Suppose it is known that the cohesion scores for adults vary normally with standard deviation $\sigma = 5$. A researcher administers FACES to a sample of 33 adults in families with a runaway teenager. The mean cohesion score in this sample is $\bar{x} = 36.9$. Give a 95% confidence interval for the mean cohesion (as rated by an adult) of families with runaway teenagers.

30. A milk processor monitors the number of bacteria per milliliter in raw milk received for processing. A random sample of ten 1-milliliter specimens from milk supplied by one producer gives the following data:

5370   4890   5100   4500   5260   5150   4900   4760   4700   4870

Suppose it is known that the bacteria count varies normally and that the standard deviation is $\sigma = 265$ per milliliter. Give a 95% confidence interval for the mean bacteria count per milliliter in this producer's milk.

31. An automatic lathe machines shafts to specified diameters as part of a manufacturing operation. Due to small variations in the operation of the lathe, the actual diameters produced follow a normal distribution with standard deviation 0.0005 inch. The shafts now being produced are supposed to have a diameter of 0.75 inch. You measure 10 such shafts and find that they have a sample diameter of $\bar{x} = 0.7505$. Give a 95% confidence interval for the true mean diameter μ of the shafts being produced.

● **32.** The upper and lower deciles of any normal distribution are located 1.28 standard deviations above and below the mean.

    a. Use this information to give a recipe for an 80% confidence interval for a population proportion p based on the sample proportion $\hat{p}$ that is accurate for large sample sizes n.

    b. Give an 80% confidence interval for the proportion of visitors to Yellowstone favoring vehicle restrictions, using the data in Exercise 13.

● **33.** The upper and lower deciles of any normal distribution are located 1.28 standard deviations above and below the mean.

    a. Use this information to give a recipe for an 80% confidence interval for the mean μ of a normal population based on the sample mean $\bar{x}$ of a simple random sample of size n.

    b. Give an 80% confidence interval for the mean FACES cohesion score in Exercise 29.

● Optional exercise.

34. Give the center line and control limits for a control chart for means $\bar{x}$ of samples of size 4 in the gold-plating process described in Exercise 24. Use 2σ limits, as in the machining operation example.

35. It is common for laboratories to keep a control chart for a measurement process based on regular measurements of a standard specimen. Suppose that you are maintaining a control chart for the laboratory scale in Exercise 27 by weighing a 5-gram standard weight three times at regular intervals. What should be the center line of your chart? What are the control limits if you decide to use 3σ limits?

36. The laboratory instrument of Exercise 28 is monitored by measuring a specimen with known concentration 50 each morning. The specimen is measured three times and an $\bar{x}$ control chart is kept. What should be the center line and 2σ control limits for this chart?

Use 3σ limits in the control charts of Exercises 37 to 39. That is, use control limits that are three standard deviations on either side of the mean μ. Also look for runs of eight or more consecutive observations on the same side of the center line. Notice that the center line and control limits are the same for all three charts.

37. In the data set below are $\bar{x}$'s from samples of size 4 with $\mu = 101.5$ and $\sigma = 0.2$. Only random variation is present. Make an $\bar{x}$ chart of these data, using the given μ and σ. Are any points out of control?

Sample	$\bar{x}$
1	101.627
2	101.613
3	101.493
4	101.602
5	101.360
6	101.374
7	101.592
8	101.458
9	101.552
10	101.463
11	101.383
12	101.715
13	101.485
14	101.509
15	101.429
16	101.477
17	101.570
18	101.623
19	101.472
20	101.531

38. The following set of $\bar{x}$'s for samples of size 4 illustrates the effect of a shift in the standard deviation. The first 10 samples have $\mu = 101.5$ and $\sigma = 0.2$, whereas the last 10 have $\mu = 101.5$ and $\sigma = 0.3$. Make a control chart for these data, using the given μ and the original σ. Are any points out of control? Is the increase in σ visible in any way on the chart?

Sample	$\bar{x}$	Sample	$\bar{x}$
1	101.602	11	101.664
2	101.547	12	101.823
3	101.312	13	101.629
4	101.449	14	101.602
5	101.401	15	101.756
6	101.608	16	101.707
7	101.471	17	101.612
8	101.453	18	101.628
9	101.446	19	101.603
10	101.522	20	101.816

39. The following set of $\bar{x}$'s for samples of size 4 illustrates the effect of a steady drift in the mean of the population. The first 10 samples have $\mu = 101.5$ and $\sigma = 0.2$, whereas the last 10 have $\sigma = 0.2$ and μ increasing by 0.04 in each successive sample, reaching 101.7 at sample 15 and 101.9 at sample 20. Make an $\bar{x}$ chart for these data. Are any points out of control? Is the upward drift in μ visible in any way on the chart?

Sample	$\bar{x}$	Sample	$\bar{x}$
1	101.458	11	101.453
2	101.618	12	101.258
3	101.507	13	101.557
4	101.494	14	101.484
5	101.533	15	101.896
6	101.334	16	101.634
7	101.547	17	101.632
8	101.695	18	101.824
9	101.351	19	101.968
10	101.555	20	101.783

▲ **40.** The U.S. government publication *Science Indicators 1980* shows that the average salary of women in all science and engineering fields is only 77% of the average salary for all male engineers and scientists. But the same source shows that in every individual field of science and engineering, the average female salary is at least 92% of the average male salary. Explain how this apparent discrepancy can come about.

41. In a study of the effect of parents' smoking habits on the smoking habits of high school students, researchers interviewed students in eight high schools in Arizona. The results appear in the following two-way table. [From S. V. Zagona (ed), *Studies and Issues in Smoking Behavior,* University of Arizona Press, Tucson, 1967, pp. 157–180.]

	Student smokes	Student does not smoke
Both parents smoke	400	1380
One parent smokes	416	1823
Neither parent smokes	188	1168

Describe the association between the smoking habits of parents and their high school children by computing and comparing several percents. Then summarize the results in plain language.

42. Here is a two-way table based on information in the 1989 *Statistical Abstract of the United States* (U.S. Bureau of the Census, Washington, D.C.).

Degrees earned in 1986 by level and sex (thousands)

	Bachelor's	Master's	Professional	Doctorate
Male	486	144	49	22
Female	502	145	25	12

a. What percent of all bachelor's degrees were earned by women?

b. How many master's degrees were awarded in 1986?

c. What percent of all degrees earned by women were doctorates?

d. Summarize in words the relation between the level of degrees and the sex of the recipient. Back your summary by computing and comparing appropriate percents.

▲ Discussion exercise.

43. The following pair of two-way tables compare the batting records of two baseball players, Bill and Will. How well a batter hits may depend on whether the pitcher is a right-hander or a left-hander, so both the type of pitcher and the result (hit or out) are recorded for each time at bat.

Bill

	Right-hander		Left-hander	
	Count	Percent	Count	Percent
Hits	40	40	80	20
Outs	60	60	320	80

Will

	Right-hander		Left-hander	
	Count	Percent	Count	Percent
Hits	120	30	10	10
Outs	280	70	90	90

a. Combine the information in these tables to make a two-way table of batter (Bill or Will) by outcome (hit or out) for all times at bat. Which player gets a hit a higher proportion of the time? (The proportion of at bats in which a player gets a hit is his batting average.)

b. Who has the higher batting average against right-handed pitching? Who has the higher batting average against left-handed pitching?

c. Explain carefully, as if talking to a skeptical baseball manager, how it is possible for one player to do better against both right-handers and left-handers and yet have a lower overall batting average. Which hitter would you prefer to have on your team?

44. A community has two hospitals. Hospital A is a large medical center, while Hospital B is a fashionable spa for prosperous patients. An article in the local newspaper claims that a higher percent of surgery patients die at Hospital A than at Hospital B. The paper says that people who need surgery should go to Hospital B. The following pair of two-way tables look at the data but add information about a hidden variable, the condition of the patients before surgery.

Hospital A

	Good condition		Poor condition	
	Count	Percent	Count	Percent
Died	6		57	
Survived	594		1443	

Hospital B

	Good condition		Poor condition	
	Count	Percent	Count	Percent
Died	8		8	
Survived	592		192	

a. Fill in the percent columns in the tables. Use these percents to show that both a higher percent of patients in good condition and a higher percent of patients in poor condition survive at Hospital A than at Hospital B.

b. Combine the information in the two tables to make a single table of patient outcome (died or survived) by hospital (A or B). Show that, as reported, a higher percent of patients survive at Hospital B.

c. Explain carefully, as if talking to a skeptical reporter, how Hospital A can have a poorer overall survival rate even though it does better than Hospital B for both classes of patients.

· III ·

Social Choice and Decision Making

A revolution currently taking place in the field of mathematics is the successful use of mathematics as a fundamental tool to study human beings — their behavior, values, interactions, conflicts, organizations, fair allocations, and decision making, as well as their interface with modern technology and complex organizations. This latter revolution could eventually prove to be as far reaching as the turning of mathematics to study physical objects and their motion some three centuries ago. As mathematics and computers play an increasingly important role in understanding our social institutions, a new profession is emerging devoted to thinking mathematically about human affairs.

In particular, human decision making is being influenced profoundly by modern mathematics, and several particular mathematical subjects have been created primarily to assist in arriving at good decisions. Many aspects involved in arriving at a decision are, of course, nonquantitative in nature. These may relate to history, past experience, instinct, judgment, morality, and so forth. As a consequence one often refers to decision making as an art rather than as a science. On the other hand, many ingredients in contemporary decision making are mathematical in nature, and one can also view this activity as a scientific subject.

A decision maker will begin by listing the options over which he or she has some control and the likely outcomes resulting from these choices. The person

Registering to vote, New York City. [Miriam White. © 1987.]

may attempt to identify all relevant variables and the relationships among them, and may associate quantitative measures when possible. Moreover, the decision maker must clarify his or her own values, identify desired goals, and spell out explicitly any limiting resources or social constraints. One then seeks the best possible result obtainable.

The situation is typically confounded by a variety of different uncertainties involving data and forecasting that typically cannot be completely resolved in advance. The effect that other decision makers may have on the outcome, and the best contingent responses to their moves, should be predetermined. Various ethical concerns such as fairness may well need consideration, and ways to ascertain group opinions may be necessary. Finally, decision makers must study the social and political context in which the decision will be implemented.

Several different mathematical subjects have been introduced since World War II, many for the purpose of assisting individuals or groups in arriving at good or equitable decisions. As an illustration, a dozen major aspects of making decisions with the corresponding

mathematical specialties are listed in the table at the bottom of this page. All of these fields, except for continuous optimization, statistics, and probability theory, have developed mostly in recent decades and in the context of mathematics applied to human actions and organization. Topics 8, 9, and 10 in the table are the subject matter of Part III, whereas topics 11 and 5 were considered in Parts I and II, respectively.

In this part of the text we will illustrate a few of the mathematical techniques available to assist decision makers. In Chapter 9 we discuss the important problem of social choice. How does a group of individuals, each with his or her own set of values, select one outcome from a list of possibilities? This problem arises frequently in any democratic society, and even in more authoritarian institutions where decisions are made by more than one person. We will learn that all voting systems have inherent flaws and that agendas designed for ascertaining the collective group will are often subject to manipulation. Group decision making is inherently a strategic encounter, and every citizen should be aware of the difficulties and pitfalls that can arise in this arena.

A decision maker's concerns	Related mathematical subjects
1. Identify and measure strategic variables	Theory of measurement
2. Understand a complex system	System analysis, graph theory
3. Quantify one's preferences	Utility theory
4. Formulate objectives and constraints	Mathematical programming
5. Acquire data and forecast results	Statistics
6. Determine the most efficient outcomes	Optimization theories
7. Deal with uncertainty	Probability theory
8. Resolve conflicts	Game theory
9. Group decision mechanisms	Social choice theory
10. Equity considerations	Fair division theory
11. Make decisions using a multidisciplinary approach	Operations research, management science
12. Make decisions in an institutional setting	Policy science

In Chapter 10 we consider decision-making bodies in which the individual voters or parties do not have equal power. In particular we will look at weighted voting systems such as stockholders in a corporation or political parties in a national assembly in which the voters cast different numbers of votes. We first observe that power in such systems is not necessarily proportional to the voters' weights. The notion of power is of fundamental importance in political science, although it is typically difficult to quantify. We will nevertheless describe one popular index for measuring power for weighted voting systems, an index that has proved useful in rulings by courts regarding local governments whose elected officials represent districts having different populations. It allows one to assign weighted votes to the legislators from constituencies of varying sizes so that the individuals in the districts are represented in an equitable manner.

A general theme throughout this part concerns the idea of fairness in decision making. In Chapter 11 this becomes most explicit. Here we describe some fair division schemes in which a group of individuals with different tastes can be assured of each receiving what he or she views as a fair share when dividing up "smooth" objects like cakes or "chunky"

goods such as estates. We then discuss the apportionment problem that is concerned with rounding fractions in an equitable manner. It arises in many fair allocation problems in which the things to be allocated must be multiples of some basic unit. This occurs, for example, in political representation, personnel assignments, and when adding capacity to a transportation system. The listing of reasonable assumptions for fair division schemes leads naturally to a discussion of axioms needed for mathematical models to always achieve a fair distribution.

Chapter 12 introduces the mathematical field called game theory, which describes situations involving two or more decision makers seeking different goals. Game theory provides a collection of models to assist in the analysis of conflict and cooperation. It prescribes optimal strategies for games of total conflict in which one's gain is another's loss. It also provides insights into purely cooperative situations as well as encounters of partial conflict that involve aspects of both competition and cooperation. Some particular games such as those known as "prisoners' dilemma" and "chicken" provide us with insights into certain social paradoxes that we routinely meet in our daily lives.

·9·

Social Choice:
The Impossible Dream

The basic question of *social choice,* of how groups can best arrive at decisions, has occupied social philosophers and political scientists for centuries. Social-choice theory arose to help explain voting and other decision-making processes. Voting is a subject that lies at the very heart of representative government and participatory democracy. In both theory and practice, voting poses difficult problems.

The fundamental problem is to turn individual preferences for different outcomes into a single choice by the group as a whole. This situation arises whenever government representatives pass a bill, stockholders decide on a course of action, a political party nominates a presidential candidate, or a community elects members to serve on the school board. Few people realize that the voting method they use can significantly affect the outcome of an election, and that the voting mechanism they use is typically subject to manipulation.

The first type of voting that often comes to mind is that of majority rule. In the case of **majority rule** each voter votes for one candidate, and the candidate receiving over half of the total votes is declared the winner. However, this method is truly effective solely in elections in which only two candidates are competing for a single office. The process is then quite simple: the candidate with the larger share of votes wins. But where there are more than two candidates, it is possible that the person with the highest tally will not actually hold a majority of all votes cast.

Most civilized societies have developed a variety of voting procedures to single out particular options from a longer list of feasible alternatives. How do we decide which of these voting schemes should be used?

All voting methods have some inherent faults, and any method of voting we use will occasionally give rise to rather paradoxical results. So no single voting method is either universally applicable or the best overall: for each situation we have to select one of the many available voting schemes, knowing that the method we choose may greatly influence the result of the election.

The surprisingly large number of different voting procedures from among which we can choose include majority rule, plurality wins, elimination and runoffs, sequential pairwise comparisons, various weighted or scoring schemes, approval voting, and a host of various other partitioning schemes that choose successively between subsets of potential outcomes. All these methods can at times produce some disconcerting results (see Spotlight 9.1, p. 245).

The outcome of an election can be affected by the type of voting method employed or by the order or formatting of the questions. The agenda itself can be a significant factor in determining which motion survives, and agendas are usually subject to manipulation. For example, insincere amendments can be introduced for diversionary purposes or to kill off popular motions at an early stage. Voters may sometimes benefit by falsifying their preferences and misleading others. The final outcome of an election can be altered by any of these strategic-voting maneuvers.

As we focus on decision-making procedures in this chapter, we will see that it is no easy matter to ascertain the "true will of the people." The problem of unifying individual preferences into one particular choice for the whole group is essentially unsolvable. We will now examine a few actual cases to illustrate these difficulties in voting.

THE ROMAN SENATE

Almost 2000 years ago, the Roman historian Pliny the Younger was grappling with a voting dilemma:

A debate arose in the Senate concerning the freedmen of the consul Afranius Dexter; it being uncertain whether he killed himself, or whether he died by the hands of his freedmen; and again, whether they killed him from a spirit of malice, or of obedience.

It appears that there were questionable circumstances surrounding the death of Consul Dexter. If his freedmen (former slaves and servants) did in fact execute him, it is not clear whether it was an act of "mercy killing," according to Dexter's own wishes, or murder. His death may also have been the result of suicide. The three possible circumstances are shown in Figure 9.1, with appropriate verdicts for the freedmen indicated in parentheses. Pliny continues in his letter:

One of the senators (it is of little purpose to tell you I was the person) declared that he thought these freedmen ought to be put to the question, and afterwards released. The sentiments of another were, that the freedmen should be banished, and of another, that they should suffer death. It was impossible to reconcile such a diversity of opinions.

Pliny describes three groups in the Roman Senate:

• Group *A* believed the freedmen were innocent and thus favored their *acquittal.*

• Group *B* considered them guilty to some extent and thought the appropriate punishment was *banishment.*

• Group *C* believed the freedmen guilty of the crime of murder and felt that they should be *condemned* to death.

We can use *A*, *B*, and *C* to represent these three groups and *a*, *b*, and *c* to denote the three actions: acquittal, banishment, and condemnation. To be more specific, let's suppose that these three groups consist of 40%, 35%, and

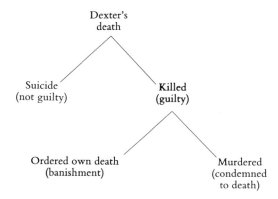

Figure 9.1   How was Dexter killed?

25% of the senators, respectively. We can arrange this information in a table:

	Group		
	A	*B*	*C*
Preferred action	*a*	*b*	*c*
Percent	40	35	25

The outcome of this trial could depend as much on the voting procedure the senators use as on the kind of information they have, as well as the strategies or arguments they employ. We'll consider five agendas in this chapter.

Method 1. In the **plurality** method of voting the candidate with the most votes is declared the winner. The winner need not have a majority of the votes cast.

If the prisoners' fate were determined by a plurality, then at first glance it looks as if they would go free: in a plurality vote, the position with the most votes is declared the winner. Our table shows that outcome *a* is favored by the largest number of the senators, 40%, including Pliny himself.

However, the senators may choose to vote differently. What if they know the distribu-

tion of potential votes in advance? Suppose they know that the numbers 40%, 35%, and 25% would be in favor of *a*, *b*, and *c*, respectively. Perhaps those in group *C* would then compromise on their hard-line position (the death penalty) and vote instead for banishment. The result would be 40% for *a* and 35% + 25% = 60% for *b*. Banishment would then carry the day.

In many cases, we can reasonably expect group *C* to vote in this way, that is, in favor of outcome *b* rather than their first choice, *c*. Such **strategic**, or **insincere, voting** is not at all uncommon.

How does strategic voting occur in this case? Suppose that we have additional information about the groups *A*, *B*, and *C*. Assume, for example, that we know their second and third choices as well. It seems reasonable that group *A* would prefer *a* over *b* and *b* over *c* and that group *C* would prefer the reverse order: *c* over *b* and *b* over *a*. It may well be that those in group *B*, who favor banishment *b*, would be divided in their second choice. Some may prefer acquittal *a* to the harsh penalty *c*, whereas others who are against the death penalty may nevertheless feel that the guilty must be punished in some manner, even if it means execution. For the sake of simplicity, however, let's assume that *all* members of group *B* favor *b* over *a* and *a* over *c*. This **schedule of preferences** is summarized as follows:

	Group		
	A	*B*	*C*
First choice	*a*	*b*	*c*
Second choice	*b*	*a*	*b*
Third choice	*c*	*c*	*a*
Percent	40	35	25

It is fairly clear that the senators in group *A* have no real choice other than to vote for acquittal. So we'll assume that *A* votes for *a*.

Group *B* is free to choose either *a*, *b*, or *c*, as is group *C*. We can represent these strategies and the resulting outcomes as follows:

		C votes		
A votes	*a*	*a*	*b*	*c*
	a	*a*	*a*	*a*
B votes	*b*	*a*	*b*	*a*
	c	*a*	*a*	*c*

In the preceding table, the three rows correspond to the three choices for group *B*, the three columns indicate the three options for group *C*, and the letters in the table itself are the outcomes when the corresponding strategies are chosen by these groups. For example, if *A* voted for *a*, *B* voted for *b*, and *C* voted for *c*, the result would be the boldfaced *a*, found in the second row and third column in this table. This result, in which each senator has voted for his most preferred outcome, is called **sincere voting.**

A closer examination of the table shows that both groups *B* and *C* will benefit if *C* votes for *b* instead of *c*. If group *C* switches its vote, *C* will achieve its second choice *b* instead of its third choice *a*, and group *B* will achieve its first choice *b*.

		C votes		
A votes	*a*	*a*	***b***	*c*
	a	*a*	*a*	*a*
B votes	*b*	*a*	***b***	*a*
	c	*a*	*a*	*c*

This analysis suggests that group *C* should vote *insincerely* and select *b*. Groups *B* and *C* will in effect have formed a *coalition* against *A*. However, neither collusion nor communication need take place to bring this about. Not only do both *C* and *B* benefit from voting for *b*, but once they have done so, there is no possible further switch that can do better for either. Thus this action is self-reinforcing since no voting group can now deviate unilaterally and expect to gain from its action. (Technically, we refer to these choices as being in *equilibrium*, a concept which arises in Chapter 12.)

Sequential Voting

One major objection to plurality voting is that when there are more than two outcomes, the final outcome may be favored by less than half of those voting — only 40% in the case of *A*. In order to make sure that the ultimate decision receives a majority vote, it may be necessary to resort to a *runoff* election or to some other type of **sequential voting** — a procedure that requires a majority vote at each step. We can now consider two such additional agendas that the Roman Senate could have used.

In the first of these agendas, the Senate will vote first between innocent and guilty; and only when a guilty verdict results will they decide the punishment, *b* or *c*. This agenda is depicted in Figure 9.2.

For this scheme, *a* loses to "not *a*" 40% to 60% (35% + 25%) on the first round, and *b* beats *c* by 75% (40% + 35%) to 25% on the second ballot. You can see that banishment

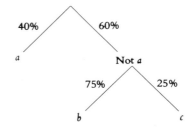

Figure 9.2 A runoff election.

SPOTLIGHT 9.1 Historical Highlights of Voting Methods

Historically, popular voting methods have often led to counterintuitive results. In the eighteenth century, certain voting paradoxes were brought to light by Jean-Charles de Borda (1733–1799) and Marie Jean Antoine Nicholas Caritat (1743–1794), the Marquis de Condorcet. Many scholars in the nineteenth and twentieth centuries also expressed concern over the lack of adequate voting methods to support the emerging representative forms of government and group decision making. For example, Lewis Carroll, whose real identity was the Oxford mathematician Reverend Charles Lutwidge Dodgson (1832–1898), appeared to deplore the tendency of voters to adopt a

principle of voting which makes an election more of a game of skill than a real tool of the wishes of the electors, and . . . my own opinion is that it is better for elections to be decided according to the wish of the majority than of those who happen to have the most skill at the game.

A review of historical developments in this area is given in *The Theory of Committees and Elections* (1958) by the English economist Duncan Black.

Kenneth Arrow's "impossibility" result in 1951 demonstrated that all attempts to arrive at a perfectly suitable technique of voting for all occasions were doomed to failure (see Spotlight 9.3). His theorem inspired three different approaches to analyzing group-decision procedures.

First, *social-choice* theory emphasizes normative models and the axiomatic approach. It is concerned with the conditions under which particular voting schemes will guarantee desirable outcomes. *Collective Choice and Social Welfare* (1970) by Oxford political economist A. K. Sen describes the social-choice perspective.

Second, the more descriptive approach called the theory of *public choice* explains the way individuals actually behave in existing group-decision forums to best obtain their aims. Among other things, public-choice theorists study the frequency of incidents involving undesirable outcomes or insincere voting. William H. Riker of the University of Rochester and James M. Buchanan of George Mason University (winner of the 1986 Nobel Memorial Prize in Economic Science) have been leading figures in these advances, and the journal *Public Choice* publishes research in the field.

A third direction is shown in the technical volume *Game Theoretical Analysis of Voting in Committees* (1984) by Bezalel Peleg. This approach accepts the strategic nature of group decision making as a given and uses existing subjects such as game theory to learn how to better compete when such instances arise.

will win if the voters vote sincerely at each decision point. There is no maneuvering or collusion that *A* or *C* can undertake in this case that will produce a better outcome for them. Whereas insincere voting was the optimal strategy in the case of plurality voting, the best strategy in this case is to vote in a sincere manner.

In the second of the sequential agendas, we assume that the Senate moves to decide *first* upon the appropriate punishment, *b* or *c*, before it addresses the question of guilt. This

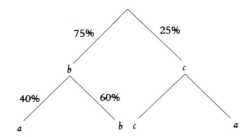

Figure 9.3 The outcome under sincere voting.

agenda is pictured in Figure 9.3. Sincere voting would result in *b* winning over *c* by 75% to 25% on the first round and then *b* winning over *a* by 60% to 40%. It appears as though *b* should win.

However, it is not clear that *A* will vote sincerely: Group *A* may well vote for outcome *c* on the first ballot, which could result in the middle position *b* being eliminated on the first vote by 65% (40% + 25%) to 35%. After eliminating *b*, group *A* would change from *c* to *a* on the second ballot; thus, preference *a* would ultimately prevail over *c* by 75% (40% + 35%) to 25% (see Figure 9.4). When forced to choose between *a* and *c*, group *B* should go with *a*, their second choice, rather than *c*, their least-preferred outcome.

On the other hand, the groups may not vote this way after all. *C* is well aware that group *A*, in an effort to eliminate *b* at an early stage, may vote insincerely on the first round. In effect, *A* and *C*, representing the extreme positions *a*

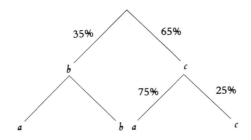

Figure 9.4 An outcome under strategic voting.

and *c*, have temporarily united to eliminate the middle position *b*. However, *C* need not go along with this ploy by *A*. To avoid eliminating *b*, *C* may actually vote for *b* instead of *c* at the first tally.

In short, in the first round *A* may vote for *c* rather than *b* and *C* may actually vote for *b* rather than *c*. This is the type of deplorable behavior that Lewis Carroll must have had in mind! (See Spotlight 9.1.) Strategically speaking, however, this type of thinking could be viewed as simple common sense.

Sequential voting can take many different forms, especially when selecting one from among several different candidates. The resulting outcome may well depend upon ordering of the issues, as well as upon strategic voting.

Pairwise Comparisons

Let's make a fourth attempt to resolve Pliny's problem. Consider what would happen if we held an election between each pair of outcomes: *a* versus *b*, *b* versus *c*, and *c* versus *a*. Under sincere voting,

- *b* beats *a* 60% to 40%
- *b* beats *c* 75% to 25%
- *a* beats *c* 75% to 25%

It seems as though *b* should be the winner because *b* beats either of the other positions when they meet head to head. Such a winner, if it exists, is called a Condorcet winner.

Method 2. A candidate who wins over every other candidate in a head-to-head ballot is called a **Condorcet winner.** (Such a winner may not exist.)

Once again, however, those in group *A* could frustrate the victory for outcome *b* by pretending to alter their "preference" schedule from

$$
\begin{array}{ccc}
\underline{A} & & \underline{A} \\
a & \text{to} & a \\
b & & c \\
c & & b
\end{array}
$$

That is, they will vote for c over b whenever this pair comes up for a vote. As a result, we wind up with

* b beats a 60% to 40%
* a beats c 75% to 25%
* c beats b 65% to 35%

So we end up in a tie, indicated by the cycle

* b beats a beats c beats b

This last result is said to violate the **law of transitivity,** which says that if x is preferred to y, and if y is preferred to z, then x is preferred to z. Our example shows the **paradox of voting,** or the **Condorcet paradox:** even if individuals hold to the law of transitivity, the voters as a group may not satisfy it. It is, however, only one of a variety of different paradoxes that occur in voting situations (see Spotlight 9.2, p. 248, and Spotlight 9.3, p. 254).

Having discussed only four of the many possible agendas that could have been used by the Roman Senate in the trial of Dexter's freedmen, we can appreciate Pliny's dilemma: the fate of the freedmen depends to a large extent on game playing!

Rank and Score

Many elections ask each voter to submit a complete ranking (preference schedule) of all the candidates. The goal is to arrive at a final group rank ordering of all the contestants that best expresses the desires of the electorate. The purpose is not only to determine the winner, say, the class valedictorian, but also to arrive at who finished second, third, and so on, as in the case of one's rank in his or her senior class. In other applications, such as an election to a hall of fame, the first few finishers each receive the award, while the remaining nominees are "also rans." One common mechanism for achieving this objective is to assign points to each voter's rankings and then to sum these for all voters to obtain the total points for each candidate. If there are ten candidates, for example, then we could assign 10 points to each first-place vote for a given candidate, 9 points for each second-place vote, 8 for each third, and so forth. The candidate with the highest total number of points is the winner. Subsequent positions are assigned to those with the next highest tallies.

Method 3. A voting method that assigns points in a descending manner to each voter's subsequent ranking and then sums these points to arrive at a group's final ranking is called a **Borda count.**

Let us return to Pliny's problem in the Roman Senate example and apply a Borda count. According to this agenda they could score 3 points for each senator's first choice, 2 points for each second choice, and 1 point for each third choice. The tally for the three outcomes is as follows:

a: $(3)(40\%) + (2)(35\%) + (1)(25\%) = 2.15$

b: $(3)(35\%) + (2)(40\%) + (2)(25\%) = 2.35$

c: $(3)(25\%) + (1)(40\%) + (1)(35\%) = 1.50$

We conclude that b (banishment) is the winner, a is second, and c is the losing position. In this situation, however, only outcome b will be implemented, and the rankings of a and c are irrelevant. We will see in what follows that

SPOTLIGHT 9.2 Unattainable Ideals

A popular view of mathematics holds it to be a collection of factual statements (theorems, propositions) that are deduced as true from more basic premises (axioms, hypotheses), plus a variety of more mechanical techniques (algorithms) used to compute actual solutions to a particular problem. In contrast to this notion, however, mathematicians also devote significant efforts to discovering illustrations (counterexamples) that show that some conjectured facts are not true. Moreover, some of the most important mathematical discoveries demonstrate that a certain presumed situation does not exist or that some highly desirable outcome cannot always be attained.

The Pythagoreans of ancient Greece knew that the number $\sqrt{2}$ was *irrational,* in the sense that it could not be written as a ratio p/q for any two integers p and q. (See Exercise 17.) Such insights often have the effect of terminating the search for some ideal state or for certainty within a subject. They may place theoretical limits on what can be achieved by means of the scientific method. Four outstanding cases from the twentieth century follow.

In 1927, the German physicist Werner K. Heisenberg (1901–1976) announced the *uncertainty principle,* which states that it is impossible to determine precisely both the position and momentum of a particle at a given time. (The product of the uncertainties in these two variables always exceeds a particular constant.) This result played havoc with the popular deterministic philosophy of the nineteenth century.

Borda counts are vulnerable to insincere voting as well as to the particular number of points assigned to each position in a ranking. So we still have not arrived at a definitive answer to Pliny's dilemma. Now let us turn to some contemporary scoring systems which will illustrate some difficulties that may arise when using Borda counts.

EXAMPLE: The Football Poll. A poll by 25 sports announcers is used to rank the football teams from among the four following universities: Miami (of Florida), Notre Dame, Penn State, and Southern California. They elect to assign 3 points to each announcer's first choice, 2 points to a second, 1 to a third, and 0 for a fourth. There are $4! = 4 \times 3 \times$ $2 \times 1 = 24$ possible rankings, but assume that only the five in the following table appear:

	Number of announcers					
Choice	8	6	5	4	2	Points
First	Mi	ND	PS	SC	ND	3
Second	ND	Mi	Mi	PS	PS	2
Third	PS	SC	ND	Mi	SC	1
Fourth	SC	PS	SC	ND	Mi	0

We calculate each team's total points as follows:

$$\text{Mi: } (3)(8) + (2)(6) + (2)(5)$$
$$+ (1)(4) + (0)(2) = 50$$

In 1931, the Austrian-American mathematician Kurt Gödel (1906–1978) published his paper on "formally undecidable propositions." (It proved that given any set of axioms, there would always be statements within the system ruled by these axioms that could be neither proved nor disproved on the basis of these axioms.) This showed that the paradoxes that had been disturbing mathematical logicians for the previous half-century were unavoidable. Gödel showed, against the hopes of some, that the totality of mathematics could not be deduced from any single system of axioms.

In 1951, Kenneth J. Arrow (1921–) listed five highly desirable properties that one would expect any reasonable voting system to possess. He then went on to prove that no possible voting method could satisfy these properties in all situations. (See Spotlight 9.3.) Every voting scheme will, at times, exhibit shortcomings. This *impossibility theorem* destroyed the dream of social philosophers who had sought fair and nonmanipulative social-choice mechanisms for more than a century. It also forced social scientists to use more rigorous methodologies in their analyses.

In 1980, Michel L. Balinski (1933–) and H. Peyton Young (1945–) showed that there is no general method for rounding a set of fractions to integers with a given sum that will always satisfy three very natural conditions. Therefore, allocating seats to states in the U.S. House of Representatives or seats to parties in parliament has no completely satisfactory solution. The two-century search by the U.S. Congress and other representative bodies was doomed from the start (see Chapter 11). More generally, attempts to apportion discrete objects in an equitable manner can result in undesirable allocations.

ND: $(2)(8) + (3)(6) + (1)(5)$
$+ (0)(4) + (3)(2) = 45$

PS: $(1)(8) + (0)(6) + (3)(5)$
$+ (2)(4) + (2)(2) = 35$

SC: $(0)(8) + (1)(6) + (0)(5)$
$+ (3)(4) + (1)(2) = 20$

The resulting ranking is (1) Miami, 50 points; (2) Notre Dame, 45; (3) Penn State, 35; and (4) Southern California, 20.

However, this poll is vulnerable to insincere voting. For example, what if three of the six voters who ranked ND over Mi over SC over PS suspected prior to the poll that Mi would edge out ND, and they decided to vote insincerely in an attempt to have ND come out on top? They could move Mi from second to fourth place in their rankings. This would take 6 points away from Mi (and add a total of 6 for SC and PS) and thus result in ND becoming the winner. However, such strategic voting may not end there. Thirteen of the seventeen voters who placed Mi over ND could be of like mind and place ND lower in their preference schedules. Many other possibilities for strategic voting appear in a poll of this type.

Another serious difficulty with Borda counts is that the outcome may well depend upon the scale of numbers selected, as is indicated in the following illustration.

EXAMPLE: The Horse Show. The four horses A, B, C, and D are finalists in a show in which they are rated evenly on four attributes W, X, Y, and Z. They place first, second, third, and fourth in each category according to the following table:

Horse	*Attributes*			
	W	X	Y	Z
A	First	Third	Second	Third
B	Second	First	Fourth	Second
C	Fourth	Fourth	First	First
D	Third	Second	Third	Fourth

If one uses a Borda count that awards 3, 2, 1, and 0 points for a first, second, third, and fourth, respectively, then we see that A and B are tied for the championship with 7 points apiece:

$$A: (3)(1) + (2)(1) + (1)(2) = 7$$
$$B: (3)(1) + (2)(2) + (1)(0) = 7$$
$$C: (3)(2) + (2)(0) + (1)(0) = 6$$
$$D: (3)(0) + (2)(1) + (1)(2) = 4$$

On the other hand, if we score the points as 5, 3, 1, and 0, respectively, then horse B is the winner with 11 points:

$$A: (5)(1) + (3)(1) + (1)(2) = 10$$
$$B: (5)(1) + (3)(2) + (1)(0) = 11$$
$$C: (5)(2) + (3)(0) + (1)(0) = 10$$
$$D: (5)(0) + (3)(1) + (1)(2) = 5$$

If points are instead assigned as 5, 2, 1, and 0, then horse C wins:

$$A: (5)(1) + (2)(1) + (1)(2) = 9$$
$$B: (5)(1) + (2)(2) + (1)(0) = 9$$
$$C: (5)(2) + (2)(0) + (1)(0) = 10$$
$$D: (5)(0) + (2)(1) + (1)(2) = 4$$

If the points are awarded as 5, 3, 2, and 1, as is often the case in such animal competitions, then the result is a three-way tie for first:

$$A: (5)(1) + (3)(1) + (2)(2) + (1)(0) = 12$$
$$B: (5)(1) + (3)(2) + (2)(0) + (1)(1) = 12$$
$$C: (5)(2) + (3)(0) + (2)(0) + (1)(2) = 12$$
$$D: (5)(0) + (3)(1) + (2)(2) + (1)(1) = 8$$

Which horse(s) should be declared the winner?

Borda counts are also used to score competitions in which several teammates can enter an event, and so more than one competitor can contribute points to the same team from a single event.

EXAMPLE: The Track Meet. The results of a track and field meet among the three schools A, B, and C are summarized in the following table:

Place	*Events*				
	Dash	Run	Hurdles	Jump	Throw
First	A	B	A	A	B
Second	A	C	B	C	C
Third	B	C	A	C	C
Fourth	C	C	C	A	C

If the finishing positions are scored 4, 3, 2, and 1, respectively, then school C (which has strong depth) is the victor with 19 points, to 18 for A and 13 for B. On the other hand, if the top places are scored 5, 3, 2, and 1, then A wins with 21 (because of more firsts) to C's 19 points, and B again is last with 15 points. There may be more than four finishers in the events and each is assigned a position in the ranking, but only the first four places count in the scoring.

Ties (or indifferences in preference) can also be incorporated into a Borda count by dividing the accumulated points evenly among the tied contestants. For example, if a remeasurement indicated that A and C were really tied for first (and second) place in the Jump, then they would each receive 3.5 points in the 4, 3, 2, 1 scoring system. Then team C would have 19.5 points and A would receive 17.5. In the 5, 3, 2, 1 scoring scheme, the final result would then be 20 for A, 20 for C, and 15 for B.

BOGUS AMENDMENTS

We can use still another simple election situation to show how diversionary amendments, when strategically introduced, can mislead some voters into acting against their own interest.

Assume that three representatives A, B, and C each have the choice of voting in favor of or against a new bill N. Voting against this new law means that the old law O will prevail. Assume that two of the three voters do prefer the proposed bill N over the existing law O, as indicated in this table of preferences:

| | Voter | | |
	A	B	C
First choice	N	N	O
Second choice	O	O	N

In a direct comparison between the outcomes N and O, N will win by a vote of 2 to 1. Nevertheless, voter C may attempt to defeat N by the following maneuver. He proposes to modify the new bill N with an amended version called M. C selects the amendment so that A prefers M most of all, whereas B prefers M least of all. This may be done, perhaps, by merely shifting some of the proposed reward in bill N from B to A. Meanwhile C pretends to

prefer O over M and M over N. The schedule of preferences now becomes:

| | Voter | | |
	A	B	C
First choice	M	N	O
Second choice	N	O	M
Third choice	O	M	N

The new agenda and voting appear in Figure 9.5. When voting between N and M is sincere at the first decision point, M wins over N by 2 to 1. At the second step, O beats M by 2 to 1. Voter C has tricked the others into defeating N and maintaining the status quo O. Voter A should have noticed this tactic and resisted the temptation to initially vote for the fleeting amendment M.

In the course of a long, complex agenda and heated debate, we must be continuously on guard to avoid being manipulated into voting against our own long-range interests. Those designing the agenda can often rig it in their own favor. For example, one contingent might stack up a larger number of popular outcomes and pit them against a *single* highly desired one in an attempt to eliminate this single outcome at an early stage of the agenda. As a general rule of thumb, it's best to enter the more preferred outcomes at a later stage of the agenda. The chances of survival may increase when there are fewer competing alternatives and fewer remaining votes to be taken.

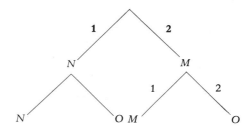

Figure 9.5 Voting on a bogus amendment.

EVERYONE WINS

We have seen how the outcome of an election may very well depend on the voting procedure or agenda as well as on strategic choices by the voters. To show an extreme case of how the method chosen might affect the results in a realistic situation, we will consider an example of a political party convention at which five different voting schemes are adopted. Assume that there are 55 delegates to this national convention, at which five of the party members, denoted by $A, B, C, D,$ and E, have been nominated as the party's presidential candidate. Each delegate must rank all five candidates according to his or her choice. Although there are $5! = 5 \times 4 \times 3 \times 2 \times 1 = 120$ possible rankings, many fewer will appear in practice because electors typically split into blocks with similar rankings. Let's assume that our 55 delegates submit only six different preference schedules, as indicated in the following table:

Number of delegates

	18	12	10	9	4	2
First choice	A	B	C	D	E	E
Second choice	D	E	B	C	B	C
Third choice	E	D	E	E	D	D
Fourth choice	C	C	D	B	C	B
Fifth choice	B	A	A	A	A	A

We see from the preceding table that the 18 delegates who most favor nominee A rank D second, E third, C fourth, and B fifth. Although A has the most first-place votes, he is actually ranked last by the other 37 delegates. Note that the 6 electors who most favor nominee E split into two subgroups of 4 and 2 because they differ between B and C on their second and fourth rankings. We will assume that our delegates must stick to these prefer-

ence schedules throughout the following five voting agendas. That is, we will not allow any delegate to switch preference ordering in order to vote in a more strategic manner.

1. *Plurality.* If the party were to elect its candidate by a simple plurality, nominee A would win with 18 first-place votes, in spite of the fact that A was favored by less than one-third of the electorate and was ranked dead last by the other 37 delegates.

2. *Sequential: Winners' runoff.* On the other hand, if the party decided that a runoff election should be held between the top two contenders, who together received a majority of the first-place votes in the initial plurality ballot, then candidate B outranks A on 37 of the 55 preference schedules and is declared the winner in the runoff.

3. *Sequential: Losers eliminated.* Another approach that could be used is holding a sequence of ballots and eliminating at each stage the nominee with the fewest first-place votes. The last to survive this process becomes the winning candidate. We see in our example that E, with only 6 first-place votes, is eliminated in the first round. E can then be deleted from our table of preferences, and *all* 55 delegates will vote again on successive votes. On the second ballot, the number of first-place votes for the 4 remaining nominees is

$$\begin{array}{cccc} A & B & C & D \\ 18 & 16 & 12 & 9 \end{array}$$

Thus, D is eliminated. Note that the 6 delegates who most favored E earlier now vote for their second choices, that is, 4 for B and 2 for C. On the third ballot the 9 first-place votes for D are reassigned to C, their second choice, giving

$$\begin{array}{ccc} A & B & C \\ 18 & 16 & 21 \end{array}$$

Thus, *B* is eliminated. On the final round, 37 of the 55 delegates favor *C* over *A*, and therefore *C* wins by this method.

4. *Borda count.* Given that they now have the complete preference schedule for each delegate, the party might instead choose to use a straight Borda count to pick the winner. This could be done, for example, by assigning 5 points to each first-place vote, 4 points for each second, 3 points for a third, 2 points for a fourth, and 1 point for a fifth. The highest total score of

$$191 = (5)(9) + (4)(18) \\ + (3)(12 + 4 + 2) \\ + (2)(10) + (1)(0)$$

is achieved by *D*, who then wins. Note that *A* has the lowest score (127) and *B* the second worst (156).

5. *Condorcet.* In the Condorcet method, each nominee is matched head-to-head with every other. There are 10 such competitions, and each candidate appears in 4 of them. Assuming sincere voting, we can easily see that *E* wins out over

- *A* by a vote of 37 to 18
- *B* by a vote of 33 to 22
- *C* by a vote of 36 to 19
- *D* by a vote of 28 to 27

In this case, the Condorcet method does produce a winner, namely *E*.

In summary, our political party has employed five different common voting procedures and has come up with five different winning candidates. We see from this illustration that those with the power to select the voting method may well determine the outcome.

Approval Voting

One voting method that shows great promise for electoral reform is called **approval voting.** It is particularly suitable for elections in which several candidates typically compete, such as the party primary elections for the President of the United States. Many existing multicandidate-election procedures should be reviewed with a mind toward adopting this simple and practical method.

In approval voting, each voter is allowed to give *one* vote to each of the candidates on the multicandidate slate. No limit is set on the number of candidates an individual can vote for: Voters can approve of as many choices as they like and show disapproval by withholding a vote on that candidate. This system replaces the traditional "one person, one vote" by "one candidate, one vote."

The winner in approval voting is the candidate who receives the largest number of approval votes. This approach is also appropriate in situations where more than one candidate or outcome may win, for example, in electing new members to an exclusive society such as the National Academy of Sciences or the Baseball Hall of Fame.

In past years, several important officials in New York State have been elected with much less than a majority of the vote. Clearly, some of these contests would have been reversed if approval voting had been used. The 1970 U.S. Senate race in New York State gave James Buckley 39%, Richard Ottinger 37%, and Charles Goodell 24% of the vote. Buckley may well have been last in approval voting or been the Condorcet loser in a head-to-head battle with either of the other two candidates. Similarly the result in 1980 was Alfonse D'Amato 45%, Elizabeth Holtzman 44%, and incumbent Jacob Javits 11%. Polls indicated that more of Javits's supporters preferred Holtzman to D'Amato and that she probably would have won in a runoff or under approval

SPOTLIGHT 9.3   Kenneth J. Arrow

Kenneth Arrow.

For centuries, mathematicians have searched for a perfect voting system. Finally, in 1951, economist Kenneth Arrow proved that finding an absolutely fair and decisive voting system is impossible. Kenneth Arrow is the Joan Kenney Professor of Economics, as well as a professor of operations research, at Stanford University. In 1972, Arrow received the Nobel Memorial Prize in Economic Science for his outstanding work in the theory of general economic equilibrium. His numerous other honors include the 1986 von Neumann Theory Prize for his fundamental contributions to the decision sciences. He has served as president of the American Economic Association, the Institute of Management Sciences, and other organizations.

Dr. Arrow talks about the process by which he developed his famous impossibility theorem and his ideas on the laws that govern voting systems:

My first interest was in the theory of corporations. In a firm with many owners, how do the owners agree when they have different opinions, for example, about the prospects of the company? I was thinking of stockholders. In the course of this, I realized that there was a paradox involved — that majority voting can lead to cycles. I then dropped that discussion because I was frustrated by it.

I happened to be working with The RAND Corporation one summer about a year or two later. They were very interested in applying concepts of rationality, particularly of game theory, to military and diplomatic affairs. That summer, I felt not like an economist but instead like a general social scientist or a mathematically oriented social scientist. There was tremendous interest in game theory, which was then new.

Someone there asked me, "What does it mean in terms of national interest?" I said, "Oh, that's a very simple matter," and he said, "Well, why don't you write us a little memorandum on the subject." Trying to write that memorandum led to a sharper formulation of the social-choice question, and I realized that I had been thinking of it earlier in that other context.

I think that society must choose among a number of alternative policies. These policies may be thought of as quite comprehensive, covering a number of aspects: foreign policy, budgetary policy, or whatever. Now, each individual member of the society has a preference, or a set of preferences, over these alternatives. I guess that you can say one alternative is better than another. And these individual preferences have a property I call *rationality* or *consistency,* or more specifically, what is technically known as *transitivity:* if I prefer a to b, and b to c, then I prefer a to c.

Imagine that society has to make these choices among a set. Each individual has a preference ordering, a ranking of these alternatives. But we really want society, in some sense, to give a ranking of these alternatives. Well, you can always produce a ranking, but you would like it to have some properties. One is that, of course, it be responsive in some sense to the individual rankings. Another is that when you finish, you end up with a real ranking, that is, something that satisfies these consistency, or transitivity, properties. And a third condition is that when choosing between a number of alternatives, all I should take into account are the preferences of the individuals among those alternatives. If certain things are possible and some are impossible, I shouldn't ask individuals whether they care about the impossible alternatives, only the possible ones.

It turns out that if you impose the conditions I just stated, there is no method of putting together the individual preferences that satisfies all of them.

The whole idea of the axiomatic method was very much in the air among anybody who studied mathematics, particularly among those who studied the foundations of mathematics. The idea is that if you want to find out something, to find the properties, you say, "What would I like it to be?" [You do this] instead of trying to investigate special cases. And I was really accustomed to this approach. Of course, the actual process did involve trial and error.

But I went in with the idea that there was some method of handling this problem. I started out with some examples. I had already discovered that these led to some problems. The next thing that was reasonable was to write down a condition that I could outlaw. Then I constructed another example, another method that seemed to meet that problem, and something else didn't seem very right about it. Then I had to postulate that we have some other property. I found I was having difficulty satisfying all of these properties that I thought were desirable, and it occurred to me that they couldn't be satisfied.

After having formulated three or four conditions of this kind, I kept on experimenting. And lo and behold, no matter what I did, there was nothing that would satisfy these axioms. So after a few days of this, I began to get the idea that maybe there was another kind of theorem here, namely, that there was no voting method that would satisfy all the conditions that I regarded as rational and reasonable. It was at this point that I set out to prove it. And it actually turned out to be a matter of only a few days' work.

It should be made clear that my impossibility theorem is really a theorem [showing that] the contradictions are possible, not that they are necessary. What I claim is that given any voting procedure, there will be some possible set of preference orders for individuals that will lead to a contradiction of one of these axioms.

But you say, "Well, okay, since we can't get perfection, let's at least try to find a method that works well most of the time." Then when you do have a problem, you don't notice it as much. So my theorem is not a completely destructive or negative feature any more than the second law of thermodynamics means that people don't work on improving the efficiency of engines. We're told you'll never get 100% efficient engines. That's a fact—and a law. It doesn't mean you wouldn't like to go from 40% to 50%.

voting. John Lindsay was reelected mayor of New York City in 1969 with only 42% of the vote. In 1977, Edward Koch beat Mario Cuomo for mayor in a runoff after they initially won only 19.8% and 18.6% of the vote, respectively. (Four others also got over 10% of the vote in that initial election.) Sequential voting or approval voting might have given quite a different view of the "will of the people" and even have selected different winners in these cases.

Approval voting may well prove to be particularly effective in presidential primaries when several contestants are entered. The results of the 1980 New Hampshire elections for the Republican Party were

Ronald Reagan	50%
George Bush	23%
Howard Baker	13%
Others	14%

An ABC News exit poll indicated that approval voting might have given this tally:

Ronald Reagan	58%
George Bush	39%
Howard Baker	41%

It is very possible that such results would have delayed the withdrawal of Senator Baker from the race. Perhaps he would have become Vice President in 1981 and President in 1989.

Approval voting, like any other voting method, is not entirely free of faults. It, too, is subject to strategic manipulation and can give rise to counterintuitive outcomes. Analyses to date indicate that it is no more vulnerable to insincere voting than other known methods. On the other hand, it is practical, simple, and easy to implement. It gives the voters greater freedom in expressing themselves without requiring more complicated ranking schemes, which have their own inherent problems. Approval voting has a lot in its favor and seems ripe for widespread implementation.

We know from **Arrow's impossibility theorem** (Spotlight 9.3, p. 254) that there can never be a perfect voting system. To select a voting system is to compromise between the different shortcomings inherent in each. Nonetheless, social-choice theorists continually strive to create (or perhaps rediscover) better voting schemes in an attempt to minimize such flaws.

REVIEW VOCABULARY

Approval voting Each voter indicates approval or disapproval for each candidate or issue on a ballot, as opposed to voting for only one candidate or issue.

Arrow's impossibility theorem The discovery by Kenneth J. Arrow that any voting system can give undesirable outcomes.

Borda count Assigning points to voters' preferences and summing the points for each candidate to determine the winner.

Condorcet (or voting) paradox Candidate A beats B, B beats C, and C beats A.

Condorcet winner A candidate who beats every other candidate in a one-on-one ballot.

Insincere voting Voting contrary to one's true preferences in an attempt to obtain a better outcome in the long run.

Law of transitivity If A wins over B and B wins over C, then A must win over C.

Majority More than half of the votes cast.

Plurality The case where a candidate with the most votes in a multicandidate race is declared the winner. The number of votes for the winner could be less than half.

Preference schedule A list of possible outcomes in the order a voter most prefers them.

Sequential voting A voting procedure in which successive ballots are taken for the purpose of eliminating some candidates or issues before the final vote.

Sincere voting   Voting in a manner consistent with one's preference schedule. One always votes for the most-preferred candidates or outcomes on each ballot.

Strategic voting   Voting insincerely on a ballot in an attempt to achieve a more preferable final outcome than could have resulted by voting sincerely.

SUGGESTED READINGS

Brams, Steven J., and Peter C. Fishburn: *Approval Voting,* Birkhäuser, Boston, 1982. This volume is a research-level work on developments in the recently popular (but rediscovered) method now called approval voting. However, the first chapter is a highly readable and superb introduction to this voting method and its uses.

Davis, Morton D.: *Mathematically Speaking,* Harcourt Brace, New York, 1980. In Chapter 6 there is an excellent introduction to the problem of voting, which includes an elementary discussion of the properties desired of any voting method (Arrow's axioms), which no voting system can achieve in general.

Farquharson, Robin: *Theory of Voting,* Yale University Press, New Haven, 1969. Some of the examples presented in this chapter, as well as other illustrations, are discussed in more detail in this elementary but historically important monograph.

Luce, R. Duncan, and Howard Raiffa: *Games and Decisions,* John Wiley, New York, 1957. (Also available as a Dover paperback.) In Chapter 14 there is a more technical introduction to Arrow's axioms and theorem, as well as a proof of his famous impossibility theorem, which is suitable for upper-division undergraduates.

Malkevitch, Joseph, and Walter Meyer: *Graphs, Models and Finite Mathematics,* Prentice-Hall, Englewood Cliffs, N.J., 1974. In Chapter 10 there is an excellent introduction to the problem of voting. It includes an elementary discussion of the properties desired of any voting method (Arrow's axioms).

Roberts, Fred S.: *Discrete Mathematical Models,* Prentice-Hall, Englewood Cliffs, N.J., 1976. In Chapter 10 there is a more technical introduction to Arrow's axioms and theorem, as well as a proof of his famous impossibility theorem, which is suitable for upper-division undergraduates.

EXERCISES

1. How many different ways can a voter:

   a. rank 3 choices (when ties are not allowed)?

   b. rank 4 alternatives (without ties)?

   c. rank n potential outcomes (without ties)?

2. How many different ways can a voter:

   a. rank 3 choices when ties are not allowed, but *incomplete* rankings can be submitted (e.g., a first choice without giving a second or third choice)?

   b. rank 3 choices when ties are allowed and *complete* rankings are required?

3. Consider the trial in the Roman Senate example when the preference schedule of the Roman senators is given by the following table. The group favoring banishment now splits into two factions B and B' with different second and third choices.

	Group of senators			
	A	B	B'	C
First choice	a	b	b	c
Second choice	b	a	c	b
Third choice	c	c	a	a
Percent	40	20	15	25

a. What verdict would result if they used the sequential agenda in Figure 9.2 in the text, and the senators voted sincerely at each step?

b. What verdict would result if they used the sequential agenda in Figure 9.3 in the text, and the senators voted sincerely at each ballot?

c. What verdict would result if they used a Borda count which assigned 3, 2, and 1 point(s) for a first, second, and third choice, respectively, and the senators voted sincerely?

d. What verdict would result if they used the Condorcet method, and the senators voted sincerely in each pairwise comparison?

●**e.** Discuss the possibilities for strategic voting in cases **a, b, c,** and **d** above.

4. What would be the outcome in the trial in the Roman Senate example if some senator moved that they first vote on whether they wanted the verdict of banishment or not? (If b did not win on this first ballot, then they would vote between a and c.)

a. Assume the senators vote sincerely at each stage.

●**b.** Assume that the voters vote strategically (insincerely) whenever it is to their advantage to do so.

5. The 10 members of a party's platform committee must pick one issue to receive the highest priority in the upcoming campaign. The three contenders are defense D, education E, and health H, and their preference schedules are as follows:

	Number of members		
	4	3	3
First choice	D	E	H
Second choice	E	H	D
Third choice	H	D	E

● Optional exercise.

a. Which issue wins if they first vote between E and H, and then vote between this initial winner and D?

b. Which issue wins if they first vote between D and E, and then vote between this initial winner and H?

c. Could those who most prefer E vote insincerely in some way to change the outcomes in **a** or **b** in a way that benefits them?

d. Which issue wins if they use a Borda count that scores 3 points, 2 points, and 1 point for each first choice, second choice, and third choice, respectively?

e. Could those who most prefer H vote insincerely in some way so as to change the outcome in **d** to their advantage?

6. One hundred voters who are to elect one of the three candidates A, B, or C have the following preference schedules:

	Number of voters			
	38	30	25	7
First choice	A	C	B	B
Second choice	B	A	C	A
Third choice	C	B	A	C

a. Which candidate wins an election using the plurality method?

b. Who wins if there is a runoff election between the top two finishers in the initial plurality ballot in **a**?

c. Who would win in **a** and **b**, respectively, if the 7 voters who prefer B over A and A over C were to switch their preference ranking to A over B over C?

d. If the 45 voters who now prefer A over B over C (after the switch made in **c**) knew everyone's preference schedule, could they vote more strategically to ensure a victory for A when the voting method in **b** is used?

7. Thirteen students decide to vote on whether to play baseball B, soccer S, or volleyball V at their picnic. Their schedule of preference is as follows:

	Number of students			
	5	2	4	2
First choice	B	S	V	V
Second choice	S	V	B	S
Third choice	V	B	S	B

a. Which sport wins if they use the plurality method?

b. Which one wins if they use a Borda count that assigns 3, 2, and 1 points to each first, second, and third choice, respectively?

c. Which one wins if they first eliminate the one with the fewest first-place votes and hold a runoff between the other two?

d. Which one wins if they first eliminate the one with the most last-place votes and have a runoff between the other two? Is the method decisive in this case?

e. Which one wins in **d** if the last two students misrepresent their preference ranking and pretend it is V over B over S rather than V over S over B as listed in the table?

f. Would there be a Condorcet winner if the students did vote sincerely?

8. One hundred sports writers with the following preference schedules are to pick the best college football team among Alabama A, Michigan M, and Washington W:

	Number of writers		
	52	38	10
First choice	W	M	A
Second choice	M	W	M
Third choice	A	A	W

a. Which team wins if the election is by a Borda count that assigns 3, 2, and 1 points to each first, second, and third choice, respectively?

b. If those who most favor Michigan suspected that Washington would win and thus voted insincerely for Alabama as their second choice, what would the outcome be?

c. If the supporters of Washington believed that the insincere voting in **b** might take place, could they still vote so as to guarantee that Washington wins?

9. Eleven students must decide whether to dine together at a Chinese, Italian, or Mexican restaurant. Their preference schedules are as follows:

	Number of students		
	5	2	4
First choice	Chinese	Mexican	Italian
Second choice	Mexican	Italian	Mexican
Third choice	Italian	Chinese	Chinese

a. What choice will the group make if they vote sincerely according to the following methods:

(1) the plurality method

(2) eliminating the restaurant with the fewest first-place votes and having a runoff between the other two

(3) eliminating the restaurant with the most last-place votes and having a runoff between the other two

b. Is any restaurant a Condorcet winner?

c. What choice will be made if they use a Borda count that assigns x points to each first choice, y points to each second choice, and z points to each third choice when

(1) $x = 3$, $y = 2$, and $z = 1$?

(2) $x = 4$, $y = 2$, and $z = 1$?

(3) $x = 5$, $y = 2$, and $z = 1$?

● **d.** Is there any way to pick the points x, y, and z in **c** with $x > y > z$ so that the Italian restaurant wins the Borda count?

10. The result of a swim meet between the four schools A, B, C, and D is given by the following table:

	Event				
Place	Sprint	Distance	Relay	Medley	Dive
First	B	A	B	A	B
Second	D	D	D	C	D
Third	D	D	C	B	C
Fourth	C	B	A	D	D
Fifth	C	B	B	C	D

a. How do the teams rank in this competition if the finishing positions are scored 5, 4, 3, 2, and 1, respectively?

b. How do the teams rank if the positions are scored 5, 3, 2, 1, and 0, respectively?

c. If the first three finishers in the Dive event were actually tied for first place, what would the teams' scores be

(1) in case **a**?

(2) in case **b**?

d. If the first two finishers in the Dive were disqualified and the new order of finish in this event is C, D, D, A, and A, then what are the teams' scores

(1) in case **a**?

(2) in case **b**?

● Optional exercise.

11. The result of a gymnastics meet between three teams A, B, and C is given by the following table:

	Event			
Place	Beam	Floor	Vault	Bars
First	A	C	C	B
Second	A	B	B	C
Third	C	A	A	A
Fourth	B	C	A	B

a. How do the teams rank in this competition if the finishing positions are scored 4, 3, 2, and 1, respectively?

b. How do the teams rank if the positions are scored 3, 2, 1, and 0, respectively?

c. If the first two finishers in the Bars event were actually tied for first place, what would the teams' scores be

(1) in case **a**?

(2) in case **b**?

d. If the first finisher in the Floor was disqualified and the new order of finish in this event is B, A, C, and A, then what are the teams' scores

(1) in case **a**?

(2) in case **b**?

12. Ten board members vote by approval voting on eight candidates for new positions on their board as indicated in the following table. An X indicates an approval vote. For example, voter 1, in the first column, approves of candidates A, D, E, F, and G, and disapproves of B, C, and H.

	Voters									
Candidates	1	2	3	4	5	6	7	8	9	10
A	X	X	X			X	X	X		X
B		X	X	X	X	X	X	X	X	
C			X					X		
D	X	X	X	X	X		X	X	X	X
E	X		X		X		X		X	
F	X		X	X	X	X	X	X		X
G	X	X	X	X	X			X		
H		X		X		X		X		X

a. Which candidate is chosen for the board if just one of them is to be elected?

b. Which candidates are chosen if the top four are selected?

c. Which candidates are elected if 80% approval is necessary and at most four are elected?

d. Which candidates are elected if 60% approval is necessary and at most four are elected?

13. The 45 members of a school's football team vote on three nominees A, B, and C by approval voting for the award of "most improved player" as indicated in the following table. An X indicates an approval vote.

Nominee	7	8	9	9	6	3	1	2
A	X			X	X		X	
B		X		X		X	X	
C			X		X	X	X	

Number of voters

a. Which nominee is selected for the award?

b. Which nominee gets announced as runner-up for the award?

c. Note that two of the players "abstained," i.e., approved of none of the nominees. Note also that one person approved of all three of the nominees. What would be the difference in the outcome if one were to "abstain" or "approve of everyone"?

14. Given that three members of a four-person committee prefer a newly proposed bill N over the old existing law O, can you suggest the type of amendment M to N that the advocate of O should propose in an attempt to defeat N (and M). Note that M or N must receive three or four of the four votes cast in order to pass, whereas the existing law wins on a tie vote of 2 to 2.

15. Consider the example in the text section "Everyone Wins," in which 55 delegates at a national convention vote on five candidates for the party's presidential nominee. Determine the winning candidate if they used a voting method that eliminates the loser at each step: at each ballot, eliminate the candidate with the most last-place votes, and then continue with successive ballots with all 55 delegates voting each time.

16. Assume that the members A, B, and C of a three-person committee have the following preference schedules over the three possible outcomes a, b, and c:

	A	B	C
First choice	a	b	c
Second choice	b	c	a
Third choice	c	a	b

Member

Each member can vote secretly for one outcome, and the majority rules. Furthermore, A is the chairman and has the power to break tie votes.

    a. What would the result be if each member voted sincerely for his or her most-preferred outcome?

    b. What do you expect to actually happen in this situation?

    c. Can you explain why this example is often referred to as *the chairman's paradox?*

● **17.** Can you prove that $\sqrt{2}$ is an irrational number? Hint: Assume to the contrary that $\sqrt{2} = p/q$, where p and q are integers and p/q is a fraction expressed in *lowest terms.* Then note that $p^2 = 2q^2$ and so $p = 2r$ is an even number, and thus $p^2 = (2r)^2 = 4r^2 = 2q^2$ and so q is also an even number.

18. Consider the following class project: Pick some upcoming election involving more than two alternatives. For example, select a few of the leading candidates for a major party's presidential nominee. Compare the class results for the following different voting methods:

    a. Vote for only one candidate and select the winner by the plurality method.

    b. If the winner in **a** does not have a majority, then hold a runoff ballot.

    c. Have each voter provide his or her preference schedule (that is, each ranks the candidates) and then select the winner by a Borda count.

    d. Use the method of approval voting where the winner is the one with the largest number of approval votes.

    e. Is there a Condorcet winner?

● Optional exercise.

·10·

Weighted Voting Systems: How to Measure Power

In many voting situations, each citizen has an equally weighted vote, called "one man, one vote." Each person clearly asserts equal influence on the outcome. Each is equally powerful. On the other hand, many voting situations occur in which the participants have different numbers of votes. Such systems are referred to as *weighted voting systems.* Here, some individuals may have greater influence than others on the results. Those individuals who cast more heavily weighted votes may or may not turn out to be more powerful, and blocs of voters, called **coalitions,** can typically increase their power (see Spotlight 10.1, p. 267).

For example, the shareholders' votes in a public corporation are weighted by the number of shares of stock they own. In the U.S. Electoral College, each state typically casts its total bloc of votes for one presidential candi-

date. The number of votes per state currently varies from 3 for states with small populations to 47 for California. Legislative bodies with strong party discipline can be viewed as weighted voting systems in which the parties, rather than the individual representatives, are the voting units.

In this chapter, we will be primarily interested in the notion of *power.* We will attempt to arrive at a mathematical measure for the power of an individual voter or of a bloc of voters in a weighted voting system. We will see that a person's influence is not always proportional to the fraction of the votes he or she casts. In a formal sense, power is a much more irregular function of the distribution of weights among the voters as well as of the fraction, or *quota,* of votes necessary to pass legislation and the total number of voters. We will first develop an index for measuring power, which is commonly referred to as the *Banzhaf power index,* and then return to several familiar institutions that actually make use of weighted voting.

HOW WEIGHTED VOTING WORKS

To see how weighted voting works, let's consider New York's Nassau County in the late 1950s, which had a county board of supervisors consisting of six representatives from five separate districts. Two of the representatives were elected at large from one district, the city of Hempstead. Because the districts had unequal populations, the representatives' votes were weighted accordingly. The names of the five districts and the number of votes for each supervisor are listed in Table 10.1.

A close look at these numbers reveals a surprising fact. The two representatives from Hempstead share a total of 18 votes. This is greater than 15, or over half of the total for the entire board. The two representatives from Hempstead could therefore pass any issue requiring only a simple majority quota, that is, 16 of the 30 votes.

TABLE 10.1 The Nassau County board in the 1950s

Municipality	1954 population	1958 number of votes
Hempstead⎤ Hempstead⎦	618,065	⎰ 9 ⎱ 9
North Hempstead	184,060	7
Oyster Bay	164,716	3
Glen Cove	19,296	1
Long Beach	17,999	1
Totals	1,004,136	30

(Actually, the charter for Nassau County at the time had a secondary condition that in fact disallowed any *one* district from passing a bill whenever it was opposed by the other four districts. Because the first two representatives both come from the same municipality, Hempstead, this provision actually prevented this particular pair from forming a winning coalition. Although this secondary rule alters the analysis presented here, including the existence of dummy voters described below, for purposes of illustration we will ignore this secondary condition in our discussion. Other weighted voting systems used by counties in New York State and elsewhere have had powerless voters with no such secondary provision to save them from this fate.)

The supervisor from North Hempstead, who has 7 votes, can join with *either one* of the two from Hempstead, and the resulting sum of 16 is also a simple majority (unaffected by the secondary provision). Thus, any two of those first three representatives have sufficient weight to pass a bill. Apart from the special condition, which we are ignoring, these three legislators do indeed share equal power.

Ironically, a close look at the figures for Oyster Bay, Glen Cove, and Long Beach reveals that these last three supervisors possess no power whatsoever. Because they share a total of only 5 votes, they cannot team up with *any* of the other representatives to turn a losing coalition into a winning one. If they joined one of the representatives from Hempstead, their votes would total 14. If they joined the representative from North Hempstead, their votes would total 12. In either case they fall short of the simple majority quota of 16 necessary to pass most issues in this county.

We can see that in the case of a simple majority, these three supervisors are left without any actual voting power, even though they may have other kinds of influence on the decision-making process; for example, they may serve on committees or draw up and defend

SPOTLIGHT 10.1 Sets and Subsets

The most primitive notion in contemporary mathematics is the idea of a set. A *set* is a collection of distinct objects of our imagination. The objects in the set are called *elements*. The great German mathematician Georg Cantor (1845–1918), who advocated this approach, described a set as

a bringing together into a whole of definite well-distinguished objects of our perception or thought—which are to be called the elements of the set.

Our concern in this chapter is with sets whose elements are voters and whose subsets are the various coalitions or blocs of voters.

It should be emphasized that one does not give definitions of the words "set" or "element." A theory must begin with some undefined terms, plus some rules of logical inference, before it proceeds to introduce new, defined terms and initial axioms and finally goes on to deduce new theorems.

We can talk about a set of physical objects in the external world, such as the set of students in the classroom, the set of cars in the parking lot, or the set of planets in our solar system. In mathematics we usually speak about mental objects, and we represent them with abstract symbols. We represent a set of particular elements by listing these elements between braces. Here are some examples of sets that arise in mathematics:

1. The set of integers from 1 through 3, which is written $\{1, 2, 3\}$.

2. The set of vertices, or corner points, of a unit square. Using the typical Cartesian coordinates in a plane, we can express these as $\{(0, 0), (0, 1), (1, 0), (1, 1)\}$.

3. The set of edges of a cube.

4. The set of roots of the equation $x^2 - 5x + 4 = 0$, which is $\{1, 4\}$.

5. The set of points in the plane that are inside the circle described by the equation $x^2 + y^2 = 4$. This can be expressed as $\{(x, y): x^2 + y^2 < 4\}$.

We also consider the set that contains no elements whatsoever. This is called the *empty set* and will be denoted by the symbol $\varnothing$.

The number of distinct elements in a set is called its **cardinality.** For example, the cardinalities of the set examples 1 through 5 are 3, 4, 12, 2, and infinity, respectively.

A set T is called a *subset* of a set S if every element in T is also an element of S. We also say that S is a *superset* of T in this case. For example, the set $\{1\}$ is a subset of the set $\{1, 2, 3\}$ in example 1 and of the set $\{1, 4\}$ in example 4 (above). Any set is considered a subset of itself, whereas the empty set $\varnothing$ is a subset of every set. However, a subset T of S is said to be a *proper* subset of S if T is not all of S and T is not $\varnothing$, the empty set. A set with only one element, that is, a set of cardinality 1, has 2 subsets: itself and the empty set. The set of all subsets of $\{1, 4\}$ is $\{\{1, 4\}, \{1\}, \{4\}, \varnothing\}$. The set $\{1, 2, 3\}$ has 8 subsets: $\{1, 2, 3\}, \{1, 2\}, \{1, 3\}, \{2, 3\}, \{1\}, \{2\}, \{3\}$, and $\varnothing$. A set with four elements has 16 subsets. In general, a set of cardinality n has 2^n subsets. If T is a subset of S, then the elements in S that are not in T form the *complementary* set $S - T$ of T in S. For example, the set $\{2, 3\} = \{1, 2, 3\} - \{1\}$ is called the *complement* of the subset $\{1\}$ in $\{1, 2, 3\}$.

potential legislation. But as it stands, they are essentially disenfranchised by the initial constitutional structure of the board's voting rules. A voter who cannot turn a losing coalition into a winning one is called a **dummy.** A dummy can likewise never cause a winning coalition to lose by singly defecting from it.

Some time later, the weights for the Nassau County board were changed (see Table 10.2). In this case, the two representatives from Hempstead and the one from Oyster Bay now share equal power in the typical case of a simple majority vote of 58. The remaining three supervisors are dummies. Note that between 1958 and 1964, the third representative, the one from North Hempstead, went from having an equal share of the power to dummy status, whereas the one from Oyster Bay had the reverse experience and became one of the three sharing full power.

By 1970 they had changed the simple-majority quota necessary to win from 58 to 63 but maintained the same weights they used in 1964. In 1976 they changed the weights to 35, 35, 23, 32, 2, and 2, respectively, and used a winning quota of 71. In these last two cases, the quota is somewhat more than a bare simple majority of the total weights. Nevertheless, it results in at least some power for each representative.

TABLE 10.2 The Nassau County board in the 1960s

Municipality	1960 population	1964 number of votes
Hempstead ⎱		31
Hempstead ⎰	728,625	31
North Hempstead	213,225	21
Oyster Bay	285,545	28
Glen Cove	22,752	2
Long Beach	25,654	2
Totals	1,275,801	115

Mathematical Notation for Weighted Voting

We can represent a **weighted voting system** involving n voters by listing $n + 1$ numbers:

$$[q: w_1, w_2, \ldots, w_n]$$

The n numbers $w_1, w_2, \ldots, w_n$ are the respective *weights* for the n voters who are indicated by $1, 2, \ldots, n$. The initial number q is the number of votes necessary to pass an issue and is called the **quota.** For example, the Nassau County Board of Supervisors for 1958 can be expressed as

$$[16: 9, 9, 7, 3, 1, 1]$$

and in 1976 as

$$[71: 35, 35, 23, 32, 2, 2]$$

A subset of voters is called a **winning coalition** whenever the sum of its weights equals or exceeds this value q. It is usually assumed that the quota q necessary to win exceeds $\frac{1}{2}w$, where w is the sum of all the weights:

$$w = w_1 + w_2 + \cdots + w_n$$

EXAMPLES: Illustrations of Weighted Voting Systems

1. Consider a small corporation owned by two people who possess 60% and 40% of the stock, respectively. If measures are allowed to pass by a simple majority, we express this as

$$[51: 60, 40]$$

Clearly, the first shareholder maintains full power and is in effect a **dictator.**

2. Let's examine a different company, where the stock is split among three individuals holding 49%, 48%, and 3%: [51: 49, 48, 3]. There is no dictator in this case. Any coalition of two or more has a simple majority, so that power is equally divided among the three stockholders. Although the third person holds only 3% of the stock, this stockholder has equal influence. His or her 3%, added to either of the other two stockholders' votes, would sway the majority.

3. In contrast, look at the case where there are four owners with 26%, 26%, 26%, and 22% of the stock: [51: 26, 26, 26, 22]. The last person, with 22% of the stock, is a dummy. This stockholder is not capable of providing a losing coalition with enough power to win. The power in this case is equally divided among the first three shareholders.

4. Finally, let's assume that the representatives to a national assembly are split along party lines according to the numbers 45, 43, 8, and 4: [51: 45, 43, 8, 4]. Any two of the first three parties can form a winning coalition, whereas the smallest party has dummy status. (It is said that the Liberal Party in Great Britain "lives for the day" when the numbers divide in such a manner.)

These examples clearly illustrate that power need not be even approximately proportional to one's share of the vote. Power has a more complicated relationship to the weights, and we will proceed to examine that relationship.

THE BANZHAF POWER INDEX

The notion of power is fundamental to understanding and explaining political events. Power is also a basic ingredient of group decision making. Because power is an illusive concept, many of its aspects are difficult to identify and measure. It is unlikely that any simple mathematical formula can capture much of the essence of power.

Nonetheless, we will attempt to arrive at a numerical index for measuring power in the abstract, at least for such highly structured voting bodies as weighted voting systems. The

resulting index should provide some useful insights for the design of equitable voting structures that arise in representative democracies (see Spotlight 10.2, p. 270).

We have already seen from our examples above that individual voters or blocs of voters cannot simply take their fraction of the total vote as a meaningful indication of their share of power. The importance of an individual or of a coalition is directly related to the ability to enforce its own will. Power relates to achieving the desired end—in short, power is winning.

An individual can frequently appear on the winning side, however, without being powerful. For example, not all professional athletes on a team that regularly wins can demand high salaries. The truly powerful, high-salaried players are those who are crucial or decisive in bringing about a win. Similarly, the real significance of a vote is whether it is essential to victory.

So one reasonable measure of formal power is the frequency with which a voter or vote is **pivotal** (or marginal or critical) to a decision. Power has to do with the number of different ways one person alone can turn defeat into victory, or vice versa. In other words, it rides on whether the switch in one vote can reverse the whole outcome.

We will therefore take as our relative measure of power the number of *different* ways an individual can join a losing coalition and thereby turn it into a winning coalition. (This is identical to the number of *distinct* winning coalitions that would lose if this person defected.) The number of ways one can swing losing to winning, or equivalently, winning to losing, determines the **Banzhaf power index.**

EXAMPLE: A Three-Person Committee.
Consider the weighted voting system [3: 2, 1, 1], with quota 3 and weights, 2, 1, and 1. This could be a three-member committee with sim-

SPOTLIGHT 10.2 Power Indices

Lloyd S. Shapley.

Martin Shubik.

John F. Banzhaf III.

James A. Coleman.

The first widely accepted numerical index for assessing power in voting structures called *simple games* was proposed by mathematician Lloyd S. Shapley (1923–) of UCLA and economist Martin Shubik (1926–) of Yale University. Their index is a special case of the well-known Shapley value, which plays a fundamental role as a fairness concept in game theory and mathematical economics. A particular voter's power in this case is proportional to the number of different *permutations* (or orderings) of all the voters in which he or she casts the pivotal vote—the vote that first turns losing into winning.

The Banzhaf power index described in this chapter was introduced independently by John F. Banzhaf III (1940–), a George Washington University lawyer and well-known consumer advocate, and social scientist James A. Coleman (1926–) of the University of Chicago. Their index is the one most often used in actual court rulings, perhaps because early cases were brought to court by Banzhaf. A voter's power is proportional to the number of different possible voting *combinations* (or arrangements) in which he or she plays a pivotal role.

ple majority rule, except that the first person, the chairperson, has veto power over any outcome. The winning coalitions are those with weights summing to 3 or 4: {1, 2}, {1, 3}, and {1, 2, 3}. The coalitions {1, 2} and {1, 3} are **minimal winning,** in the sense that each of the members is essential to remaining winning. The nonwinning coalitions are **losing:** {2, 3}, {1}, {2}, {3}, and ∅, the empty set.

Voter 1 can join a losing coalition and turn it into a winning coalition in three distinct ways:

1 joins {2} to obtain {1, 2}

1 joins {3} to obtain {1, 3}

1 joins {2, 3} to obtain {1, 2, 3}

Thus, the power index of voter 1 is 3. Voter 2 can turn a loss into a win in only one way:

2 joins {1} to obtain {1, 2}

Similarly, voter 3 can also change losing to winning in just one way:

3 joins {1} to obtain {1, 3}

Thus, voters 2 and 3 each have a power index of 1.

These five pivotal cases can be pictured as follows:

$$\{1, 2\} \qquad \{1, 3\} \qquad \{1, 2, 3\}$$

$$\uparrow 1 \qquad \uparrow 1 \qquad \uparrow 1$$

$$\{2\} \qquad \{3\} \qquad \{2, 3\}$$

$$\{1, 2\}$$

$$\uparrow 2$$

$$\{1\}$$

$$\{1, 3\}$$

$$\uparrow 3$$

$$\{1\}$$

We summarize the Banzhaf index for this example as (3, 1, 1). Note that voter 1, the chairperson, has an index three times that of the other committee members. It is not the absolute magnitude of these numbers that matters, but rather their relative values. One could also express the Banzhaf power index for this committee as $(6, 2, 2)$, $(\frac{3}{5}, \frac{1}{5}, \frac{1}{5})$, or $(\frac{6}{8}, \frac{2}{8}, \frac{2}{8})$ for reasons that will become clear later.

Computing the Power Index

It is not difficult to calculate the power index for a weighted voting system when there are only a few voters. We merely list all the theoretically possible ways that the individuals can vote, that is, all the different combinations of yes and no votes. In the case of n voters there will be 2^n such combinations. We would then examine each combination in turn to see which votes are critical.

Consider the three-person committee [3: 2, 1, 1] presented above. Table 10.3 lists the eight distinct combinations of voters 1, 2, and 3, according to whether they vote yes, denoted by Y, or no, indicated by N. Whether the issue will pass (P) or be defeated (X) is indicated in the outcomes below each of the combinations.

We must examine each column in the table to see which voters are pivotal to the outcome. This means checking each yes or no vote in each combination to determine whether a

TABLE 10.3 Each combination of votes either passes or fails

Voter	Weight	Combinations of votes							
1	2	Y	Y	Y	N	Y	N	N	N
2	1	Y	Y	N	Y	N	Y	N	N
3	1	Y	N	Y	Y	N	N	Y	N
Total Y weights		4	3	3	2	2	1	1	0
Outcome: Pass (≥3)		P	P	P					
Fail (<3)					X	X	X	X	X

switch of the one vote will cause a change in the result.

For example, the first combination

$$
\begin{array}{c}
Y \\
Y \\
\underline{Y} \\
P
\end{array}
$$

results in the issue passing by a unanimous vote. But if the first voter changes his or her vote from yes to no

$$
\begin{array}{cc}
Y \longrightarrow N \\
Y \quad\quad Y \\
\underline{Y} \quad\quad \underline{Y} \\
P \quad\quad X
\end{array}
$$

then the outcome changes to "defeat." We will indicate that this vote is pivotal by circling the Y in the first row of the first column of the table:

$$
\begin{array}{c}
ⓎY \\
Y \\
\underline{Y} \\
P
\end{array}
$$

On the other hand, if only voter 2 switches his or her vote from yes to no in this first combination, the end result remains the same — the issue still passes by a weight of 3 "for" to 1 "against":

$$
\begin{array}{cc}
Y \quad\quad Y \\
Y \longrightarrow N \\
\underline{Y} \quad\quad \underline{Y} \\
P \quad\quad P
\end{array}
$$

Similarly, voter 3 is not pivotal in this first combination:

$$
\begin{array}{cc}
Y \quad\quad Y \\
Y \quad\quad Y \\
\underline{Y} \longrightarrow \underline{N} \\
P \quad\quad P
\end{array}
$$

Let's now consider the second column in our table:

$$
\begin{array}{c}
Y \\
Y \\
\underline{N} \\
P
\end{array}
$$

If voter 1 changes from yes to no, the outcome goes from pass to defeat:

$$
\begin{array}{cc}
Ⓨ \longrightarrow N \\
Y \quad\quad Y \\
\underline{N} \quad\quad \underline{N} \\
P \quad\quad X
\end{array}
$$

Likewise, if only voter 2 switches his or her vote, the result again goes from pass to defeat:

$$
\begin{array}{cc}
Y \quad\quad Y \\
Ⓨ \longrightarrow N \\
\underline{N} \quad\quad \underline{N} \\
P \quad\quad X
\end{array}
$$

So voters 1 and 2 are pivotal to the second combination, but voter 3 is not pivotal; the outcome remains the same even if voter 3 decides to switch:

$$
\begin{array}{cc}
Y \quad\quad Y \\
Y \quad\quad Y \\
\underline{N} \longrightarrow \underline{Y} \\
P \quad\quad P
\end{array}
$$

Accordingly, we can proceed to each column in Table 10.3, checking whether each voter is pivotal and circling the corresponding Y and N. For example, in the third combination, 1 and 3 are critical:

$$
\begin{array}{cccc}
Ⓨ \longrightarrow N & Y & Y \\
N & N & N & N \\
\underline{Y} & \underline{Y} & Ⓨ \longrightarrow N \\
P & X & P & X
\end{array}
$$

In the fourth column, only voter 1 is pivotal, because his or her vote changes defeat to pass:

$$
\begin{array}{ccc}
Ⓝ & \longrightarrow & Y \\
Y & & Y \\
\dfrac{Y}{X} & & \dfrac{Y}{P}
\end{array}
$$

In the fifth column, 2 and 3 are pivotal:

$$
\begin{array}{cccc}
Y & Y & Y & Y \\
Ⓝ \longrightarrow Y & & N & N \\
\dfrac{N}{X} & \dfrac{N}{P} & Ⓝ \longrightarrow \dfrac{Y}{X} & \dfrac{Y}{P}
\end{array}
$$

In the sixth and seventh columns, only 1 is pivotal:

$$
\begin{array}{cccc}
Ⓝ \longrightarrow Y & & Ⓝ \longrightarrow Y \\
Y & Y & N & N \\
\dfrac{N}{X} & \dfrac{N}{P} & \dfrac{Y}{X} & \dfrac{Y}{P}
\end{array}
$$

In the last column, no voter is pivotal.

In summary, our original table now takes the form shown in Table 10.4.

Finally, if we total up the number of circles in each voter's row, we arrive at the power index of (6, 2, 2). This result of 6, 2, and 2 corresponds to the previously obtained indices of 3, 1, and 1, because our second approach counts each pivotal vote twice: it counts the number of times a voter can turn winning into losing (P into X) as well as the times this voter can turn losing into winning (X into P). This double counting, however, leaves the *relative* magnitude, or ratios, of the power indices the same.

This procedure for computing the power index is straightforward when the number of voters n is rather small. But as n increases, the number of combinations 2^n grows rapidly. We would need to enlist the help of a computer to calculate the power index. For still larger values of n, this method of computation may

TABLE 10.4 The pivotal voters in each combination

Voter	Weight	Combinations of votes							
1	2	Ⓨ	Ⓨ	Ⓨ	Ⓝ	Y	Ⓝ	Ⓝ	N
2	1	Y	Ⓨ	N	Y	Ⓝ	Y	N	N
3	1	Y	N	Ⓨ	Y	Ⓝ	N	Y	N
Total Y weights		4	3	3	2	2	1	1	0
Outcome: Pass (≥3)		P	P	P					
Fail (<3)					X	X	X	X	X
Pivotal voters:		1	1 2	1 3	1	2 3	1	1	

become prohibitive, even with the help of a computer.

EXAMPLE: A Four-Person Corporation. Consider the weighted voting system [51: 40, 30, 20, 10]. This could be four shareholders in a corporation who have 40%, 30%, 20%, and 10% of the stock, respectively, where a simple majority (taken here as 51%) is necessary to win on any ballot.

To find the Banzhaf power index for this case, one begins by listing the $2^4 = 16$ distinct combinations of yes (Y) and no (N) votes for the four shareholders, as is indicated in Table 10.5. The total percent of yes votes for each such combination (row in the table) is indicated to its right. An issue passes or fails, indicated by P or X, when the percent of yes votes exceeds the quota of 51%. One then must examine each vote in each combination to determine whether or not it is pivotal to the result. Will the change of just this one yes vote Y to a no vote N cause the outcome to switch from pass (P) to fail (X)? Or else, will the change of a single N to a Y cause the outcome to switch from X to P? Each such pivotal vote is circled in Table 10.5. Counting the number of circles for each shareholder results in the Banzhaf power index (10, 6, 6, 2). It is interesting to

TABLE 10.5 The Banzhaf power index for the corporation

Voter:	1	2	3	4			
Weight (%):	40	30	20	10			
Combinations					%Y	P	X
	Y	Y	Y	Y	100	P	
	(Y)	Y	Y	N	90	P	
	(Y)	(Y)	N	Y	80	P	
	(Y)	N	(Y)	Y	70	P	
	N	(Y)	(Y)	(Y)	60	P	
	(Y)	(Y)	N	N	70	P	
	(Y)	N	(Y)	N	60	P	
	(N)	Y	Y	(N)	50		X
	Y	(N)	(N)	Y	50		X
	(N)	Y	(N)	Y	40		X
	(N)	(N)	Y	Y	30		X
	Y	(N)	(N)	N	40		X
	(N)	Y	N	N	30		X
	(N)	N	Y	N	20		X
	N	N	N	Y	10		X
	N	N	N	N	0		X
No. of pivots:	(10,	6,	6,	2)	= Banzhaf index		

observe that the second and third shareholders have different amounts of stock but equal power in the actual voting.

Note again that our procedure of listing all combinations and counting each pivotal vote actually does a "double count." For each combination in which a voter is pivotal by changing Y to N, there is the resulting combination in which this voter's change from N to Y will also be pivotal by switching X back to the P they started from. So one needs to count only the Y to N changes that result in a P to X switch; or else just the N to Y changes that result in an X to P switch. For example, in Table 10.5, both the former and latter counts result in (5, 3, 3, 1). Either this or the previously obtained (10, 6, 6, 2) can be taken as the Banzhaf index, since the relative size or

ratios among these numbers rather than their absolute magnitudes is what is important.

Finding the Weights

Many existing voting situations are not explicitly described as *weighted* voting systems. In many of these cases, however, one can find appropriate weights $w_1, w_2, \ldots, w_n$ and a quota q so as to express these as weighted voting systems. The following example provides an illustration of this. On the other hand, well-defined voting rules can exist in practice that cannot be represented as weighted voting systems. Three examples of this possibility appear in Exercises 15, 16, and 17.

EXAMPLE: Australia. Some decisions in Australia have been made by means of the following voting procedure. Each of the six states can cast one vote, the federal government (FG) has two votes, and the majority rules. Furthermore, the FG can break a tie in its favor.

There are two types of minimal winning coalitions in this example: either five states without the FG, or two states plus the FG. This particular voting system is equivalent, for example, to the weighted voting system

$$[5: 3, 1, 1, 1, 1, 1, 1]$$

where the first voter, the FG, has a weight of $w_1 = 3$ that is three times that of any state. We see that the two types of minimal winning coalitions have a total weight equal to the quota 5.

One can compute the Banzhaf power index for this system as in the previous two examples by listing the $2^7 = 128$ combinations of yes and no votes and by examining each of the seven votes in each such combination to see whether or not this vote is pivotal. (One may wish to design a computer program to undertake this task.)

On the other hand, we observe that each of the six states appears in a symmetrical position, and thus they must all have the same Banzhaf index. This allows one who is familiar with some of the counting techniques that one learns in algebra and probability courses to arrive at the number of combinations in which each voter is pivotal. The FG can join a losing coalition and turn it into a winning one whenever it joins a coalition of any two, three, or four of the six states. The number of distinct ways it can do this is

$$\frac{6!}{4!2!} + \frac{6!}{3!3!} + \frac{6!}{2!4!} = 50$$

(Similarly, the FG can turn a winning coalition into a losing one in 50 ways.) A particular state can join a losing coalition L and thus turn it into a winning one in two different ways:

1. L consists of four *other* states.

2. L consists of one *other* state plus the FG.

The number of combinations in which case 1 can occur is $5!/4!1! = 5$. Similarly, case 2 can happen in five ways. This gives a total of 10 pivots for a given state. (This state can likewise turn winning into losing in 10 ways.)

We conclude that the ratio of the Banzhaf power index of the FG to a state is $100 \div 20$. In summary, the FG has two votes to one for a state, plus the power to resolve ties. This could be represented by a weighted voting system with weights 3 to 1 (see Exercise **6a**). The power ratio, however, is 5 to 1.

Veto Power

Another type of coalition worth noting is one that can prevent the remaining players from winning. A coalition has **veto power** if its complementary coalition is losing. In the

three-person committee example the coalitions $\{1, 2, 3\}$, $\{1, 2\}$, $\{1, 3\}$, $\{2, 3\}$, and $\{1\}$ all have veto power since their respective complementary coalitions $\varnothing$, $\{3\}$, $\{2\}$, $\{1\}$, and $\{2, 3\}$ are all losing. A *minimal veto-power coalition* is one that has veto power, but has no proper subset with veto power. In our example, $\{1\}$ and $\{2, 3\}$ are the minimal veto-power coalitions. (One could also analyze power by focusing on the minimal veto-power coalitions rather than the minimal winning coalitions, but we will not do so in this chapter.)

A Probabilistic Interpretation

If one were to make the assumption that each one of the 2^n combinations in an n-person voting system (weighted or not) were equally likely to occur, then the Banzhaf power indices, when divided by 2^n, could be interpreted as the probabilities that the voters will be pivotal in a given vote. In the Australia example the probability that the FG will be pivotal is $100/128 = 0.781$. The probability that a given state is pivotal is $20/128 = 0.156$. (Note that these probabilities need not sum to one, because none, one, or more than one voter might be pivotal in any particular combination.)

The assumption that every combination is equally likely to occur is rather unrealistic when considering any existing voting system. Differences in ideology may cause certain voters frequently to vote alike or to be often in opposition. Such bloc voting would make the appearance of some combinations more frequent than others. On the other hand, if one is designing a new constitution for an organization, it may well be a reasonable goal to assume all combinations as equally likely and to create a voting system whose Banzhaf index is used to measure fairness for the voting entities, whether individuals or multiperson voting units.

Optional A PICTORIAL VIEW OF WEIGHTED VOTING

We have seen that power depends more on which coalitions are winning than on the particular weight the voter holds. A voting scheme is merely a rule for dividing or partitioning the collection of all coalitions into two classes: those coalitions that win and those that lose. This viewpoint leads us to a highly geometric way to visualize weighted voting systems and the Banzhaf power index. First, however, we will describe a geometric way of viewing the subsets of a set.

One can picture the subsets of a set with cardinality 1, 2, 3, or 4 with the help of Figure 10.1. The *vertices* in these diagrams correspond

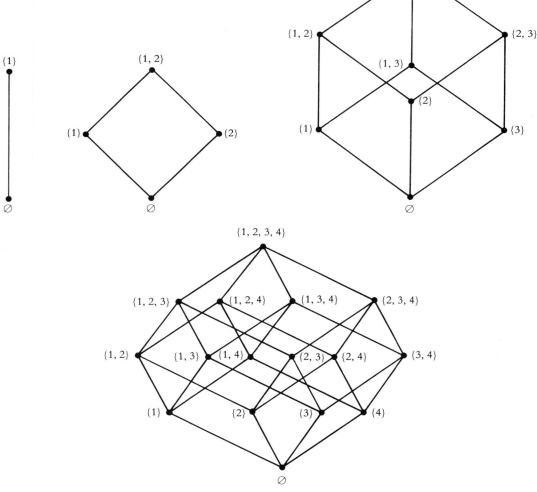

Figure 10.1 The subsets of sets with 1, 2, 3, and 4 elements.

to the subsets, and the relationship "is a subset of" is indicated by the *edges* in the diagrams. For example, the eight subsets of {1, 2, 3} can be represented by a cube. The edge arising from {1, 3} indicates that it is a subset of {1, 2, 3} and the two edges descending from {1, 3} to {1} and {3} show that the former is a superset of the latter.

In the case of three voters designated 1, 2, and 3, there are 2^3, or 8, subsets, as illustrated in the cube in Figure 10.1. A three-person voting rule divides these 8 subsets into the winning coalitions (typically the larger ones), which appear toward the top of the diagram, and the losing coalitions, which appear near the bottom. We assume that the grand coalition {1, 2, 3} is always winning. Such is the case of *unanimity,* whereas the empty set ∅ always loses.

Let's return to the three-person committee example [3: 2, 1, 1] with quota 3 and weights

2, 1, and 1. Recall that it was characterized by listing the winning coalitions

$$\{1, 2\}, \{1, 3\}, \text{ and } \{1, 2, 3\}$$

and losing coalitions

$$\{2, 3\}, \{1\}, \{2\}, \{3\}, \text{ and } \emptyset$$

This division of the subsets into two classes is indicated by the red curve in Figure 10.2, which we can call the **quota curve,** or **cut.** In other words, a voting system corresponds to removing just enough edges of the cube so that the vertices fall into two parts, as indicated in Figure 10.3. That is, the losing coalitions are disconnected from the winning ones.

The power index can be determined directly from Figure 10.4. Each edge in the cube corresponds to exactly one voter. For example, the right-hand edge, running from vertex {2, 3}

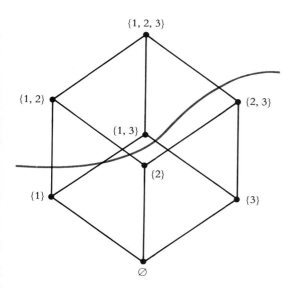

Figure 10.2 The red curve separates winning coalitions from losing coalitions.

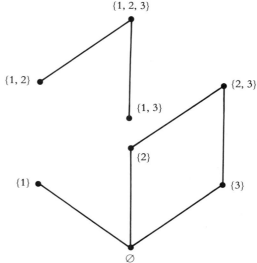

Figure 10.3 The cube after removing the edges connecting the winning coalitions to losing coalitions.

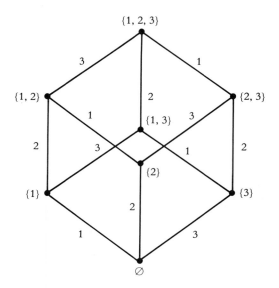

Figure 10.4 Each edge in the cube is labeled with its corresponding voter.

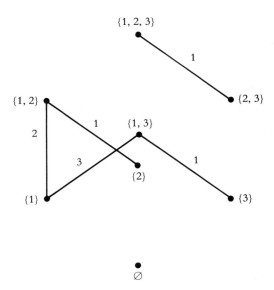

Figure 10.5 The edges connecting winning coalitions to losing coalitions.

down to vertex {3}, corresponds to voter 2. We can label each edge in our cube accordingly.

Furthermore, the edges that were removed in Figure 10.3 are precisely the ones that are pivotal. They are the edges connecting losing to winning coalitions in Figure 10.5. We need only count the number of times a given voter appears on such crucial edges to arrive at his or her power index.

For example, voter 1 appears on the critical edges joining {2} to {1, 2}, {3} to {1, 3}, and {2, 3} to {1, 2, 3}, for an index of 3. Voter 2 appears only once, on the left-hand edge connecting {1} to {1, 2}. Similarly, voter 3 appears only on the edge joining {1} to {1, 3}. Again, we arrive at the Banzhaf power index (3, 1, 1).

These latticelike diagrams, depicting the collection of all possible coalitions of voters, provide a highly pictorial way to describe voting systems, as well as a way to compute power indices. However, such representation becomes rather difficult when we consider more than three voters. In this case we must visualize the vertices and edges of higher-dimensional cubes, the so-called **hypercubes,** or ***n*-cubes,** which are described in the following section. •

HYPERCUBES, COORDINATES, AND COMBINATIONS

Multidimensional Cubes

The cube is one of the most commonly occurring and regular of geometric objects. There is a natural way to extend the idea of a cube to dimensions besides three. We can define a point as a *zero-dimensional cube.* A line segment can be thought of as a *one-dimensional cube;* it is built up by moving a point a unit distance in some direction. Similarly, a square in the plane can be viewed as a *two-dimensional cube.* A square can be obtained from a line segment of length 1, by moving it a unit distance perpendicular to its length. The regular cube in three-dimensional space can be called a *three-dimensional cube,* or a *3-cube.* The cube can be obtained from a square, or *2-cube,* by moving it into the third dimension.

Similarly, we can move a 3-cube a unit distance into a fourth dimension, generating a *four-dimensional cube,* or *4-cube.* (Cubes of dimensions 1, 2, 3, and 4 are illustrated in Figure 10.1.) We can continue in this fashion to generate cubes of still higher dimensions, the so-called multidimensional cubes, or *n-cubes*; these are also called *hypercubes.* However, our ability to visualize *n*-cubes geometrically fails us as the dimension *n* continues to increase.

Coordinates for *n*-Cubes

In analytic geometry we use ordered pairs of real numbers (x, y) to represent points in the plane. Similarly, we can use triples of numbers (x, y, z) or (x_1, x_2, x_3), to indicate points in three-dimensional space. We can continue in this manner, thinking of four-dimensional space as the set of all 4-tuples of numbers (x_1, x_2, x_3, x_4). In fact, one can generalize in this way to consider *n*-dimensional space as merely the *n*-tuples $(x_1, x_2, \ldots, x_n)$ of real numbers $x_1, x_2, \ldots, x_n$, where *n* can be any integer from 1 onward.

Using this technique, we can represent the vertices of a unit cube of any dimension in terms of coordinates that make use only of the numbers 0 and 1. We can represent the zero-dimension cube (a point) by the number, or coordinate, (0); the vertices of a 1-cube (the endpoints of a line segment) by the coordinates (0) and (1); the vertices of a 2-cube (a square) by the coordinates (0, 0), (1, 0), (0, 1), and

(1, 1); and the vertices of a 3-cube by the eight 3-tuples (0, 0, 0), (1, 0, 0), (0, 1, 0), (0, 0, 1), (1, 1, 0), (0, 1, 1), (1, 0, 1), and (1, 1, 1). We can continue in this fashion, representing the vertices of *n*-cubes in higher dimensions by means of 2^n *n*-tuples of coordinates using only 0 or 1.

Voting Combinations

If abstentions are not allowed, there are eight different ways that three voters (1, 2, and 3) can each record either a yes or no vote. Each voting arrangement, called a **combination,** is shown in the following table, where *Y* indicates a yes vote and *N* a no:

Voter	Combination of votes							
1	Y	Y	Y	N	N	N	Y	N
2	Y	Y	N	Y	N	Y	N	N
3	Y	N	Y	Y	Y	N	N	N

In each such combination, the set of voters partitions into the subset of yes voters and the complementary subset of no voters. In the case of a secret ballot, however, we can observe only four different outcomes:

Voting outcome		Number of
Y	N	combinations
3	0	1
2	1	3
1	2	3
0	3	1

In the case of *n* voters, there are 2^n different voting combinations that can result in $n + 1$ possible outcomes. Using the shorthand factorial notation $n! = n(n - 1)(n - 2) \ldots (2)(1)$, these combinations can be enumerated as

TABLE 10.6 Types of voting combinations

| Voting outcome | | Number of |
Y	N	combinations
n	0	$\dfrac{n!}{n!0!} = 1$
$(n-1)$	1	$\dfrac{n!}{(n-1)!1!} = n$
$(n-2)$	2	$\dfrac{n!}{(n-2)!2!} = \dfrac{n(n-1)}{2}$
$(n-3)$	3	$\dfrac{n!}{(n-3)!3!} = \dfrac{n(n-1)(n-2)}{6}$
$\vdots$	$\vdots$	$\vdots$
1	$(n-1)$	$\dfrac{n!}{1!(n-1)!} = n$
0	n	$\dfrac{n!}{0!n!} = 1$

shown in Table 10.6. The number of different combinations that give rise to each particular outcome corresponds to a binomial coefficient in the expansion

$$2^n = (1+1)^n = \sum_{k=0}^{n} \frac{n!}{(n-k)!k!}$$

where $\displaystyle\sum_{k=0}^{n}$ indicates the sum of the $n+1$ terms when $k = 0, 1, 2, 3, \ldots, n-1, n$.

Summary

Our analysis shows that there is a very natural *one-to-one correspondence* between the following four mathematical objects:

1. The vertices of an n-cube
2. The n-tuples of 0s or 1s
2. The combinations of Y or N votes
4. The subsets of a set of n elements

This correspondence is illustrated for the case when $n = 2$, as follows:

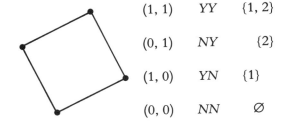

(1, 1)	YY	$\{1, 2\}$
(0, 1)	NY	$\{2\}$
(1, 0)	YN	$\{1\}$
(0, 0)	NN	$\varnothing$

We can interpret a coordinate 1 to mean that the corresponding element is in the subset and interpret a 0 to mean that it is not in the subset. Likewise, a 1 corresponds to a Y and a 0 to an N. This relationship of the binary coordinates 0 and 1 to subsets and combinations is useful in undertaking large-scale computer calculations when n is large. These computations arise in such situations as calculating power indices for voting systems and calculating various probabilities for statistical events.

A crucial ingredient of modern mathematics is to discover that patterns or structures that initially appear different are really the "same." One can then use any one of these systems to represent any other. The more detailed analysis of only one such system can then efficiently replace the study of them all. This ascending to a higher level of mental abstraction is the essence of theoretical and applied mathematics today. •

APPLYING THE BANZHAF INDEX

The Banzhaf index for measuring power is useful both in analyzing current voting systems and in designing new ones. Where today's voting systems are concerned, this index can provide insights into existing inequities. (See also Spotlight 10.3, p. 282.)

The U.S. Electoral College is used to elect the President of the United States. The power indices in this case show a slight bias in favor of the large states, relative to their populations, when we view the *states* themselves as 51 voters casting weighted votes, according to the number of Congress members in each state, with three additional votes allowed for the District of Columbia.

A more in-depth analysis of the power of the *individual voter* in such presidential elections shows a much greater inequity. A voter from California is more than three times as likely to be crucial in electing the President as a voter from the District of Columbia. On the other hand, several of the schemes proposed for reforming the Electoral College give undue influence to small states, according to Banzhaf's index. These results seem counterintuitive in view of the way most states have acted regarding Electoral College reform. The need for a more equitable Electoral College in the United States becomes a very popular issue nearly every presidential election year, and especially so in those years in which there is a third party that could conceivably carry a state.

In order to create a new federal law in the United States, a bill must be passed by a simple majority in both the House of Representatives and the Senate, and then signed by the President. Or, in the case of a presidential veto, it must be approved by a two-thirds majority in both houses. Using a different index, Lloyd Shapley and Martin Shubik calculate that each legislative body as a whole is about 2.5 times as powerful as the President. On the other hand, the President has almost 40 times the power of an individual senator and 175 times that of a sole representative in the House.

Power indices also prove useful when new voting systems are designed. Consider the case of a local representative body, such as a county board or a board of education for a unified school district. Two desired objectives are often in conflict. The region may partition in a natural way into submunicipalities with common interests. The first desired objective is that these municipalities be represented by their own elected official. However, this natural division into existing communities may not give rise to districts with populations of similar size, so that the second desired objective, the popular principle of "one man, one vote," cannot be met because larger districts would be underrepresented. One way to avoid redistricting into equal-size districts that cut across current submunicipality boundaries is to resort to weighted voting.

Weighted voting has been used in many local governments. One cannot, however, merely weight a representative's vote in direct proportion to the number of constituents he or she represents, because such weighting can sometimes lead to strange results, as we saw in the case of Nassau County. Instead, the weights for the representatives should be determined so that their resulting Banzhaf power indices are nearly proportional to their respective populations. Several court rulings in New York State have approved this approach. For example, Table 10.7 gives the weighted voting system introduced in 1982 by Tompkins County, New York, for their 15-member board of representatives for the case of a simple majority vote.

More than half of the 63 counties in New York State have actually used weighted voting or seriously considered this possibility over the past 25 years. Several school boards have implemented such systems, especially in the state of Maine. A few governing bodies in New York City have employed weighted voting, and it is currently being proposed for some others (see Exercise 13). This movement toward weighted voting systems at the local level owes much to lawyer Banzhaf, who initiated court cases in the 1960s in an attempt to arrive at governing bodies that represented their constituents in a more equitable manner than the ones existing at the time.

TABLE 10.7 Board of representatives, Tompkins County, New York (1982)

District number	Name of municipality	1980 Census population	Relative population	Assigned weights	Relative weights	No. of crucial combinations	Relative power	Discrepancy*
6	Town of Lansing	8317	1.433	404	1.515	4747	1.415	0.0122
14	Town of Dryden, East	7604	1.310	333	1.249	4402	1.316	0.0047
8	Towns of Enfield/Newfield	6776	1.167	306	1.148	3934	1.176	0.0076
3	City of Ithaca, Ward 3	6550	1.128	298	1.118	3806	1.138	0.0085
4	City of Ithaca, Ward 4	6002	1.034	274	1.028	3474	1.039	0.0045
11	Town of Ithaca, Southeast	5932	1.022	270	1.013	3418	1.022	0.0000
1	City of Ithaca, Ward 1	5630	0.970	261	0.979	3218	0.962	0.0080
2	City of Ithaca, Ward 2	5378	0.926	246	0.923	3094	0.925	0.0015
10	Town of Ithaca, Northeast	5235	0.902	241	0.904	3022	0.903	0.0019
9	Town of Groton	5213	0.898	240	0.900	3006	0.899	0.0007
7	Towns of Caroline/Danby	5203	0.896	240	0.900	3006	0.899	0.0027
5	City of Ithaca, Ward 5	5172	0.891	238	0.893	2978	0.890	0.0007
12	Town of Ithaca, West	4855	0.836	224	0.840	2798	0.836	0.0002
15	Town of Ulysses	4666	0.804	214	0.803	2666	0.797	0.0084
13	Town of Dryden, West	4552	0.784	210	0.789	2622	0.784	0.0003
Totals:		87,085	15.000	3999	15.000	50,191	15.000	0.0000
Quota necessary to win:				2000				

* Discrepancy = |relative power − relative population| ÷ relative population.

SPOTLIGHT 10.3 Binary Systems and Their Applications

A great number of different types of systems have inputs of a binary nature, which in turn determine an output of a binary type. In this chapter we are concerned with voting systems in which individuals vote yes or no and the resulting combination of votes determines whether an issue passes or fails. This system is illustrated by the following diagram:

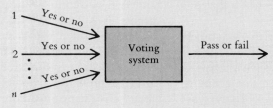

There are many other instances in which the inputs are either one of two alternatives and the resulting output or consequence is also twofold. Some other common examples of **binary systems** follow:

1. An examination is passed or failed depending on which questions are answered correctly or incorrectly. The questions may in turn be of a true or false category.

2. A machine runs or stalls out, depending on which component parts function properly or break down.

3. A message gets through or is cut off, depending on which channels are functioning or impeded.

4. Whether a vehicle successfully reaches its destination or is halted depends on which routes are passable or blocked. This may depend on which drawbridges are up or down along the river that divides a city.

5. A battle is won or lost, depending on which supply links are usable or interdicted. In turn, "victory" or "defeat" in a war depends on which battles are won or lost.

6. A facility is open for business or shut down, depending on which employees are present or absent.

7. A project is feasible or impossible, depending on what talent or equipment is available or tied up.

8. An electrical signal is present or absent, depending on whether various contacts are open or closed.

9. A security system is safe or breachable, depending on which detectors are sensing or inactivated—for example, which guards are alert or sleeping.

The techniques developed in this chapter for measuring power in voting systems can be applied to other types of binary systems. We can use them to arrive at some measure of the system's reliability, the importance of a particular component in it, or the probability that the system will or will not function. A space shuttle with its great number of components, for example, must be known to be highly reliable before it "flies." The owner of a nuclear power plant must be able to assure that there is an extremely low probability that a terrorist could divert any potentially dangerous (and precious) material from the large and complex facility.

Advances in engineering and materials have led in turn to major progress in the ability of modern digital computers to perform complex calculations rapidly. The design, electronics, inner logic, networking, and programming activities related to these computing machines rely heavily upon *discrete mathematics*. In particular, the mathematics of binary systems arises frequently in the construction and use of computers. These discrete forms of mathematics contrast with the more continuous mathematics of calculus, advanced by Sir Isaac Newton (1642–1727) and Gottfried W. Leibniz (1646–1716), which has proved so useful to the physical sciences and traditional engineering.

If modern technology had instead taken us in the direction of some sort of analog computing device, then the more continuous approaches of classical mathematics might have played a more fundamental role in arriving at today's computer science and computer engineering as well.

REVIEW VOCABULARY

Banzhaf power index A numerical measure of power for individuals in certain voting situations. It is given by the number of different combinations of yes and no votes in which a person's vote is pivotal.

Binary system A rule for assigning one of two possible outcomes given the input variables, each of which is also one of two types (for example, the numbers 0 or 1, or a vote in favor of or against an issue).

Cardinality The number of elements in a set.

Coalition A subset of a set of voters.

Combination A partitioning of a set into a subset and its complementary subset, for example, a list of those voters who voted in favor of and against an issue.

Cut (quota curve) A partitioning of all the subsets of a set into two disjoint classes, for example, those coalitions of voters that win and those coalitions that lose.

Dictator A person who can pass any issue without the aid of any other voters.

Dummy A person with no power, in the sense that his or her vote can never be pivotal to any outcome.

Hypercube (*n*-cube) The natural extension of the concept of a square in the plane and a cube in space to the analogous configurations in higher-dimensional space. Its vertices can be represented by the coordinate n-tuples $(x_1, x_2, \ldots, x_i, \ldots, x_n)$, where each x_i is either 0 or 1.

Losing coalition A coalition of voters that does not have enough votes to pass an issue, that is, a nonwinning coalition.

Minimal winning coalition A winning coalition that is as small as possible in the sense that if any member were to leave the coalition it would become a losing coalition.

Pivotal A person is pivotal in a particular voting combination if a change in just his or her vote will alter the outcome of the ballot.

Quota The smallest number of votes necessary to pass an issue.

Veto power A coalition has veto power if its complementary coalition is a losing coalition.

Weighted voting system A voting situation in which various individuals can have different numbers of votes. It can be represented by the symbol $[q: w_1, w_2, \ldots, w_n]$, where the numbers $w_1, w_2, \ldots, w_n$ represent the numbers of votes held by the n individuals and where q is the quota necessary to win.

Winning coalition A coalition of voters that has enough votes to pass an issue, that is, a nonlosing coalition.

SUGGESTED READINGS

Several of the books listed under Suggested Readings for Chapter 9 have short chapters on power indices. Many other texts on finite or discrete mathematics as well as some on mathematics in the social sciences, written over the past 30 years, also have a chapter on weighted voting. Most of these, however, stress the original Shapley-Shubik power index, which counts pivots in permutations rather than in combinations as in the Banzhaf index. The two books by Brams and by Goldberg are examples.

One of the most popular institutions to which power analysis has been applied is the federal government of the United States of America, and in particular to the U.S. Electoral College for electing the President. For basic references on this subject, see the cited works by Banzhaf, Shubik, and Owen.

Banzhaf, John F., III: "One Man, 3.312 . . . Votes: A Mathematical Analysis of the Electoral College," *Villanova Law Review,* 13:304–332, 333–346 (Winter 1968); and 14:86–96 (Fall 1968).

Barrett, Carol, and Hanna Newcombe: "Weighted Voting in International Organizations," *Peace Research Reviews,* vol. II, April 1968, Canadian Peace Research Institute, Oakville, Ontario. A long list of international organizations that make use of weighted voting systems is given in this article.

Brams, Steven J.: *Paradoxes in Politics: An Introduction to the Nonobvious in Political Science,* Free Press, New York, 1976.

Brams, S. J., W. F. Lucas, and P. D. Straffin, Jr. (eds.): *Political and Related Models. Modules in Applied Mathematics,* vol. 2, Springer-Verlag, New York, 1983. Chapters 9 and 11 by Lucas and Straffin, respectively, provide an extensive survey of both the Banzhaf-Coleman and the Shapley-Shubik power indices, including computational methods as well as many illustrations, exercises, and references. Chapter 10 proposes an alternate measure of power. A similar but shorter survey appears in the article by Lambert.

Goldberg, Samuel: *Probability in Social Science,* Birkhäuser, Boston, 1983.

Lambert, John P.: "Voting Games, Power Indices, and Presidential Elections," *UMAP Journal,* 9(3):213–267, 1988.

Owen, Guillermo: *Game Theory,* 2d ed. Academic Press, Orlando, 1982, Chapter 10.

Shubik, Martin (ed.): *Game Theory and Related Approaches to Social Behavior,* Wiley, New York, 1964, especially Part 3, "Political Choice, Power, and Voting."

EXERCISES

1. Give the cardinality of:

  a. The set {1,3}

  b. The set {1, 2, 4}

  c. The set of edges in a cube

  d. The set of faces in a cube

  e. The set of all subsets of the set {1, 2, 3}

  f. The set of all subsets of the set {1, 2, 4} that contain the element 4

  g. The set of all subsets of the set {1, 2, 4} that do not contain the element 4

  h. The set of positive even integers

  i. The set of vertices in a 5-cube

  j. The set of all roots of the equation $x^2 - 4x + 3 = 0$

  k. The set of all points on the graph of the equation $y = 2x + 3$

2. Prove that any set S with cardinality n has 2^n subsets (including S itself and the empty set $\varnothing$).

3. For each of the following weighted voting systems describe the following:

(1) All the winning coalitions

(2) All the minimal winning coalitions

(3) All the losing coalitions

(4) Any dummy voters

(5) All of the coalitions with veto power

a. [51: 52, 48]

b. [2: 1, 1, 1]

c. [3: 2, 2, 1]

d. [8: 5, 4, 3]

e. [51: 45, 43, 8, 4]

f. [51: 28, 27, 26, 19]

g. [16: 10, 10, 10, 1]

h. [21: 10, 10, 10, 10, 1]

i. [51: 28, 24, 24, 24]

j. [6: 4, 3, 2, 1]

k. Nassau County in New York in 1958

l. Nassau County in New York in 1964

m. [4: 3, 1, 1, 1, 1]

4. a. List the 16 possible combinations for how four voters 1, 2, 3, and 4 can each vote either yes Y or no N on an issue.

b. List the 16 subsets of the set $\{1, 2, 3, 4\}$.

c. Show a natural one-to-one correspondence between the lists in parts **a** and **b**.

d. In how many of the combinations in part **a** is the vote (1) 4 Y to 0 N? (2) 3 Y to 1 N? (3) 2 Y to 2 N?

5. Determine the Banzhaf power index for each of the weighted voting systems given in Exercise 3.

6. Consider the voting system for the Australia example given in the text.

a. Show that this can be represented as a weighted voting game of the form $[q: w_1, 1, 1, 1, 1, 1, 1]$ where w_1 is any number in the range $2 < w_1 \leq 3$ and $q = w_1 + 2$ (where w_1 need not be an integer).

b. Describe the minimal veto-power coalitions in the Australia example.

c. Assume that three of the six states in this voting system always vote the same way on any issue. Consider this bloc of three states as just one voter and express this as a five-person weighted voting system.

d. Determine the Banzhaf power index for the five-person system in part **c.**

e. Have the states that formed a bloc in part **c** increased their power? By how much?

7. Consider a four-person voting system composed of four voters denoted by 1, 2, 3, and 4 in which the winning coalitions are {1, 2, 3, 4}, {1, 2, 3}, {1, 2, 4}, {1, 3, 4}, and {1, 2}.

a. List the minimal winning coalitions for this voting system.

b. Express this as a weighted voting system.

c. Give the Banzhaf power index for this voting system.

8. Consider the four shareholders in a corporation with 48%, 23%, 22%, and 7% of the stock and where a simple majority rules.

a. How much stock can the first shareholder (who has 48%) sell, without changing anyone's power index, to

(1) the second shareholder?

(2) the third shareholder?

(3) the fourth shareholder?

(4) a new fifth shareholder?

b. How much stock can the fourth shareholder (who has 7%) sell, without changing anyone's power index, to

(1) the first shareholder?

(2) the second shareholder?

(3) the third shareholder?

(4) a new fifth shareholder?

c. How much stock can the fourth shareholder sell, before becoming a dummy, to

(1) the first shareholder?

(2) the second shareholder?

(3) the third shareholder?

(4) a new fifth shareholder?

d. How much stock can the second shareholder sell to the third without changing anyone's power index?

9. Show that the voting scheme for the United Nations Security Council can be expressed as the weighted voting system [39: 7, 7, 7, 7, 7, 1, 1, 1, 1, 1, 1, 1, 1, 1, 1].

10. Consider a nine-person committee in which each member has one vote and a simple majority wins (for example, the U.S. Supreme Court).

a. How is the power distributed among the nine voters?

b. If a coalition of five of these voters made a secret pact to first vote only among themselves with majority rule, and then to vote all the same way in the full committee, then how is power distributed among the nine members?

c. If a three-person subgroup of the five-person coalition in part **b** were also to form an agreement such as in part **b**, then how would the power be distributed among the nine committee members?

11. Show that the weighted voting systems for Nassau County in New York in 1970 and 1976 had no dummies.

12. In some international boxing competitions the winner of a match is determined as follows: There is a primary panel of five judges; if they rule 5 to 0 or 4 to 1 in favor of one competitor, then he is declared the winner. However, a ruling of 3 to 2 by the first panel is considered indecisive, and the matter is then referred to a second panel of five judges. This secondary panel can reverse a 3 to 2 vote by the primary panel only when they vote 1 to 4 or 0 to 5. But if the secondary panel votes 5 to 0, 4 to 1, 3 to 2, or 2 to 3, then the 3 to 2 decision of the primary panel is upheld. Can this be written as one weighted voting game involving all 10 judges? If so, indicate how.

13. The New York City Board of Estimates has had different actual or proposed weighted voting systems as indicated.

| | Number of votes | | |
| | System | | |
	A	B	C
Citywide officials			
Mayor	3	4	35
Controller	3	4	35
Council president	3	4	35
Borough presidents			
Brooklyn	2	2	11.3
Manhattan	2	2	7.3
Queens	1	2	9.6
Bronx	1	2	6.0
Staten Island	1	2	1.8
Total weight w	16	22	141.0
Quota q	9	12	71

a. Describe the minimal winning coalitions in each of the three cases A, B, and C.

b. Describe the minimal veto-power coalitions in each case.

c. Determine the Banzhaf power index in each case.

d. Can you provide a simpler set of numbers for the weights of the borough presidents in case C that serve the same purpose?

e. The weights in case *C* are related to the boroughs' populations. Do the resulting power indices from part **c**, or your new weights in part **d**, reflect these differences in populations?

14. The European Community Council of Ministers has had weighted voting systems as follows:

Country	1958	1973	1981
		Weights	
France	4	10	10
Germany	4	10	10
Italy	4	10	10
Belgium	2	5	5
Netherlands	2	5	5
Luxembourg	1	2	2
England		10	10
Denmark		3	3
Ireland		3	3
Greece			5
Total weight *w*	17	58	63
Quota *q*	12	41	45

a. Show that Luxembourg is a dummy in 1958, but not in 1973 when its vote increased relatively less than the other old (1958) members.

b. Compute the Banzhaf power index for

 (1) 1958

● (2) 1973

● (3) 1981

● **c.** Show from (2) and (3) in part **b** that Luxembourg's relative power index increased again from 1973 to 1981.

15. Certain well-defined voting systems cannot be represented as weighted voting systems. Can you prove that there is no way to pick positive numbers q, w_1, w_2, w_3, w_4 such that $[q: w_1, w_2, w_3, w_4]$ gives the four-person voting system in which the winning coalitions are $\{1, 2\}, \{3, 4\}, \{1, 2, 3\}, \{1, 2, 4\}, \{1, 3, 4\}, \{2, 3, 4\}$, and $\{1, 2, 3, 4\}$? Recall that for every coalition S and every voter i in S we must have $\Sigma_{i \text{ in } S} w_i \geq q$ when S is winning and $\Sigma_{i \text{ in } S} w_i < q$ when S is losing. (Note that this voting scheme is "improper" in the sense that the two *disjoint* coalitions $\{1, 2\}$ and $\{3, 4\}$ are both winning.)

● Optional exercise.

●**16.** Consider the seven-person voting system in which the voters are indicated by 1, 2, 3, 4, 5, 6, and 7, and the minimal winning coalitions are {1, 2, 3}, {3, 4, 5}, {1, 5, 6}, {1, 4, 7}, {2, 5, 7}, {3, 6, 7}, and {2, 4, 6}. Any superset of a winning coalition is also winning. In the following figure, the seven voters correspond to seven points, and the seven minimal winning coalitions correspond to the six lines and the one circle. Can you prove that this voting scheme cannot be expressed as a weighted-voting system? (Note that this voting scheme is "constant sum" in the sense that the complementary coalition {1, 2, 3, 4, 5, 6, 7} − S is losing whenever the coalition S is winning.)

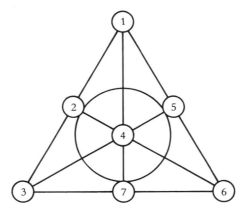

●**17.** In the early 1980s Canada adopted a method for amending the Canadian constitution that gives most of the power to the federal government. Another proposal under consideration for many years would also have involved a vote by the 10 provinces. This proposal can be described briefly in terms of minimal veto-power coalitions as follows: Veto power would be held by Ontario, by Quebec, by any 3 of the 4 Atlantic provinces, by the 3 prairie provinces together, and by British Columbia along with any one of the prairie provinces. Show that this voting scheme for the 10 provinces cannot be expressed as a weighted voting system.

18. The most common voting system is that in which each of the n voters has an equally weighted vote and a simple majority is necessary to win. This can be expressed as the n-person weighted voting system [q: 1, 1, . . . , 1] where the quota q can be given as the integer:

$$q = \frac{n+1}{2} \qquad \text{when } n \text{ is odd}$$

$$q = \frac{n+2}{2} \qquad \text{when } n \text{ is even}$$

a. Assuming that each combination of yes and no votes is equally likely, what is the probability that an individual voter will be pivotal when

(1) $n = 1$?   (3) $n = 3$?   (5) $n = 5$?   ●(7) n is arbitrary, but odd?

(2) $n = 2$?   (4) $n = 4$?   (6) $n = 6$?   ●(8) n is arbitrary, but even?

● Optional exercise.

b. As the number of voters doubles

(1) from $n = 2$ to $n = 4$ ● (3) or from any n to $2n$

(2) or from $n = 3$ to $n = 6$

does the probability that a voter will be pivotal decrease to half?

● **c.** Discuss the ratio of the probabilities in going from n to $2n$ in part **b** (3) as n becomes large.

▲ **19.** Discuss how many "really different" voting systems (weighted or not) there can possibly be when there are merely two voters.

● **20. a.** How many "really different" and "reasonable" (see below) voting systems can there possibly be involving just three voters?

It seems "reasonable" to assume that every three-person voting system should have the following three properties:

(1) The grand coalition {1, 2, 3} of the three voters 1, 2, and 3 should always be winning (and the empty coalition ∅ should always be losing).

(2) If a coalition of players T is winning, then every superset S of T is also winning (and equivalently, if a coalition of players S is losing, then every subset T of S is losing).

(3) If two coalitions are *disjoint* subsets of {1, 2, 3} (i.e., they have no voters in common), then they cannot both be winning.

b. List the winning coalitions for each system in part **a**.

c. List the minimal winning coalitions for each case in part **a**.

d. List the minimal veto-power coalitions for each case in part **a**.

e. For each system in part **a**, draw the cut (quota curve) in the three-dimensional cube as was done in Figure 10.2 for the three-person committee example.

▲ **21.** Discuss why the British Parliament in early 1974 was "unstable." The parties and their number of seats were as follows:

Party	Seats
Labour	301
Conservatives	296
Liberals	14
Irish Unionists	11
Scottish Nationalists	7
Welsh Plaid Cymru	2
Irish Catholics	1
Others	3
Total	635

● Optional exercise.
▲ Discussion exercise.

● **22.** Consider some national parliament with a multiparty system such as those in Italy or Portugal, which have been somewhat unstable in recent years. Show how the power of the different parties has changed over time. Discuss how alliances of parties have managed to maintain temporary stability.

23. The Vice President of the United States is allowed to vote to break ties in the U.S. Senate. Discuss how his Banzhaf power index compares with that of an individual senator.

24. Consider some country or city that has a representative form of government consisting of more than two parties. Consider the parties to be the individual voting units and present this government as a weighted voting system. Then compute the Banzhaf power index for each party.

● **25.** The power index of Banzhaf as well as that of Shapley and Shubik have been employed to study power in the U.S. federal government (see Banzhaf, 1968; Shubik, 1964; and Owen, 1982, in the Suggested Readings). It would make an interesting project to do a similar power analysis of the representative forms of government that have existed in eastern Europe, especially since the increase in democracy in several such countries and their "states" since the late 1980s.

● Optional exercise.

·11·

Fair Division and Apportionment

Achieving fairness is an important aspect of decision making. How can a group of students divide up a pizza so that each perceives his or her share as fair? How should an estate be divided equitably among the heirs? Can the head of a department fairly allocate such tasks as teaching assignments and committee activities? Can the number of legislators in a representative assembly be assigned in an impartial way? The goal in such fair-allocation problems is to have all persons feel that they obtain a fair and unbiased share of the available benefits or losses, in light of the various limitations present.

In practice we often appeal to someone with authority or experience to answer such questions. Mother will surely divide the cake into fair portions. Managers, we assume, will assign work in a just manner. Labor and management

will call in an expert arbitrator to detail the resolution of a dispute. Failing agreement between a husband and wife, a judge may dictate the division of the assets in divorce proceedings.

Another common approach to fairness is first to devise some appropriate measure of inequity and then attempt to find an allocation that minimizes inequity — either the largest individual inequity or the total group inequity. Still other approaches take a more statistical view of the problem and seek to achieve even distributions in some average sense, for example, over repeated allocations.

Our approach in this chapter differs from all these. We wish to arrive at methods for fair division that involve only the participants themselves and that satisfy individuals with different value systems. We do not want to assume that each person assigns the same worth or utility to objects or tasks, nor do we want some third party to guess what another's values might or should be. The items to be divided need be neither uniform (homogeneous) nor symmetrical in character. We also want to avoid talking about equity in terms of statistical notions. For example, it is hardly fair to suggest that we flip a coin, giving you all the cake if it is heads and me the whole cake if it is tails, even though we each have an "expected value of half a cake."

In general, a **fair-division problem** consists of n individuals, called **players,** whom we indicate by the numbers $1, 2, \ldots, i, \ldots, n$. They must partition some set S of goods (or losses) into n disjoint parts $S_1, S_2, \ldots, S_i, \ldots, S_n$. The objective is to find subsets S_i so that each person i considers the share S_i as fair in his or her own personal value system.

We will illustrate some **fair-division schemes** for three different cases. First, we will examine the **continuous case,** when the S to be divided is *finely divisible,* like a cake. Second, we will consider the more difficult **discrete case,** where S consists of various *indivisible objects.* For example, an estate's house, furniture, silverware and china, artworks, and automobiles are objects that cannot be further subdivided. Finally, we will describe a discrete type of allocation known as the **apportionment problem.** It arises when the individual shares must be integers, for example, in assigning the number of representatives each state will be allocated in the U.S. Congress.

THE CONTINUOUS CASE

Let us consider first the fair division of objects that are finely divisible into a great variety of different parts. Examples include cake, land, money, and a pile of sand. We will begin with the case of only two participants.

Two Players

A traditional technique for dividing a finely divisible object S in a fair manner between two players 1 and 2 is "one cuts, the other chooses":

1. Player 1 divides the set S into two pieces S_1 and S_2.

2. Player 2 picks either piece, S_1 or S_2.

3. Player 1 is given the piece not selected by player 2.

Because the roles of players 1 and 2, the cutter and the chooser, are asymmetrical, a player may prefer one role, say, the chooser, over the other. It is not uncommon to flip a coin at the start to decide whether player 1 or 2 will be the cutter.

This method (with or without the coin flip) assures that each of the players can realize a fair piece, provided that two axioms are met:

1. Each player is able to divide the set S into two parts so that *either* one of the pieces is acceptable by that player as fair.

2. Given *any* division of S into two parts, each player will find at least one of the pieces acceptable.

These assumptions apply to the positions of cutter and chooser, respectively. Axiom 1 guarantees that the cutter will be satisfied with whatever piece is remaining, whereas axiom 2 assures the chooser of finding at least one piece agreeable, no matter how the cutter divides object S.

If these assumptions are not in fact realized, then a fair result may not be reached using this method. For example, if one player will accept only the full cake as fair and the other wants a (nonempty) piece, then no method can arrive at a fair split. Any fair-division scheme must presume some reasonability conditions similar to axioms 1 and 2.

It is important to note that our axioms and our method assume neither that the players can assign numerical measures to pieces of S nor that they must delineate all acceptable pieces before the cut is made. The method requires only that the operational requirements in axioms 1 and 2 are met and each player is able to decide whether a given piece is fair. This is an example of applying mathematical concepts to practical problems without the explicit use of numbers.

Multiple Players

A number of fair-division schemes have been proposed for more than two players, along with corresponding axioms to assure that acceptable distributions can always be reached. One such "divide and choose" approach for the case of *n* participants is called the **last-diminisher** method. Let's see how this method works for seven students who want to divide a chocolate cake:

1. Student 1 cuts any piece he or she views as fair from the cake.

2. Student 2 can "pass" on this piece, or diminish the piece cut if he or she views it as too large.

3. Students 3, 4, 5, 6, and 7 each have the right — but not the obligation — to further diminish the remaining piece as their turns come.

4. The piece is assigned to the *last* player who elected to diminish it, who then exits from the game. This is student 1 if all the other players passed, that is, if no one chose to challenge the piece as unfair.

5. The above steps are repeated with the remaining six players, using the original cake less the piece that was assigned to the last diminisher.

6. This process is repeated for five players, then four players, then three, until only two players remain in the game.

7. The final two players can continue in the same manner or decide to use the "cut and choose" method.

The reader may want to list the assumptions under which this method will guarantee that a fair division can always result. For example, it might be essential that the values of all the deleted parts always add up to the total value of the cake. (The whole is equal to the sum of its parts.) That is, the act of dismembering the whole into many pieces has not lessened any student's appetite for the part he or she eventually receives. In the case of a crumbly cake and actual slicing, this assumption may well fail. On the other hand, if the object being divided is land, the preliminary cuts may only be made by marking on a map, which would hardly devalue the land itself. As in all mathematical models, the conclusions may not apply if the assumptions are not satisfied.

Figure 11.1 A fair-division scheme can help us decide how to best divide a cherry pie among any number of individuals. [Photo by John Paul Endress.]

THE DISCRETE CASE

In many fair-division problems, including some inheritances, some of the objects to be allocated cannot be further subdivided into smaller parts. One approach in such cases is to attempt to assign numerical values, such as dollar amounts, to the objects and then divide the total sum into fair ratios. The final allocation can then be achieved by assigning either the objects themselves or the dollar equivalents. This will typically require monetary side payments between the players. A great number of auction and bidding schemes have been introduced to force participants to honestly reveal their individual monetary values for specific objects.

Two Players

In the case of two players dividing one object into equal shares, we can arrive at a fair division as follows:

1. Players A and B enter sealed bids of amounts a and b, representing their respective honest evaluations of the object.

2. The object is awarded to the higher bidder, say B, whose bid was $b > a$.

3. Then the higher bidder B pays the lower bidder A the amount of

$$\frac{a}{2} + \frac{b-a}{4} = \left(\frac{a}{2} + \frac{b}{2}\right) \div 2$$

As a consequence, player B realizes an amount that B values as

$$b - \left(\frac{a}{2} + \frac{b-a}{4}\right) = \frac{b}{2} + \frac{b-a}{4}$$

It follows that each player receives half of his or her own evaluation (which is his or her fair share) plus a surplus of $(b-a)/4$. Paradoxi-

cally, when the participants in a fair division assign different values, one can arrive at a split that gives each of them *more* than a fair share.

EXAMPLE: A Two-Person Inheritance. Consider the case where Alice and Barbara inherit a house; they bid $100,000 and $150,000, respectively. They see Alice's fair share as $50,000 and Barbara's as $75,000. Our method assigns the house to Barbara and has her pay $62,500 to Alice. Barbara ends up with a house worth $150,000 to her, less $62,500 paid out, for a net gain of $87,500. Similarly, Alice receives her fair share ($50,000) plus $12,500, for a total of $62,500. So each receives her fair share plus $12,500.

Multiple Players

EXAMPLE: A Four-Person Inheritance. We will illustrate one of the possible ways the discrete fair-division method described above can be extended to more than two persons. Consider the case of four children with equal shares, Ann, Bob, Carol, and Don, who inherit their parents' estate consisting of three objects, a house, a summer cabin on the lake, and a boat. Assume that each heir enters a secret bid (in dollars) on each of the objects as indicated by the top three lines of Table 11.1.

The next row in this table, denoted as line 1, gives the sum of each heir's three bids. These amounts are taken as their respective evaluations for the whole estate. The numbers in line 2 are one-fourth of those in line 1. They represent the fair shares due the heirs in monetary terms in this case of equal division. Note that these numbers are not all the same! They do, nevertheless, represent one-fourth of the total estate's value for each heir according to his or her own evaluation, and we do not wish to impose some other evaluation upon any of these individuals. In line 3 we denote the items that are awarded to the highest bidders. (One

TABLE 11.1 Fair division of an estate

	Ann	Bob	Carol	Don
Bids on:				
House	120,000	200,000	140,000	180,000
Cabin	60,000	40,000	90,000	50,000
Boat	30,000	24,000	20,000	20,000
Fair division:				
1. Sum of the bids:	210,000	264,000	250,000	250,000
2. Fair shares:	52,500	66,000	62,500	62,500
3. Objects awarded:	Boat	House	Cabin	—
At the high bids:	30,000	200,000	90,000	0
4. Remaining claims:	22,500	−134,000	−27,500	62,500
5. Total surplus: 76,500				
6. Share of surplus:	19,125	19,125	19,125	19,125
7. Final settlements:	Boat	House	Cabin	
	+41,625	−114,875	−8,375	+81,625

could use a random device to pick the awardees if necessary to break any ties at this point.) Line 4 indicates the differences between line 2 and the declared values of the goods awarded in line 3. This represents the additional value due each person and is negative whenever the value of the goods on line 4 exceeds the fair share on line 2. Such negative numbers are payments to be made *into* the "kitty" and become available for further distribution to the heirs. The sum, −76,500, of the numbers on line 4 is never positive, and the absolute value of this sum is called the *surplus* and is recorded on line 5. We distribute this surplus in four equal shares of 19,125 to each of the participants as indicated on line 6. Line 7 sums up the goods awarded in line 3 plus the final amounts of money the heirs receive. A negative amount here indicates a side payment to compensate for goods obtained whose worth exceeds one's fair share plus his or her surplus. Observe that the sum of all the numbers on this final line is zero. Note that each heir ends up with $19,125 in value in addition to what he or she declared as a fair share. This fair-division scheme shows again that whenever some participants have different evaluations of some objects, there is an allocation in which everyone of them obtains *more* than a fair share.

Exercises 18 to 26 show some variations and extensions that can be made in this particular fair-division scheme.

APPORTIONMENT

Another important class of fair-division problems is the *apportionment problem.* It arises when we are required to round fractions so that their sum is maintained at some given constant value (see Spotlight 11.1, p. 298, regarding fractions and their rounding). This constraint appears in many situations, for example, in cases where the fractions correspond to numbers of persons (who must be assigned in whole numbers) or in statistical tables where we may wish to round percentages while we maintain the sum at 100%.

SPOTLIGHT 11.1 Fractions and Approximations

Number Systems

The concept of number is one of the most fundamental notions in mathematics. One starts with the *natural numbers* 1, 2, 3, 4, 5, 6, 7, . . . , which are also called the *whole* numbers, the *counting* numbers, or the *positive integers*. The famous nineteenth-century German mathematician Leopold Kronecker (1823 – 1891) attests to the fundamental role of the natural numbers in his much quoted quip: "God made the integers: all else is the work of man."

One important extension of the natural numbers is the *positive fractions.* These are ratios p/q of natural numbers p and q. The number p is called the *numerator* and q the *denominator* of the fraction p/q. These ratios are also referred to as *positive rational numbers.* Any such fraction p/q can be divided by another fraction r/s according to the rule

$$\frac{p}{q} \div \frac{r}{s} = \frac{p \times s}{q \times r}$$

and the quotient will again be a rational number. In other words, the positive fractions are said to be closed under the operation of division, as well as under addition and multiplication.

Fractions can also be expressed in *decimal notation,* as in these examples:

$$\frac{1}{4} = \frac{25}{100} = 0.25$$

$$\frac{5}{4} = 1\frac{1}{4} = 1\frac{25}{100} = 1.25$$

$$\frac{8}{3} = 2.666666 \ldots$$

$$\frac{7}{12} = 0.583333 \ldots$$

$$\frac{1}{7} = 0.142857142857 \ldots$$

Note that the decimal representations of some fractions, such as one-fourth (0.25), terminate, whereas others, such as eight-thirds (2.6666 . . .), go on without end. However, all rational numbers expressed in their decimal form will either terminate or else become infinitely repeating. For example, one-seventh will continue to repeat the block of digits 142857 unendingly.

Use of the Hindu-Arabic number system, along with the decimal notion, was a great leap forward in the history of mathematics. It greatly simplified performing the elementary operations of arithmetic.

However, as even the ancient Greeks knew, fractions are not sufficient to represent all "real" measurements. For example, some ratios that arise naturally cannot be represented as rational numbers. The ratio of the diagonal of a square to one of its sides is $\sqrt{2} = 1.4142146. \ldots$ The ratio of the circumference of a circle to its diameter, denoted by the Greek letter pi, is $\pi = 3.1415927. \ldots$ These are examples of irrational numbers. They cannot be represented either as terminating decimals or as repeating decimals.

Rounding Fractions and Its Consequences

In many practical applications of arithmetic we are forced to *round* fractions. To "round off " a fraction is to approximate it by another fraction that uses fewer digits in its decimal representation. For many purposes, it may be sufficient to approximate the fraction $\frac{1}{3}$ by 0.33333, or $\frac{1}{7}$ by 0.14286.

For example, if your savings account at the local bank grows at the annual rate of $6\frac{2}{3}\% = 0.066666 \ldots$, compounded quarterly, then the interest that accrues after 3 months on an investment of $1000 is

$$\frac{1}{4} \times 6\frac{2}{3}\% \times (\$1000) = \frac{1}{4} \times \frac{20}{3} \times \frac{1}{100} \times \$1000$$

$$= \frac{5}{3} \times \$10$$

$$= \frac{\$50}{3}$$

$$= \$16.666666 \ldots$$

The bank will typically credit $16.66 to your account and keep the remaining fraction of a cent for itself. Customers rarely complain when the bank keeps their two-thirds of a penny, although such fractions from many different accounts may add up to a nontrivial sum for the bank. On the other hand, if the bank were computing interest on a loan, it would normally round up and charge you $16.67 in interest for the quarter.

In most practical applications, the rounding of fractions is done routinely and its impact on any resulting calculation is trivial. However, when one is adding or multiplying a very large string of numbers, the accumulation of many small errors may eventually create a significant error. In most cases where it may matter (for example, in computer programs), one can usually incorporate additional decimal places into the calculations to reduce the potential final error to a negligible level.

In some situations, however, the rounding of fractions can have major impact. For example, the election of Rutherford B. Hayes as President of the United States in 1876 depended, among other things, on the way in which fractions had been rounded. Had a different procedure been employed for rounding fractions in the U.S. Congress and the Electoral College, including the method they were supposed to be using according to the law at the time, Samuel J. Tilden would have been declared President instead.

The method used for allocating fractions may at times prove significant, especially if we are concerned with fairness. The surprising conclusion is that there is no one method for apportioning fractions that is entirely free of all undesirable outcomes. Every method must on occasion demonstrate some counterintuitive anomalies or undesirable behavior.

TABLE 11.2 University enrollment by colleges

College	Number of students	Exact percent	Rounded percent	Integer percent	Apportioned percents	
Arts & Sciences	6716	33.580	33.6	34	33.6	34
Engineering	4832	24.160	24.2	24	24.1	24
Agriculture	4093	20.465	20.5	20	20.5	20
Business	3211	16.055	16.1	16	16.0	16
Law	852	4.260	4.3	4	4.3	4
Architecture	296	1.480	1.5	1	1.5	2
Total	20,000	100.000	100.2	99	100.0	100

EXAMPLE: College Enrollments. Consider a university with 20,000 students and six colleges. The student populations are displayed in Table 11.2. The exact percents appear in the third column. The fourth and fifth columns show the percents rounded to fewer decimal places. We use the normal rounding procedure — rounding fractions up when they are greater than one-half and down when they are less than one-half — to round these percentages to one decimal place or to integer values. Note that in both cases the total percentages do not sum to 100%. Alternate apportionments that realize the sum of 100% for these fractions appear in the last two columns.

Apportionment problems often appear in allocation situations. College officials assign faculty to colleges or departments. A typical school district, for example, must schedule courses to classrooms; place teachers in schools; assign students to schools, to classrooms, or to different sections of a course; and assign school buses to various routes or schools. The U.S. Navy must allocate personnel with various qualifications and ranks to a large variety of positions and tasks. A traffic engineer has to assign subway cars or buses to different trains or routes to best meet the expected demand. Planners everywhere must apportion their scarce resources to optimize the quality of service.

Political Apportionment

The apportionment problem arises frequently in determining political representation in democratic institutions. In this case, a fair representative share may involve a fraction, whereas the representatives themselves are individuals with one vote apiece. In some national government bodies with a representa-

Figure 11.2 A university must resolve many apportionment problems, such as arise in class scheduling. [Photo by Robert Cohen. © 1987 Photographic Services, University of Delaware. All rights reserved.]

tive form of government, the number of seats in the parliament that are assigned to a particular party is directly proportional to the number of votes the party received, with the total size of the assembly held constant. A party's proportional share, which we call its **quota,** typically will not be a whole number; so it must be replaced by a nearby integer value.

For example, if a party obtains 41.23% of the votes in a national election for a parliament with 120 seats, its quota is given by the number q in the formula

$$\frac{q}{120} = 41.23\%$$

or

$$q = (120)(0.4123) = 49.476$$

However, the number of individual representatives assigned to this party must be a whole number, such as 49 or 50, or perhaps even some other integer value.

The apportionment problem also arises in a federal form of government, where the number of representatives from a given region is proportional to the population of that region. For example, the U.S. House of Representatives, in recent years, has had 435 voting members, where the number from each state is intended to be proportional to the state's population. The population of California, for example, was 23,668,562, according to the 1980 census, whereas the population of all 50 states at that same time was 225,867,174. So California's *quota* is given by q where

$$\frac{q}{435} = \frac{23,668,562}{225,867,174}$$

which yields

$$q = 45.5835 \ldots$$

The current method for apportioning seats in the House, called the **Hill-Huntington method,** gave California 45 seats from 1983 to 1992.

Figure 11.3 The United States House of Representatives in session in Washington, D.C.

The U.S. House of Representatives is one of the best-known and most frequently studied cases of political apportionment. After each decennial census, this matter is typically a topic of intense debate. Some half-dozen different methods have been seriously considered for use in the U.S. Congress, and four of these have been implemented at various times. These are referred to in U.S. political history as the methods of Hamilton, of Jefferson, of Webster and Willcox, and of Hill and Huntington; each is named after famous U.S. statesmen or distinguished mathematical scientists who argued for its use. John Quincy Adams also proposed a method, but it was never implemented. The first presidential veto in U.S. history occurred when George Washington vetoed an apportionment bill advocated by Alexander Hamilton in favor of one supported by Thomas Jefferson.

The fascinating history of apportionment in the U.S. Congress is told in a delightful and very important book, *Fair Representation: Meeting the Ideal of One Man, One Vote*, by Michel L. Balinski and H. Peyton Young. After describing the classical methods of apportionment in some detail, the authors recommend as most appropriate for political apportionment the method advocated by Senator Daniel Webster in the nineteenth century and by statistician Walter F. Willcox throughout most of this century. The authors' suggestion differs from the currently used Hill-Huntington method, which became the law of the land in 1941.

It should be noted that the Electoral College, which formally elects the President of the United States, gives to each state a number of votes equal to its number of congress members, including both senators and all House members. Thus, the method of apportionment used for Congress can also affect who is elected President, as was the case in 1876.

APPORTIONMENT METHODS

An apportionment problem arises when there are *n states,* denoted by

$$1, 2, \ldots, i, \ldots, n$$

with respective *populations*

$$p_1, p_2, \ldots, p_i, \ldots, p_n$$

along with a given integer-valued *house size h.* An *apportionment* then consists of *n* nonnegative integers

$$a_1, a_2, \ldots, a_i, \ldots, a_n$$

which sum to *h.* Each state's fair share, or quota, q_i, is given by the formula

$$\frac{q_i}{h} = \frac{p_i}{p} \qquad \text{or} \qquad q_i = h \frac{p_i}{p}$$

where $p = p_1 + p_2 + \cdots + p_n$ is the total population of all *n* states. The aim of an apportionment is to arrive at *n* integers a_i that sum to *h*

Figure 11.4 In the 1970s mathematicians Michel L. Balinski *(left)* and H. Peyton Young *(right)* analyzed apportionment methods and recommended use of the Webster-Willcox method.

and are "close" to the respective fractional values q_i.

In our example on college enrollments, assume that there is a student representative assembly with 100 seats called the Student Senate. How many seats should be assigned to each college? The quota q_i for the ith college is given by

$$q_i = \frac{hp_i}{p} = \frac{100p_i}{20,000} = \frac{p_i}{200}$$

where p_i is the number of students in the ith college. Since $h = 100$, these numbers q_i are the same as the exact percents listed in Table 11.2. Our objective is to select six integers a_i that sum to 100 and approximate the respective q_i's .

Figure 11.5 The apportionment method of largest fractions, also known as the Hamilton method, was named for Alexander Hamilton. [*Alexander Hamilton,* John Trumbull; National Gallery of Art, Washington; Andrew W. Mellon Collection.]

Hamilton's Method

One very natural way to do this apportionment is called the **method of largest fractions,** or the **Hamilton method** in U.S. history. First, one assigns to each college i the *largest integer* in its quota q_i. This is called the *lower quota*. For this example, these numbers are

33, 24, 20, 16, 4, and 1

However, this accounts for only 98 of the 100 seats. (The six fractional parts we dropped add up to two seats.) So two more seats remain to be allocated. Second, one can assign the remaining seats, one each, to those colleges with the largest fractional parts in their quotas. These are the first and sixth colleges with fractional parts .580 and .480. The Hamilton method results therefore in the apportionment

34, 24, 20, 16, 4, and 2

which was given in the final column of Table 11.2. Note that $q_6 = 1.480$ was rounded up to 2 even though its fractional part was less than 0.5.

This straightforward apportionment method of Hamilton can give rise to one undesirable property referred to as the *Alabama paradox*. This is illustrated by the following financial problem.

EXAMPLE: Salary Increments. The chairman of a small college mathematics department supervises three other faculty, whose current salaries are $43,100, $42,150, and $10,000. (Because the third person works only part time, his salary is low.) The chairman is instructed by her dean to award salary increments for the coming year at approximately 5%, up to a total salary pool of $100,000, and to round all salaries to multiples of $1000.

The chairman increases everyone's salary proportionally by 5% and using the Hamilton method rounds to multiples of $100 and $1000, so as to sum to exactly $100,000. These figures are shown in Table 11.3.

The chairman is displeased with the result, however, because she had to round the first two professors' salaries downward. In an attempt to alleviate the problem, the dean re-

TABLE 11.3 Apportioning salaries

Professor	Current salaries	Increased by 5%	Rounded to 100s	Rounded to 1000s
A	43,100	45,255	45,200	45,000
B	42,150	44,257	44,300	44,000
C	10,000	10,500	10,500	11,000
Totals	95,250	100,012	100,000	100,000

sponds by allocating an additional $1000 to the salary pool, bringing it up to a total of $101,000. The chairman then recomputes her figures, using 6% instead of 5% (see Table 11.4).

In this case, as the salary pool increased from $100,000 to $101,000, the third professor's salary decreased from $11,000 to $10,000. The added condition that salaries be in multiples of 1000 caused him to receive no increase whatsoever, even though his 6% fair share of $600 brings him closer to $11,000 than to $10,000. This illustration of the Alabama paradox plagues the Hamilton apportionment method and many of its variants.

Divisor Methods

In addition to the Hamilton method of largest fractions, there is another whole class of apportionment schemes called **divisor methods.** Any divisor method consists of two parts. First, it focuses on a particular number d called a **divisor** and determines how often d divides into the respective populations $p_1, p_2, \ldots, p_i, \ldots, p_n$. Each resulting quotient p_i/d will have an integer part, plus a fractional remainder. Second, a particular divisor method must specify the conditions under which each remainder is to be rounded down or rounded up. That is, the remainder is either deleted from the quotient, or else it causes the integer part of the quotient to increase by one. The three simplest and most popular divisor methods are the following:

1. The **Jefferson method** discards all fractional parts and thus takes the integer parts of the quotients p_i/d for the apportionment numbers a_i.

2. The **Webster-Willcox method** rounds the fractional remainders either down or up, according to whether they are less than or greater than one-half.

3. The **Adams method** rounds any positive fraction upward, to arrive at the *smallest* integers that are greater than or equal to the quotients p_i/d.

TABLE 11.4 Apportioning larger salaries

Professor	Current salaries	Increased by 6%	Rounded to 100s	Rounded to 1000s
A	43,100	45,686	45,700	46,000
B	42,150	44,679	44,700	45,000
C	10,000	10,600	10,600	10,000
Totals	95,250	100,965	101,000	101,000

Figure 11.6 Thomas Jefferson favored a method of apportionment biased in favor of states with large populations. [The Bowdoin College Museum of Art.]

problem. A few good guesses and use of the "trial and error, and try again" approach, with the aid of a calculator, will usually result in a useable value of d without excessive effort.

Let us return to our example on college enrollments. A first guess for a divisor might be taken as the *average* number of students represented by each delegate to the Student Senate. This is given by

$$d = \frac{p}{h} = \frac{20,000}{100} = 200$$

This number will be too large for the Jefferson method and too small for the Adams method (unless all the quotients p_i/d are integers), but d is a reasonable first guess for the Webster-Willcox method. When any such choice d does not lead to a_i's that add to h, then one appropriately decreases or increases the value

Intuitively, we see that the divisor corresponds to some desirable unit size. For example, the U.S. Constitution requires that congressional districts average at least 30,000 "people." In a transportation allocation problem the divisor may be the number of seats in a subway car. In a school district situation the divisor may correspond to the ratio of students to teachers or to some desired average (or minimal or maximal) class size.

Given a particular apportionment problem to solve and a specific (divisor type) apportionment method to use, the task is to determine a workable value of the divisor d. The number d must be selected so that the resulting quotients p_i/d and appropriate roundings produce numbers a_i that do in fact sum to the given house size h. There is an interval of numbers d, rather than a unique solution, that will solve this problem, and d need not be an integer. One does not typically rely on any algorithm to produce an appropriate value of d for a given

Figure 11.7 Statesman and orator Daniel Webster (1782–1852) *(left)* and statistician and social scientist Walter F. Willcox (1861–1964) *(right)* advocated their method for apportioning the U. S. House of Representatives during much of the nineteenth and twentieth centuries. [*left:* Hood Museum of Art, Dartmouth College, Hanover, N.H.; purchased through the Julia L. Whittier Fund. *right:* Department of Manuscripts and University Archives, Cornell University Libraries.]

of d. One guesses again and again until a satisfactory value of the divisor d is eventually found for the particular apportionment method in use.

The values of p_i/d for $d = 200$ in this example are the same as the quotas q_i, and the exact percents in the third column of Table 11.2. For this value of d the appropriate roundings of Jefferson, Webster-Willcox, and Adams produce the sums of 98, 99, and 104, respectively. Table 11.5 displays the values of p_i/d for the six colleges using the selected values of $d = $ 194, 198, 202, and 206, as well as the value 200.

1. For $d = 194$ the integer parts sum to 100 and this provides for a Jefferson apportionment.

2. For $d = 198$ and rounding up for the first and third colleges, one arrives at a Webster-Willcox apportionment.

3. For $d = 206$ and rounding up for all six colleges, one obtains an Adams apportionment.

These three apportionments are indicated in the last three columns of Table 11.5. The first two methods produce the same results in this case. The Adams method favors the smaller colleges.

An Application to Scheduling

Let us see how our apportionment methods work when applied to a practical allocation problem.

A small senior high school has only one mathematics teacher who teaches five classes each day. One hundred students preregister to take one of three mathematics courses: 51 students sign up for tenth-grade geometry, 30 for eleventh-grade algebra, and 19 for twelfth-grade calculus. This yields the following apportionment problem: Given a total of five sections, how many sections of each subject should be offered?

If there are 100 students and five sections, the average class size will be 20. However, we cannot realize this average in each class be-

TABLE 11.5 Values of p_i/d for different d

Colleges i	Populations p_i	Value of divisor d					Apportionment		
		$d = $ 194	$d = $ 198	$d = $ 200	$d = $ 202	$d = $ 206	Jefferson	Webster-Willcox	Adams
1	6716	34.619	33.919	33.580	33.248	32.602	34	34	33
2	4832	24.907	24.404	24.160	23.921	23.456	24	24	24
3	4093	21.098	20.672	20.465	20.262	19.869	21	21	20
4	3211	16.552	16.217	16.055	15.896	15.587	16	16	16
5	852	4.392	4.303	4.260	4.218	4.136	4	4	5
6	296	1.526	1.495	1.480	1.465	1.437	1	1	2
Jefferson sum		**100**	98	98	96	94	100		
Webster-Willcox sum		104	**100**	99	98	97		100	
Adams sum		106	104	104	102	**100**			100

TABLE 11.6 Apportioning course sections

Grade	Course	Number of students	Quota q	Integer part	Fractional part	Fraction rounded	Hamilton apportionment	Course average
10	Geometry	51	2.55	2	0.55	1	3	17
11	Algebra	30	1.50	1	0.50	0	1	30
12	Calculus	19	0.95	0	0.95	1	1	19
Totals		100	5.00	3	2.00	2	5	20

cause 20 does not divide evenly into the numbers 51, 30, and 19. Instead, we compute the ideal fair share of sections for each subject — the quotas. For example, the quota for geometry will be $q = \frac{51}{100} \times 5 = 2.55$. Similarly, the quotas for algebra and calculus are 1.50 and 0.95, respectively. Table 11.6 summarizes this information.

The apportionment problem consists in rounding these three quotas to whole numbers while maintaining the sum at 5. The Hamilton method first assigns to each subject the largest integer in its quota, its *lower quotas*. For this example, they are the numbers 2, 1, and 0, respectively, and are shown in the fifth column of Table 11.6. They account for only three of the five sections. Second, assign the remaining two sections to those subjects with the largest fractional parts in their quotas. Because the calculus course has the largest remainder, with fractional part 0.95, and geometry has the next highest, with 0.55, the resulting Hamilton apportionment is 3, 1, and

1. The resulting average class size for the three individual subjects are then 17, 30, and 19, respectively.

Suppose now that when the term begins, the actual enrollments in the three math courses have changed to 52, 33, and 15. The new quotas, integer parts, fractional parts, and apportionment are given in Table 11.7.

Comparing Tables 11.6 and 11.7, you observe that even though the number of students taking geometry has increased from 51 to 52, the number of geometry sections decreased from 3 to 2 because one section formerly allocated to geometry has been switched to algebra, which had a still larger increase in enrollment. This particular anomaly, the failure of **quota monotonicity,** is inherent in any reasonable apportionment method: one group's quota may increase, only to have its apportionment decrease.

Let us now consider our three divisor methods. First, consider a divisor equal to the average class size of 20. The number of times

TABLE 11.7 Apportioning course sections with revised enrollments

Grade	Course	Number of students	Quota q	Integer part	Fractional part	Fraction rounded	Apportionment	Course average
10	Geometry	52	2.60	2	0.60	0	2	26
11	Algebra	33	1.65	1	0.65	1	2	16.5
12	Calculus	15	0.75	0	0.75	1	1	15
Totals		100	5.00	3	2.00	2	5	20

TABLE 11.8 Apportioning classes with divisor 20

Course	Number of students	Quotient		Apportionment		
		Integer	Fraction	Jefferson	Webster-Willcox	Adams
Geometry	52	2	$\frac{12}{20}$	2	3	3
Algebra	33	1	$\frac{13}{20}$	1	2	2
Calculus	15	0	$\frac{15}{20}$	0	1	1
Totals	100	3	2	3	6	6

that 20 divides the preregistration figures of 52, 33, and 15 is given in Table 11.8, along with the total number of sections apportioned, using this divisor, according to the methods of Jefferson, Webster-Willcox, and Adams. Because none of these totals of 3, 6, and 6 meets the requirement of precisely five sections, none of these apportionment schemes provides a suitable answer for this particular divisor.

So we will consider other divisors. For example, a divisor of 26 yields an Adams apportionment of two sections each of geometry and algebra, and one of calculus (see Table 11.9). This divisor of 26 can be viewed as the maximum class size for any single section, although it does not necessarily follow that we must assign precisely 26 students to each of the two sections in geometry.

Going in the other direction, a divisor of 16 (see Table 11.10) results in a Jefferson apportionment of 3, 2, and 0. In this case, the divisor can be viewed as the minimal allowable class size, although here we are likely to overrule this minimum requirement of 16 per section and schedule a class for the 15 students wanting calculus, even though the added section would have to be taken away from one of the other courses. A similar minimal condition arises in the U.S. House of Representatives because the Constitution requires at least one seat for each state. Of course, apportionment methods can be redefined to take account of various minimum or maximum conditions.

Finally, a divisor of 22 yields the apportionment shown in Table 11.11, where the fraction $\frac{11}{22}$ has been rounded upward under the Webster-Willcox method. Thus, for a divisor

TABLE 11.9 Apportioning classes with divisor 26

Course	Number of students	Quotient		Apportionment		
		Integer	Fraction	Jefferson	Webster-Willcox	Adams
Geometry	52	2	$\frac{0}{26}$	2	2	2
Algebra	33	1	$\frac{7}{26}$	1	1	2
Calculus	15	0	$\frac{15}{26}$	0	1	1
Totals	100	3	$\frac{22}{26}$	3	4	5

TABLE 11.10 Apportioning classes with divisor 16

Course	Number of students	Quotient		Apportionment		
		Integer	Fraction	Jefferson	Webster-Willcox	Adams
Geometry	52	3	$\frac{4}{16}$	3	3	4
Algebra	33	2	$\frac{1}{16}$	2	2	3
Calculus	15	0	$\frac{15}{16}$	0	1	1
Totals	100	5	$\frac{20}{16}$	5	6	8

of 22—or 21 if you wish—the Webster-Willcox method assigns two sections each of geometry and algebra and one of calculus.

UNDESIRABLE OUTCOMES

Given several different methods for apportioning fractions, we would naturally ask whether one method is better than the others. Is there any method that is completely fair in the sense that it satisfies all of the criteria we most desire? Perhaps surprisingly, the answer is no.

Every *known* apportionment method will, in some particular cases, demonstrate some undesirable property. Furthermore, research done in the 1970s by Balinski and Young has proved that *any possible* apportionment method, whether known or not yet discovered, must in fact produce some unpleasant result in some instances. These difficulties are unavoidable. In particular, we can list three properties we would want to hold for any fair apportionment method, and then prove that *no* method can possibly exist that always satisfies all three conditions simultaneously. Any attempt to avoid these difficulties is doomed to failure. Balinski and Young's result is another good illustration of an *impossibility theorem,* which states that some desired outcome cannot in general be realized (see Spotlight 9.2). Such discoveries may well appear, at least initially, as paradoxical or counterintuitive. Let's look at each of three desired conditions in turn, seeing how it may be violated.

TABLE 11.11 Apportioning classes with divisor 22

Course	Number of students	Quotient		Apportionment		
		Integer	Fraction	Jefferson	Webster-Willcox	Adams
Geometry	52	2	$\frac{8}{22}$	2	2	3
Algebra	33	1	$\frac{11}{22}$	1	2	2
Calculus	15	0	$\frac{15}{22}$	0	1	1
Totals	100	3	$\frac{34}{22}$	3	5	6

1. The apportionment methods of Jefferson and of Adams violate what is known as the **quota condition** — a standard requiring that the apportioned assignment be one of the two whole numbers nearest to the true quota. For example, the census of 1830 listed New York State's population as 1,918,578 and the total population of the United States as 11,931,000. At that time, the House of Representatives had 240 seats, so the quota for New York State was determined by the formula

$$q = \frac{1,918,578}{11,931,000} \times 240 = 38.593 \ldots$$

The Jefferson method of apportionment, which is biased in favor of large states, gave New York an apportionment of 40, whereas the Adams method, which is biased in favor of small states, would have assigned only 37 seats. Both these methods violated the quota condition because they produced an apportionment differing by a full integer or more from the quota. Both the Webster-Willcox method and the Hill-Huntington method would have assigned 39 seats to New York in 1830, satisfying the quota condition. Even these two methods can also violate the quota condition in other cases, although they are statistically less likely to do so.

2. The apportionment method of Hamilton will always satisfy the quota condition. However, it often demonstrates an undesirable property called the **Alabama paradox**, as we already witnessed in our example on salary increments. Using the census of 1880, the Hamilton method would have given the state of Alabama 8 seats if the U.S. House of Representatives had a total of 299 seats, whereas this method would assign Alabama only 7 seats if the house size were raised to 300. Thus, an increase in the house size would cause Alabama to lose a seat. Both Texas and Illinois were to gain a seat. (A method that produces this result is said to violate a condition called **house monotonicity**.)

In an attempt to discover an apportionment method that would never violate either the quota condition or the condition of house monotonicity, Balinski and Young invented the **quota method.** It is a variant of the Jefferson method, which assigns seats one at a time, but it is modified so that the quota condition is verified at each step.

3. Unfortunately, the quota method may violate a third desirable property, called **population monotonicity**: using this method, a state may increase its population and yet end up with fewer seats! For example, using the quota method, five states with the populations

| 122 | 35 | 17 | 16 | 10 |

would receive

| 4 | 2 | 0 | 0 | 0 |

seats, respectively, in a house of 6 seats. However, if the populations were changed to

| 122 | 39 | 17 | 16 | 10 |

then the resulting apportionment by the quota method would be

| 4 | 1 | 1 | 0 | 0 |

You can see that the second state increased in population but nonetheless lost a seat.

(The latter properties called house monotonicity and population monotonicity should not be confused with the term quota monotonicity introduced earlier.) Balinski and Young then showed that no apportionment method will *always* satisfy the three desirable properties of quota condition, house monotonicity, and population monotonicity. (Their definition of "house monotone" is more restrictive than that indicated by the example above.) The selection of an apportionment method is necessarily a compromise between these three drawbacks, in which one must pick the drawback one is willing to accept.

REVIEW VOCABULARY

Adams method The apportionment scheme of John Quincy Adams is the divisor method that rounds all positive fractions upward to the next integer.

Alabama paradox An apportionment method exhibits the Alabama paradox if an increase in the size of a legislative body can cause an individual state to lose a representative.

Apportionment problem The apportionment problem arises when one must round a list of fractions to integers in a way that preserves the sum of the original fractions. It occurs in assigning representatives to parties or states in an assembly, in assigning teachers to courses in classrooms, in rounding percentages to integers that sum to 100% in a statistical table, and in many other instances.

Continuous (divisible) case A fair-division problem in which the object to be divided has no indivisible components and can be finely divided into parts; examples include time, land, money, or sand.

Discrete (indivisible) case A fair-division problem in which some parts of the objects to be divided, such as the house and cars of an estate, cannot be finely divided into arbitrarily small parts in any manner.

Divisor A number, representing a desirable unit size, which is basic to any particular divisor method of apportionment.

Divisor method An apportionment scheme in which the resulting apportionment is obtained in two steps. First, the populations are divided by a suitable constant value d, called a *divisor,* to obtain integer values plus fractional remainders. Second, a rule is given to determine whether the remainders will be dropped or rounded up in order to arrive at the final apportionment. The methods of Adams, Jefferson, and Webster-Willcox are divisor methods, as is the Huntington-Hill method currently used by the U.S. Congress for the House of Representatives.

Fair-division problem To divide up some gains or losses into n separate parts so that each

of n people considers the part he or she receives as a fair allocation.

Fair-division scheme A method for solving a fair-division problem. Each participant in the procedure must have a way to realize a piece that he or she views as fair in his or her own value system. Any such scheme must be based on certain reasonability assumptions in order to guarantee that a fair division will exist.

Hamilton method (method of largest fractions) The apportionment method of Alexander Hamilton assigns to each state the integer part (lower quota) of its quota and then adds 1 (to get upper quota) for those states with the largest fractional parts in their quota. The numbers of 1s so added is enough to reach the given house size.

Hill-Huntington method A divisor apportionment method (not explained in this chapter) named after early twentieth-century mathematicians Joseph A. Hill and Edward V. Huntington. It has been used since 1930 as the method for apportioning the U.S. House of Representatives after each decennial census.

House monotonicity (See **Monotone**.)

Jefferson method Jefferson's apportionment scheme is the divisor method that rounds all fractions down to zero and takes only the integer parts as the apportionment.

Last-diminisher method A fair-division method for dividing a continuous object among any number of players.

Monotone A function of a variable is monotone if this function will not decrease in value whenever the variable increases in value. Several types of monotonicity are desirable for apportionment schemes: (1) An apportionment method is *house monotone* (avoids the Alabama paradox) if no state can lose a seat in the house when the size of the house increases. (2) An apportionment method is *population monotone* when no state can lose a seat when only its population increases. (3) An apportionment method is *quota monotone* when no state can lose a seat whenever its quota increases. Property 3 is rarely achieved in an apportionment scheme. Properties 1 and 2 hold for all divisor methods, although property 2 fails for the popular Hamilton method.

Player   A participant in a fair division.

Population monotonicity   (See **Monotone**.)

Quota   A state's quota in an apportionment problem is given by the formula $q = hp/P$, where h is the size of the house (the total number of representatives), p is the population of the state, and P is the total population of all the states. The quota represents the state's fair share, but it is typically a fraction that must be rounded to an integer value. The largest integer less than or equal to q is called the *lower quota;* the smallest integer greater than or equal to q is called the *upper quota.*

Quota condition   An apportionment method satisfies the quota condition if it always results in an apportionment that rounds each quota q_i either to the largest integer less than or equal to q_i (lower quota) or else to the smallest integer greater than or equal to q_i (upper quota).

Quota method   An apportionment method (not described in detail in this chapter) that fails to have the property of population monotonicity.

Quota monotonicity   (See **Monotone**.)

Webster-Willcox method   The apportionment scheme of statesman Daniel Webster and statistician Walter F. Willcox is the divisor method that rounds fractions greater than $\frac{1}{2}$ upward and fractions less than $\frac{1}{2}$ downward.

SUGGESTED READINGS

Balinski, M. L., and H. P. Young: *Fair Representation: Meeting the Ideal of One Man, One Vote,* Yale University Press, New Haven, 1982. In the 1970s Balinski and Young were the first mathematical scientists to analyze apportionment methods in depth and from the point of view of desirable properties (axioms). Their book is a highly readable introduction. It provides the very interesting history for the case of the U.S. House of Representatives as well as the basic theory of apportionment in the first and second halves of the book, respectively.

Balinski, M. L., and H. P. Young: "The Webster Method of Apportionment," *Proceedings of the National Academy of Sciences, U.S.A.,* 77:1–4, (1980). Balinski and Young present a technical outline of desirable properties and their impossibility theorem. Also included is their argument in support of the Webster-Willcox method.

Balinski, M. L., and H. P. Young: "The Quota Method of Apportionment," *American Mathematical Monthly,* 82:701–730 (1975). Balinski and Young describe their quota method of apportionment (since discounted) that satisfies the quota condition and house monotonicity.

Dubins, L. E., and E. H. Spanier: "How to Cut a Cake Fairly," *American Mathematical Monthly,* 68:1–17 (1961). This article gives some extensions of the simple fair-division concepts introduced in this chapter.

Huntington, E. V.: "The Apportionment of Representatives in Congress," *Transactions of the American Mathematical Society,* 30:85–110 (1928). This is a fairly elementary introduction to the Hill-Huntington method of apportionment containing many illustrations. They arrive at their method by minimizing a "measure of inequity" between pairs of states. The author also derives the rank function for their (divisor) method which simplifies actual calculations for particular apportionments. (The Hill-Huntington method rounds quotas q down or up depending upon whether they are less than or greater than the "geometric mean" $\sqrt{n(n + 1)}$ of the lower quota n and upper quota $n + 1$.)

Kuhn, H. W.: "On Games of Fair Division," in Martin Shubik (ed.), *Essays in Mathematical Economics,* Princeton University Press, Princeton, 1968, pp. 29–37. This article provides extensions of and additional references for the fair-division concepts presented in this chapter.

Lucas, W. F.: "The Apportionment Problem," in S. J. Brams, W. F. Lucas, and P. D. Straffin, Jr. (eds.), *Political and Related Models,* Springer-Verlag, New York, 1983, pp. 358–396. This chapter is a teaching module at a somewhat more advanced level than the presentation in the text. It gives alternate definitions for the popular apportionment methods. It also describes the rank functions, which every divisor method must have, that provide an alternate and more direct way for actually calculating apportionments for the commonly used divisor methods.

Steinhaus, H.: *Mathematical Snapshots,* Oxford University Press, Oxford, 1960. A brief but significant introduction to both fair division and apportionment is provided in this popular book on interesting mathematical topics.

EXERCISES

1. Give a precise definition of "grade point average" (GPA), as it is computed at your college.

2. In order to graduate summa cum laude at Mount LAX College, students need to have a 3.75 GPA in their last two years. If Mary has 228 quality points for 61 credit hours, will she receive this honor?

3. In order to graduate from Mud Tech, where "+" and "−" after grades count in the averages, a student must have an overall GPA of 2.000. If John has 253.95 quality points for 127 credit hours, will he graduate on time?

4. a. Entering the last day of the 1941 baseball season, Boston Red Sox slugger Ted Williams had 179 hits in 448 times at bat. If Williams did not play that day, would he have averaged .400 for the year?

b. Williams got 6 hits in 8 appearances during a double header that day to be the last batter in nearly 50 years to exceed .400. What was his batting average when rounded to three decimal places?

5. Who won the American League batting championship in 1945 when George Stirnweiss of New York got 195 hits in 632 times at bat and Tony Cuccinello of Chicago went 124 for 402?

6. Who won the American League batting championship in 1949 when Ted Williams of Boston got 194 hits in 566 times at bat and George Kell of Detroit went 179 for 522?

7. Who won the American League batting championship in 1970 when Carl Yastrzemski of Boston got 186 hits in 566 times at bat and Alex Johnson of California went 202 for 614?

8. Who won the National League batting championship in 1931 when Bill Terry of New York hit 213 out of 611, Jim Bottomly of St. Louis hit 133 out of 382, and Chick Hafey of St. Louis hit 157 out of 450?

Exercises 9 to 14 refer to the following box score of an NBA basketball game played between the Detroit Pistons and the Los Angeles Lakers.

Pistons 108, Lakers 97

Detroit	Min:Sec	FG-A	FT-A	R	A	P	T	Los Angeles	Min:Sec	FG-A	FT-A	R	A	P	T
Aguirre	29:50	10–16	5–5	3	2	4	25	Green	37:42	1–8	7–8	5	0	5	9
Edwards	32:21	6–10	5–5	2	2	4	17	Worthy	46:38	7–22	5–6	13	2	4	19
Laimbeer	42:14	5–10	5–5	10	2	4	15	Thompson	43:45	2–5	6–8	5	0	1	10
Dumars	39:52	9–18	8–11	5	5	1	26	E. Johnson	39:50	10–16	7–8	7	10	5	28
Thomas	44:41	4–18	2–2	6	16	4	10	Scott	32:39	5–9	4–4	2	4	6	15
Salley	23:07	2–4	0–0	3	0	6	4	Cooper	32:33	4–7	2–3	1	1	4	10
V. Johnson	14:00	0–4	1–2	2	0	3	1	Drew	18:31	1–5	0–0	1	2	3	2
Rodman	38:55	5–7	0–0	14	0	3	10	Divac	13:22	0–1	4–4	0	0	2	4
Totals	265:00	41–87	26–30	45	27	29	108	Totals	265:00	30–73	35–41	34	19	30	97

	1st	2d	3d	4th	OT	Totals
Pistons	15	27	28	27	11	108
Lakers	26	23	27	21	0	97

9. a. What percent of their field goals (FG) of those attempted (A) did each team make?

   b. What percent of their free throws (FT) did each team make?

10. What players made three-point field goals?

11. Round the "minutes:seconds" column (Min:Sec) to just minutes using the apportionment method of Hamilton.

12. Round the "minutes:seconds" column to just minutes using the apportionment methods of

   a. Jefferson

   b. Webster-Willcox

   c. Adams

13. Compute the number of points scored (T) per minute played for each player using

   a. the Min:Sec column for minutes played

   b. the rounded "minutes" played given in Exercise 11

   c. the rounded "minutes" played given in Exercise 12**c**

14. Repeat Exercise 13 for

   a. rebounds (R)

   b. assists (A)

   c. personal fouls (P)

15. If Sheila bids $1.50 and Jean bids $1.25 for one Frisbee, how would you reach a fair division of this object?

16. If John bids $28,225 and Mary bids $32,000 on their aging parents' old classic car which they no longer drive, how would you reach a fair division?

17. John and Mary inherit their parents' old house and classic car. John bids $28,225 on the car and $55,900 on the house. Mary bids $32,100 on the car and $59,100 on the house. How should they arrive at a fair division?

18. Can you modify your fair division scheme in Exercise 17 so that both John and Mary receive one of the two objects while still considering the allocation as fair?

19. a. Describe a fair division for three heirs A, B, and C who inherit a house in the city, a small farm, and a valuable sculpture and who submit sealed bids (in dollars) on these objects as follows:

	A	B	C
House	145,000	149,999	165,000
Farm	135,000	130,001	128,000
Sculpture	110,000	80,000	127,000

b. Describe a fair division for the three heirs A, B, and C if their shares are $\frac{5}{10}$, $\frac{3}{10}$, and $\frac{2}{10}$, respectively.

c. Describe a fair division if their shares are $\frac{1}{2}$, $\frac{1}{4}$, and $\frac{1}{4}$, respectively.

20. a. Describe a fair division for three children E, F, and G who inherit equal shares of their parents' classic car collection and who submit sealed bids (in dollars) on these five cars as follows:

Cars	E	F	G
Duesenberg	18,000	15,000	15,000
Bentley	18,000	24,000	20,000
Ferrari	16,000	12,000	16,500
Pierce-Arrow	14,000	15,000	13,500
Cord	24,000	18,000	22,000

b. Describe a fair division for E, F, and G if their shares are $\frac{1}{2}$, $\frac{1}{4}$, and $\frac{1}{4}$, respectively.

c. Describe a fair division for E, F, and G if their shares are $\frac{3}{6}$, $\frac{2}{6}$, and $\frac{1}{6}$, respectively.

Exercises 21 to 26 apply to the following alternate fair-division method for dividing discrete objects (which differs from the one presented in the text): Each of the n participants makes a sealed bid on each object. Each object is then assigned to the highest bidder. Each participant will finally obtain either objects or cash equal in value to $\frac{1}{n}$th of the *sum* of the *highest* bids for the objects.

21. Do Exercise 15 using this alternate fair-division method.

22. Do Exercise 16 using this alternate fair-division method.

23. Do Exercise 17 using this alternate fair-division method.

24. Do Exercise 18 using this alternate fair-division method.

25. Do Exercise 19 using this alternate fair-division method.

26. Do Exercise 20 using this alternate fair-division method.

27. Discuss what assumptions (axioms) are necessary in order that the fair-division scheme presented in the text for continuous objects and multiple players (the last-diminisher method) will always result in a fair division.

28. Consider the following scheme used to divide a large submarine sandwich among three children:

STEP 1. Mother passes the knife slowly from left to right over the top of the sandwich. The first child who says "Stop" is given the piece to the left of the knife and leaves the contest.

STEP 2. Mother continues to move the knife. The next child to say "Stop" receives the piece to the left and exits the game.

STEP 3. The piece to the right of the knife in step 2 is given to the remaining child.

a. Is this a reasonable fair-division scheme?

b. What assumptions need to be made in order for this to be a fair method?

c. Can this approach be extended to more than three children?

29. Pablo Picasso left his enormous estate to six heirs with shares as follows:

Person	Relationship	Share
Jacqueline	Wife	$\frac{11}{32}$
Claude	Child	$\frac{3}{32}$
Paloma	Child	$\frac{3}{32}$
Maya	Child	$\frac{3}{32}$
Bernard	Grandson	$\frac{6}{32}$
Marina	Granddaughter	$\frac{6}{32}$

Can you suggest a reasonable fair-division scheme for the heirs that takes into account the huge numbers of objects involved?

30. It makes for interesting class projects to implement the continuous and discrete fair-division schemes discussed in this section in the classroom using a (nonhomogeneous or irregularly shaped) cake, a large pizza, a can of mixed nuts, a bag of mixed candies, or an assortment of candy bars.

31. Consider a small college with three divisions as follows:

Arts	690
Science	435
Business	375
Total	1500

a. How would you apportion the five seats in the student senate to the three divisions?

b. If the student numbers the next year are 555, 465, and 480, respectively, how would you apportion the five seats?

c. Has any division gained students but lost representation in going from **a** to **b**?

32. Reconsider the college example given in the text (p. 300).

College	Number	Percent	Percent rounded
Arts & Sciences	6716	33.580	34
Engineering	4832	24.160	24
Agriculture	4093	20.465	20
Business	3211	16.055	16
Law	852	4.260	4
Architecture	296	1.480	2
Total	20,000	100.000	100

The last column gives the rounded percentages apportioned according to the Hamilton method of largest fractions.

a. How would the last column appear if one used the apportionment methods of (1) Jefferson, (2) Webster-Willcox, and (3) Adams?

b. Give values of the divisor d that work for each of the three methods in **a**.

33. Consider a state with 11 counties with populations (in thousands) of 8785, 126, 125, 124, 123, 122, 121, 120, 119, 118, and 117.

a. What is the best apportionment for an assembly with 100 seats?

b. What would the apportionment be when you use the methods of (1) Hamilton, (2) Jefferson, (3) Webster-Willcox, and (4) Adams?

c. For which method in **b** is the quota condition violated? That is, when does some state receive more than its upper quota or less than its lower quota?

34. Consider a state with six counties with populations of 9215, 159, 158, 157, 156, and 155. Answer all the questions in the previous exercise.

35. Consider a country with six states with populations: 27,774, 25,178, 19,947, 14,614, 9225, and 3292 and a house size of 36 seats. Find the apportionment using the methods of (1) Hamilton, (2) Jefferson, (3) Webster-Willcox, and (4) Adams. (You may wish to use a hand calculator.)

36. For an interesting project, take some national assembly in which the number of representatives a party receives is proportional to the number of votes it receives in an election, then calculate the number of seats each party receives according to the four different apportionment schemes listed in Exercise 35. Israel, Japan, and several European countries have such systems.

Exercises 37 to 40 relate to the classroom scheduling example discussed in this chapter. The problem is to determine the number of class sections to be assigned to each subject, given the number of students listed in the exercise. Solve each exercise using the apportionment methods of (1) Hamilton, (2) Jefferson, (3) Webster-Willcox, and (4) Adams.

37. geometry 43, algebra 42, and calculus 10

38. geometry 55, algebra 25, and calculus 20

39. geometry 67, algebra 23, and calculus 5

40. geometry 76, algebra 19, and calculus 20

● **41.** Apportionment methods can also be characterized in terms of minimizing the distance (or the inequity) between the quota point $q = (q_1, q_2, \ldots, q_s)$ and the integer apportionment point $a = (a_1, a_2, \ldots, a_s)$, where $a_1 + a_2 + \cdots + a_s = h = q_1 + q_2 + \cdots + q_s$ for a house size of h seats and s states.

a. Can you show that the apportionment scheme that minimizes the Euclidean distance

$$\min_a \left[\sum_{i=1}^{s} (a_i - q_i)^2 \right]^{1/2}$$

gives the Hamilton method?

b. Can you show that the apportionment scheme that minimizes the Manhattan distance

$$\min_a \left[\sum_{i=1}^{s} |a_i - q_i| \right]$$

gives the Hamilton method?

● **42. a.** What is the cardinality of the set $\{1, 2, 3, \ldots\}$ of natural numbers?

b. What is the cardinality of the set $\{2, 4, 6, \ldots\}$ of even positive integers?

c. Which of the sets in **a** and **b** has the larger number of elements? In what sense?

● Optional exercise.

d. Georg Cantor (1845–1918) argued that the set {1, 2, 3, . . .} of natural numbers has as many elements as the set of positive rational numbers by showing a one-to-one correspondence between the set {1, 2, 3, . . .} and all the fractions:

$$\frac{1}{1} \quad \frac{1}{2} \quad \frac{1}{3} \quad \cdot \cdot \cdot \quad \frac{1}{p} \quad \cdot \cdot \cdot$$

$$\frac{2}{1} \quad \frac{2}{2} \quad \frac{2}{3} \quad \cdot \cdot \cdot \quad \frac{2}{p} \quad \cdot \cdot \cdot$$

$$\frac{3}{1} \quad \frac{3}{2} \quad \frac{3}{3} \quad \cdot \cdot \cdot \quad \frac{3}{p} \quad \cdot \cdot \cdot$$

$$\vdots \quad \vdots \quad \vdots \quad \quad \vdots$$

$$\frac{q}{1} \quad \frac{q}{2} \quad \frac{q}{3} \quad \cdot \cdot \cdot \quad \frac{q}{p} \quad \cdot \cdot \cdot$$

$$\vdots \quad \vdots \quad \vdots \quad \quad \vdots$$

Can you show such a relationship?

·12·

Game Theory: The Mathematics of Competition

Conflict is a central theme in human history and literature. It arises naturally whenever two or more individuals try to control the outcome of events. People compete in such situations because they have both freedom of choice and different values.

Game theory is a serious mathematical subject created to study situations involving conflict and cooperation. It is a new approach in that it brings scientific methods and the powerful tools of mathematics to bear on the topic. Game theory began in earnest in 1944 with the publication of *Theory of Games and Economic Behavior* by John von Neumann and Oskar Morgenstern. (See Spotlight 12.2, p. 332.)

A game situation arises when two or more individuals, called **players,** are each able to act freely and to select from a list of available options. These options are referred to as **strategies.** These choices in turn lead to various out-comes, called **payoffs.** Each player has various *preferences* among the resulting rewards or penalties. The theory of games, then, is concerned with notions such as selection of the best strategies, equilibrium outcomes, bargaining and negotiations, formation and stability of coalitions, fair division, and resolution of conflict. This subject deals with the rules of the game, individual and coalition values, side payments and repeated play, as well as various kinds of uncertainty and chance events. Game theory differs from the traditional subjects of statistics and probability in that it treats two or more individuals with different goals or objectives.

Many confrontations are primarily *noncooperative,* for example, those between combatants in warfare or competitors in sports. In these encounters, the adversaries' ultimate objectives are typically at cross-purposes: a gain for one means a loss for the other. Other social activities, such as those in economics or politics, typically have a large cooperative component. However, most human interactions involve a delicate mix of cooperative and noncooperative behavior. In business, for example, people cooperate to maintain a healthy economy even as they compete for shares in the marketplace.

In the following sections we will present six simple examples of noncooperative *two-person* games. The first four will be games of complete or total conflict, and the last two will be games of only partial conflict.

A Location Game

Two young entrepreneurs, Henry and Lisa, plan to locate a new restaurant at a main-route intersection in the nearby mountains. They can agree on all aspects except one. Lisa likes low elevations, whereas Henry wants greater heights — the higher up, the better. In this one regard, their preferences are diametrically opposed.

The layout for their location problem is shown in Figure 12.1. You can see that three

routes, Avenue A, Boulevard B, and Country Road C, run in the east-west direction and that three highways, numbered 1, 2, and 3, run in the north-south direction. The altitudes at the nine corresponding intersections are given in the following table in thousands of feet and are illustrated in Figure 12.2.

		Highways	
Routes	1	2	3
A	10	4	6
B	6	5	9
C	2	3	7

Henry and Lisa agree to turn their decision into a competitive game, as follows: Henry will select one of the three routes, A, B, or C, and Lisa will simultaneously pick one of the

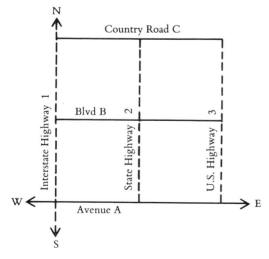

Figure 12.1 The road map for the location problem.

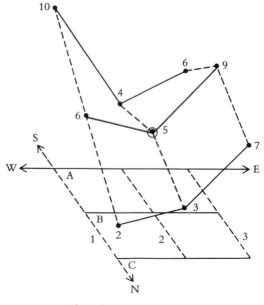

Figure 12.2 Three-dimensional road map showing Henry's and Lisa's selections (color). Underneath, the slanted map shows the map in Figure 12.1 in perspective.

three highways, 1, 2, or 3. The restaurant will then be located at the resulting intersection.

Henry is rather pessimistic and considers the lowest altitude along each of the routes A, B, and C. He gets the numbers 4, 5, and 2, which are the respective row minima, indicated in the right-hand column of the next table. He notes that the highest of these values is 5. He can elect the corresponding route, B, and guarantee himself an altitude of at least 5000 feet.

		Lisa			
		Highways			
	Routes	1	2	3	Row minima
Henry	A	10	4	6	4
	B	6	5	9	5
	C	2	3	7	2
	Column maxima	10	5	9	

Lisa likewise does a worst-case analysis and lists the highest — for her, the worst — elevations for each highway. These numbers, 10, 5, and 9, are the column maxima and are listed in the bottom line of this table. The best of these outcomes from her point of view is 5. If she picks State Highway 2, then she is assured of an elevation of no more than 5000 feet.

So Henry has a choice that will result in 5 or higher and Lisa can choose so as to hold him down to 5 or less. The resulting height 5 at the intersection of Route B and Highway 2 is, simultaneously, the lowest value along Boulevard B and the highest on State Highway 2. Such an outcome for a game is called a **saddle point**, or **mountain pass**. The reason for using these terms should be clear from the saddle shape of the "payoff surface" in Figure 12.2.

The resolution of this contest is for Henry to pick B and Lisa to elect 2; then each settles for the resulting elevation of 5. This number 5 is called the **value** of the game. The selections B and 2 are the **optimal** (pure) **strategies,** and, along with the value, are referred to as the **solution** of the game.

There is no need for secrecy in the case of a game with a saddle point. Even if Henry were to reveal his choice of B in advance, Lisa would be unable to use this knowledge to exploit him. In fact both players can use the given information to compute the best strategies for their opponents as well as for themselves. In games with saddle points (in pure strategies) it turns out that players' worst-case analyses lead to the best possible solution.

MIXED STRATEGIES

Most competitive games do not have a saddle point with optimal *pure* strategies as was the case in the location game just discussed. Players more typically must maintain secrecy about their intentions. They must take care not to reveal beforehand what particular pure strategy they will select. This concealment can be achieved by employing the notion of a mixed strategy. A **mixed strategy** is a particular "randomization" over a player's list of *pure* strategies. The specific pure strategy to use in a given play of the game can be selected by means of some probabilistic mechanism. (Note that a *pure* strategy is a special case of a *mixed* strategy: The total probability of 1 is assigned to just the one pure strategy.) When a player resorts to such mixed strategies, however, the resulting outcome of the game is no longer predictable in advance, and it must be described in terms of the statistical concept of "average" or "expected value."

EXAMPLE: Matching Pennies. In the two-person version of matching pennies each of the two players simultaneously shows either a head H or a tail T. If the two coins match, with either two heads or two tails, then the first player (Player I) receives both coins. If the coins do not match, that is, if either one is an H while the other is a T, then the second player (Player II) lays claim to the two coins.

This game can be represented by the following table, called the **game matrix:**.

		Player II	
		H	T
Player I	H	1	-1
	T	-1	1

The two rows correspond to Players I's two pure strategies: H and T. The two columns likewise give Player II's two pure strategies: H and T. The numbers in the table are the corresponding *winnings* for Player I and *losses* for Player II. If two H's or two T's are played, Player I wins 1¢ from Player II. When an H and a T are selected, Player I pays out 1¢ to Player II.

It is usually fruitless for one player to attempt to outguess the other in this game. They should instead resort to mixed strategies and use "expected values" to estimate their likely gains or losses. We saw in Chapter 7 that if one of the n payoffs $s_1, s_2, \ldots, s_n$ will occur with the probabilities $p_1, p_2, \ldots, p_n$, respectively, then the **expected value** E is given by

$$E = p_1 s_1 + p_2 s_2 + \cdots + p_n s_n$$

where $p_1 + p_2 + \cdots + p_n = 1$ and where each $p_i \geq 0$.

The best thing for Player I to do is to randomly select H half of the time and T half of the time. This mixed strategy $(p_1, p_2) = (1 - p, p) = (\frac{1}{2}, \frac{1}{2})$ could be realized in practice by a flip of the coin. Player I's resulting expected value is

$$E = E(p, H) = \tfrac{1}{2}(1) + \tfrac{1}{2}(-1) = 0$$

whenever Player II plays H, and

$$E = E(p, T) = \tfrac{1}{2}(-1) + \tfrac{1}{2}(1) = 0$$

when Player II plays T. Player I's average outcome of 0 is called the *value* of the game, but now it must be understood in a statistical sense. In a given play of the game, Player I will either win 1¢ or lose 1¢. However, his or her expectation over many plays of this **fair** game is 0. The best mixed strategy for Player II is likewise a half-half mix H and T.

Player II gains nothing by knowing that Player I is using the *optimal* mixed strategy $(\frac{1}{2}, \frac{1}{2})$. However, Player I must not reveal to Player II whether an H or T will be displayed in any given play of the game, before Player II is committed to his or her own choice of an H or T. On the other hand, if Player II knew that Player I was using a particular *nonoptimal* mixed strategy $(1 - p, p)$ where $p \neq \frac{1}{2}$, then Player II could take advantage of this knowledge and increase his or her average winnings over time to above the value zero. (See Exercise 23.)

EXAMPLE: Nonsymmetrical Matching. Players I and II can each show either heads H or tails T. When two H's appear, Player II pays $5 to Player I. When two T's appear, Player II pays $1 to Player I. When one H and one T are displayed, then Player II collects $3 from Player I.

This game is given by the following game matrix which displays the payoff from Player II to Player I:

		Player II	
		H	T
Player I	H	5	-3
	T	-3	1

A worst-case analysis, like that which solved our initial location game, is of little help here. Player I may lose $3 whether he plays H or T. Player II can keep her losses down to $1 by always playing T (and thus avoiding the $5 penalty when two H's appear). However if Player II sticks to T and Player I knows this, then Player I will also play T in each game and always collect $1 from Player II. Can Player II do better than lose $1 in each play of the game?

Consider the situation where Player I uses a mixed strategy $(1 - p, p)$, which says play H with probability $1 - p$ and play T with probability p where $0 \leq p \leq 1$. Against Player II's pure strategy H, Player I's expected value is

$$E = E(p, H) = (5)(1 - p) + (-3p) = 5 - 8p$$

Against Player II's T, Player I's expectation is

$$E = E(p, T) = (-3)(1 - p) + 1p = -3 + 4p$$

These two linear equations in the two variables E and p are sketched in Figure 12.3. Note that the four "boundary points" where these two lines meet the vertical lines $p = 0$ and $p = 1$ are just the four payoffs appearing in the game matrix.

Player I's goal is to select p so as to achieve the highest value of E, against *both* strategies H and T by Player II, simultaneously. The best value of p, which will maximize Player I's return in both eventualities, occurs at the intersection of these two straight lines. We can thus solve these two equations in two unknowns to obtain the point $p = \frac{2}{3}$ and $E = -\frac{1}{3}$. The *optimal* mixed strategy $(1 - p, p)$ for Player I is to pick H and T with probabilities $\frac{1}{3}$ and $\frac{2}{3}$, respectively. The expected value for Player I is $-\frac{1}{3}$.

A similar calculation for Player II results in the same optimal mixed strategy $(\frac{1}{3}, \frac{2}{3})$ and expected value $-\frac{1}{3}$. Recall that the payoffs for Player II are *losses*, so that this $-\frac{1}{3}$ means she

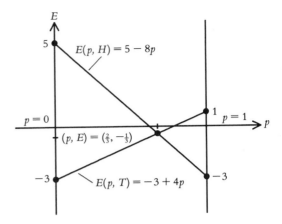

Figure 12.3 Solution for the nonsymmetrical matching game.

gains $\frac{1}{3}$. So this game is *unfair*. It favors Player II, who will win $33\frac{1}{3}$ cents *on average* for each time the game is played.

The last two examples are illustrations of what are called **zero-sum games.** The payoff to one player is the negative of the corresponding payoff to the other. The sum of their payoffs is zero. What one wins the other loses. These are also referred to as *strictly competitive* or *antagonistic* games. The two-person, zero-sum games are called **matrix games,** because they can be completely described by a square or rectangular table of numbers. These numbers represent the payoffs to Player I, while their negatives are the resulting payoffs to Player II.

The solution technique used in our last example works for any matrix game in which each player has only two viable strategies. We must use a slightly more involved method when only one of the two players has more than two strategies, as illustrated in the next game. However, first check to see whether a game has a saddle point in pure strategies before employing any solution method for finding optimal mixed strategies.

A Duel Game

Consider the strategic encounter that takes place between the pitcher and batter in the game of baseball. First, let's assume that the pitcher has enough control to throw the ball to whatever part of the strike zone he wishes. To simplify matters, let's suppose that he will select one of three pitches:

- HI: a fastball high and inside

- MD: a fastball down the middle

- LO: a fastball low and outside

These alternatives, HI, MD, and LO, are the pitcher's three strategies and are depicted in Figure 12.4.

Next, let's assume that a particular batter is known to average

- .300 against high inside pitches

- .400 against pitches over the middle

- .200 against low outside pitches

A batting average of .300 means that the batter will hit safely about 3 times out of 10.

However, the batter also has the option of outguessing the pitcher. He can choose to guess HI and step back from the plate as he swings. In this case his average becomes

- .400 against HI

- .200 against MD

- .000 against LO

Similarly, he can guess LO and average

- .000 against HI

- .300 against MD

- .400 against LO

So the batter also has the three strategies that are designated HI, MD, and LO.

This information can be summarized in the following game matrix:

		Pitcher		
		HI	MD	LO
Batter	HI	.4	.2	.0
	MD	.3	.4	.2
	LO	.0	.3	.4

We assume that all this information is known to both players. The numbers represent the corresponding probability that the batter will hit safely; they can serve as a measure of his likely reward. Whereas the batter's objective is to maximize the resulting payoff, the pitcher has the opposite goal. He aims to minimize the likelihood that the batter will hit safely. What is good for the batter is bad for the pitcher, and vice versa. They are complete adversaries. This again is an example of a strictly competitive game. Since the batter's expected gain is the same as the pitcher's expected loss, and vice versa, this is a zero-sum game.

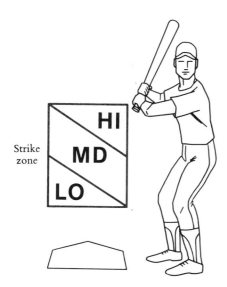

Figure 12.4 The three strike zones, which correspond to the pitcher's three strategies.

Clearly, this model is an oversimplification of real baseball. We have excluded from our analysis pitches outside the strike zone, foul balls, bases on balls, and other possible strategies and outcomes. We could enlarge the game to make it more realistic, but that would only complicate our calculations without altering the nature of the analysis.

To solve this game we must answer the question, Which choices are best for the two players? One approach to this problem, if thinking only in terms of *pure strategies,* is to reason along the following lines:

Batter (to himself): If I select MD, which has the best overall average for me, I am protected against the zero payoff in all cases.

Pitcher (to himself): If the batter does choose MD, then I should throw LO and limit him to a .200 average.

Batter: The pitcher knows that I am thinking MD, and he is thus considering a LO pitch. Therefore, I should really guess LO and hit .400.

Pitcher: But I know he is considering LO; so, on second thought, I should surprise him and aim HI.

Batter: In light of the preceding chain of thought, I should expect a HI pitch and swing accordingly.

Pitcher: In light of his last reasoning about HI, I should really throw LO.

Batter: Then I should guess LO.

Pitcher: So I should pitch HI.

Batter: I'd better guess HI after all.

Pitcher: I will surprise him with LO.

This form of cyclical reasoning can continue on without end. It provides no resolution to the decision problem. Clearly, there is no one pitch that is best in all instances. In fact, the batter can do better than to rely on the safe strategy MD, despite the overall protection it seems to provide.

The answer to the players' problems lies in our notion of a mixed strategy. Both players must maintain an element of surprise and must not give away their choices in advance. They can guarantee surprise by randomly selecting their strategies. That is, the pitcher should select his pitches at random, and the batter should do so as well. This can be done in a way that ensures an optimal mixed strategy for each player. In short, the players can improve their average performance over the long run if they play the odds and randomize properly.

Let's look more closely at the pitcher's problem. He must determine the best probabilities for each of his three pitches. These probabilities tell the pitcher the ratio of different types of pitches he should select. For example, if he determined a probability of 0.3 for LO, then three-tenths of his pitches should be LO.

An old baseball adage says, "Pitch to the corners." That means that a pitcher should avoid throwing down the middle. (A mathematical analysis of the game matrix verifies this adage, showing that the pitcher should indeed assign probability zero to the MD. See Exercises 17 and 18.) So the pitcher should randomly choose between HI and LO only. Eliminating this middle column MD reduces the size of our game matrix to 3 by 2.

Next, consider the three outcomes that may occur, depending upon whether the batter guesses HI, MD, or LO:

1. If the batter guesses HI, then a random mix of HI and LO pitches will cause the batter to average somewhere between .400 and .000. This is indicated by the line labeled HI between the heights .400 and .000 in Figure 12.5. The height of the sloping line indicates the expected batting average of the batter when he guesses HI, and the distance along the horizontal axis indicates the probability p with

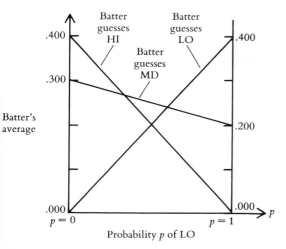

Figure 12.5 A graph of the batting averages, which depend on the batter's guesses (the three lines) and the pitcher's strategy given by p.

the graph, which are indicated by the red curve in Figure 12.6. He should select his probability p so as to obtain the lowest point on this top curve. This can be done by finding the point at which the MD line and the LO line intersect.

The mathematical solution to this problem gives the result $p = 0.6$. The pitcher should throw LO with a probability $p = 0.6$ and HI with probability $1 - p = 0.4$. In other words, six-tenths of his pitches should be LO and four-tenths should be HI. This will result in holding the batter to an expected average of .240 or less.

A similar mathematical analysis would show that the batter has an optimal mixed strategy if he guesses

◆ HI with probability 0

◆ MD with probability $q = 0.8$

◆ LO with probability $1 - q = 0.2$

This will guarantee him an average of .240 over the long run.

which the pitcher throws LO. The equation of the HI line is

$$E(\text{HI}, p) = (0.4)(1 - p) + 0p = 0.4 - 0.4p$$

2. If the batter guesses MD, his average will fall somewhere between .300 and .200, as shown in Figure 12.5. The equation of this MD line is

$$E(\text{MD}, p) = (0.3)(1 - p) + 0.2p = 0.3 - 0.1p$$

3. Similarly, if the batter guesses LO, his average will be between .000 and .400, as also shown in Figure 12.5. The equation of this LO line is

$$E(\text{LO}, p) = (0)(1 - p) + 0.4p = 0.4p$$

The pitcher wants to keep the batting average as low as possible against all three batting eventualities simultaneously. The pitcher is therefore concerned with the highest points in

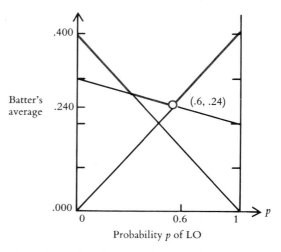

Figure 12.6 The pitcher's optimal mixed strategy is to select p so as to obtain the lowest point (circle) on the top curve (color).

The number .240 is the value of this game. The pair of optimal mixed strategies, $p = 0.6$ for the pitcher and $q = 0.8$ for the batter, along with the resulting value .240, gives us the *solution* for this game.

It is usually essential for the players to act in an unpredictable manner — they must maintain secrecy. If the baseball pitcher were to "telegraph" his pitches and the batter were able to detect these signals in advance and react in time, then the batter could raise his batting average. On the other hand, if the batter were not mixing his guesses in an optimal way, then the pitcher could make use of this information to improve his own performance by throwing to the zone that the batter was least likely to guess. (See Exercises 19 and 20.)

We have seen that the batter has a mixed strategy (0, 0.8, 0.2), which assures him of a batting average of .240, and that the pitcher has a mixed strategy (0.4, 0, 0.6), which will hold the batter's average down to this *same* number .240, called the value of the game. This result is an illustration of the famous **minimax theorem** of John von Neumann. It states that for any matrix game there is a *unique* value that either player can realize by using an optimal mixed strategy. (Recall, however, that this value is a gain for Player I and a loss for Player II.)

Solving Matrix Games

Given any matrix game we should first go through the simple check to see whether it has a saddle point in *pure* strategies. List the minimum number from each row and list the maximum number from each column, as we did in the location game. If the maximum of the row minima is equal to the minimum of the column maxima, then the resulting value and the corresponding pure strategies (which may not

be unique) provide a solution for the game. The value will appear in the game matrix as the lowest number in its row and the largest in its column.

Next, we may be able to discard some strategies (i.e., eliminate some rows and columns in the matrix), because these strategies are bad ones and should never be played. If all the payoffs in one row of the game matrix are greater than or equal to the respective payoffs in another row, then the former strategy is said to **dominate** the latter strategy and the latter row can be eliminated. Similarly, if all the payoffs in one column are less than or equal to the corresponding payoffs in another column, then the former dominates the latter and the latter column can be deleted. For example, in our location game, Route B dominates Route C, and Highway 2 dominates Highway 3. Domination can also take more complicated forms. In our baseball game, for example, the pitcher did not consider using his middle pitch MD. A half-half mix of his pitches HI and LO results in average payoffs of .200, .250, and .200, which are less than or equal to, respectively, the .2, .4, and .3 realized by his MD pitch. Eliminating dominated rows and columns reduces the size of the matrix game to be solved.

If we can use domination to reduce the size of a game matrix to n rows by two columns where $n > 2$, then we can solve for Player II's optimal mixed strategy $(1 - p, p)$, and the value of the game, by the same method used in our 3 by 2 baseball game. This geometric approach works when there are n lines as well as 3 lines (in Figures 12.5 and 12.6). We still find the value p that realizes the lowest point on the "upper curve" formed by the n lines involved. This requires solving a corresponding pair of two linear equations in two unknowns E and p, a topic that also arose in Chapter 4 on linear programming. Furthermore, there will then always be an optimal mixed strategy $(p_1,$

$p_2, \ldots, p_n$) for Player I that only involves (one or) *two* nonzero probabilities q_i and $q_j = 1 - q_i$, which are viable strategies to use against Player II's $(1 - p, p)$. (Analogous methods can be used to solve games with 2 rows and n columns.)

Several general algorithms have been developed since 1945 to solve any larger matrix games. Furthermore, the more recently developed subject of linear programming (see Chapter 4) is equivalent to the theory of matrix games. Any of the linear programming algorithms, such as the simplex method of G. B. Dantzig or the more recent method of N. K. Karmarkar, can also be used to solve matrix games. (See Suggested Readings for more details and references.)

Practical Applications

The element of surprise is essential in many encounters. Example of the use of mixed strategies include various inspection procedures and auditing schemes. These should employ randomness to keep potential cheaters off guard as well as for statistical purposes.

Individual investigators or regulatory agencies often monitor certain accounts or actions to check for faults, errors, or illegal activities. The investigators include bank auditors, customs agents, insurance investigators, and quality control experts. The National Bureau of Standards is responsible for monitoring proper measuring instruments and for maintaining reliable standards. The Nuclear Regulatory Agency demands an accounting of dangerous (and expensive) nuclear materials as part of its safeguards program. The Internal Revenue Service wishes to identify those cheating on taxes. The military or intelligence service may wish to sneak in or intercept a weapon or secret agent that is in the midst of many decoys. Because it is prohibitively expensive to check out

all cases, statistical methods must be used to check for violators. Many such encounters can be modeled as a competitive game, with the goal to obtain optimal mixed strategies for the inspector and the violator.

The notion of a "bluff," as in the game of poker, is also a viable strategy. For bluffing, game theory assigns an optimal probability with which one should bluff, given a particular situation. For example, in labor negotiations, a threat to strike is effective only if it is believed to some extent. (See Exercise 45.)

NONZERO-SUM GAMES

The four matrix games just presented all had the zero-sum property: One player's gain was equal to the other player's loss. The sum of their joint payoffs was always zero. Such games involve total conflict. We will now consider two examples of games that are nonzero-sum. The players' goals are only in partial conflict. There is some mutual gain to be realized by both players if they could cooperate. Such cooperation typically involves elements of communication, trust, and the threat of enforcement. If these elements are lacking, however, we are in the realm of noncooperative games. The players' individual self-interest can then lead to less-than-optimal payoffs. We will consider the following two games when played in this noncooperative mode.

The Prisoner's Dilemma

We will first look at a game that has come to be known as the **prisoner's dilemma.** This elementary two-person game provides a simple explanation of the forces at work behind arms races, price wars, costly advertising campaigns, and many other similar escalations. The prisoner's dilemma illustrates a kind of social paradox that we frequently confront in the course of our everyday lives.

The term *prisoner's dilemma* was first assigned to this game by the Princeton mathematician Albert W. Tucker (1905–) in 1950. The game involves the scenario of two suspects in crime who are held *incommunicado*. Each is given one of two choices: to steadfastly maintain their individual or mutual innocence or to sign a confession accusing the partner of committing the crime. It is usually in the individual's self-interest to confess, that is, implicate one's partner and receive a reduced sentence. Yet when both confess, they each reach a bad outcome, that is, they are both found guilty. What is good for the prisoners as a pair — steadfast denial by both — is frustrated by their pursuit of individual rewards.

We can use this simple model for another crucial international problem: the arms race. Assume that there are two nations called Red and Blue. Each of these superpowers can independently select one of two policies:

(A) Noncooperation: arm heavily in preparation for any possible war contingency.

(D) Cooperation: disarm, or at least agree to a partial ban on armaments.

There are four possible outcomes:

(D, D) Both Red and Blue choose to disarm. Viewed as a whole, this is the most preferred social outcome for them, even in light of certain risks.

(A, A) Both nations elect to arm, which is taken as the worst possibility from the global perspective.

(A, D) Red decides to arm, whereas Blue elects to disarm. This amounts to unilateral disarmament by Blue, which is the most preferred of all outcomes to Red but the least desired by Blue.

(D, A) Red disarms, whereas Blue arms. This is considered the worst result for Red and the best outcome for Blue.

This situation can be modeled by means of the following game matrix:

		Blue	
		A	D
Red	A	Arms race	Favors red
	D	Favors blue	Disarm

Here, Red's choice amounts to picking one of the two rows, whereas Blue's options correspond to the two columns. It may prove helpful to assign numerical payoffs to the four outcomes, as follows:

		Blue	
		A	D
Red	A	(2, 2)	(5, 0)
	D	(0, 5)	(4, 4)

For example, the pair of numbers (0, 5) in the second row and first column signifies a payoff of 0 to the row player Red and a payoff of 5 for the column player Blue. The least desired outcome is 0; the most preferred is 5. Only the relative magnitude, not the actual values, of these payoffs is essential for our analysis. The numbers are not intended to measure the absolute worth of unilateral disarmament or of a disarmed world compared with an armed one. They are used only to suggest the preferences of the players: we simply assume that a player prefers a larger numerical payoff to a smaller one.

SPOTLIGHT 12.1 Repeated Play of Prisoner's Dilemma

Robert Axelrod, professor of political science and public policy at the University of Michigan, is a well-known authority on the problem of the prisoner's dilemma. In the following interview he discusses the effects of playing the prisoner's dilemma game repeatedly.

Robert Axelrod, University of Michigan.

If you play the game only once, there's no future to your interaction and so you might as well take the short-term gains. In one play there's no chance to reward or punish a defection by the other player: there's no hope that you'll get a mutual cooperation going.

A very important feature of the evolution of cooperation is that there is a long-term relationship. When the prisoner's dilemma is iterated, it gives the players an opportunity to base their current choices on the previous interactions they've had. So if the other player seems willing to cooperate, you can cooperate yourself. If the other player has rarely been willing to cooperate, then it probably doesn't pay to cooperate. Repeating the game allows you to do things like base your own strategy on reciprocity. And therefore it allows you to try to mold the other player's behavior, to encourage him to cooperate with you.

I got the idea of a tournament because I was interested in determining a good way of playing prisoner's dilemma, because it captures some important features of the real world. No one seems to know exactly what's the best strategy. So I invited experts from a variety of fields — people who had written about prisoner's dilemma or game theory — and asked them what strategy they would use in this game. Then I played each with the other to see how well they would do. So I had something analogous to a computer chess tournament.

I was really surprised by the way it came out because the simplest of all the strategies submitted was the one that did best. That was "tit for tat" by Anatol Rapoport. This rule simply says to cooperate on the first move and then do whatever the other player did on the previous move. If the other player cooperated, you cooperate. If the other player defected, you defect. It works best for several reasons.

What the analysis shows is that an effective strategy is not to start defecting: never be the first to defect. But if the other side defects, it pays to be provokable. It also pays to be forgiving after you've been provoked, so as to keep the conflict as short as possible. It pays to respond promptly if someone does something you don't like.

I titled my book *The Evolution of Cooperation* to capture the analogy from biological evolution. It's an evolutionary study of cooperation, asking how it could get started in a world where there isn't any, how it can sustain itself, and how it can grow after it gets started. People are likely to continue to use strategies that are effective and to drop or change them if a strategy they use is not very effective. So in fact things tend to evolve toward more effective strategies.

SPOTLIGHT 12.2 Historical Highlights

John von Neumann. [Photo courtesy of The Institute for Advanced Study.]

Oskar Morgenstern. [Princeton University Archives.]

As early as the seventeenth century, such outstanding scientists as Christian Huygens (1629–1695) and Gottfried W. Leibniz (1646–1716) proposed the creation of a discipline that would make use of the scientific method to study human conflict and interactions. Throughout the nineteenth century, several leading economists created simple mathematical examples to analyze particular illustrations of competitive encounters. The first general mathematical theorem in this subject was proved by the distinguished logician Ernst Zermelo (1871–1956) in 1912. It stated that any finite game with *perfect information* such as checkers or chess has an optimal solution in *pure* strategies;

Let's examine this arms race more closely. Should Red select strategy *A* or *D*? Red can see what will happen if Blue selects his first column *A*: Red will then receive a payoff of 2 for arming and 0 for disarming, so he will arm (first row). Similarly, Red notes the consequences if Blue were to select his second column *D*. In this case, Red will receive 5 for arming or 4 for disarming. Again Red decides in favor of his first row *A*. In either case, Red's first row gives him the more desired result. We say that the payoffs to Red in the first row

dominate those in the second. There is always some advantage to Red to arm, whether Blue arms or disarms.

A similar argument leads Blue also to choose *A*, that is, to pursue a policy of arming. When each nation strives to maximize its own payoff independently, the pair is driven into the outcome (*A, A*), with payoffs (2, 2). The better outcome (*D, D*), with payoffs (4, 4), appears unobtainable when this game is played noncooperatively.

The outcome (*A, A*) is said to be in **equilib-**

that is, no randomization or secrecy is necessary. A game is said to have perfect information if at each stage of the play, every player is aware of all past moves by himself and others as well as all future choices that are allowed. This theorem is an example of an *existence theorem:* it demonstrates that there must exist a best way to play such a game, but it does not provide a detailed plan for actually playing a complex game in order to achieve victory.

The famous mathematician F. E. Emile Borel (1871–1956) introduced the notion of a *mixed,* or randomized, strategy when he investigated some elementary duels around 1920. The fact that every two-person *zero-sum* game must have optimal mixed strategies and an expected value for the game was proved by John von Neumann (1903–1957) in 1928. Von Neumann's result was extended to the existence of equilibrium outcomes in mixed strategies for multiperson *general-sum* games by John F. Nash Jr. (1931–) in 1951.

Modern game theory dates from the publication in 1944 of *Theory of Games and Economic Behavior* by the Hungarian-American mathematician John von Neumann and the Austrian-American economist Oskar Morgenstern (1902–1977). They introduced the first general model and solution concept for multiperson cooperative games, which are primarily concerned with coalition formation (economic cartels, voting blocs, and military alliances) and the resulting distribution of gains or losses. Several other suggestions for a "solution" to such games have since been proposed. These include the value concept of Lloyd S. Shapley (1923–), which relates to fair allocation and economic prices and serves as well as an index of voting power.

The French artist Georges Mathieu designed a medal for the Paris Musée de la Monnaie in 1971 to honor game theory. It was the seventeenth medal to "commemorate 18 stages in the development of Western consciousness." The first medal was for the Edict of Milan in 313 AD. Game theory also has a mascot, the tiger, arising from the Princeton University tiger and the Russian abbreviation of the term "game theory" (ТЕОРИЯ ИГР).

rium because if either nation alone were to deviate from its choice of *A,* then it would be punished with the lower payoff 0, rather than 2. The forces involved prohibit only one nation from moving away from equilibrium *(A, A)*.

Even if both nations agree in advance to jointly pursue the globally optimal solution *(D, D),* this outcome is unstable because if either nation alone reneges on the agreement and secretly arms, it will benefit. Each would thus be tempted to go back on its word and

select *A.* After all, if you have no confidence in the trustworthiness of your opponent, you may be well advised to cover yourself against such a defection.

In real life, however, people often manage to avoid the noncooperative outcome in the prisoners' dilemma. The game is usually played within a larger context, where other incentives are at work. Moreover, the game is typically played on a repeating basis — it is not a one-time affair. Elements such as reputation and trust also play a role. The players realize the mutual advantages in cooperation and may arrive at this point by slowly phasing down over time. They may also resort to other helpful measures, such as better communications channels, more reliable inspection procedures, truly binding agreements, or promptly enforced penalties for violators (see Spotlight 12.1, p. 331).

The prisoner's dilemma nicely pinpoints the dynamics behind a frequently occurring social paradox. The resulting standoff, or noncooperative equilibrium outcome, is not as satisfactory a solution as were the optimal strategies derived in our previous constant-sum games. The cooperative outcome, or one that does not take into account immediate self-interest, is obviously the preferred solution in the long run.

The Game of Chicken

Let us look at one more two-person game of partial conflict, known as **chicken,** which leads to troublesome outcomes. Two drivers are approaching each other at high speeds. Each must decide at the last minute whether to swerve to the right or not swerve. There are several possible consequences:

1. Neither driver swerves, and the cars collide head-on. We assign this least preferred outcome a value of 0.

2. Both players swerve. Each loses some prestige by backing off at the brink, but they do remain alive. We give this outcome an intermediate value of 3.

3. One of the drivers swerves and badly loses face, whereas the other does not swerve and is viewed as the winner. Let us select the numerical payoffs of 1 for swerving and 5 for not swerving in this case.

These options can be summarized in the following game matrix:

		Driver 2	
		Swerve	Not swerve
Driver 1	Swerve	(3, 3)	(1, 5)
	Not swerve	(5, 1)	(0, 0)

If both players persist in their attempts to obtain the maximum payoff 5, then the resulting outcome is mutual disaster: the lowest payoff 0 for each. It is surely better for both drivers to simultaneously back down and obtain 3 each. But neither opponent wants to be in the position of being intimidated into swerving (for a payoff of 1) while the other does not give in (and appears as the winner with a payoff of 5).

Many superpower conflicts, prolonged labor disputes, and other power confrontations have elements in common with the game of chicken. We may find some comfort, however, in knowing that of the 78 essentially different two-by-two games of partial conflict, only chicken and the prisoners' dilemma give rise to such disturbing results, where players' following their immediate self-interests leads to such suboptimal results for the society as a group.

In this chapter we have seen how two-person, zero-sum games have a natural solution. It consists of an (expected) value that each player can realize by playing an optimal (mixed) strategy. In the case of nonzero-sum games, however, their resolution can be more complicated. Disconcerting and rather paradoxical results can arise, as in the case of the last two games. The subject of game theory also studies situations involving multiple players, with many strategies, acting in either a cooperative or a noncooperative manner.

REVIEW VOCABULARY

Chicken A common two-person symmetric game in which each player has two strategies: to swerve to avoid a collision or confrontation or else to continue straight ahead and cause a collision if the opponent has not chosen to swerve in the meantime. If both players refuse to swerve, disaster results; if only one backs down, the other wins.

Dominate One of a player's strategies dominates another strategy when the first one results in higher (or equal) payoffs than the latter one, against any choice made by the opposing player.

Equilibrium A set of strategies, one for each player, is in equilibrium if no one player can unilaterally alter his or her strategy to obtain a better payoff.

Expected value If one of the n payoffs s_1, s_2, . . . , s_n will occur with the respective probabilities p_1, p_2, . . . , p_n, then the expected value E is

$$E = p_1 s_1 + p_2 s_2 + \cdots + p_n s_n$$

where $p_1 + p_2 + \cdots + p_n = 1$ and where each $p_i \geq 0$.

Fair A zero-sum game is fair when the (expected) value of the game, obtained by using optimal strategies, is zero.

Game matrix A rectangular array of numbers. The rows and columns correspond to the strategies for two players, respectively, and the numerical entries represent the resulting payoffs when particular strategies are selected.

Matrix game A two-person, zero-sum game can be described by a matrix and is called a matrix game.

Minimax theorem The fundamental theorem for two-person, zero-sum games, stating that there always exist optimal mixed (randomized) strategies that enable both players to obtain the optimal expected value in the game.

Mixed strategy A strategy chosen in a probabilistic manner from a list of options (called pure strategies).

Optimal strategy A particular strategy for a player (pure or mixed) that guarantees that the resulting payoff is the best one that this player can expect to achieve against all possible choices by the opposition.

Payoffs The potential outcomes of a game, typically expressed as numbers.

Players The participants who make strategic choices in a competitive encounter.

Prisoner's dilemma A frequently occurring two-person symmetrical game in which each player has two strategies: cooperate or defect. The best outcome for the pair taken together occurs when they both cooperate. In the case where one cooperates and the other defects, the resulting payoffs are the worst and best possible, respectively.

Pure strategy Each possible way for a player to play through a game is called a strategy, or pure strategy, to distinguish it from a mixed strategy.

Saddle point (mountain pass) In a two-person, zero-sum game, a pair of strategies, one for each player, that are in equilibrium. Neither player alone can change strategy and achieve a higher payoff. The strategies and resulting value at a saddle point provide a solution for the game.

Solution An optimal strategy for each player, along with the resulting value of the game.

Strategy One of the possible ways to play a game; strategies are mixed or pure, depending on whether they are selected in a probabilistic manner (mixed) or not (pure).

Value The payoff of a game, usually expressed numerically, that results when the players play optimally.

Zero-sum game A game in which the payoff to one player is the opposite (or negative) of the payoff to the opposing player.

SUGGESTED READINGS

Brams, S. J., W. F. Lucas, and P. D. Straffin, Jr. (ed.): *Political and Related Models,* Springer-Verlag, New York, 1983. Chapter 4 by W. F. Lucas and L. J. Billera gives an elementary introduction, with many illustrations, to the multiperson cooperative games (the coalitional games) and related ideas on fair division.

Hamburger, Henry, "*N*-person Prisoner's Dilemma," *Journal of Mathematical Sociology,* 3:27–48(1973). This article shows how the fascination of the two-person prisoner's dilemma increases as one goes to more than two players.

Luce, R. Duncan, and Howard Raiffa: *Games and Decisions,* Wiley, New York, 1957; Dover, 1989. This venerable survey of most of early game theory presents the two-person zero-sum and nonzero-sum games in Chapters 4 and 5. Several different algorithms for solving the zero-sum case are mentioned in Appendix 6. The minimax theorem and its equivalence to the "duality theorem" in linear programming are given in Appendixes 2 and 5.

Rapoport, Anatol, Melvin Guyer, and David Gordon: *The 2 × 2 Game,* University of Michigan Press, Ann Arbor, 1976. This volume provides a detailed review of the 78 "different" two-by-two nonzero-sum games of which the prisoner's dilemma and chicken are the two most interesting and troublesome cases.

Williams, John D: *The Compleat Strategyst* (sic), McGraw-Hill, New York, 1954; revised edition, 1966; Dover, 1986. This gem, which contains many simple illustrations, is a humorous primer on the two-person, zero-sum games. The first and second editions provide two different solution procedures discovered by L. S. Shapley and R. Snow, and A. W. Tucker, respectively. There are many other texts or chapters in books devoted to matrix games.

Game theory has found a great number of applications in a wide variety of fields. A mere glimpse into the uses of this subject in the respective areas of biblical studies, international relations, economics, biology, business, and political science can be found in the following (mostly popular) books. Brams's books in particular give many examples of two-person, nonzero-sum games and their applications.

Brams, Steven J.: *Biblical Games,* MIT Press, Cambridge, 1980.

———: *Superpower Games,* Yale University Press, New Haven, 1985.

Case, James H.: *Economics and the Competitive Process,* New York University Press, New York, 1979.

Dawkins, Richard: *The Selfish Gene,* Oxford University Press, Oxford, 1976.

McDonald, John: *The Game of Business,* Doubleday, Garden City, N.Y., 1975; Anchor 1977.

Ordeshook, Peter J. (ed.): *Games Theory and Political Science,* New York University Press, New York, 1978.

EXERCISES

Consider the following eight two-person, zero-sum games, where the payoffs represent gains to the row player I and losses to the column player II.

1. $\begin{bmatrix} 6 & 5 \\ 4 & 2 \end{bmatrix}$ **2.** $\begin{bmatrix} 0 & 3 \\ -5 & 1 \\ 1 & 6 \end{bmatrix}$ **3.** $\begin{bmatrix} 3 & 6 \\ 5 & 4 \end{bmatrix}$ **4.** $\begin{bmatrix} -2 & 3 \\ 1 & -2 \end{bmatrix}$ **5.** $\begin{bmatrix} -1 & 3 \\ 2 & 0 \end{bmatrix}$

6. $\begin{bmatrix} 13 & 11 \\ 12 & 14 \\ 10 & 11 \end{bmatrix}$ **7.** $\begin{bmatrix} -10 & -17 & -30 \\ -15 & -15 & -25 \\ -20 & -20 & -20 \end{bmatrix}$ **8.** $\begin{bmatrix} 6 & 5 & 6 & 5 \\ 1 & 4 & 2 & -1 \\ 8 & 5 & 7 & 5 \\ 0 & 2 & 6 & 2 \end{bmatrix}$

a. Which of these games have saddle points?

b. Find the optimal strategy for Player I and for Player II, and the value for those games given in **a.**

c. List any bad strategies in these games, that is, ones the players should avoid because the resulting payoffs are dominated by the payoffs for some alternate strategy.

Solve the following three games of batter versus pitcher in baseball, where the pitcher can throw one of two pitches and the batter can guess either of these two pitches. The batter's batting averages are given in the game matrix.

9.

		Pitcher	
		Fastball	Curve
Batter	Fastball	.300	.200
	Curve	.100	.600

10.

		Pitcher	
		Fastball	Knuckleball
Batter	Fastball	.500	.200
	Knuckleball	.200	.250

11.

		Pitcher	
		Blooperball	Knuckleball
Batter	Blooperball	.400	.200
	Knuckleball	.250	.250

12. A businessman has the choice of either not cheating on his income tax or cheating and making $1000 if not audited. If caught cheating he will pay a fine of $2000 in addition to the $1000 he owes. He feels good if he does not cheat and is not audited (worth $100). If he does not cheat and is audited, he evaluates this at $-$100 (for the lost day). If he is willing to assume that this is a two-person, zero-sum game between himself and the tax agency, then what are the optimal strategies for each player and the expected value of the game?

13. When it is third down and short yardage to go for a first down in American football, the quarterback can decide to run the ball or pass it. Similarly, the other team can commit itself to defend more heavily against a run or a pass. This can be modeled as a 2 by 2 matrix game where the payoffs are the probabilities of obtaining a first down. Find the solution for this game.

		Defense	
		Run	Pass
Offense	Run	.5	.8
	Pass	.7	.2

14. Consider the game played between the opposing goalie and a soccer player who after a penalty is allowed a free kick. The kicker can elect to kick toward one of the two corners of the net or else aim for the center of the goal. The goalie can decide to commit in advance (after the kicker's decision) to either one of the sides or else remain in the center until he sees the direction of the kick. This zero-sum game can be represented as follows, where the payoffs are the probability of scoring a goal:

		Goalie		
		Breaks left	Remains center	Breaks right
Kicker	Kicks left	.5	.9	.9
	Kicks center	1	0	1
	Kicks right	.9	.9	.5

If we assume that decisions between the left or right side are made symmetrically (i.e., with equal probabilities), then this game can be represented by a 2 by 2 matrix as follows, where $.7 = (\frac{1}{2})(.5) + (\frac{1}{2})(.9)$:

		Goalie	
		Remains center	Breaks side
Kicker	Kicks center	0	1
	Kicks side	.9	.7

Find the optimal strategies for the kicker and goalie and the value of this game.

15. You have the choice of either parking illegally on the street or else parking in the lot and paying $16. Parking illegally is free if the police officer is not patrolling, but you receive a $40 parking ticket if she is. However, you are peeved when you pay to park in the lot on days when the officer does not patrol, and you are willing to assess this outcome as costing $32 ($16 for parking plus $16 for your time, inconvenience, and grief). It seems reasonable to assume that the police officer ranks her preferences in the order (1) give you a ticket, (2) not patrolling with you parked in the lot, (3) patrolling with you in the lot, and (4) not patrolling with you parked illegally.

    a. Describe this as a matrix game, assuming that you are playing a zero-sum game with the officer.

    b. Solve this matrix game for its optimal strategies and its value.

    c. Discuss whether it is reasonable or not to assume that this game is zero-sum.

    d. Assuming that you play this parking game each working day of the year, how do you implement an optimal mixed strategy?

16. Describe how a pure strategy for a player in a matrix game can be considered as merely a special case of a mixed strategy.

Exercises 17 to 22 are concerned with the batter versus pitcher game described in the text.

17. Show that the pitcher should avoid throwing to the middle zone. For example, show that his (nonoptimal) mixed strategy $(1 - p, 0, p) = (\frac{1}{2}, 0, \frac{1}{2})$ of throwing both HI and LO half of the time is better on average than using his pure strategy MD.

18. Show that if the pitcher uses his optimal mixed strategy $(1 - p, 0, p) = (0.4, 0, 0.6)$, then he does better on average than using his pure strategy MD, that is, $(0, 1, 0)$.

19. What strategy should the batter use if he knew that the pitcher was playing the (nonoptimal) mixed strategy $(0.5, 0, 0.5)$, that is, throwing a half-half mix of HI and LO?

20. What strategy should the batter use if he knew that the pitcher was playing the (nonoptimal) mixed strategy $(0.2, 0, 0.8)$?

21. Show that the batter should never select his pure strategy HI if he knew that the pitcher was using his optimal mixed strategy $(1 - p, 0, p) = (0.4, 0, 0.6)$.

22. What strategy should the pitcher use if he knew that the batter was playing the (nonoptimal) mixed strategy $(0, 1 - q, q) = (0, 0.5, 0.5)$?

23. In the matching pennies example consider the case where Player I favors heads H over tails T. For example, assume that Player I plays H three-fourths of the time and T only one-fourth of the time — a nonoptimal mixed strategy. What should Player II do if he or she knew this?

24. Assume that in the nonsymmetrical matching example that Player II is using the (nonoptimal) mixed strategy $(p, 1 - p) = (\frac{1}{2}, \frac{1}{2})$; that is, she is playing H and T with the same frequency. What should Player I do in this case if he knew this?

Find the optimal mixed strategies for both players and the value for the following eight *n* by 2 matrix games:

25. $\begin{bmatrix} 0 & 6 \\ 2 & 4 \\ 3 & 0 \end{bmatrix}$

28. $\begin{bmatrix} -4 & 2 \\ -2 & 4 \\ 1 & 3 \\ 1 & 1 \end{bmatrix}$

31. $\begin{bmatrix} 4 & -1 \\ 3 & 1 \\ 1 & 1 \\ 0 & 2 \\ -1 & 3 \end{bmatrix}$

26. $\begin{bmatrix} -2 & 2 \\ 4 & -4 \\ 2 & 0 \end{bmatrix}$

29. $\begin{bmatrix} -2 & 1 \\ -1 & 0 \\ 1 & -1 \end{bmatrix}$

27. $\begin{bmatrix} 6 & 8 \\ 4 & 10 \\ 4 & 5 \\ 5 & 2 \end{bmatrix}$

30. $\begin{bmatrix} 3 & 12 \\ 9 & 6 \\ 10 & 2 \end{bmatrix}$

32. $\begin{bmatrix} 4 & -6 \\ -2 & -6 \\ -2 & 0 \\ -4 & 0 \\ -6 & 4 \end{bmatrix}$

The matrix games in the following four exercises have multiple optimal strategies for some players. Use the geometric approach for finding optimal mixed strategies and the value for a game, and describe the resulting range of optimal strategies in each case.

33. $\begin{bmatrix} 4 & 0 \\ 2 & 2 \end{bmatrix}$

34. $\begin{bmatrix} 4 & 0 \\ 2 & 2 \\ -3 & 1 \end{bmatrix}$

● **35.** $\begin{bmatrix} 2 & -2 \\ 1 & -1 \\ -1 & 1 \end{bmatrix}$

● **36.** $\begin{bmatrix} 0 & 6 \\ 8 & 2 \\ 2 & 5 \end{bmatrix}$

37. You plan to manufacture a new product for sale next year, and you can decide to make either a small quantity, in anticipation of a poor economy and few sales, or a large output, hoping for brisk sales. Your expected profits are indicated in the following table:

		Economy	
		Poor	Good
Quantity	Small	$500,000	$300,000
	Large	$100,000	$900,000

If you want to avoid risk and believe that the economy is playing an optimal mixed strategy against you in a two-person, zero-sum game, then what is your optimal mixed strategy and expected value? Discuss some alternative ways that you may go about making your decision.

● Optional exercise.

38. On an overcast morning, deciding whether to carry your umbrella can be viewed as a game between yourself and nature as follows:

		Weather	
		Rain	No rain
You	Carry umbrella	Stay dry	Lug umbrella
	Leave it home	Get wet	Hands free

Let's assume that you are willing to assign the following numerical payoffs to these outcomes and that you are also willing to make decisions on the basis of expected values (that is, average payoffs):

$$\text{(Carry umbrella, rain)} = -2$$
$$\text{(Carry umbrella, no rain)} = -1$$
$$\text{(Leave it home, rain)} = -5$$
$$\text{(Leave it home, no rain)} = 3$$

a. If the weather forecast says there is a 50% chance of rain, should you carry your umbrella or not? What if you believe there is a 75% chance of rain?

b. If you are conservative and wish to protect against the worst case, what pure strategy should you pick?

c. If you are rather paranoid and believe that nature will pick an optimal strategy for this two-person, zero-sum game, then what strategy should you choose?

d. Another approach to this decision problem is to assign payoffs to represent what your *regret* will be after you know nature's decision. In this case, each such payoff is the best payoff you could have received under that state of nature, minus the corresponding payoff in the previous table.:

		Weather	
		Rain	No rain
You	Carry umbrella	$0 = (-2) - (-2)$	$4 = 3 - (-1)$
	Leave it home	$3 = (-2) - (-5)$	$0 = 3 - 3$

What strategy should you select if you wish to minimize your maximum possible regret?

Consider the following six two-person, nonzero-sum games and discuss the players' possible behavior when these games are played in a noncooperative manner (i.e., with no prior communication or agreements). The first payoff is for the row player; the second, for the column player.

39.

	Player II	
Player I	(5, 5)	(1, 4)
	(3, 0)	(2, 2)

40.

	Player II	
Player I	(3, 5)	(4, 4)
	(1, 2)	(3, 1)

41. Battle of the sexes:

He buys a ticket for:	*She buys a ticket for:*	
	Boxing	Ballet
Boxing	(4, 0)	(0, 0)
Ballet	(1, 1)	(0, 4)

42.

	Player II	
Player I	(3, 5)	(2, 4)
	(1, 2)	(2, 1)

43.

	Player II	
Player I	(1, 3)	(4, 1)
	(0, 6)	(3, 5)

44.

	Player II	
Player I	(2, 4)	(1, 6)
	(4, 2)	(0, 1)

●**45.** Consider the following miniature poker game with two players, I and II. Each antes $1. Each player is dealt either a high card H or a low card L, with probability one-half. Player I then folds or bets $1. If I bets, then Player II either folds, calls, or raises $1. Finally, if II raises, I either folds or calls.

Most choices by the players are rather obvious, at least to anyone who has played poker: if either player holds H, that player always bets or raises if he or she gets the choice. The question remains of how often one should bluff, that is, continue to play while holding a low card in the hope that one's opponent also holds a low card.

This poker game can be represented by the following matrix game where the payoffs are the *expected* winnings for Player I (depending upon the random deal) and the dominated strategies have been eliminated:

		Player II (when holding L)		
		Folds	Calls	Raises
Player I	Folds initially	−0.25	0	0.25
(when holding L)	Bets first and folds later	0	0	−0.25
	Bets first and calls later	−0.25	−0.25	0

a. Are there any strategies in this matrix game that a player should avoid playing?

b. Solve this game.

c. Which player is in the more favored position?

d. Should one ever bluff?

●**46.** For an interesting class project, discuss strategies and expected outcomes in the case of a three-person pistol duel. Assume that three players I, II, and III are an equal distance apart and that they shoot with accuracies 0.8, 0.6, and 0.4, respectively. Before *each* individual shoots, the players decide by lot which one of the three will take the next shot. When a player is hit he is out of the contest; the lone surviving player is the winner.

a. Discuss this game when it is played in a noncooperative manner.

b. Discuss this game when it is played in a cooperative manner; when coalitions of two can form against the third player, at least until he is eliminated.

● Optional exercise.

·IV·

On Size and Shape

Mathematics is the study of patterns and relationships. In some applications of mathematics, this truth is obscured by the details of the problem being solved. Consider as examples the spiral growth structure of a sunflower; the spiral distribution of stars, gas, and dust in the Andromeda galaxy; the intricate symmetries of old pottery designs; and the patterns of growth exhibited by human populations or by bank accounts.

Mathematicians instinctively search for geometrical and numerical patterns and for symmetry. Their discoveries of patterns and symmetries often enable us to better understand practical problems. Geometry grew as a body of knowledge because of people's need to explain and control the world around them. The set of tools created has enabled people to range further and further from daily experience.

From determining the circumference of the earth to digging straight tunnels, the ancient Greeks used the tools of geometry to make seemingly impossible measurements. More modern astronomers and mathematicians succeeded in uncovering the secrets of planetary motion. And new geometries, which have led to a much deeper understanding of the nature of mathematics itself, paved the way for the discoveries of Einstein, the basis of our present view of the workings of our universe.

Perhaps no other subject has intrigued people through the centuries as much as geometry. We perceive the symmetries and patterns of nature and search to understand the intrinsic order and beauty we observe.

The geometric symmetry of natural phenomena is strikingly revealed in this sunflower (see p. 477). [Photo by Dwight Kuhn.]

·13·

Growth and Form

Fantasy films have made us familiar with assorted giant creatures, including King Kong, Godzilla, and the 50-foot-high grasshoppers in *The Beginning of the End.* We also find supergiants in literature, such as the giant of "Jack and the Beanstalk," Giant Pope and Giant Pagan of *The Pilgrim's Progress,* and the Brobdingnagians of *Gulliver's Travels.*

Much as we appreciate those stories, even from an early age we don't really believe in monsters and giants. But could such beings ever exist? What problems would their enormous size cause them? How would they have to adapt in order to cope? (see Figure 13.1.)

Every species survives by adapting to its environment. In particular, it faces the **problem of scale:** how to adapt and survive at the different sizes from the beginning of life to the final size of a mature adult.

Certainly, there have been large land mammals (mammoths) and huge sea mammals (the blue whale) — not to mention the dinosaurs. But the tallest humans have been only 9 to 10

Figure 13.1 Could King Kong actually exist? [The Museum of Modern Art/Film Stills Archive.]

feet tall; the largest mammoth was 16 feet at the shoulder (about twice as tall as an elephant); and even the tallest dinosaur, *Supersaurus,* stood only 40 feet high.

But what about supergiants and utterly huge monsters? That they have never existed suggests that there are physical limits to size. In fact, with a few simple principles of geometry, we can show not only that lizards and apes of such size are impossible, but also that none of the living beings and objects in our world could exist, unchanged in shape, on a vastly different scale, larger or smaller.

GEOMETRIC SIMILARITY

The powerful mathematical idea that we will use is *geometric similarity.* By geometric standards, two objects are **similar** if they have the same shape, regardless of the materials of which they are made. They may even be of different sizes. Corresponding angles must be

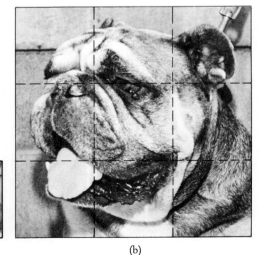

(a)   (b)

Figure 13.2 Two geometrically similar photographs. [Photo by Travis Amos.]

equal, and corresponding dimensions must all have the same factor of proportionality.

For example, when a photo is enlarged, it is enlarged by the same factor in both the horizontal and vertical directions — in fact, in any direction whatever (such as a diagonal). We call this enlargement factor the **scaling factor.** In the photos in Figure 13.2, the scaling factor is 3: the enlargement is three times as wide and three times as high as the original. We notice, however, that the enlargement can be divided into $3 \times 3 = 9$ rectangles, each the size of the original. Hence, the enlargement has $3 \times 3 = 3^2 = 9$ times the area of the original. More generally, if the scaling factor is some general number M (not necessarily 3), the resulting enlargement will have an area $M \times M = M^2$ ("M squared") times the area of the original. Thus, the *area* of a scaled-up object goes up with the *square* of the scaling factor.

What about enlarging three-dimensional objects? If we take a cube and enlarge it by a scaling factor of 3, it becomes 3 times as long, 3 times as high, and 3 times as deep as the original (see Figure 13.3).

We observe, however, that the area of each face (side) of the enlarged cube is $3^2 = 9$ times as large as that of a face or our original cube, just as the area of the photo enlarged by a factor of 3 has 9 times the area of the original. Since this fact is true for all 6 faces, the total surface area of the enlarged cube is 9 times as much as the original.

More generally, for objects of any shape whatever, the total *surface area* of a scaled-up object goes up with the *square* of the scaling factor. Thus, the surface area of an object scaled up by a factor of M is M^2 times the surface area of the original; and this feature holds true even for irregular shapes.

What about volume? The enlarged cube has 3 layers, each with $3 \times 3 = 9$ little cubes, each the same size as the original. Thus, the total volume is $3 \times 3 \times 3 = 3^3 = 27$ times as much

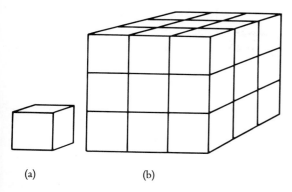

(a)   (b)

Figure 13.3   Cube (b) is made by enlarging cube (a) by a factor of 3.

as the original cube. In general, the *volume* of a scaled-up object goes up with the *cube* of the scaling factor. Thus, for an object enlarged by a scaling factor of M, the enlargement will have M^3 ("M cubed") $= M \times M \times M$ times the volume of the original. Like the relationship between surface area and M^2, this relationship holds even for irregularly shaped objects, such as science fiction monsters.

We will now discuss the terminology you should use to describe the increases and decreases that result from scaling factors or other causes.

THE LANGUAGE OF GROWTH, ENLARGEMENT, AND DECREASE

Erin's Own Irish sphagnum moss peat. It enriches your soil and makes your growing easier. Compressed to $2\frac{1}{2}$ times normal volume. — Printed on sacks of the product [*New Scientist* (May 8, 1986): 80].

Poor or imprecise use of language can often confuse a reader or listener. There is particular occasion for such confusion in discussions of growth or decrease, since the speaker or writer may want to exaggerate.

Let us take a photograph with area 1 and enlarge it three times, so that it is three times as wide and three times as high. The enlargement has area 9. What are some ways you might compare this area to the original area? Fill in the blank in the sentence: "The enlargement's area is _____ the area of the original."

"Nine times," "nine times as large as," "nine times as great as" are all simple, descriptive, correct, and unconfusing alternatives.

However, "nine times greater than," "nine times more than," "nine times larger than" are confusing alternatives. These phrases sound even grander, because they use the comparative form "more" in addition to the word "times." But "more," "larger," and "greater" refer to quantity over and above the original; so the phrases are properly interpreted as meaning "10 times as large" (the original plus nine times more). The problem involved is easiest to see if we ask what "one times greater than" could possibly mean.

This problem is not solved by speaking of percent instead. In all the incorrect phrases we have considered, "nine times" could be replaced by "900%" and the same criticism would hold.

Percent has even further hazards for the careless user of language. An increase from one to nine is an increase of 800%, not 900%. In discussions of percent we also need to distinguish percent from percentage points: if support for the President has decreased from 60% to 30%, it has dropped 30 *percentage points* but decreased 50% (because the drop of 30 percentage points is 50% of the original 60 percentage points).

There are also perils associated with speaking of decreases. For a reduction from nine to one, the reduced quantity may be described incorrectly as "nine times less than" or "900% less than." Correct terminology is that the new amount is "one-ninth as much," "eight-ninths less than," "11% as much," and "89% less than" the original.

EXAMPLE: What about That Bag of Peat Moss? The writer of the ad meant that the volume was 2.5 times as much before packaging. So 1 finished bag has the same amount as 2.5 bags before compression. Since 2.5 bags has been compressed to 1 bag, the new volume is $\frac{1}{2.5} = 0.4$ times as much as before. We could also say that it has been compressed to 40% of the original volume, or that the compressed volume is 60% less than the original volume.

Before pursuing the scaling of real three-dimensional objects, we discuss the units in which mass, weight, area, and volume are measured.

MEASURING LENGTH, AREA, VOLUME, AND WEIGHT

We give here an introduction to the common units in which various physical quantities are measured, together with a handy table of conversion factors and examples of how to convert successfully from one system of units to another.

You are no doubt familiar with the common units of the *U.S. Customary System* of measurement and their abbreviations. But please pay close attention to the systematic way to convert from one unit to another, and to the expression of approximate numbers in scientific notation:

Distance:
$$1 \text{ mile (mi)} = 5280 \text{ feet (ft)}$$
$$1 \text{ foot (ft)} = 12 \text{ inches (in)}$$

Area:
1 square mile (sq mi)
$$= 1 \text{ mi} \times 1 \text{ mi}$$
$$= 5280 \text{ ft} \times 5280 \text{ ft}$$
$$= 27,878,400 \text{ ft} \times 1 \text{ ft}$$
$$= 28 \times 10^6 \text{ sq ft, approximately}$$
$$= 27,878,400 \times 1 \text{ ft} \times 1 \text{ ft}$$
$$= 27,878,400 \times 12 \text{ in} \times 12 \text{ in}$$
$$= 4,014,489,600 \times 1 \text{ in} \times 1 \text{ in}$$
$$= 4,014,489,600 \text{ sq in}$$
$$= 4 \times 10^9 \text{ sq in, approximately}$$

We also have
$$1 \text{ square mile} = 640 \text{ acres}$$
with
$$1 \text{ acre} = 43,560 \text{ sq ft}$$

Volume:
1 cubic mile (cu mi)
$$= 1 \text{ mi} \times 1 \text{ mi} \times 1 \text{ mi}$$
$$= 5280 \text{ ft} \times 5280 \text{ ft} \times 5280 \text{ ft}$$
$$= 147,197,952,000 \times 1 \text{ ft} \times 1 \text{ ft} \times 1 \text{ ft}$$
$$= 147 \times 10^9 \text{ } 1 \text{ ft} \times 1 \text{ ft} \times 1 \text{ ft,}$$
$$\text{approximately}$$
$$= 147 \times 10^9 \times 12 \text{ in} \times 12 \text{ in} \times 12 \text{ in}$$
$$= 147 \times 10^9 \times 12 \times 12 \times 12 \text{ cu in}$$
$$= 2.5 \times 10^{14} \text{ cu in, approximately}$$

For liquid measure, the customary unit in the United States is
$$1 \text{ U.S. gallon} = 231 \text{ cu in, exactly}$$

Weight:
$$1 \text{ ton (t)} = 2000 \text{ pounds (lb)}$$

There are other units (rods, light-years, bushels, ounces, etc.), but we will not consider them here.

The metric system was first proposed in France by Gabriel Mouton, Vicar of Lyons, in 1670, and was adopted in France in 1795. The fundamental unit of length, the *meter,* was originally defined to be one ten-millionth of the distance from the North Pole to the Equator, as measured on the meridian through Paris. Later the meter was redefined as the distance between two lines marked on a platinum-iridium bar kept at the International Bureau of Weights and Measures, near Paris, when the bar is kept at a temperature of 0°C. Finally, in 1960 the meter was redefined in terms of a standard reproducible in any laboratory, namely, 1,650,763.73 times the wavelength of the orange-red light emitted by atoms of the gas krypton-86 when an electrical charge is passed through them. All other units of length, area, and volume are *defined* in terms of the meter; for example, a centimeter is a hundredth of a meter. The metric unit of weight, the *kilogram,* is still defined as the weight of a platinum-iridium standard.

In the metric system, we have

Distance:
1 kilometer (km) = 1000 meters (m)
= 100,000 centimeters (cm)
= 1×10^5 cm

1 meter (m) = 100 centimeters (cm)

Area:
1 square meter (sq m, or m²)
= 1 m × 1 m
= 100 cm × 100 cm
= 10,000 sq cm (cm²)
= 1×10^4 cm²

(Land is usually measured in a larger unit: 1 hectare = 10,000 m².)

Volume:
1 cubic meter (cu m, or m³)
= 1 m × 1 m × 1 m
= 100 cm × 100 cm × 100 cm
= 1,000,000 cu cm (cm³)
= 1×10^6 cm³

(For liquid measure, the unit is 1 liter = 1000 cm³.)

Weight:
1 kilogram (kg) = 1000 grams (g)

The metric system also has other units (angstrom, metric tonne) that we will not consider here.

What are the conversions between the U.S. Customary System and the metric system? Since 1960, the fundamental units of the U.S. Customary System, the yard (for length) and the pound (for weight), have been *defined* in terms of metric units, so that we have

1 yd = 0.9144 m, exactly

1 lb = 0.45359237 kg, exactly

The conversions of other units are

Distance:
1 in = 2.54 cm, exactly

1 ft = 12 in = 12 × 1 in = 12 × 2.54 cm
= 30.48 cm, exactly

1 mi = 5280 ft = 5280 × 1 ft
= 5280 × 30.48 cm
= 160,934.4 cm, exactly
= 1.61×10^5 cm, approximately
= 1.61 × 1 km
= 1.61 km, approximately

We can also go the other way:

1 m = 100 × 1 cm

But how much is 1 cm in terms of inches? Since

1 in = 2.54 cm

we can divide both sides of this equation by 2.54, getting

$\frac{1}{2.54}$ in = 1 cm

or, performing the division and reading from right to left,

1 cm = 0.393701 in, approximately
= 0.4 in, approximately

Hence

$$1 \text{ m} = 100 \times 1 \text{ cm} = 100 \times 0.393701 \text{ in}$$
$$= 39.3701 \text{ in, approximately}$$

For *weight* we have

$$1 \text{ lb} = 0.45359237 \text{ kg}$$

Hence

$$1 \text{ kg} = \frac{1}{0.45359237} \text{ lb}$$
$$= 2.205 \text{ lb, approximately}$$

The exercises offer practice in converting area, volume, and weight back and forth between the U.S. Customary and metric systems.

SCALING REAL OBJECTS

Real three-dimensional objects are, of course, made of matter, which has extension (volume) and substance (mass). **Mass** is the aspect of matter that is affected by forces, according to physical laws. For example, the mass of you reacts to the gravitational force of the earth by staying close to it (and your mass exerts an equal force on the earth that tends to keep the earth close to you). We perceive the mass of an object when we try to move it (as in throwing a ball). When we try to lift an object, we perceive its mass as weight, due to the gravitational force that the earth exerts on it. As we will see, gravity exerts an enormous effect on the size and shape that objects and beings can assume.

In Figure 13.4, we consider two cubes. The first is a cube of steel 1 foot on a side. The bottom face supports the weight of the entire cube. **Pressure** is force per unit weight. So the pressure exerted on the bottom face by the weight of the cube is equal to the weight of the cube divided by the area of the bottom face. A cubic foot of steel weighs about 500 pounds (we say it has a **density** of 500 pounds per cubic foot), and the area is 1 square foot; so the pressure exerted on the bottom face is therefore 500 pounds per square foot.

The second cube in Figure 13.4 is made of the same steel but is 2 feet on a side. The area of the bottom face has increased with the square of the scaling factor, so it is $2^2 \times 1 = 4$ square feet. Since volume goes up with the cube of the scaling factor, this larger cube has a volume of $2^3 \times 1 = 8$ cubic feet. Because both cubes are made of the same steel, the larger cube has eight times as much steel as the smaller; hence it weighs eight times as much as the smaller cube, or $8 \times 500 = 4000$ pounds.

When we divide this weight by the area of the bottom face (4 square feet), we find that the pressure exerted on the bottom face is 1000 pounds per square foot, or twice the pressure on the bottom face of the original cube. This makes sense because over each 1 square foot area stands 2 cubic feet of steel.

If we scale the original cube of steel up to a cube 3 feet on a side, we observe *triple* the pressure on the bottom face. In general, if the scaling factor for the cube is M, the pressure on the bottom face will be M times as much.

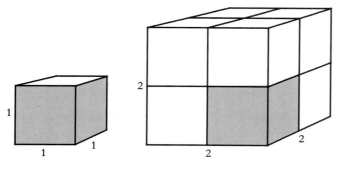

Figure 13.4 A cube of side *n* and a cube of side 2*n*.

EXAMPLE: What about a 10-Foot Cube?
If we scale the original cube of steel up to a cube 10 feet on a side, then the dimensions are

$$10 \text{ ft} \times 10 \text{ ft} \times 10 \text{ ft}$$

The total volume is

$$
\begin{aligned}
V &= \text{length} \times \text{width} \times \text{height} \\
&= 10 \text{ ft} \times 10 \text{ ft} \times 10 \text{ ft} \\
&= 1000 \text{ ft} \times 1 \text{ ft} \times 1 \text{ ft} = 1000 \text{ ft}^3 \\
&= 1000 \text{ cu ft}
\end{aligned}
$$

The weight of the cube is

$$
\begin{aligned}
W &= V \times \text{density} \\
&= 1000 \text{ cu ft} \times 500 \text{ lb/cu ft} \\
&= 500{,}000 \text{ lb}
\end{aligned}
$$

The area of the bottom face is

$$
\begin{aligned}
A &= \text{length} \times \text{width} \\
&= 10 \text{ ft} \times 10 \text{ ft} \\
&= 100 \text{ ft} \times 1 \text{ ft} \\
&= 100 \text{ ft}^2 = 100 \text{ sq ft}
\end{aligned}
$$

The pressure on the bottom face is

$$P = \frac{W}{A} = \frac{500{,}000 \text{ lb}}{100 \text{ sq ft}} = 5000 \text{ lb/sq ft}$$

This is 10 times — not "10 times more than" — the pressure on the bottom face of the original 1-foot cube.

At some scale factor the pressure on the bottom face will exceed the steel's ability to withstand that pressure — and the steel will deform under its own weight. That point for steel is reached for a cube about 3 miles on a side — the pressure exerted by the cube's weight exceeds the resistance to crushing (ability to withstand pressure, or **yield strength**) of steel, which is about 7.5 million lb/sq ft. Since a mile is 5280 feet, a 3-mile-long cube of steel would be more than 15,000 times as long as the original 1-foot cube; that is, the scaling factor is more than 15,000. The pressure on the bottom face of the cube would therefore be more than 15,000 times as much as for the 1-foot cube, or $15{,}000 \times 500$ lb/sq ft = 7.5 million lb/sq ft.

SORRY, NO KING KONGS

Unfortunately, the resistance of bone to crushing is not nearly as great as that of steel. This fact helps to explain why there couldn't be any King Kongs (unless they were made of steel!). A creature scaled up by a factor of 20 would weigh $20^3 = 8000$ times as much. Though the weight increases with the cube of the scaling factor, the ability to support the weight — as measured by the cross-sectional area of the bones, like the area of the bottom face of the cube — increases only with the square of the scaling factor.

These simple consequences of the geometry of scaling apply to other objects, natural and artificial, not only to supermonsters. For example, we can estimate how high the tallest mountains could be. Three hundred and fifty years ago, Galileo was able to give an accurate estimate of how high the tallest trees could be (see Spotlight 13.1, p. 356).

EXAMPLE: How High Can a Mountain Be? Real mountains, of course, aren't made out of steel, and they don't come in the shape of a cube. Mountains differ from one to another in composition and shape, and some assumptions about those features will be necessary in order to do any calculating. We want to make our assumptions as realistic as we can and still be able to calculate easily an estimate of how high a mountain can be. In effect, we build a simple mathematical model of a mountain.

Let's suppose that the mountain is made entirely of granite, a common material in many mountains, and we assume that the granite has uniform density. Relevant facts about granite are that it weighs 165 lb/cu ft and it has a yield strength of about 30 million lb/sq ft.

In the interests of both realism and simplicity, we assume that our model mountain is in the shape of a cone whose width at the base is the same as its height. Let's model Mount Everest: the tallest earth mountain, it is about 6 miles high. The base, then, is a circle with a distance across (or diameter) of 6 miles. The radius of the circle is half the diameter, so our model Everest has a radius of 3 miles measured at the base (Figure 13.5). Since we are taking such a round number for the height of Everest, we will record as significant only the first two digits of the results of our calculations.

What does our model Everest weigh? The relevant formula is

$$\text{Weight} = \text{density} \times \text{volume}$$

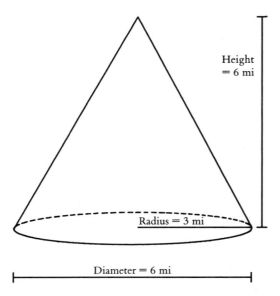

Figure 13.5 Model of Mt. Everest as a cone of granite.

We already know the density of granite (165 lb/cu ft), so to find the weight we are going to need to know how to calculate the volume of a cone. The formula is

$$\text{Volume} = \pi \times (\text{radius})^2 \times \frac{\text{height}}{3}$$

For our Everest, the radius is 3 miles and the height is 6 miles; π (pi) is about 3.14. Using those values in the formula, we find that our model Everest has a volume of about 57 cubic miles.

To find the weight of 57 cubic miles of granite, we need to do some conversion of units, since the density is given in pounds per cubic foot. Let's convert to units of feet:

$$
\begin{aligned}
1 \text{ cu mi} &= 1 \text{ mi} \times 1 \text{ mi} \times 1 \text{ mi} \\
&= 5280 \text{ ft} \times 5280 \text{ ft} \times 5280 \text{ ft} \\
&= 1.5 \times 10^{11} \text{ ft} \times 1 \text{ ft} \times 1 \text{ ft,} \\
&\qquad\qquad\qquad\qquad \text{approximately} \\
&= 1.5 \times 10^{11} \text{ cu ft, approximately}
\end{aligned}
$$

and

$$
\begin{aligned}
57 \text{ cu mi} &= 57 \times 1 \text{ cu mi} \\
&= 57 \times 1.5 \times 10^{11} \text{ cu ft,} \\
&\qquad\qquad\qquad\qquad \text{approximately} \\
&= 8.6 \times 10^{12} \text{ cu ft,} \\
&\qquad\qquad\qquad\qquad \text{approximately}
\end{aligned}
$$

So we have

$$
\begin{aligned}
\text{Weight of mountain} \\
&= 165 \text{ lb/cu ft} \times 8.6 \times 10^{12} \text{ cu ft} \\
&= 1.4 \times 10^{15} \text{ lb} \\
&= 1.4 \text{ quadrillion lb}
\end{aligned}
$$

Now that we know the weight of the mountain, we want to find out what the pressure is on the base of the cone and compare that with the yield strength of granite. (Everest is standing, so if our model is any good, that pressure will be below the yield strength.) Physics tells us that the weight of the mountain is spread evenly over the base of the cone (we are oversimplifying the geology underlying mountains). Since

$$\text{Pressure} = \frac{\text{weight}}{\text{area}}$$

we need to calculate the area of the base of the cone. The shape is a circle, and the familiar formula

$$\text{Area} = \pi \times (\text{radius})^2$$

gives an area of 28 square miles for a radius of 3 miles.

Once again, we will need to convert to units of feet in order to express the pressure in pounds per square foot, the units in which we are given the yield strength. We get

$$
\begin{aligned}
\text{Area} &= 28 \text{ sq mi} \\
&= 28 \times 1 \text{ mi} \times 1 \text{ mi} \\
&= 28 \times 5280 \text{ ft} \times 5280 \text{ ft} \\
&= 8 \times 10^8 \text{ sq ft, approximately}
\end{aligned}
$$

Then

$$
\begin{aligned}
\text{Pressure} &= \frac{\text{weight}}{\text{area}} \\
&= \frac{1.4 \times 10^{15} \text{ lb}}{8 \times 10^8 \text{ sq ft}} \\
&= 1.7 \times 10^6 \text{ lb/sq ft} \\
&= 1.7 \text{ million lb/sq ft}
\end{aligned}
$$

This number is safely below the yield strength of granite, 30 million pounds per square foot. For a mountain to come close to the limitation of the yield strength of granite, it would have to be about 18 ($= 30$ million/1.7 million) times as high as Everest, or about 100 miles high. From other physical considerations we can determine that the maximum possible height of a mountain on earth would be closer to 15 miles. That no present mountains are that high is a consequence of the earth's high amount of volcanic activity and structural deformation of the earth's crust.

What about mountains made of other materials — glass, ice, wood, old cars? They couldn't be nearly as high; the pressure would cause glass to flow, ice to melt, and old cars to compact. What about mountains on another planet? Their potential height also depends on the gravity of the planet.

SOLVING THE PROBLEM OF SCALE

A large change in scale forces a change in either materials or form. A major manifestation of the scaling problem is the tension between weight and the need to support it. For example, a real building or machine must differ from a scale model; the balsa wood or plastic of the model would never be strong enough to use for the real thing, and the materials in the scaled-up version must be aluminum, steel, or reinforced concrete. So one way to compensate for the problem of scale is to use stronger materials in the scaled-up object (see Spotlight 13.2, p. 358).

The other way to compensate is to redesign the object so that its weight is better distributed. Let's go back to our original cube. It supports all its weight on its bottom face. In the version scaled up by a factor of 3, each small cube of the bottom layer has a bottom face that is supporting that cube's weight plus the weight of the other two cubes piled on top of it.

Now, let's redesign the scaled-up cube, concentrating for simplicity only on the front face, with its nine small cubes. We take the three cubes on top and move them to the bottom, alongside the three already there. We take the three cubes on the second level, cut each in half, and put a half cube over each of

SPOTLIGHT 13.1 Galileo and the Tallest Trees

Galileo Galilei (1564–1642) was the first to describe the problem of scale, in 1638, in his *Dialogues Concerning Two New Sciences* (in which he also introduced the idea of the earth revolving around the sun):

You can plainly see the impossibility of increasing the size of structures to vast dimensions either in art or in nature; likewise, the impossibility of building

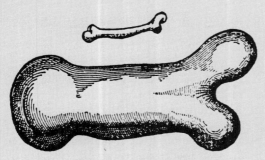

One bone, with another three times as long and thick enough to perform the same function in a scaled-up animal. [Illustration from Galileo's *Dialogues Concerning Two New Sciences.*]

ships, palaces, or temples of enormous size in such a way that their oars, yards, beams, iron-bolts, and, in short, all their other parts will hold together; nor can nature produce trees of extraordinary size because the branches would break down under their own weight, so also would it be impossible to build up the bony structures of men, horses, or other animals so as to hold together and perform their normal functions if these animals were to be increased enormously in height; for this increase in height can be accomplished only by employing a material which is harder and stronger than usual, or by enlarging the size of the bones, thus changing their shape until the form and appearance of the animals suggest a monstrosity.

To illustrate briefly, I have sketched a bone whose natural length has been increased three times and whose thickness has been multiplied until, for a correspondingly large animal, it would perform the same function which the small bone performs for

the six ground-level cubes (see Figure 13.6). We have the same volume and weight that we started with, but now there is less pressure on the bottom face of each small cube. Of course, our new design is not geometrically similar to the object we started with — it's no longer a

cube. By changing the proportions, we have given up the precise scaling of geometrical similarity, but we have managed to compensate for the scaling problem.

We observe in nature both strategies for adaptation to scaling: change of materials and change of form. Smaller animals generally do not have bony internal skeletons; larger animals generally do. Those animals made of sim-

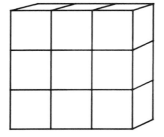
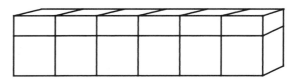

Figure 13.6 Nine small cubes rearranged to support greater weight.

its small animal. From the figures here shown you can see how out of proportion the enlarged bone appears. Clearly then if one wishes to maintain in a great giant the same proportion of limb as that found in an ordinary man he must either find a harder and stronger material for making the bones, or he must admit a diminution of strength in comparison with men of medium stature; for if his height be increased inordinately he will fall and be crushed under his own weight. Whereas, if the size of a body be diminished, the strength of that body is not diminished in proportion; indeed the smaller the body the greater its relative strength. Thus a small dog could probably carry on his back two or three dogs of his own size; but I believe that a horse could not carry even one of his own size.

(Translated by Henry Crew and Alfonso De Salvo, and published by Macmillan, 1914, and Northwestern University, 1946)

Galileo suggested 300 feet as the limiting height for a tree. Galileo's estimate has proved to be roughly correct. The world's tallest trees are giant sequoias, which grow only on the West Coast of the United States and hence were unknown to Galileo. They grow to 360 feet.

Even these giant sequoias may grow no taller than their form and materials allow. [Larry Ulrich.]

ilar materials but differing greatly in size, such as a mouse and an elephant, will most certainly differ in shape. If a mouse were scaled up to the size of an elephant, its legs could no longer support it. It would need the disproportionately thicker legs of the elephant, as well as the elephant's thick hide to contain its tissue.

Some dinosaurs, like *Supersaurus* (which weighed 30 tons, as much as a tank), had special adaptations to lighten their weight, such as hollow bones, just as some birds have. (Hollow bones also turn out to be stronger, a paradox that Galileo analyzed. Of two bones of the same weight and length, the hollow one will be wider across at its midpoint, because of the air it contains; and the greater the width, the greater the resistance to fracture.)

FALLS, DIVES, JUMPS, AND FLIGHTS

The need to support weight can be thought of as a tension between volume and area. As we scale up an object, its volume and weight go up together, as long as we maintain a constant density (for example, no air bubbles introduced into our steel to make it into a Swiss cheese!). At the same time, the ability to support the weight goes up with the cross-sectional area, just as our steel cube had to be supported by its bottom face.

Area-volume tension has many other practical consequences. It brings our child-

SPOTLIGHT 13.2 Mile-High Buildings?

Architect Hugh Stubbins discusses skyscrapers:

Scale is a relative thing. If you think of a human being standing beside the Great Pyramid at Giza, you can imagine the scale of a human being against that big mass, or even against the Empire State Building in New York. Now, scale in the past and ancient times was made so it would impress people; they made temples to impress people. Today I don't think we build big buildings to impress people; they're big because they have to be big. The requirement for this building was to have a million and a quarter square feet of usable space. Now, that's a lot of square footage! If the building were high and solid, then nobody would be close to a window. It would be just a cavern of space, which would not be very pleasant to be in. And one of the main purposes of architecture is to develop spaces that are pleasant for people to work in, to live in, and to spend their time in.

Well, I think buildings can get too big. William LaMeasure has devised a building that could be built a mile high. Years ago Frank Lloyd Wright said he had designed a building that would reach a mile high. Well, LaMeasure checked into that and found out that Wright's scheme was absolutely impossible — it could never be built. But LaMeasure is such a genius at engineering that he has devised a mile-high structure that could be built. There are four buildings actually, at four corners of a square, which are about 400 feet apart. They go up, and then there are horizontal buildings that run between these four posts. And they're open every other floor so the air can get through it. He says that such a structure can be built 5000 feet high. But who would ever want to do it? I can't imagine anybody wanting to do that. It takes too long to get from the ground up to where you're going.

hood fantasies into sharp but disappointing terms. We can forget about humans "leaping tall buildings in a single bound," "soaring like an eagle," diving miles below the sea, and jumping from airplanes without parachutes.

EXAMPLE: Falls. Area-volume tension affects how creatures respond to falling, another of gravity's effects. A mouse may be unharmed by a 10-story fall, a cat by a two-story fall, but a human may well be injured just by falling from his or her own height.

What is the explanation? The energy acquired in falling is proportional to the weight of the falling object, hence to its volume. This energy must be absorbed either by the object or by what it hits or must be otherwise dissipated at impact — for example, as sound. The fall is absorbed over part of the surface area of the object, just as the weight of the cube was distributed over its base. With scaling up, volume — hence weight, hence falling energy — goes up much faster than area. As volume increases, the hazards of falling from the same height increase.

EXAMPLE: Dives. Whales can hold their breath and stay under water for as long as 20 minutes. Why can't we? Basically, because we aren't as large as whales. A mammal's breath-holding ability depends on two things: the volume of oxygen carried in the lungs, which volume is proportional to the volume of the mammal and hence to the cube of its length; and the rate at which oxygen is absorbed by the surface area of the lungs, which area is proportional to the square of the length of the mammal. We would therefore expect the limits of duration of dive to be proportional to the lung volume divided by the lung area, thus to the length of the animal; and we would be right. Although special adaptations may play a larger role for a particular species, and exceptional individuals can outperform the average, this case does illustrate a straightforward proportional relationship. Blue whales are about 16 times as long as human adults and can hold their breath about 16 times as long.

EXAMPLE: Jumps. A flea can jump about 2 feet vertically, many times its own height. Many people believe that if a flea were as large as a person, it could jump a thousand feet into the air. Imagining — against our earlier arguments — that there could be so large a flea, we know its limits: a scaled-up flea could jump about the same height as a small flea. The strength of a muscle is proportional to its cross-sectional area (see Spotlight 13.3, p. 360). A jump involves suddenly contracting the muscle through its length, so that it turns out that the ability to jump is proportional to the volume of muscle. But the volume of the flea and the volume of its leg muscles go up in proportion. Let's say that a real flea's leg muscles account for 1% of its body. If we scale the flea up to the size of a person (without any change in its form), the enlarged flea's leg muscles would still make up 1% of its body. For either flea, each bit of muscle has the same power: in a jump, it propels 100 times its own weight, and it can do so to the same height. Both the weight of the flea and the power of its legs go up proportionately.

EXAMPLE: Flight. Wouldn't it be nice to be able to fly? Well, you have to be able to stay up. The power necessary for sustained flight is proportional to the **wing loading,** which is the weight supported divided by the area of the wings. We know that in scaling up, weight grows with the cube of the length of the bird or plane, and wing area with the square of the length. So the wing loading will be proportional to the length of the flying object.

For example, if a bird or plane is scaled up proportionally by a factor of 4, it will weigh $4^3 = 64$ times as much but have only $4^2 = 16$ times as much wing area. So each square foot of wing must support 4 times as much weight.

Second, you have to keep moving. To stay level, an airborne object must fly fast enough to maintain the lift on the wings. The minimum necessary speed is proportional to the square root of the wing loading. Combining this fact with our first consideration above, we conclude that the minimum speed goes up with the square root of the length. Our bird that was scaled up by a factor of 4 must fly $\sqrt{4} = 2$ times as fast.

Take, for instance, a sparrow, whose minimum speed is about 20 miles per hour. An ostrich is 25 times as long as a sparrow, so the minimum speed for an ostrich would be $\sqrt{25} \times 20 = 100$ miles per hour. Have you seen any flying ostriches lately? Heavy birds have to fly fast or not at all!

Of course, ostriches are not just scaled up sparrows, nor are eagles. The larger flying birds have disproportionately larger wings than a sparrow, to keep the wing loading down. The largest animal ever to have taken to the air was *Quetzlcoatlus northropi,* a flying dinosaur of 65 million years ago, with a wingspan of 36 feet and a weight of about 100 pounds.

You have to stay up, you have to keep moving — and you have to get up there. Here basic aerodynamics imposes further limits.

SPOTLIGHT 13.3 "Take That, King Richard!"

Shakespeare, following the propaganda of the Tudor historians, painted Richard III as a humpbacked Machiavellian monster. Did Richard have an advantage in armored combat because he was short? That suggestion was made some years ago by one of the leading modern historians of the Tudor era, Garrett Mattingly.

Did wearing armor give an advantage to the shorter warrior? [All rights reserved, the Metropolitan Museum of Art.]

Between a short man and a tall man, height increases by the linear dimension—from 5 feet 2 inches, say, to 6 feet—while the surface of the body increases as the square. Since it's . . . the surface of the body that the armorer must plate with steel, the armor of a short warrior, like Richard, would be lighter than a tall warrior's by a lot more than the few inches' difference in height would indicate. So Richard's notorious deadliness in battle would have been possible at least in part—or so Mattingly's speculation ran—because his armor, while protecting him as well as the big man's, left him less encumbered.

After the lecture, someone said to Mattingly that he had grasped the right idea—but by the wrong end. Muscle power, the listener claimed, is a matter of bulk—and physical volume goes up by the cube, whereas the surface to be protected goes up by the square. So the large warrior should have more strength left over than the little guy after putting on his armor. And the large warrior, swinging a bigger club, can deliver a far more punishing blow—because the momentum of the club depends on its weight, which goes up with its volume, which means by the cube. Richard was at a terrible disadvantage.

But wait a minute, a second listener said. That's true about the club—but not about the muscles. The strength of a muscle is proportional not to its bulk but to the area of its cross section. And since the cross section of muscles obviously increases by the square, just as the surface of the body does, the big guy, plated out, has no more, or less, advantage over the little guy than if both were naked.

But hang on, a third person interjected—an engineer. That's right about the muscles, but it's not right about the armor. The weight of the armor increases not simply with the increase in the surface area that it must cover but slightly faster. The reason is that to obtain the same strength with a larger area of metal, there must be reinforcing ribs. Or else the metal must be significantly thicker overall. So maybe Richard had an advantage after all.

To maximize protection within the weight, the armorer adopted two strategies: variable thickness and deflection. The unexpected fact is that armor was made as thin as possible. Thickness, reinforcement, and structural stiffening of a large surface were concentrated where opponents' weapons were likely to hit. From these strong, shaped places, the metal tapered away, until the sheet steel was as thin as the lid of a coffee can at the sides of the rib cage beneath the arms, or across the fingers, or at the cheek of a helmet.

(Quoted from Horace F. Judson, *The Search for Solutions,* Johns Hopkins Univ. Press, 1987, pp. 54–56.)

Paleontologists originally had thought that *Quetzlcoatlus northropi* weighed 200 pounds and had a 50-foot wingspan. Even though that works out to just about the same wing loading as for 100 pounds and a wingspan of 36 feet, other considerations from aerodynamics show that the larger size flying dinosaur wouldn't have been able to get off the ground.

KEEPING COOL (AND WARM)

Area-volume tension is also of crucial importance to a creature's maintenance of thermal equilibrium. Both warm-blooded and cold-blooded animals gain or lose heat from the environment in proportion to body surface area. A warm-blooded animal usually is losing heat; its equilibrium consumption, or food intake needed to maintain body heat, depends primarily on the amount of its surface area, the temperature of its ambient environment, and the insulation provided by its coat or skin. Other factors being equal, a scaled-up mammal scales up its food consumption by surface area (proportional to the square of the scaling factor), not by volume (proportional to its cube).

Mammals regulate their metabolism and maintain a constant internal body temperature. Cold-blooded animals, such as alligators or perhaps most dinosaurs, have a somewhat different problem. They absorb heat from the environment for energy, but they must also dissipate any excess heat to keep their temperature below unsafe levels. The amount of heat that must be gained or lost is proportional to total volume, because the entire animal must be warmed or cooled. But the heat is exchanged through the skin, so the rate is proportional to surface area.

Dimetrodon was a large mammal-like reptile that was the most common dinosaur roaming

Figure 13.7 *Dimetrodon* may have evolved a sail to absorb and dissipate heat efficiently. (The animal to the lower left is an *Edaphosaurus,* which also had a sail.) [Courtesy, Field Museum of Natural History.]

present-day Texas and Oklahoma 280 million years ago (see Figure 13.7). *Dimetrodon* had a great "sail" or fan on its back. As an individual grew, and as the species evolved, the sail grew. But it did not grow according to geometric similarity, the kind of growth we refer to as **proportional growth.** Instead, the area of the sail grew precisely in proportion to the volume of the animal, a fact that strongly suggests to paleontologists that the sail was a temperature-regulating organ that was able to absorb or radiate heat. So, an individual *Dimetrodon* twice as long would have eight ($= 2^3$) times as much weight and volume and also a sail with eight times as much area. If it had grown according to geometric similarity, the sail would have been twice as high and twice as wide, and hence would have had only four times as much sail area. Larger specimens of *Dimetrodon* didn't look quite like scaled-up smaller ones; we would say that the sail grew disproportionately large compared to the rest of the animal.

SIMILARITY AND GROWTH

Although a large change of scale forces adaptive changes in materials or form, within narrow limits — perhaps up to a factor of 20 — creatures can grow according to a law of similarity. They grow in such a way that their shape is preserved. A striking example of such growth is that of the chambered nautilus (*Nautilus pompilius*). Each new chamber that is added onto the nautilus shell is larger but the same shape as the previous chamber, and the shape of the shell as a whole — an *equiangular,* or *logarithmic,* spiral — remains the same (see Figure 13.8).

Most living things grow over the course of their lives by a factor greater than 2. We've seen with *Dimetrodon* that a big specimen was not just a scaled-up small one. Nor is an adult human simply a scaled-up baby. Relative to the length of the body, a baby's head is much larger than an adult's. Even the proportions of facial features are different: in a baby, the tip of the nose is about halfway down the face, while in an adult, the nose is about two-thirds of the way down. The arms of the baby are disproportionately shorter than an adult's. In the

Figure 13.8 A chambered nautilus shell. [Photo by Nancy Rodger.]

growth from baby to adult, the body does not scale up as a whole. But different parts of the body scale up, each with a different scale factor. That is, a baby's eyes grow at one rate to perhaps twice their original size, while the arms grow at another rate, to about four times their original size.

Although the laws for growth can be much more complicated than proportional growth (or even the allometric growth we discuss in the next section), more sophisticated mathematics — for example, differential geometry, the geometry of curves and surfaces — permits analysis of complex and interlocking scalings. For a model of the process in which a baby's head changes shape to grow into an adult head, we can use graph paper: we put a picture of the baby's skull on graph paper, then determine how to deform the grid until the pattern matches an adult skull (see Figure 13.9 and Spotlight 13.4, p. 365).

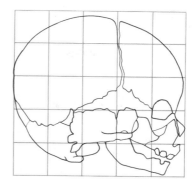

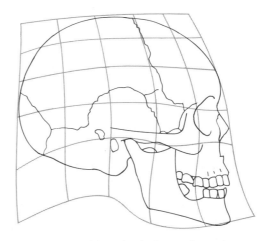

Figure 13.9 Modeling the changes in shape of a human head from infancy to adulthood. [From Richard C. Lewontin, "Adaptation." ©*Scientific American* 239(3):220, 1978. All rights reserved.]

Optional **Allometry**

If we measure the arm length or head size for humans of different ages and compare these measurements with body height, we observe that humans do not grow in a way that maintains geometric similarity. The arm, which at birth is one-third as long as the body, is by adulthood closer to two-fifths as long.

Ordinary graphing provides a way to test for differential growth. We can plot body height on a horizontal axis and arm length on the vertical axis. After doing so, we get a curve, which indicates that the height is not increasing the same amount each year. If the growth were proportional, that is, according to geometric similarity, we would have gotten a straight line, indicating the same amount of growth every year.

Is there an orderly law by which we can relate arm length to height? Let's plot again, this time using **log-log paper** — graph paper on which each axis is marked off evenly, not in units but in orders of magnitude: 1, 10, 100, 1000, and so on (see Figure 13.10). On this graph the data plot closely to a straight line. We actually can discern two different straight lines: a steeper one that fits early development,

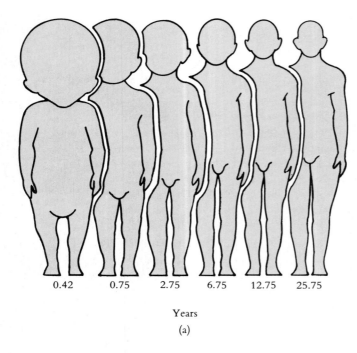

Years

(a)

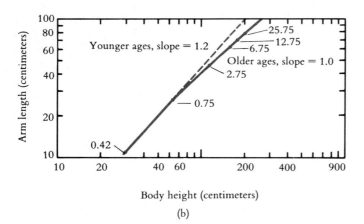

Body height (centimeters)

(b)

Figure 13.10 (a) The proportions of the human body change with age. (b) A graph of human body growth on log-log paper. The numbers shown beside the points indicate the age in years; they correspond to the stage of human development shown in part a. [From Thomas A. McMahon and John Tyler Bonner, *On Size and Life*, Scientific American Library, 1983.]

with slope 1.2, and a less steep one that fits development after 9 months of age, with slope 1.0. The change from one line to another at 9 months indicates a change in pattern of growth. The pattern after 9 months, characterized by the straight line with slope 1, is indeed proportional growth (sometimes called **isometric growth**). For the pattern before 9 months, we know from the slope (1.2) being greater than 1 that arm length is increasing relatively faster than height. That earlier growth also follows a definite pattern, called allometric growth.

Allometric growth is the growth of one feature at a rate proportional to a power of another. For the infant before 9 months, arm length grows allometrically with height. We have seen that in geometric scaling, area grows according to the square (second power) and volume according to the cube (third power) of length, so we can say that they grow allometrically with length.

If we denote arm length by y and height by x, a straight line fit on log-log paper corresponds to the algebraic relation

$$\log_{10} y = B + a \log_{10} x$$

where a is the slope of the line and B is the point where the graph crosses the vertical axis. If we raise 10 to the power of each side, we get

$$y = bx^a$$

where $b = 10^B$. This equation describes a **power curve**: y as a constant multiple of x raised to a certain power.

For a slope $a = 1$, we get $y = bx$, which is a linear relationship describing proportional growth, that is, growth according to geometric similarity. On ordinary graph paper, proportional growth appears as a straight line, allometric growth as a curve. On log-log paper, both patterns appear as straight lines. •

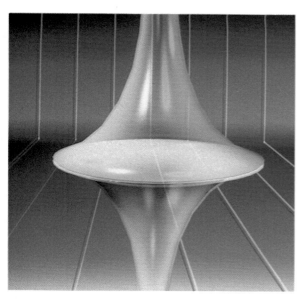

COLOR PLATE 16 A pseudosphere, a model of the hyperbolic plane. (See pp. 452–454.)

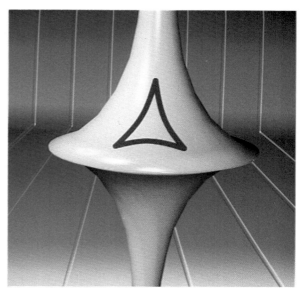

COLOR PLATE 17 A triangle on the pseudosphere, illustrating that in a hyperbolic geometry the sum of the angles in any triangle is less than 180°. (See pp. 452–454.)

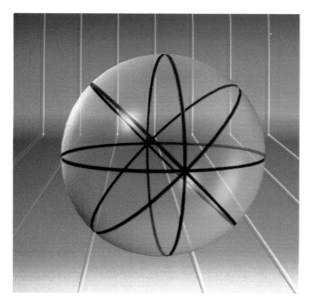

COLOR PLATE 18 In spherical geometry the shortest path between two points lies along a great circle. In an elliptic geometry (such as spherical geometry), every two "lines" (great circles) intersect. (See pp. 454–456.)

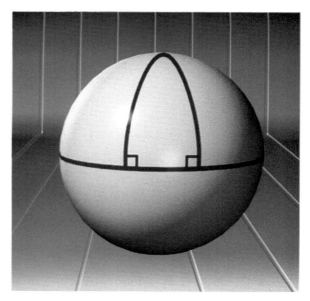

COLOR PLATE 19 In spherical geometry, a triangle on the sphere can have two or even three right angles. In an elliptic geometry (such as spherical geometry), the sum of the angles of any triangle is greater than 180°. (See pp. 454–456.)

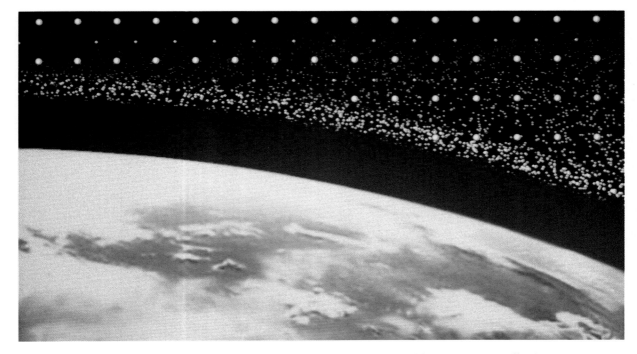

COLOR PLATE 20 The continual development of better algorithms and faster computers is allowing scientists to devise more realistic models of complex systems such as the earth's atmosphere. (See Chapter 18.)

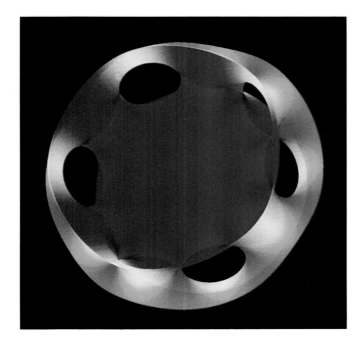

COLOR PLATE 21 This surface represents an example in the infinite sequence of embedded minimal surfaces discovered by Hoffman and Meeks. A surface is called "minimal" when each small piece of it is curved like a soap film, which seeks to minimize its area. The example has genus 4. Genus is related to the number of holes in the surface. [Mathematics by David Hoffman and William H. Meeks III. Computer graphics by Jim Hoffman. Image produced at the University of Massachusetts using a Ridge computer and a Raster Technologies graphics controller.] (See Chapters 18 and 20.)

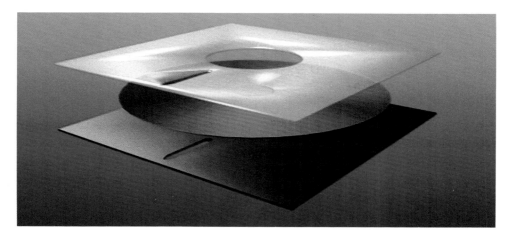

COLOR PLATE 22 An exploded view of a floppy diskette in which information is stored on the circular magnetic media. When the diskette is placed in a disk drive, a magnetic head (similar in principle to that found in a tape recorder) can "read" the information stored on the spinning disk by sensing magnetic states (either positive or negative) and sending the patterns electronically to the computer. Data can be written to the disk by altering existing magnetic states. (See Chapter 19.)

COLOR PLATE 23 A totally artificial computer-generated fractal mountain and valley created by Benoit B. Mandelbrot and Richard F. Voss. [Copyright 1982 by B. B. Mandelbrot and R. F. Voss.] (See Spotlight 20.1.)

(a)

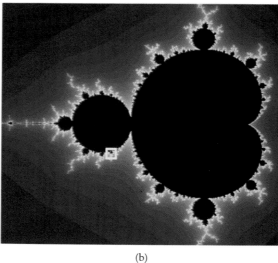

(b)

COLOR PLATE 24 (a) Computer-generated enlargement of part of the boundary of the Mandelbrot set, whose members are complete numbers graphed as points in the complex plane. The points in the set are shown in black; the colored regions of the image represent numbers outside the set that, following a repeated operation, flee the boundary. The boundary of the Mandelbrot set is a fractal. (b) The Mandelbrot set. Note the area in the small square, which is seen enlarged in part (a). [From H.-O. Peitgen and P. H. Richter, *The Beauty of Fractals,* Heidelberg, Springer-Verlag, 1986.] (See Spotlight 20.1.)

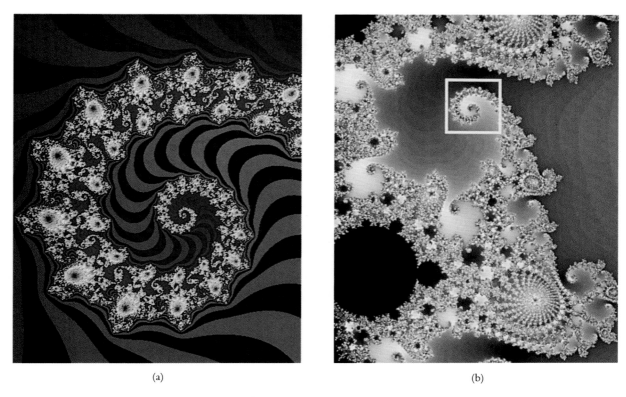

(a)

(b)

COLOR PLATE 25   (a) Enlargement of the Mandelbrot set known as the "Tail of the Seahorse." (b) Boundary region showing area (in square) enlarged in part (a). [From H.-O. Peitgen and P. H. Richter, *The Beauty of Fractals,* Heidelberg, Springer-Verlag, 1986.] (See Spotlight 20.1.)

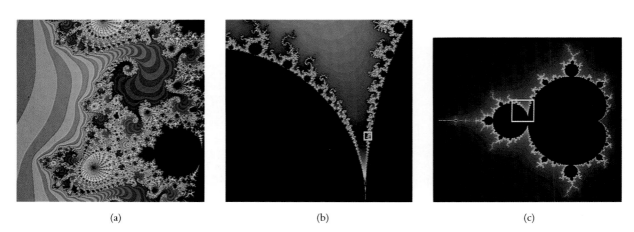

(a)

(b)

(c)

COLOR PLATE 26   (a) Enlargement of a boundary portion of the Mandelbrot set known as "Seahorse Valley." (b) The small square represents the region enlarged in part (a). (c) The Mandelbrot set showing the region (in the square) enlarged for part (b). [From H.-O. Peitgen and P. H. Richter, *The Beauty of Fractals,* Heidelberg, Springer-Verlag, 1986.] (See Spotlight 20.1.)

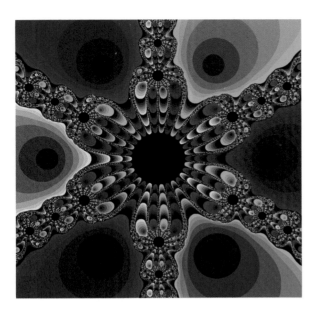

COLOR PLATE 27 The six basins of attraction of the solutions of $z^6 - 1 = 0$ as solved by Newton's method, plotted on the complex plane (z-plane). All regions of a common color represent starting points that converge to a common solution (red = 1.0, green = 0.5 + 0.866i, blue = −0.5 + 0.866i, yellow = −1.0, magenta = −0.5 − 0.866i, cyan = 0.5 − 0.866i). Different shades of each color represent different iteration counts required for convergence (to within a distance of 0.1 of a solution), and they cycle from dark to light with increasing iteration count. [Illustration: Scott A. Burns, University of Illinois at Urbana-Champaign.] (See Spotlight 20.2.)

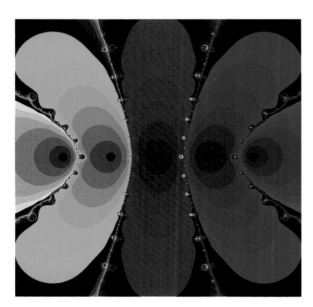

COLOR PLATE 28 Five basins of attraction of the solutions of $[(z-2)(z-1)(z)(z-1)(z-2)] = 0$ as solved by Newton's method. All five solutions occur on the real axis {−2, −1, 0, 1, 2} and are assigned the colors {yellow, green, red, magenta, blue}, respectively. Near each "boundary" between basins of attraction, all five colors are tightly intermingled, indicating a sensitive dependence on initial conditions in these regions. [Illustration: Scott A. Burns, University of Illinois at Urbana-Champaign.] (See Spotlight 20.2.)

COLOR PLATE 29 A close-up of the basins of attraction of the solutions of $z^{10} - 1 = 0$ as solved by Newton's method. The boundary region between two of the ten solutions (0.309 + 0.951i and −0.309 + 0.951i) is shown, revealing the fractal nature of the basins' boundaries. The black region at the lower edge of the image is the top of a large black hole centered at the origin, similar to the smaller one seen in the $z^6 - 1 = 0$ image, which results from a limit on the maximum number of iterations allowed before the iteration is terminated. Many preimages of the black hole are visible in the boundary region. [Illustration: Scott A. Burns, University of Illinois at Urbana-Champaign.] (See Spotlight 20.2.)

COLOR PLATE 30 Texture-mapping. In texture-mapping, a two-dimensional pattern, either derived from a digitized photograph or produced with a paintbox program, is "wrapped" around a three-dimensional computer-graphics object. The checked pattern was created with a paintbox program and then texture-mapped onto the spiral. The sky and clouds are from a black-and-white photograph that was colored with a paintbox program. [Computer graphics by Symbolics, Inc.] (See Chapter 20.)

COLOR PLATE 31 Bump-mapping. The etched appearance of the spiral was produced with a technique called "bump-mapping." In bump-mapping, the angles of the surface normals used in shading are varied so as to produce the effect of light reflecting from an irregular surface. [Computer graphics by Symbolics, Inc.] (See Chapter 20.)

COLOR PLATE 32 Three-dimensional computer-generated solid image of a pear. [Computer graphics by Philip Zucco on equipment by Symbolics, Inc.] (See Chapter 20.)

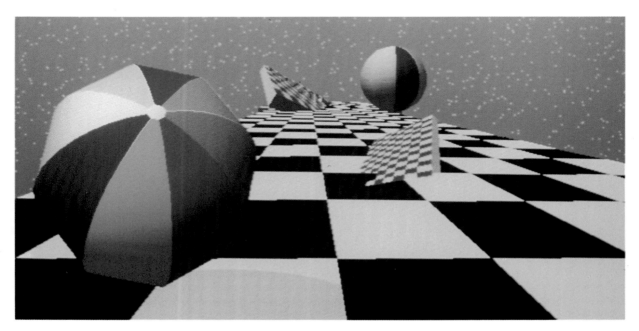

COLOR PLATE 33 Computer-generated image. [Computer graphics by Philip Zucco on equipment by Symbolics, Inc.] (See Chapter 20.)

SPOTLIGHT 13.4 Helping to Find Missing Children

It can be valuable to be able to predict what a developing organ will look like in the future. For example, what does a child look like now who was kidnapped 2 years ago, at age 3?

At Face Systems, Inc., in New York, a computer and a more sophisticated version of our graph-paper technique are used to answer such questions. Staff transfer to the computer screen a photo of the missing child, taken right before the disappearance, and a photo of an older sibling or of one of the parents as an older child. Then they superimpose the two images and wrap the image of the missing child's face to fit the shape of the other face. The result is a rough idea of what the missing child may look like. As mathematicians and biologists refine their models of how faces change over time, this technique will improve. It may even become possible for a child to gain an idea of how he or she may look at age 40 or 65.

CONCLUSION

We have examined the problem of scale and noted that a large change in scale forces a change in either materials or form. A particular instance of the problem of scale is area-volume tension, and we have seen how an animal's size and geometric shape affect its abilities to move and to keep itself warm or cool.

In this chapter we have explored the limitations of life in three dimensions; in Chapter 17 we will see that dimensionality also imposes surprising limits on artistic creativity in devising patterns.

REVIEW VOCABULARY

Allometric growth A pattern of growth in which one feature grows at a rate proportional to a power of another feature.

Area-volume tension The fact that in scaling up, volume increases faster than area.

Density Weight per unit volume.

Inch Unit of measure now defined to be exactly 2.54 centimeters.

Isometric growth Proportional growth.

Log-log paper Graph paper on which both the vertical and the horizontal scales are logarithmic scales, that is, the scales are marked in orders of magnitude 1, 10, 100, 1000, . . . , instead of 1, 2, 3, 4, ...

Mass The aspect of matter that is affected by forces, according to physical laws.

Power curve A curve described by an equation $y = cx^n$, so that y is proportional to a power of x.

Pressure Weight divided by area.

Problem of scale As an object or being is scaled up, its area and its volume increase at different rates, forcing adaptations of materials or shape.

Proportional growth Growth according to geometric similarity.

Scaling factor The number by which each linear dimension of an object is multiplied.

Similar   Two objects are geometrically similar if they have the same shape, regardless of the materials they are made of. They need not have the same size. Corresponding linear dimensions must have the same factor of proportionality.

Wing loading   Weight supported divided by wing area.

Yield strength   The maximum ability of a substance to withstand pressure without crushing or deforming.

SUGGESTED READINGS

Gould, Steven Jay: "Size and Shape." Chapter 21 in *Ever Since Darwin,* Norton, New York, 1977.

Haldane, J. B. S.: *On Being the Right Size and Other Essays,* Oxford University Press, New York, 1985. Succinctly surveys area-volume tension, flying, the size of eyes, and even the best size for human institutions.

McMahon, T. A., and J. T. Bonner: *On Size and Life,* Scientific American Library, New York, 1983. Astonishingly beautiful and informative book on the effects of size and shape on living things.

Stevens, Peter S.: *Patterns in Nature,* Atlantic Monthly Press, Boston, 1974. Splendid treatment of the problem of scale and other physical phenomena in nature: flows, meanders, branching, trees, soap films, cracking, and packing.

Thompson, D'Arcy: *On Growth and Form,* Cambridge University Press, 1917, 1961. "A discourse on science as though it were a humanity" (J. T. Bonner), this was the first book to describe in quantitative terms the processes of growth and shaping of biological forms.

EXERCISES

Most of the exercises below require a calculator; one that offers square roots will suffice.

1. Suppose you are printing photographs from negatives of so-called 35-millimeter film, whose frames actually measure 24 by 36 millimeters, which is just under 1 inch by $1\frac{1}{2}$ inches.

a. First you make some contact prints, which are exactly the same size as the negatives. What is the scaling factor of a contact print?

b. One enlargement you want to make is to be three times as high and three times as wide as the negative. What is the scaling factor for this print? How does its area compare with the area of the negative?

c. Considering the negative as measuring approximately 1 inch by $1\frac{1}{2}$ inches, what is the approximate scaling factor for a 4 by 6 print, that is, one that is 4 by 6 inches? What is the area of the print?

d. The size of so-called 3 by 5 prints can vary, depending on whether the print has a border or not. For a common commercially made print, the size is about $3\frac{1}{16}$ by $4\frac{19}{32}$ inches. For such a print, what is the scaling factor of the enlargement from the negative?

e. The cost of raw photographic paper is pretty close to exactly proportional to the area of the paper. Suppose you are comparing the cost of getting 3 by 5 enlargements versus 4 by 6 enlargements, and let's assume for the sake of simplicity that the prints are exactly 3 by 5 inches and 4 by 6 inches. The smaller prints cost 17 cents each, and the larger cost 50 cents each. From what you know about scaling factors and their role in areas, what can you say about the relative cost of the two kinds of prints?

f. Based on the amount of paper used, what would you expect a 7 by 10 print to cost, considering the cost of the 3 by 5 prints in part **e**? Considering the cost of the 4 by 6 prints in part **e**?

2. The area of a circle can be expressed in terms of the diameter (the distance across the center from one side to the other, or twice the radius) as

$$\text{Area} = \pi \times (\text{radius})^2 = \pi \times \left(\frac{\text{diameter}}{2}\right)^2$$

If we apply a scaling factor M to the diameter of a circle, then — as in the case of the square we considered in the text — the area of the scaled circle changes with M^2, the square of the scaling factor. A natural application of this idea, of course, is to your local pizza parlor and the prices on its menu. The actual prices at the pizza restaurant closest to Beloit College are $5, $6, $6.95, and $7.95, respectively, for small (10-inch), medium (12-inch), large (14-inch), and extra large (16-inch) cheese pizzas.

a. What is the scaling factor for an extra large pizza compared to a small one?

b. How many times as large in area is the extra large pizza compared to the small one?

c. How much pizza does each size give per dollar? What "hidden" assumptions are you making about how the pizzas are scaled up?

d. The corresponding prices for a pizza with "the works" are $8.25, $9.75, $11.95, and $13.95. Is there any size of these for which you get more pizza per dollar than some size of the cheese pizzas?

3. Toy trains, sometimes called model trains, come in various sizes or gauges. Not all toy trains are exact scale models of real trains, but some are.

a. HO-gauge toy trains are usually built to an exact scale of 1 to 87, meaning that a part 1 foot long on the real train will be one eighty-seventh of a foot long on the toy train. What is the scaling factor of an HO-gauge toy train?

b. How does the volume of a real boxcar compare with the volume of an HO-gauge scale model?

c. O-gauge toy trains are built to a scale of approximately $\frac{1}{4}$ inch to a foot, meaning that a part 1 foot long on the real train will be one-fourth of an inch long on the toy train. (In fact, O-gauge trains tend to be a little shorter than exact scale would demand, and their wheels are oversized compared to exact scale.) What is the scaling factor of an O-gauge toy train?

4. Doll houses and their furnishings are customarily built to a scale of exactly 1 inch to 1 foot, meaning that an item 1 foot long in a real house is 1 inch long in a doll house.

a. What is the scaling factor for a doll house?

b. If a doll house were made of the same materials as a real house, how would their weights compare?

For Exercises 5 to 9, refer to the following: Consider the square *PQRS* and the point *C* in the figure below. *P′* is the image of the point *P* on the ray *CP* and such that the length of the line segment *CP′* is twice the length of the line segment *CP*. Similarly, the images *Q′*, *R′*, and *S′* are determined. Further, the image of *any* point on the square *PQRS* is so determined. Such a transformation is called an *expansion*.

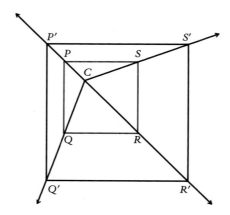

Now, consider the Δ*DEF* and the point *C* in the figure below. *D′* is the image of the point *D* on the ray *CD* and such that the length of the line segment *CD′* is one half the length of the line segment *CD*. Similarly, the images *E′* and *F′* are determined. Further, the image of *any* point on Δ*DEF* is so determined. Such a transformation is called a *contraction*.

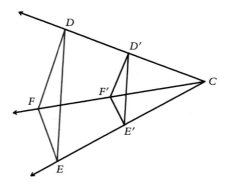

Definition: A *dilation* with *center C* and *scale factor k* (*k* > 0) is a transformation that maps *C* onto itself and any other point *P* onto a unique point *P′* such that *P′* lies on the ray *CP* and such that the length of the line segment *CP′* is *k* times the length of the line segment *CP*. If *k* > 1, the dilation is an expansion. If 0 < *k* < 1, the dilation is a contraction. If *k* = 1, the dilation is the identity transformation.

5. For this exercise, use the figure below for the required dilations.

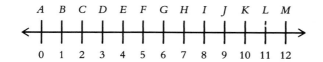

a. Determine the image of D with center at A and scale factor of 2.

b. Determine the image of G with center at A and scale factor of 1.

c. Determine the image of D with center at A and scale factor of 3.

6. For this exercise, use the figure in Exercise 5 for the required dilations.

a. Determine the image of K with center at A and scale factor of $\frac{1}{2}$.

b. Determine the image of J with center at A and scale factor of $\frac{1}{3}$.

c. Determine the image of A with center at A and scale factor of 6.

7. $\triangle ABC$, in the xy-plane, has vertices at $A(-2, 4)$, $B(0, -5)$, and $C(5, 6)$. Determine $\triangle A'B'C'$, the image of $\triangle ABC$, under a dilation with center at the origin and scale factor of $\frac{1}{2}$.

8. $\triangle DEF$, in the xy-plane, has vertices at $D(-5, -6)$, $E(-1, 0)$, and $F(4, -2)$. Determine $\triangle D'E'F'$, the image of $\triangle DEF$, under a dilation with center at the origin and scale factor of 3.

9. Identify each of the following statements as being true or false:

a. Dilations preserve distances.

b. Dilations preserve betweeness of points.

c. Dilations preserve collinearity of points.

d. Dilations preserve angle measure.

e. Dilations preserve parallelism of lines.

f. Dilations preserve perpendicularity of lines.

g. Dilations preserve orientation of points.

10. Two geometric figures are *similar* if they have the same shape but not necessarily the same size. Indicate whether the geometric figures described below are always, sometimes, or never similar:

a. Two squares

b. Two isosceles triangles

c. Two equilateral triangles

d. Two pentagons

e. Two regular pentagons

f. Two rectangles

    g. A square and a rectangle

    h. Two circles

    i. A regular pentagon and a regular hexagon

    j. Two angles

11. Identify each of the following statements as being true or false:

    a. Similar figures are also congruent.

    b. Congruent figures are also similar.

    c. Every polygon is similar to itself.

    d. If polygon A is similar to polygon B and polygon B is similar to polygon C, then polygon A is similar to polygon C.

    e. Corresponding interior angles of similar polygons are congruent.

12. The Susan B. Anthony dollar coin was a failure with the U.S. public, who found it too small and light. Suppose you have been put in charge of designing a new dollar coin that is to be made of the same material as the current U.S. 25-cent piece ("Liberty quarter") and weigh four times as much. A quarter can be described geometrically as a circular cylinder approximately $\frac{15}{16}$ inch in diameter and $\frac{1}{16}$ inch thick. Since your new dollar should weigh four times as much, it will need to have four times the volume of a quarter. (You may find it helpful to know that the formula for the volume of a cylinder is $\pi \times \dfrac{\text{diameter}}{2} \times \text{height.}$)

    a. A member of your public advisory panel suggests that the requirements will be fulfilled if you just double the diameter and double the thickness. What do you tell this individual, in the most diplomatic terms?

    b. If you go along with the member's suggestion to double the diameter, how thick does the coin need to be?

    c. Another member of the board feels that the resulting coin would be too large in diameter to be convenient and proposes instead that you scale up the quarter proportionally (she took a course from the first edition of this book). What would the dimensions be for this new dollar?

13. One of the famous problems of Greek antiquity was the *duplication of the cube.* Our knowledge of the history of the problem comes down to us from Eratosthenes of Cyrene (ca. 284 to ca. 192 B.C.), who is famous for his estimate of the circumference of the earth (see Chapter 15). According to him, the citizens of Delos were suffering from a plague. They consulted the oracle, who told them that to rid themselves of the plague, they must construct an altar to a particular god that would be geometrically similar to the existing one but double the volume.

    a. How would the volume of the new altar compare with the old if each of its linear dimensions were doubled?

    b. What should the scaling factor be for the new altar?

(The actual problem intended by the oracle was to construct with straightedge and compasses a line segment with this scaling factor as its length, a task that was shown in the nineteenth century to be impossible. Eratosthenes relates that the Delians interpreted the problem in this sense, were perplexed, and went to ask Plato about it; Plato told them that the god didn't really want an altar of double the volume but wished to shame them for their "neglect of mathematics and their contempt for geometry.")

14. Criticize the following claims, which were cited in the *New York Times* of 9/25/87 and 10/21/87:

   a. A new dental rinse "reduces plaque on teeth by over 300%."

   b. An airline working to decrease lost baggage has "already improved 100% in the last six months."

   c. "If interest rates drop from 10% to 5%, that is a 100% reduction."

15. The principle that area scales with the square of length, and volume with the cube, has important consequences for the depiction and interpretation of data in graphic form. Suppose we wish to indicate in an artistic way that the weekly income of a U.S. carpenter is twice that of a carpenter in (mythical) Rotundia. We draw one money bag for the Rotundian and another one "twice as large" for the American. [Illustration from Darrell Huff, *How to Lie with Statistics*, Norton, 1954, p. 69.]

What's the problem? Well, first, people tend to respond to graphics by comparing *areas*. Since the larger moneybag is twice as high and twice as wide as the smaller one, the image of it on the page has four times the area. Second, we are used to interpreting depth and perspective in drawings in terms of three-dimensional objects. Since the larger bag is also twice as thick as the smaller, it has eight times the volume. The shading and other artistic aspects of the drawing help leave the subconscious impression that the U.S. carpenter earns *eight* times as much, instead of twice as much.

With these ideas in mind, evaluate the following depictions of data. [Illustrations reproduced or adapted from Edward R. Tufte, *The Visual Display of Quantitative Information*, Graphics Press, 1983, pp. 55, 57, and 70.]

a.

Comparative Annual Cost per Capita for care of Insane in Pittsburgh City Homes and Pennsylvania State Hospitals.

Pittsburgh Civic Commission, *Report on Expenditures of the Department of Charities* (Pittsburgh, 1911), p. 7.

b.

This line, representing 18 miles per gallon in 1978, is 0.6 inches long.

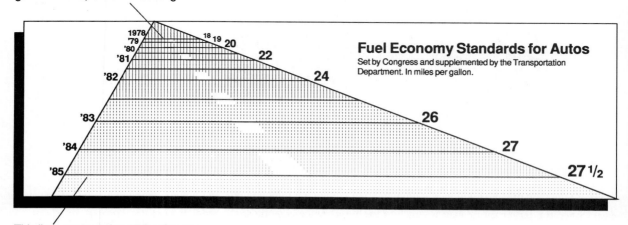

This line, representing 27.5 miles per gallon in 1985, is 5.3 inches long.

New York Times, August 9, 1978, p. D2

c.

1958 — EISENHOWER

1963 — KENNEDY

1968 — JOHNSON

Purchasing
Power
of the
Diminishing
Dollar

Source: Labor Department

1973 — NIXON

1978 — CARTER

Washington Post, October 25, 1978, p. 1.

16. As in Exercise 15, evaluate the depictions below. [Illustrations reproduced or adapted from Edward R. Tufte, *The Visual Display of Quantitative Information*, Graphics Press, 1983, pp. 62 and 69.]

a.

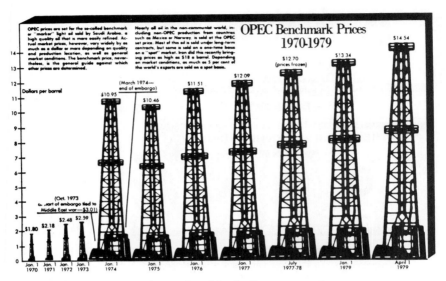

Washington Post, March 28, 1979, p. A-18.

b.

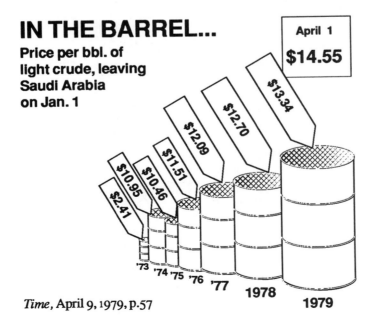

Time, April 9, 1979, p. 57

c.

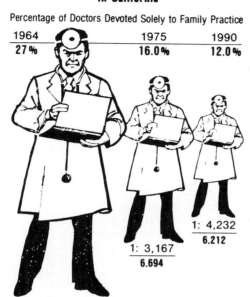

THE SHRINKING FAMILY DOCTOR
In California

Percentage of Doctors Devoted Solely to Family Practice

1964	1975	1990
27%	16.0%	12.0%

1: 4,232
6.212

1: 3,167
6.694

1: 2,247 RATIO TO POPULATION
8.023 Doctors

Los Angeles Times, August 5, 1979, p. 3.

---

17. Here's one just for fun. A humorous and impractical measure of speed is "furlongs per fortnight." A furlong is an eighth of a mile, and a fortnight is 14 days.

   a. How many furlongs per fortnight is 55 miles per hour?

   b. The speed of sound in dry air at sea level is about 1090 feet per second. How much is that in furlongs per fortnight?

   c. The speed of light is about 3.00×10^8 meters per second. How much is that in furlongs per fortnight?

18. A *light-year* is a measure of distance: the distance that light travels in a year.

   a. How long is a light-year in kilometers?

   b. In miles?

   c. In angstroms? (1 angstrom $= 10^{-10}$ m)

19. Consider a real locomotive that weights 88 tons and an HO-gauge scale model of it. (See Exercise 3 above.)

   a. How much would an exact scale model weigh, in tons?

   b. What assumptions are involved in your answer to part **a**?

   c. How much would an exact scale model weigh, in pounds?

   d. In kilograms?

   e. In metric tonnes? (1 metric tonne $= 1000$ kg)

20. Gasoline is sold in the United States by the U.S. gallon and in Canada by the liter. (1 U.S. gallon = 231 cu in; 1 liter = 1000 cm³) What is the equivalent cost, in U.S. dollars per U.S. gallon, for gasoline in Canada priced at 50.5 Canadian cents per liter, if one Canadian dollar exchanges for 85.5 cents the U.S.?

21. What is the equivalent cost, in dollars per pound, for ground beef in Spain selling for 900 pesetas per kilo, when a U.S. $1 converts to 109 pesetas?

22. An ad for a software package for data analysis on the Apple II included a data set on tropical rain forests and deforestation. The data were given in hectares and were accompanied by the statement, "A hectare equals 10,000 square miles or 2471 acres." What conversion factors should have appeared instead?

23. In connection with Eratosthenes' measurement of the circumference of the earth, Chapter 15 discusses the Greek measuring unit of a *stadium*. In *The American Heritage Dictionary,* Second College Edition (Houghton-Mifflin, Boston, 1982), we read for the second meaning of *stadium:* " An ancient Greek measure of distance . . . equal to about 185 kilometers, or 607 feet." The name of the unit came from the length of a racecourse that was a bit less than an eighth of a mile long. If the numbers in the definition are correct, what are the correct units that should have appeared?

24. According to classical Greek sources, Pythagoras (sixth century B.C.) used geometric scaling to model the height of Hercules, the most heroic figure in classical mythology, in the epic poems of Homer. Pythagoras compared the lengths of two racecourses, one (according to tradition) paced off by Hercules and the other by a man of average height. Both were 600 "paces" long, but the one established by Hercules was longer because of Hercules' longer stride (600 "Herculean" paces vs. 600 paces by a normal man). A normal man in the time of Pythagoras would have been about 5 ft tall.

   a. If the distance paced off by Hercules was 30% longer than the other racecourse, how tall was Hercules? What does your calculation assume?

   b. In fact, the ancient sources do not give the original data but only the two conflicting answers that Hercules was 4 cubits tall and 4 cubits and 1 foot tall. A cubit was supposed to be the distance between a person's elbow and the tip of the middle finger of the person's outstretched arm, much as a "foot" was originally the length of a person's foot. So the measurement depended on the person, though there was some attempt at standardization. Although the length of a Greek cubit is not known precisely, most estimates place it between 17 and 22 inches. What range does this give for the height of Hercules, in feet and inches? In centimeters?

25. Goliath (of David and Goliath, as related in the Bible in I Samuel 17:4) was "six cubits and a span." A span was originally the distance from the tip of the thumb to the tip of the little finger when the hand is fully extended, about 9 inches. What range of heights for Goliath would this indicate, in feet and inches? In centimeters?

26. The weight of a 1-foot cube of steel is 500 pounds. What is the pressure on the bottom face in

   a. Pounds per square inch?

   b. Atmospheres? (1 atmosphere = 14.7 pounds per square inch)

27. In an article on adding organic matter to soil, the magazine *Organic Gardening* (March 1983) said, "Since a 6-inch layer of mineral soil in a 100-square-foot plot weighs about 45,000 pounds adding 230 pounds of compost will give you an instant 5% organic matter."

a. What is the density of the mineral soil, according to the quotation?

b. How does this density compare with that of steel?

c. How do you think the quotation should be revised to be accurate?

28. A mature gorilla weights 400 pounds and stands 5 feet tall.

a. Give an estimate of its weight when it was half as tall.

b. What assumptions are involved in your estimate?

c. A mature gorilla's two feet together have a combined area of about 1 square foot. When the gorilla is standing on its feet, what is the pressure on its feet, in pounds per square inch?

29. Suppose King Kong is a gorilla scaled up with a scaling factor of 10.

a. How much does the King weigh?

b. What is the pressure on the King's feet, in pounds per square inch?

30. A human infant at birth usually weighs between 5 and 10 pounds and has a height (length) between 1 and 2 feet, with the shorter babies having the lesser weight. Considering the weight and height of an adult human, give an argument that human growth must not be just proportional growth.

31. In the children's story *Peter Pan,* Peter and Wendy can fly. We can suppose that they are 4 feet tall, so they are about 12 times as tall as a sparrow is long. What should their minimum flying speed be?

32. Icarus of Greek legend escaped from Crete with his father, Daedalus, on wings made by Daedalus and attached with wax. Against his father's advice, Icarus flew too close to the sun; the wax melted, the wings fell off, and he fell into the sea and drowned. What must have been his minimum cruising speed? What assumptions does your answer involve?

33. Recent years have seen the beginnings of human-powered controlled flight, in the *Gossamer Condor* and other superlightweight planes. The *Gossamer Condor* is far longer than an ostrich, but it flies at only 12 miles per hour. How can it?

34. What would an individual *Quetzlcoatlus northropi* weigh that had half the wingspan of an adult?

35. If an individual *Quetzlcoatlus northropi* weighed 50 pounds, what would have been its wingspan?

36. Jonathan Swift's Gulliver also traveled to Lilliput, where the Lilliputians were human-shaped, but only about 6 inches tall. In other words, they were geometrically similar in shape to ordinary human beings but only one-twelfth as tall. What would a Lilliputian weigh?

37. Are Lilliputians ruled out by the size-shape and area-volume considerations in this chapter? If you think they are, what considerations do you find convincing? If not, why not?

38. Use algebra to demonstrate that for proportional growth, "area scales as volume to the two-thirds power."

●**39.** Consider the following data for various mammals:

	Weight (kilograms)	Calories per kilogram
Guinea pig	0.7	223
Rabbit	2	58
Human	70	33
Horse	600	22
Elephant	4,000	13
Whale	150,000	17

Graph these data on log-log paper. What can you conclude from your graph? [Note that you may have to "create" your own log scale on the weight axis, as the weights span six orders of magnitude. Commercial log paper is usually limited to 2, 3, or perhaps 5 orders of magnitude ("cycles").] If you don't have log-log paper available, use your calculator to take the logarithms of all the numbers in the table and graph these values on ordinary graph paper.

●**40.** Listed in the table below are the winning times in the 1983 World Rowing Championships for rowing shells with one, two, four, and eight oars. The men's times are for 2000 meters, the women's for 1000 meters. Convert the times to speed. For men and women separately, plot speed versus number of oarsmen on ordinary graph paper, and then on log-log paper. Is the relationship proportional? Allometric?

Fit the best line you can through the log-log data and estimate its slope. Compare your results with the theoretical discussion and data in McMahon and Bonner's *On Size and Life,* Scientific American Library, 1983, pp. 42–47 and 67.

Event	Number of oars	Men (2000 meters)	Women (1000 meters)
Single sculls	1	6:49.75	3:36.51
Pairs without coxswain	2	6:35.85	—
Fours without coxswain	4	6:14.83	3:26.68
Eights (with coxswain)	8	5:34.39	2:56.22

As in Exercise 39, if you don't have log-log paper, you can calculate the logarithms of the numbers and graph those.

● Optional exercise.

·14·

The Size of Populations

Many of the problems we face relate to populations and their changes over time. We all have a stake in the problems associated with human population growth, such as hunger and disease. Our food supplies are affected by the growth and behavior of nonhuman populations such as bacteria, locusts, and rats. Even inanimate populations affect us. The growing "populations" of household refuse and nuclear waste pose disposal and storage issues, with accompanying environmental questions. Similarly, but more favorably, a growing population of dollars in a bank account can provide resources that enrich our lives.

To apply the ideas and methods of mathematics to the study of populations, we must clarify and refine familiar notions. We will therefore pose and answer some fairly precise questions about populations:

♦ How are its members described?

♦ How big is the population?

♦ Where is it located and how is it distributed?

♦ How fast is it growing or shrinking?

♦ How is its structure or makeup changing?

The term **population structure** refers to the divisions of a population into subgroups. For example, human populations are frequently described according to age structure, as in U.S. census data. For a specific problem it may be advantageous to break a population down according to economic, social, or educational criteria. We will use the term **population growth** to refer to both increases (positive growth) and decreases (negative growth) in population size.

In this chapter we will pay particular attention to the size of a population and to the way its size changes over time. We will investigate models of population growth of two different kinds. In one kind, the population grows at a rate proportional to its current size, so that, for example, when it is twice as large, it is growing twice as fast. In the other kind of population, the increase in the population is the same in each time interval. For examples, we will focus on two seemingly different kinds of populations—financial and biological—in order to illustrate the broad application of our models.

GEOMETRIC GROWTH AND FINANCIAL MODELS

We begin our study with a population whose structure and behavior are relatively simple—the population of dollars in a bank account. We deposit money in a savings account; our primary concerns are the safety and the growth of such savings. Suppose that we deposit $1000 in an account that, we are told, "pays interest at a rate of 10%, compounded and paid annually." Assuming that we make no other deposits or withdrawals, how much is in the account after 1, 2, or 5 years?

The $1000 is usually called the **initial balance** or the **principal** of the account. At the end of one year, **interest** is added. The amount of interest is 10% of the principal, $100 in this case. So the balance at the beginning of the second year is $1100. During the second year the interest is also 10%—not of the initial balance of $1000 but of the new balance of $1100—so at the end of the second year, 10% of $1100, or $110, is added to the account.

Notice that during the second year we earn interest on both the initial balance of $1000 and on the $100 interest earned during the first

year. Interest that is paid on both the principal and on the accumulated interest is known as **compound interest.** The simple but remarkable consequence is that we receive more interest during the second year than during the first, that is, the account grows by a greater amount during the second year. At the beginning of the third year the account contains $1,210, so at the end of the third year we receive $121 in interest. Again this is larger than the amount we received at the end of the preceding year. Moreover, the increase during the third year

$$\text{Third-year interest} - \text{second-year interest} = \$121 - \$110 = \$11$$

is larger than the increase during the second year

$$\text{Second-year interest} - \text{first-year interest} = \$110 - \$100 = \$10$$

Thus, not only is the account balance increasing each year, but the amount added also increases each year. This type of growth is called **geometric growth** or **exponential growth.**

There is another way to pay interest, called **simple interest.** In this method, interest is paid only on the original balance, no matter how much interest has accumulated. With simple interest, for an account with $1000 and a 10% interest rate, we receive $100 interest at the end of the first year; so at the beginning of the second year, our account contains $1100, as before. But at the end of the second year, we again receive only $100; so at the beginning of the third year, our account contains $1200. In fact, at the end of each year we receive just $100 in interest. Clearly this method yields less than if interest were compounded.

Although the simple-interest method is seldom used in today's competitive financial markets, we frequently observe this type of growth, called **simple growth** or **arithmetic growth,** in other contexts. If you withdraw the interest paid on an account at compound interest, then there is no difference — as far as the amount remaining in the account is concerned — between compound interest and simple interest. (Of course, the practical difference to you is that you get to spend all that extra compound interest!)

The amounts in accounts paying interest at the rate of 10% per year with compound and simple interest are shown in Table 14.1 and in the graph in Figure 14.1. The impressive growth associated with compound interest is dramatically illustrated by these figures, as is the distinction between geometric and arithmetic growth. It is this distinction that led the demographer and economist Thomas Malthus (1766–1834) to his famous theory that human populations grow geometrically,

TABLE 14.1 The growth of $1000: compound interest versus simple interest

Years	Amount in account from compounded interest	Amount from simple interest
1	1100.00	1100.00
2	1210.00	1200.00
3	1331.00	1300.00
4	1464.10	1400.00
5	1610.51	1500.00
10	2593.74	2000.00
20	6727.50	3000.00
50	117,390.85	6000.00
100	13,780,612.34	11,000.00

whereas food supplies grow arithmetically (see Spotlight 14.1, p. 383).

Populations that grow by the same amount in each time interval follow the arithmetic growth model. Nuclear waste generated by a power plant generally follows an arithmetic growth model because the amount of waste added each year depends on the fixed size of the power plant, not on the growing size of the waste dump. The population of medical doctors in the United States grows arithmetically, since the fixed number of medical schools each graduate about the same total number of doctors each year. On the other hand, general human populations tend to grow geometrically because the number of children born increases as the population increases.

THE MATHEMATICS OF GEOMETRIC GROWTH

Suppose we have the option of depositing our $1000 in a bank that pays interest at the rate of 10% per year but compounds the interest

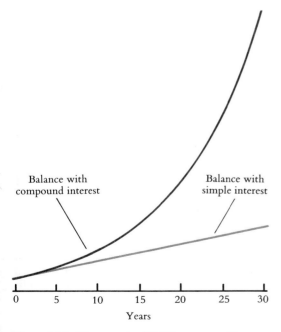

Balance with compound interest

Balance with simple interest

0   5   10   15   20   25   30

Years

Figure 14.1 The growth of $1000: compound interest and simple interest.

quarterly, that is, four times per year. With an interest rate of 10% per year, we are entitled to one-fourth, or 2.5%, each quarter. In this case, the quarter (3 months) is the **compounding period,** the time that elapses before interest is calculated on the account.

Let's see what happens during the first year. At the end of the first quarter, we have the original balance plus $25 interest, so the balance at the beginning of the second quarter is $1025. During the second quarter we receive interest equal to 2.5% of $1025, or $25.63, so our new balance at the end of the second quarter is $1050.63. Continuing in this manner, we find that the balance at the end of the first year is $1103.81.

Note that even though the account was advertised as paying 10% interest, the interest we receive over the course of the year, $103.81, amounts to more than 10% of our principal — in fact, to 10.381%. The 10% rate is known as the **(nominal) annual rate;** the 10.381% rate is the **(effective) annual yield.**

If interest were compounded monthly (12 times per year) or daily (365 times per year), the balance would be still larger. A comparison of yearly, quarterly, monthly, and daily compounding for an interest rate of 10% is shown in Table 14.2.

We summarize our results in a general formula before we consider other examples. From now on, we will express all interest rates as fractions. For example, 10% will be expressed as 0.10; to convert a percentage to a fraction, we divide the percentage by 100,

which means moving the decimal point two places to the left. An interest rate of 0.10 is the same as 10%, or $\frac{10}{100}$; an interest rate r is the same as 100r%, or (the rate in %)/100.

We generalize from our observations for annual and monthly compounding. For annual compounding, we found that at the end of one year we have

$$\begin{aligned} \text{Initial balance} + \text{interest} \\ = \$1000 + \$1000(0.10) \\ = \$1000(1 + 0.10) \end{aligned}$$

and for quarterly compounding we have at the end of the first quarter

$$\begin{aligned} \text{Initial balance} + \text{interest} \\ = \$1000 + \$1000(0.025) \end{aligned}$$

and at the end of the second quarter

$$\begin{aligned} \text{Initial balance} + \text{interest} \\ = \$1000 + \$1000(0.025) \\ + [\$1000 + \$1000(0.025)]\,(0.025) \\ = [\$1000 + \$1000(0.025)] \\ \times (1 + 0.025) \\ = \$1000(1 + 0.025)(1 + 0.025) \\ = \$1000(1 + 0.025)^2 \end{aligned}$$

The pattern continues in this way, so that we have $1000(1 + 0.025)^4$ at the end of the fourth quarter.

In the more general setting, with an initial balance of P and an interest rate $r\ (= 100r\%)$

TABLE 14.2 Comparing compound interest: the value of $1000, at 10% annual interest, if interest is compounded

Years	Compounded yearly	Compounded quarterly	Compounded monthly	Compounded daily	Compounded continuously
1	1100.00	1103.81	1104.71	1105.16	1105.17
5	1610.51	1638.62	1645.31	1648.61	1648.72
10	2593.74	2685.06	2707.04	2717.91	2718.28

SPOTLIGHT 14.1 Thomas Malthus

Thomas Malthus. [The Bettman Archive/ BBC Hulton.]

Thomas Malthus (1766–1834), a nineteenth-century English demographer and economist, based a well-known prediction on his perception of the differences in the rates of growth of two populations, the human population and the "population" of food supplies.

His belief was that human populations increase at a much faster rate than do food supplies. He concluded, however, that over the long run there would be restrictions on the natural growth of human populations too, including war, disease, and starvation — hardly an optimistic forecast and, doubtless, responsible for the dreary image associated with his views.

per compounding period, we have at the end of the first compounding period

$$P + Pr = P(1 + r)$$

This amount can be viewed as a new starting balance. Hence, in the next compounding period the amount $P(1 + r)$ grows to

$$P(1 + r) + P(1 + r)r = P(1 + r)(1 + r)$$
$$= P(1 + r)^2$$

The pattern continues, and we reach the following conclusion:

If a principal P is deposited in an account that pays interest at the rate r per compounding period, then after n compounding periods the account contains $P(1 + r)^n$.

Given an annual interest rate, we can determine the amount in an account with interest compounded quarterly, monthly, or according to any other compounding period, by using

our formula with an interest rate r per compounding period and with the total elapsed time expressed in terms of the number n of compounding periods.

EXAMPLE: Compounding Interest for 10 Years. Suppose we have a principal of $P = 1000$ with interest at an annual rate of 10%. Using the formula $P(1 + r)^n$, we can determine the amount in the account after 10 years using several different compounding periods:

♦ *Annual compounding.* If the compounding is done once a year, then the interest rate of 10% per year gives $r = 0.10$, and we determine the amount in the account after 10 years to be

$$1000(1 + 0.10)^{10} = 1000(1.10)^{10}$$
$$= 1000(2.59374)$$
$$= 2593.74$$

♦ *Quarterly compounding.* If the interest is compounded every quarter, then $r = 0.10/4 =$

0.025, and after 10 years (or 40 quarters) the account contains

$$1000\left(1 + \frac{0.10}{4}\right)^{40} = 1000(1.025)$$
$$= 1000(2.68506)$$
$$= 2685.06$$

• *Monthly compounding.* Finally, if compounding is done monthly, then $r = 0.10/12 = 0.008333$. The amount in the account after 10 years, or 120 months, is

$$1000\left(1 + \frac{0.10}{4}\right)^{120} = 1000(1.008333)^{120}$$
$$= 1000(2.70704)$$
$$= 2707.04$$

These entries are found in the last row of Table 14.2. (See Spotlight 14.2.)

e — A VERY SPECIAL NUMBER

The results given in Table 14.2 illustrate a general trend: for a fixed interest rate, more frequent compounding results in larger ending balances. Thus, as you move from left to right in any row of the table, the amounts increase. However, the amounts in an account that result from more and more frequent compounding do not grow indefinitely; instead, they get closer and closer to a number that can be determined in advance from the interest rate. This number is shown in the far right column in each row.

What happens when we increase the frequency of compounding? We will first suppose the absurdly high interest rate of 100% per year compounded n times per year, and then examine interest rates closer to the ones we usually see in stable economies. For an initial balance of $1, the amount we have at the end of one year is $\$(1 + 1.00/n)^n$. As n increases, this term, which is just $(1 + 1/n)^n$,

gets closer and closer to a special number called e. This is illustrated in the following table, where the dots (ellipses) indicate that more decimal places follow.

n	$(1 + 1/n)^n$	n	$(1 + 1/n)^n$
1	2.0000000...	100	2.7048138...
5	2.4883200...	1,000	2.7169239...
10	2.5937425...	10,000	2.7181459...
50	2.6915879...	100,000	2.7182682...

The value of e is 2.71828. . . . Similarly, for any number x, the quantity $(1 + x/n)^n$ gets closer and closer to the number e^x as n becomes large.

EXAMPLE: Continuous Compounding of Interest. For $1000 at an annual interest rate of 10%, compounded n times in the course of a single year, the balance at the end of the year is $(1 + 0.10/n)^n$. This quantity gets closer and closer to $e^{0.1} = 1.10517$ . . . as the number of compoundings n is increased. No matter how frequently interest is compounded — daily, hourly, every second, infinitely often ("continuously") — the original $1000 at the end of one year cannot grow beyond $1105.17. The corresponding values for other years are shown in the last column of Table 14.2.

If a principal P is deposited in an account that pays interest at the (nominal) annual rate of $r = 100r\%$ compounded continuously, then after 1 year the account contains Pe^r; after m years, it contains Pe^{rm}.

In addition to its fundamental importance in banking and growth, the number e occurs naturally in several other common contexts as well, especially those involving matching.

EXAMPLE: The "Guy Gift" Problem. A custom in some families is for each family

SPOTLIGHT 14.2 Long-term Investments

The sums that result from small initial deposits and relatively modest interest rates over long time periods are truly astonishing. The miracle of compound interest is frequently illustrated with the following example.

Myth and legend have it that Manhattan Island was "purchased" in 1650 for the paltry sum of $24. We consider the fate of a single dollar from that time. If it had been deposited then in an account paying interest at 5% compounded yearly, then the single dollar would have grown to more than $16 million by 1990!

The effect of small changes in interest rates can also be illustrated. If the assumed interest rate were 6% rather than 5%, then the value of the account by 1990 would have been $402 million.

member to buy a holiday gift for just one other family member (colloquially, "guy"). In advance of the holiday, all the members' names are put into a hat, and each member draws out a name at random. If anyone draws his or her own name, the drawing is annulled and is redone. What is the probability that the first drawing is successful in assigning each person someone else's name? (This problem is often called the *hat-check problem*, after a whimsical imaginary situation in which men who check their hats at a theater checkroom are returned hats at random.) For a large family, the answer is approximately $1/e \approx 0.37$. (The situation is complicated further with the usual additional condition that the drawing is annulled if a husband or wife draws a spouse's name.)

EXAMPLE: The Secretary Problem. An employer interviews applicants one after another, in a random order. At the end of an interview, the employer either hires the applicant or else (we suppose) loses the applicant forever. The employer wants to maximize the chance of hiring the best person available.

Suppose, for example, that there are 100 applicants. Many people would think that because the applicants are interviewed in random order, the probability of selecting the best one

must be 1 in 100. The following strategy, however, increases this probability to greater than 1 in 4: Interview, but do not hire, the first 50 applicants. Then pick the first applicant from the second group of 50 who is better than all of the first 50. This strategy certainly selects the best candidate if the best candidate is in the second group and the second-best candidate is in the first group, which together happen with probability one-fourth. Of course, if the best candidate is in the first group, which will happen with probability one-half, the strategy fails to hire anyone.

In fact, with a variation on this strategy, the employer can maximize the probability of hiring the best applicant by interviewing a fraction $1/e \approx 0.37$ of the pool and hiring the first applicant thereafter who is better than all the ones interviewed so far. The probability of hiring the best person available is then $1/e \approx 0.37$.

GROWTH MODELS FOR BIOLOGICAL POPULATIONS

We can now use a *geometric* growth model to make rough estimates about sizes of human populations. Indeed, projections reported in the popular media are based on such a model; they use for the growth rate *r* the difference between the birth rate and the death rate, for which the technical term is the **rate of natural increase.** In the terminology we have been using for financial models, this is the effective annual yield, so we may think of it as a growth rate that is compounded annually.

A shortcoming of such a model is that birth and death rates rarely remain constant for very long, so projections must be made with extreme care. In addition, we exclude the effect of net migration. In the short run, however, predictions based on the model may provide useful information. Let's apply this model to two questions about the U.S. population.

EXAMPLE: Predicting the U.S. Population.

The population of the United States was about 250 million at the beginning of 1990. It was increasing then at an average growth rate of 0.7% per year. What is the anticipated size of the U.S. population at the beginning of the year 2000? What will it be if the rate of natural increase turns out instead to be 0.4% per year, or 1.0% per year?

To answer these questions, we apply our geometric model with the initial population size equal to 250 million. In the first case $r = 0.007$ (because 0.7% equals 0.007). Using a year as the compounding period and the formula $P(1 + r)^n$, where $n = 10$, we find that the projected population size in the year 2000 is

Population in 2000
= (population in 1990) $(1 + \text{growth rate})^{10}$
= $250{,}000{,}000 \, (1 + 0.007)^{10}$
= $250{,}000{,}000 \, (1.007)^{10}$
= $250{,}000{,}000 \, (1.072247)$
= $268{,}000{,}000$, approximately

(The result of our calculation can't be any more precise than the ingredients. Because our estimates of population and growth rate are rough approximations, we don't copy down all the digits our calculator gives but instead round off.)

In the same way, we find that with a growth rate of 0.4% per year, we predict a population in the year 2000 of 260 million, while a growth rate of 1.0% per year yields a predicted population of 276 million. It is clear that an uncertainty of three-tenths of one percentage point, or 0.003, in the growth rate of a population has major implications, even over fairly short time horizons. The presence or absence of 8 million people would have a significant impact on our social and economic systems. According to our model, the increase would be children (as opposed to adult immigrants), who would need schools built and teachers trained. Indeed, a great deal of concern over the long-range funding of the Social Security programs results from uncertainties over birth rates. Figure 14.2 gives a graph of the structure of the U.S. population expected in the year 2000, structured by age and sex.

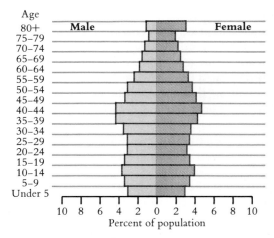

Figure 14.2 Graph of the estimated population of the United States in the year 2000 grouped by age and sex and shown as a percent of the total population. This estimate is based on a geometric growth model.

Rates of natural increase in most Third World countries are well above those experienced by the United States and other industrialized nations. It is not uncommon to find growth rates of 3% per year (and more) in developing nations. With its growth rate of 2.9%, Nigeria's population of 118 million in mid-1990 will become 157 million by the middle of the year 2000, a one-third increase. If such growth rates were to be maintained, the population would double in about 25 years. Projections of this sort are at the root of worldwide concern over our ability to provide sufficient food and other resources for all people.

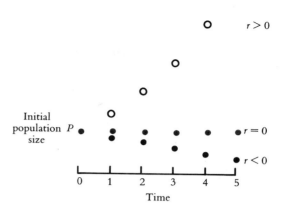

Figure 14.3 Projected population over five time units assuming a geometric growth model with growth rate r.

LIMITATIONS ON GEOMETRIC GROWTH

Let's now think more carefully about the implications of using a geometric growth model to describe the growth of a population.

Examining Figure 14.3, we see that for a positive growth rate r, the size of the population increases as time increases, at least for all times shown in the graph. In fact, no matter how long a time span we consider, the model predicts that the population continues to grow, even to astronomical numbers.

Such predictions are clearly unreasonable for many situations. For instance, no biological population can continue to increase without limit. Its growth is eventually constrained by the availability of resources such as food, shelter, and psychological and social "space." As the population grows, it eventually reaches a level at which there are no resources for new members. A geometric growth model cannot describe forever the growth of such a population.

Let's now think about the way actual biological populations behave. Their percentage growth rate is likely to depend on the size of the population and may actually decrease as population size increases. (This happened fairly steadily to the U.S. population from 1865 to 1945.) For a sufficiently large population, the "growth" may even be negative. How can we account for a decline in the growth rate? One way this could occur is if, as the population size increases, the resources per individual decrease, and thus the energy available for growth and reproduction decreases. There may in fact be a maximum population size that can be supported by the available resources. Such a population size is called the **carrying capacity** of the environment. In one model for growth that takes into account the carrying capacity, we reduce the growth rate r by a factor that indicates how close the population size P is to the carrying capacity M:

$$\text{Growth rate} = r \left(1 - \frac{\text{population size}}{\text{carrying capacity}} \right)$$
$$= r \left(1 - \frac{P}{M} \right)$$

This method is known as the **logistic model** of population growth. We can check that this model has the properties we want. For small population sizes, that is, for values of the initial population P that are small relative to the carrying capacity M, the quantity P/M is small; therefore $1 - P/M$ is close to 1 and the growth rate is close to r.

EXAMPLE: Predicting the U.S. Population Using the Logistic Model. The U.S. population from 1790 to 1950 closely followed a logistic model with $r = 0.031$, $P =$ population in $1790 = 3,900,000$ and $M = 201$ million. For the early decades after 1790, the population was a small fraction of the "carrying capacity," and it grew at close to the rate r of 3.1% per year (a rate similar to many Third World countries today). By 1920 the U.S. population had reached 106 million, a little more than half of the "carrying capacity," and indeed the growth rate had slowed by about one-half, to 1.5% per year.

Obviously, the 1990 U.S. population of 250 million exceeds the hypothesized "carrying capacity" of 201 million. Predicting the 1990 population on the basis of the 1950 population by using the logistic model would have resulted in great error. The postwar baby boom and increased immigration are two factors in the difference between prediction and fact. The predominant factor, however, is that the structure of the U.S. population changed, from a large proportion of people making their living on family farms to a highly urbanized society. Since the structure of the population changed, the model based on the prior structure is no longer valid under the new circumstances.

For a logistic model, as the size of the population increases, the growth rate decreases — because the term containing the population P has a negative sign. For a population size equal to the carrying capacity, that is, when $P = M$, the growth rate is zero.

If at any time the population size exceeds the carrying capacity, then the growth rate becomes negative (because $P > M$ in the formula for growth rate used in the logistic model) and the population size decreases. The carrying capacity refers to long-range capacity to support the population, so the population could exceed it for short periods of time. This could happen either because the population grows very rapidly and surges above the carrying capacity, or because of a sudden decrease in the food supply, thus lowering the carrying capacity, as happens to deer and other animals in winter.

The logistic model provides excellent predictions for the growth of some populations, particularly in laboratory environments. For example, the graph in Figure 14.4 shows the growth of a population of fruit flies in a glass enclosure with a limited food supply. On the

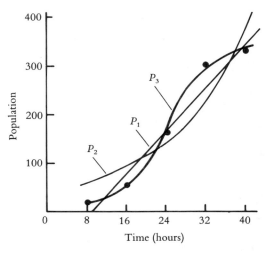

Figure 14.4 The growth of a population of fruit flies: P_1 is the best approximation by a simple growth model; P_2 is the best approximation by a geometric growth model; and P_3 is the best approximation by a logistic model. The red points show the actual values. [Adapted from Daniel Maki and Maynard Thompson, *Mathematical Models and Applications*, Prentice-Hall, 1973, p. 431.]

same coordinate system we show the predictions of an arithmetic population model (labeled P_1), a geometric population model (labeled P_2), and a logistic population model (labeled P_3). Predictions based on the logistic model come closest to the actual growth.

RENEWABLE RESOURCES

A **renewable natural resource** is a resource that tends to replenish itself if we allow it to evolve without intervention. Obvious examples are fish, wildlife, and forests. Petroleum and coal are examples of nonrenewable resources. We would like to predict how much of a resource we can harvest and still allow the resource to replenish itself.

We will concentrate on the subpopulation of individuals that have a commercial value. In the instance of a forest, the subpopulation might be trees of a commercially useful species and appropriate size. Because trees of different sizes have different commercial values, we will measure the size of the population in common units of equal value. For example, we measure a forest not by counting the trees but by estimating the number of board feet of usable timber. Similarly, we might measure the size of a fish population in terms of pounds rather than numbers of fish. When a population is measured in this way—in common units of equal value—we say that we are considering the **biomass** of the population.

REPRODUCTION CURVES

We use a figure called a **reproduction curve,** which predicts next year's population based on this year's population. A typical reproduction curve is shown in Figure 14.5. The reproduc-

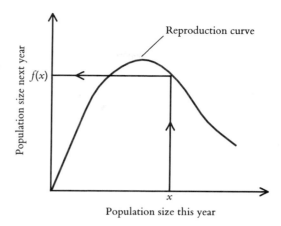

Figure 14.5 A typical reproduction curve.

tion curve represents the total change in the population's biomass from one year to the next, including the growth of continuing members, plus the addition of new members, minus losses due to death and other factors. Although the precise shape of the curve will vary from one population to another, reasonable biological conditions result in a curve of the general shape shown.

Let's take a closer look at the reproduction curve in Figure 14.5. The size of the population in the current year is measured on the horizontal axis; let x be a typical value. For a population of size x this year, its size *next* year is given by the height of the curve above the horizontal axis. This value is denoted by $f(x)$. (You may think of f as standing for "forthcoming.")

Figure 14.6 shows the same reproduction curve, only this time with the addition of a broken line inclined at a 45° angle with the horizontal axis. If you experiment with various choices for x, you will see that whenever the reproduction curve is above the broken line, then the next year's population is larger than this year's. In fact, the vertical distance from the broken line to the curve is the gain in population, or **natural increase,** which in al-

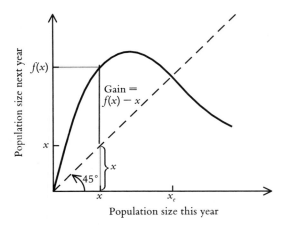

Figure 14.6 Depiction of the natural increase (gain) in population from one year to the next. The population size x_e is the equilibrium population size, for which the population 1 year later is the same, or $f(x_e) = x_e$.

on the horizontal axis corresponding to the second year's population. Proceeding vertically from there to the curve then yields a height that is the population in the third year.

Figure 14.7 shows the results of several of these traces for the same reproduction curve, each starting from a different size initial population, that is, from a different point on the horizontal axis. The resulting variation is quite surprising — it can even be "chaotic" in a very specific mathematical sense, and is an example of how what would otherwise appear as random behavior in fact follows a pattern corresponding to a very simple deterministic mathematical model.

SUSTAINED-YIELD HARVESTING

gebraic terms is $f(x) - x$. Whenever the reproduction curve is below the broken line, then the next year's population is smaller than this year's and $f(x) - x$ is negative. Note that for the special population size labeled x_e, the population at which the reproduction curve crosses the broken line, the population is exactly the same next year as this year. The population size x_e is known as the **equilibrium,** or **steady state, population size.** (Notice that if you project to both axes lines from where the reproduction curve and the broken line cross, the resulting figure is a square.)

The broken line provides a convenient way to trace the evolution of the population over a period of several years (see Figure 14.7). Begin with the first year's population on the horizontal axis, go up vertically to the curve; the height is the population in the second year. Proceed horizontally from the point on the curve over to the broken line; this point on the broken line is located directly above the point

Many biological populations are harvested by predators (including humans). **Yield** is the amount harvested at each harvest. For the present discussion we will focus on a **sustained-yield harvesting policy,** that is, a harvesting policy that if continued indefinitely will maintain the same yield. Sustained-yield harvesting policies are obviously important to timber companies and other enterprises that plan to use a natural resource over a long period of time. Often a key question is to determine the best or optimal harvesting policy, and the answer is often a sustained-yield policy.

Under a sustained-yield harvesting policy, the population after each year's harvest will be the same. To achieve this stability, the amount h that is harvested must exactly equal the amount by which the population naturally increases each year. We recall that the population increases from x to $f(x)$ in 1 year, so the amount of natural growth is $f(x) - x,$ which,

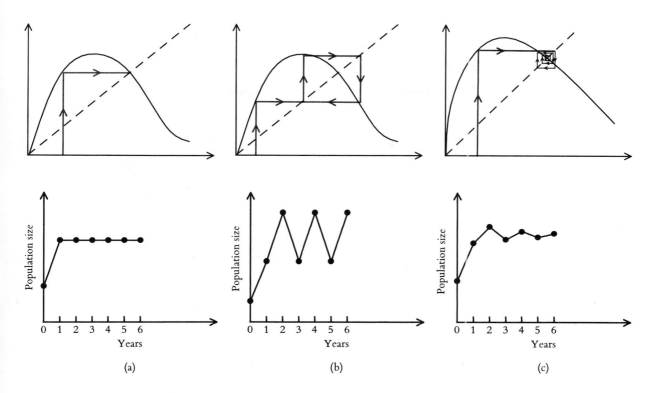

Figure 14.7 Examples of the dynamics over time for the same reproduction curve but different starting populations. (a) The population goes in one year to the equilibrium population and stays there year after year. (b) After initial adjustment, the population cycles between values over and under the equilibrium population. (c) The population spirals in toward the equilibrium population.

to achieve sustained yields, must equal h. So for sustained yield, we must have $h = f(x) - x$, or equivalently $x = f(x) - h$ (see Figure 14.6).

Observe that depending on x, the value of h, which is the vertical distance from the broken line to the curve, can vary all the way from 0 (for $x = 0$; or for $x = x_e =$ the equilibrium population, in the absence of harvesting) up to some maximum value (for an x somewhere between 0 and x_e). A common problem facing

a timber company or a fishery is to determine a harvesting policy that results in the largest possible sustainable yield, or **maximum sustainable yield.** In terms of our diagram, the goal is to select x so that the sustainable harvest is as large as possible. Without a description of the reproduction curve in numeric terms, the best we can do is estimate the value from our graph. The value of the population size that achieves maximum sustainable yield is shown

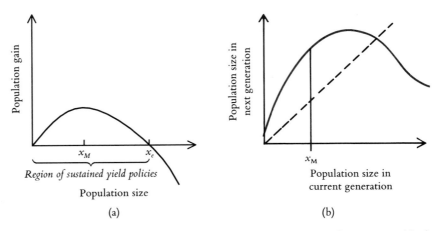

Figure 14.8 Determining the maximum sustainable yield x_M using a reproduction curve. (a) The natural increase, or sustainable yield, for each population size, with the maximum sustainable yield determined by the high point on the curve. The corresponding population size is x_M. (b) The reproduction curve, with the population size x_M corresponding to the maximum sustainable yield. The maximum sustainable yield is the greatest vertical distance from the 45° line to the reproduction curve.

as x_M in Figure 14.8. At this point, the vertical distance between the broken line and the reproduction curve — which represents the harvest — is as large as possible.

CONSIDERATIONS FROM ECONOMICS

Let's now consider how the costs of harvesting may complicate our analysis. We'll consider two models: one for a cattle ranch, and one for either a fishing boat or a tree farm.

In our models we assume that the price we receive for our harvest is the same for each harvested unit and does not depend on the size of our harvest. In effect, we are assuming that our operation is so small a part of the total market for the population in question that the size of our harvest will not substantially affect overall supply and hence price. We will let p denote the price that we obtain for each harvested unit, so that the total revenue from our operation is price per unit times number of units harvested, or ph.

Remember that we are concerned with determining the optimal harvesting level for a sustained-yield policy. We want to stay in business; therefore, we are not going to extinguish our resource for quick profits. In particular, for any given population size, we will harvest exactly the gain in population from one year to the next.

EXAMPLE: Cattle Ranching. For the cattle ranch, assume that the cost of harvesting a unit of the population is the same for each unit and does not depend on how many units we harvest. We include in the cost of harvesting all the costs of raising the cattle and delivering them to market. Let us say that it costs c to harvest each unit. Since our cost does not de-

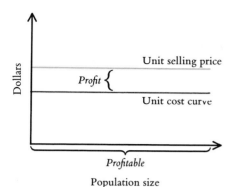

Figure 14.9 The unit cost, unit revenue, and unit profit of harvesting one unit, as a function of population size, for the cattle ranch.

pend on the population size, our cost curve is a horizontal line (Figure 14.9).

Now, if we harvest h units, the associated cost is ch. Then our net profit P from harvesting h units from a population consisting of x units is

Net profit P
= revenue − cost
= value of harvested units − harvest cost
= $ph − ch = (p − c)h$

As long as the selling price per unit is higher than the harvest cost per unit, we make a positive profit. The points of view of economics and biology agree, since the maximum profit occurs for the maximum sustainable yield.

EXAMPLE: Fishing and Logging.

Our key assumption for this model is that the cost of harvesting a unit of the population depends on how abundant the population is. This assumption incorporates the familiar principle of **economy of scale:** the cost of harvesting one unit decreases as the size of the population increases. For example, the same fishing effort yields more fish when fish are more abundant. Similarly, a logger's costs of harvesting one tree are less when that tree is in a stand of many harvestable trees than when it is in one with few; this is the logger's motivation for wanting to "clear-cut" large stands of trees. Since we assume that the cost goes down as the size of the population goes up, our cost curve looks something like Figure 14.10. The size of a population from which one unit is harvested is shown on the horizontal axis; the cost of harvesting a single unit is measured on the vertical axis. The fact that the curve slopes downward and to the right is a reflection of the basic principle that the cost of harvesting a single individual is less in a large population than in a small population.

Our analysis proceeds as before, except now our unit cost is no longer constant but depends on x. So we write it as $c(x)$, and our net profit is

$$\text{New profit } P = [p − c(x)] \times h$$

An optimal harvesting policy will depend on the relation between price and costs. There are two cases (Figure 14.11).

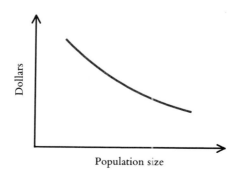

Figure 14.10 The unit cost, as a function of population size, for fishing or logging.

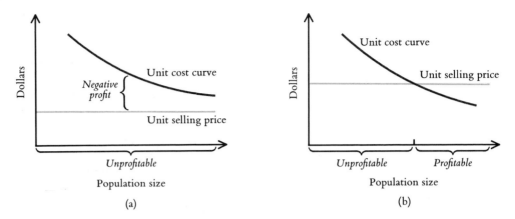

Figure 14.11 The unit cost, unit revenue, and unit profit of harvesting one unit, as a function of population size, for fishing or logging. (a) The market price is below harvesting cost, for all population sizes. (b) The operation is profitable for populations above a certain minimum size.

First, if the price we receive for a harvested unit is less than the cost of harvesting that unit for all population sizes, then it is impossible to make a positive net profit. The best we can do is to have a net profit of zero, by harvesting no units from our population.

The second case is of greater interest: Above a certain population size, the price we receive for one unit of the population is more than the cost of harvesting one unit. Now it is possible to generate a positive net profit (Figure 14.12). At the same time, the harvest is smaller for population sizes that are very large. There must be some population size, call it x_Q, that gives a maximum net profit. Using calculus, it can be shown that x_Q is actually larger than x_M, the population that gives the maximum sustainable yield, as shown in Figure 14.11.

But our models fail to take into account a very critical feature of a modern economy that we have already concentrated on earlier in this chapter: the time value of money, as measured by the interest capital can earn. In the next section, we see one important explanation why biological populations are susceptible to overexploitation and even extinction.

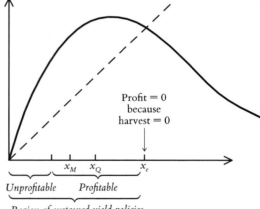

Figure 14.12 Region of profitability for sustained-yield policy, with the economically optimal population size x_Q marked.

WHY ELIMINATE A RENEWABLE RESOURCE?

We might well ask why anyone would want to eliminate a renewable resource. In at least some instances, such as the case of the now-extinct passenger pigeon, populations have been completely harvested (see Spotlight 14.3, p. 396). We can now apply our approach to understand why.

Sustained-yield policies operate under the assumption that some of the revenues will be received at a fairly distant future time. It is reasonable that the value of these revenues should be discounted to reflect the loss of income that would be earned if the funds were available for investment today. Rather than their current value, we should consider what economists call the **present value** of revenues to be received in the future. After all, a dollar today is worth more than a dollar you expect to receive a year from now.

To be specific, if funds are invested at an interest rate of $100r\%$ per year, the present value V of an amount A to be received n years in the future is related to A by the formula $A = V(1 + r)^n$. Hence $V = A/(1 + r)^n$. In this model, our goal is to maximize the sum of the present values of all future receipts of a sustained-yield harvesting policy. We refer to this sum as the present value of our return. The optimal harvesting policy will depend on the price p per unit harvested, the cost function $c(x)$, and the additional consideration of the interest rate r.

We don't delve into the details of the calculations here, but instead just give the results of the analysis.

Again there are several cases to consider. In the first case, suppose that the cost of harvesting $c(x)$ exceeds the price for all population sizes x. Then it is impossible to have a positive net profit, and the best we can do is have a return of zero. The optimal policy in this case is to harvest nothing.

For the second case, suppose that there is a population size x such that $c(x) = p$. Then there is a population size between x and x_e (the equilibrium population size) for which the present value of the total return is maximized.

In particular, the population is maintained at that level (after harvesting) and is not completely harvested.

For the third case, suppose that p is larger than $c(x)$ for *all* population sizes x. Then the conclusion depends on the value of the interest rate r. If r is small, then the situation is the same as in the second case. On the other hand, if r is large, then it may be that the optimal harvesting policy is not to sustain the yield but to harvest the entire population immediately. This conclusion certainly corresponds to our intuition: if the price is high enough and if the proceeds can be invested at a sufficiently high rate of return (interest), then the most profitable course of action is to harvest everything — that is, extinguish the resource — and invest the proceeds.

Let's put this in the simplest and starkest terms. Suppose you own a population of a valuable resource, such as a forest, and the cost of harvesting is small relative to the value of the resource. If the rate at which the population is increasing is greater than the current interest rate on investments, it pays you to let the forest keep on growing.

On the other hand, if the forest is growing more slowly than the interest rate, your economically optimal policy is to cut down all the trees now and invest the money. You could then start raising cattle on the land — and right there you have the scenario that is resulting in deforestation all over the world.

The sobering fact is that *very few economically significant renewable resources can sustain annual growth rates over 10%.* Many, like whales and most forests, have growth rates in the 4% to 5% range. These values — even a growth rate of 10% — are far below the return investors expect on their investment. For example, Wisconsin electric power utilities — far from being an exciting growth industry — are guaranteed by the state a profit of 14.25%; and venture capital firms expect to exceed 25% profit.

SPOTLIGHT 14.3 Extinction of the Passenger Pigeon

Although once numbering in the billions, passenger pigeons are now extinct. [Courtesy, Field Museum of Natural History, Chicago.]

Historically, the utilization of a renewable resource has followed a characteristic pattern. First comes a stage of expanding harvests, perhaps based on a new use of the resource or on new harvesting technology. This is followed by concern for overutilization. Conservation measures are then adopted and the industry either stabilizes or collapses.

In some cases, the population has actually become extinct. For example, the passenger pigeon was once considered to be the world's most abundant land bird. Over a century ago, they numbered from 3 to 5 billion, traveling and nesting in huge flocks, mostly in eastern North America. But by 1914, the last remaining passenger pigeon had died at the Cincinnati Zoo.

The demise of the passenger pigeon can clearly be traced to expanding harvests, brought about by new technology—in this case, the development of the eastern railroad network and the telegraph. To understand how these developments were able to severely diminish such an abundant species, we can look at the ecological characteristics that were once key to the passenger pigeon's earlier success: colonization and nomadism.

Passenger pigeons were not solitary. They nested in colonies containing millions of pairs (the entire population consisted of perhaps fewer than a dozen flocks). The flocks were sometimes so immense that they were reported to have obscured the sun. (The largest flight ever recorded was estimated to contain 2.23 billion birds.) By traveling and nesting in such large groups, pigeons were able to literally "shield" themselves from predators. No matter where they nested, there were not enough local predators to significantly reduce their numbers. This concept is known as "predator satiation."

Passenger pigeons fed on large crops of nuts found in the deciduous forests of eastern (and occasionally midwestern) North America. Because the location of crops large enough to accommodate their numbers varied from year to year, passenger pigeons rarely nested in the same place two years in a row, with nesting sights ranging from New York and Pennsylvania to Michigan or Wisconsin. Thus, it was difficult to predict their location from one nesting season to the next.

By all accounts, the final decline of the passenger pigeon was rapid. The arrival of flocks of passenger pigeons had always meant food to the local people, but it was probably not until pigeons were actually harvested for market that the population began to markedly diminish. (Market harvesting began before 1800 but was not a major industry until 1840.)

While we can never know for certain, it is believed that the technological developments of the nineteenth century — namely, the railroad and the telegraph — increased the efficiency and scope of market harvesting to the point where it was ultimately responsible for the extinction of the passenger pigeon. By the time of the Civil War, the railroad network through America, east of the Mississippi, was complete. This network allowed the professional pigeoners, who numbered about 1000 in their heyday, rapid access to all major nesting colonies. It also provided a fast means of shipping barrels of pigeons to the big city markets in the east and midwest. The telegraph was able to keep professional pigeoners informed of the locations of nesting colonies. In fact, the entire operation was organized so efficiently that word of any pigeon nestings spread rapidly for hundreds of miles. Since the railroads benefited from the pigeon harvest, it is likely that they, too, helped to see that this information was transmitted.

The fact that passenger pigeons nested in gigantic colonies — which at one time had assured their safety from predators — now made them especially accessible to harvesting for market. People did not understand that such an abundant resource could ever be severely diminished. They also did not allow for undisturbed nesting sites so that the pigeons could replenish their numbers. Instead, harvests at the nesting sites were so efficient and complete that there were no successful nesting colonies for a period of over 10 years. The last known colonial nesting attempt occurred in 1887 in Wisconsin, but the site was rapidly abandoned by the birds, probably because of disturbances.

If the harvest had not occurred at the nesting colonies, it is unlikely that the adult population could ever have been exterminated. Or, if only the adults had been harvested, the species might have survived. But because the fat nestlings were especially prized, and because many birds were driven away from nesting sites by the violent hunting methods sometimes used (shooting, setting trees on fire), the adults could not replace themselves, and the fate of the passenger pigeon was sealed.

Passenger pigeons were once a renewable resource, but within a period of about 20 years — twice an individual pigeon's lifetime — they became extinct. Certain other species, such as the bison, have also been reduced to numbers below a level of economic significance.

REVIEW VOCABULARY

Arithmetic growth   Growth by a constant amount in each time period.

Biomass   A measure of a population in common units of equal value.

Carrying capacity   The maximum population size that can be supported by the available resources.

Compound interest   The method of paying interest on both the principal amount and the accumulated interest in an account.

Compounding period   The interval that elapses before interest is calculated on an account.

Economy of scale   Costs per unit decrease with increasing volume.

Effective annual yield   A yearly rate of interest that is compounded only once per year.

Equilibrium population size   The population size for which the size next year will be exactly the same as the size this year.

Exponential growth   Geometric growth.

Geometric growth   Growth by a constant proportion in each time period.

Initial balance   Initial deposit in a bank account.

Interest   Money earned on a bank account.

Logistic model   A particular population model that begins with near geometric growth but then tapers off toward a limiting population (the carrying capacity).

Maximum sustainable yield   The largest sustainable yield.

Natural increase   The growth of a population that is not harvested.

Nominal annual rate   The stated annual rate of interest, which is then subject to compounding.

Population growth   Change in population, whether increase (positive growth) or decrease (negative growth).

Population structure   The division of a population into subgroups.

Present value   The value today of money to be received in the future.

Principal   Initial balance.

Rate of natural increase   Birth rate minus death rate, the annual rate of population growth without taking into account net migration.

Renewable natural resource   A resource that tends to replenish itself if it is allowed to evolve without external intervention; examples are fish, forests, wildlife.

Reproduction curve   A curve that shows population size in the next year plotted against population size in the current year.

Simple growth   Arithmetic growth.

Simple interest   The method of paying interest on only the initial balance in an account and not on any accrued interest.

Steady state   The population size for which the size next year will be exactly the same as the size this year. The amount harvested is exactly equal to the natural increase.

Sustained-yield policy   A harvesting policy that can be continued indefinitely while maintaining the same yield.

Yield   The amount harvested at each harvest.

SUGGESTED READINGS

Cherfas, Jeremy: "What Price Whales," *New Scientist* 5:36–40 (June 1986). Offers history of the whaling industry and concrete facts about both the economics and the biology of whaling.

Clark, Colin: "Some Socially Relevant Applications of Calculus," *The Two-Year College Mathematics Journal* 4(2):1–15 (Spring 1973). Gives a mathematical approach to animal resource economics.

———— "The Mathematics of Overexploitation," *Science* 181:630–634 (August 17, 1973).

Kleinbaum, David G., and Anna Kleinbaum: Adjusted Rates: The Direct Rate. UMAP Modules in Undergraduate Mathematics and Its Applications: Module 330. COMAP, Inc., Arlington, Mass., 1980. Reprinted in *The UMAP Journal* 1(1)(1980) 49–80, and in *UMAP Modules: Tools for Teaching 1980,* Birkhäuser, Boston, 303–334. A beginning exploration into the structure of populations, which explains, for instance, the paradox of how a Third World country can have a lower mortality rate than the United States, yet have a higher mortality rate for every age group.

Lindstrom, Peter A.: Nominal vs. Effective Rates of Interest. UMAP Modules in Undergraduate Mathematics and Its Applications: Module 474. COMAP, Inc., Arlington, Mass., 1988.

Reprinted in *UMAP Modules: Tools for Teaching 1988,* edited by Paul J. Campbell, COMAP, Inc., Arlington, Mass., 21–53. A bank account at *nominal* rate of 5% interest compounded daily actually earns an *effective* rate of 5.13%. A learning module, requiring no more background than this chapter, that teaches about the difference between nominal and effective rates of interest and how to calculate them.

Maki, D.P., and M. Thompson: *Finite Mathematics,* 2d ed., McGraw-Hill, New York, 1983, Chapter 11.

Population Reference Bureau, Inc., Annual World Population Data Sheet, 777 14th St. NW, Suite 800, Washington, D.C. 20005.

Schwartz, Richard H: *Mathematics and Global Survival,* Ginn Press, Needham Heights, Mass., 1989.

EXERCISES

Many of the exercises below require a calculator; a few require a scientific calculator [with buttons for exponential (EXP) and natural logarithm (LN) functions].

1. You deposit $1000 at 8% annual rate of interest. What will the balance be at the end of 1 year, and what is the effective annual yield, if the interest paid is

   a. simple interest?

   b. compounded annually?

   c. compounded quarterly?

   d. compounded continuously?

2. In some cases it is important to look at problems that are in a sense the reverse of those discussed in this chapter. In a financial setting such a question might be: How much do you need to deposit today in an account that pays interest at a known rate in order to have a specified amount at a specified time in the future? This question is crucial in certain financial planning considerations, for instance, in planning for a major purchase in the future.

Suppose that we ask for a specified amount, say A dollars, at a specified time in the future, say n years, and we know that the account pays interest at the rate $100r\%$ per year. The unknown quantity—namely, the amount that must be deposited today—will be denoted by P. Using our basic formula, we know that A, P, r, and n are related by

$P = A/(1 + r)^n$. The quantity P is known as the *present value* of an amount A to be paid n years in the future.

Suppose that you will need $15,000 to pay for a year of college 8 years in the future, and you can buy a certificate of deposit whose interest rate of 10% compounded quarterly is guaranteed for that period. How much do you need to deposit?

3. As in Exercise 2, except that the interest rate is only 8%. How much do you need to deposit?

4. The situation described in Exercise 2 gives the rationale for the pricing of the so-called *zero-coupon bonds.* These securities pay no current interest but are sold at a substantial discount from redemption value. The difference between purchase price and redemption value provides income to the bondholder at the time of redemption or resale. If the interest rate in the economy is now 7%, what should be the price of a zero-coupon bond that will pay $10,000 8 years from now? (Use daily compounding.)

5. In times of economic inflation, prices behave like populations undergoing exponential growth.

  a. Suppose inflation proceeds at a level rate of 5% per year from 1990 through 1995. Find the cost in 1995 of a basket of goods that cost $1 in 1990.

  b. During price inflation, the value of the dollar behaves like a population undergoing exponential decay; all our formulas work, with a little adjustment. If we let i represent the effective annual rate of inflation, then what costs $1 now will cost $(1 + i)$ this time next year, and a dollar then will buy only $1/(1 + i) = 1 - i/(1 + i)$ times as much. The quantity $d = -i/(1 + i)$ behaves like an interest rate, even though it is negative; the value of P dollars m years from now is given by $P(1 + d)^m$. Supposing 5% annual inflation from 1990 through 1995, what will be the value of a dollar in 1995 in constant 1990 dollars?

  c. Depreciation of the value of equipment is just like population growth, except that the rate of growth is negative. If you bought a car in 1990 for $10,000 and its value in current dollars depreciates steadily through 1995 at a rate of 15% per year, what will be its value in 1995 in 1995 dollars?

  d. If there is also 5% annual inflation from 1990 through 1995, what will be the value in 1995 of the car in part **c** in "inflation-adjusted" (constant 1990) dollars?

6. The *rule of 72* is a rule of thumb for finding how long it takes money at interest to double: If $100r$% is the annual rate of interest, then the doubling time is approximately $72/100r$ years.

  a. Calculate the balance at the end of the predicted doubling time for each $1000, with annual compounding, for the small growth rates of 3%, 4%, and 6%.

  b. As in part **a,** for the intermediate interest rates of 8% and 9%.

  c. As in part **a,** for the larger interest rates of 12%, 24%, and 36%.

  d. What do you conclude about the rule of 72?

7. (For this exercise you will need a calculator with **EXP** and **LN** buttons.) More frequent compounding yields greater interest, but with diminishing returns as the frequency of compounding is increased. For small interest rates, there is little difference in yield for compounding annually, quarterly, monthly, daily, or continuously. Investigating doubling times with continuous compounding leads to understanding why the rule of 72 of Exercise 6 works. Recall that for continuous compounding at a (nominal) annual rate r, the balance A at the end of m years is Pe^{rm} for an initial principal of P. Let D be the number of years that it takes for the initial principal to double. Then we have $2P = A = Pe^{rD}$, so $e^{rD} = 2$. Taking the "natural logarithm" of both sides yields $rD = \ln 2$, where "ln" stands for the natural logarithm. The natural logarithm is represented on a calculator by a button marked either "ln" or "LN" (not "log," which stands for a different kind of logarithm). Using the button gives $\ln 2 = 0.693$. So we have $rD = 0.693$, from which we can determine D if we know r.

a. Calculate the doubling times for continuous compounding at 3%, 6%, and 9%, and compare them with those predicted by the rule of 72. What do you conclude? Why do you think people prefer a rule of 72 rather than a rule of 69.3?

b. Using the analysis for doubling as a model, devise a "rule of_____" for the time it takes money to triple. Use your rule to predict how long it will take $600 at 5% interest to triple.

8. In some cases we know the principal, the current balance, and the interval of time, and we want to learn the interest rate. For example, money market funds typically report earnings to investors each month, based on interest rates that vary from day to day. The investor may be interested in knowing some average rate of interest for the month, but usually no such figure is reported by the fund (nor are each of the daily rates given). For our purposes, we will simplify and assume that the same interest rate holds for each day of the month and that compounding is done daily. The key tool will be our general compounding formula, that the balance A is given by $P(1 + r)^n$, where P is the original principal, r is the interest rate per compounding period, and n is the number of compounding periods. For our purposes here, the compounding period is one day. We have $A = P(1 + r)^n$, so $A/P = (1 + r)^n$. Taking the nth root of each side gives

$$1 + r = \left(\frac{A}{P}\right)^{1/n} \quad \text{so} \quad r = \left(\frac{A}{P}\right)^{1/n}$$

We illustrate with an example. Suppose the monthly statement from the fund reports a beginning balance (P) of $7373.93 and a closing balance (A) of $7416.59 for 28 days (n). We thus have

$$r = \left(\frac{7416.59}{7373.93}\right)^{1/28} - 1 = (1.005785246)^{0.035714286} - 1 = 0.000206042$$

Thus the daily rate of interest is 0.0206042%. We obtain the annual effective yield by compounding at this rate 365 times: $(1 + 0.0206042)^{365} = 1.0780972$, for an effective annual yield of 7.81%.

Suppose that the preceding month the initial balance had been $7331.35, and again the report had been for 28 days. Calculate the effective annual yield.

9. A 1990 advertisement reads, "If you had put $100 per month in this fund starting in 1980, you'd have $37,747 today." How much money was deposited during this period? Assume that deposits were made on the first day of the month, starting on January 1, 1980, through December 1, 1989, and that interest is paid monthly on the last day of the month. How much money was deposited during this period? What annual rate of interest, compounded monthly, would lead to the result described in the advertisement? Hint: This problem can be solved on a calculator by tediously adding deposits and crediting interest, or by writing a short computer program to do the same. But there is also a relevant formula. For any real number x and any positive integer n, the following relation holds:

$$1 + x + x^2 + x^3 + \cdots + x^n = \frac{x^{n+1} - 1}{x - 1}$$

In this formula the dots mean that all powers of x up to and including the nth (i.e., through x^n) are included in the sum on the left-hand side.

10. Parents commonly use savings accounts to accumulate funds for the college education of their children. Frequently the funds are deposited over a period of several years, and the funds together with accumulated interest are used to pay the costs of education.

Suppose that $1000 is to be deposited each year into an account that pays interest at a rate of 8% per year compounded annually. If the first payment is made when a child is 2 years old and the last is made when the child is 17 (16 payments), how much is available when the child begins college at the age of 18? (You can make use of the formula in Exercise 9; but be careful and think through the problem, as the formula itself will not give the final answer.)

11. The total population of the less-developed countries (excluding China) was 3.0 billion in mid-1990, and the growth rate was 2.4% per year (this is an effective annual yield, so you may think of it as compounded annually). If this growth rate continues until mid-2000, what will be the size of the population then?

12. If the growth rate of the less-developed countries of Exercise 11 changes suddenly to 2% in 1995, what will be the size of the population in mid-2000?

13. If the growth rate of the less-developed countries of Exercise 11 decreases by one-twentieth of a percentage point per year, what will be the size of the population in mid-2000?

14. In mid-1990 Zimbabwe had a population of a little over 10 million. If the population grows at a rate of 1% per year for 10 years and then grows at a rate of 4% for another 10 years, what is the final population size?

15. What is the final population size for a population of 10 million if the population grows at a rate of 4% per year for 10 years, and then grows at a rate of 1% per year for another 10 years?

16. Can you now, on the basis of Exercises 14 and 15, propose a general conclusion about the effects on final population size of the order in which high and low growth years occur in a population? (The actual rate of growth of the population of Zimbabwe in mid-1990 was 3.5% per year, among the highest in the world.)

17. Apply the rule of 72 of Exercise 6 to estimate the doubling times for the following populations (figures are for mid-1990):

    a. China, 1.1 billion, 1.4%

    b. the world as a whole, 5.3 billion, 1.7%

    c. Africa, 660 million, 2.9%

18. As in Exercise 17, but for

a. Europe, 500 million, 0.3%

b. United States, 250 million, 0.7%

19. Kenya was one of the most rapidly growing nations in the world in 1990, with an estimated growth rate of 4.1% per year. The populations of Kenya for the decades 1950 through 1990 are as shown in the table below. The projected population sizes for 2000 and 2010 are also shown. What assumptions about the rate of growth of the Kenya population led to these projections?

Year	Population (millions)
1950	6.0
1960	8.1
1970	11.2
1980	16.7
1990	25.2
2000	38.3
2010	58.2

20. Which population pyramid in the accompanying figure shows the highest birth rate?

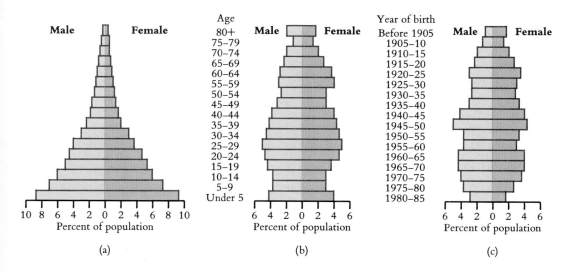

(a)   (b)   (c)

21. Venezuela in mid-1990 had a population of almost 20 million, and its growth rate was about 2.4% per year for the previous 10 years. What was its population in 1980?

22. In 1985 the population of Brazil was 150 million and the growth rate was 2.0% per year. In the same year the population of the United States was 250 million and the growth rate was 0.7% per year (both rates exclude net migration). If these growth rates were to continue into the future, when would the population of Brazil be the same as the population of the United States? (Hint: Begin by finding the population sizes at 20, 40, and 60 years in the future. If the population of Brazil is less than that of the United States at one time and greater than that of the United States at a later time, then the two populations must be equal at some time in between.)

For Exercises 23 and 24, refer to the following: Radioactive elements emit particles and decrease in quantity at a predictable rate. For such elements, the length of time that it takes to decrease to half its original mass is called the *half-life* of the element. For instance, the radioactive isotope radium-226 has a half-life of approximately 1600 years. This means that if a substance contains 1000 grams of radium-226 now, then 1600 years from now, the substance will contain 500 grams of the isotope. Further, in 3200 years, the substance will contain 250 grams of the isotope, and in 4800 years, only 125 grams of the isotope will remain in the substance.

23. a. The isotope carbon-14 has a half-life of approximately 5700 years. If a substance contains 20,000 grams of the isotope now, in how many years will there be only 5000 grams of the isotope remaining in the substance?

b. The isotope plutonium-230 has a half-life of approximately 24,000 years. If a substance contains 10,000 grams of the isotope now, in how many years will there be only 1250 grams of the isotope remaining in the substance?

24. a. The half-life of the isotope uranium-238 is approximately 4.5 billion years. In how many years will 40 grams of the isotope be reduced, through emissions, to 5 grams of the isotope?

b. The isotope einsteinium-254 has a half-life of approximately 276 days. If a substance contains 30,000 grams of the isotope now, approximately what portion of the isotope will remain in the substance 4 years from now?

For Exercises 25 and 26, refer to the following: Carbon dating is used to determine the age of a fossil. A living body radiates approximately 900 rays from carbon-14 per gram per hour. By measuring the number of rays emitted per gram per hour, the age of a fossil can be estimated. If a fossil is giving off 28 rays of carbon-14 per gram per hour, the estimated age of the fossil could be determined by working backward as follows:

Age of fossil (in years)	Rays emitted per gram per hour
0	900
5,700	$\frac{1}{2}(900) = 450$
11,400	$\frac{1}{2}(450) = (\frac{1}{2})(\frac{1}{2})(900) = (\frac{1}{2})^2(900) = 225$
17,100	$\frac{1}{2}(225) = (\frac{1}{2})^3(900) = 112.5$
22,800	$\frac{1}{2}(112.5) = (\frac{1}{2})^4(900) = 56.25$
28,500	$\frac{1}{2}(56.25) = (\frac{1}{2})^5(900) = 28.125$

Hence, the 900 rays would be decreased to approximately 28 rays in approximately 28,500 years, the estimated age of the fossil. (Note: An age of 0 for the fossil denotes the time of death of the living body.)

25. A fossil is emitting approximately seven rays of carbon-14 per gram per hour. Determine the approximate age of the fossil.

26. A fossil is determined to be approximately 5100 years old. Approximately how many rays of carbon-14 is the fossil emitting per gram per hour?

27. Suppose that a population of size P is growing at a rate r given by $r = P(100 - P)$. Estimate the value of P corresponding to the largest value of r.

28. Suppose that a population of size P grows by an amount

$$Pk\left(1 - \frac{P}{100}\right)$$

between observations. We view k as an intrinsic growth rate (the rate of population growth without resource constraints) and the number 100 as a carrying capacity of the environment.

a. For an initial population of 10 and $k = 0.8$, find the sizes of the population for the next 10 observations.

b. As in part **a,** but for an initial population of 110. What differences do you observe between the results for these two initial populations?

c. As in part **a** and **b,** but with $k = 1.8$.

d. On the basis of your analysis of the situations, what can you say about the dependence of the population growth on the parameter k?

29. Suppose that a population of size P grows by the amount

$$Pk\left(1 - \frac{P}{M}\right)$$

between observations, where k is the intrinsic growth rate and M is the carrying capacity of the environment. Suppose also that the carrying capacity is 100 for the first 5 observations and then drops to 70. (There is an environmental catastrophe at that time; for instance, a flood wipes out much of the food supply).

a. For an initial population of 20 and $k = 0.9$, find the population sizes for the first 10 observations.

b. As in part **a,** but suppose also that the carrying capacity M is increasing steadily. This might be the case if, for example, the food supply were increasing steadily. Suppose that in the nth year the carrying capacity is $100 + 5n$. For an initial population of 10 and $k = 0.7$, find the population sizes for the first 10 observations.

c. As in part **b,** but for an initial population of 10 and $k = 2$. Find the population sizes for the first 10 observations.

d. What do you conclude? Describe what is happening in terms of the setting of the problem.

30. Suppose a reproduction curve for a certain population is as in the accompanying figure, where the units are in thousands.

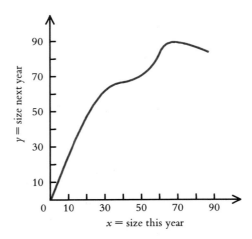

a. Estimate the sustainable yield corresponding to a population of size 10 remaining after the harvest.

b. Estimate the maximum sustainable yield.

31. A reproduction curve for a population is shown in the figure below. Estimate the equilibrium population size and the maximum sustainable yield.

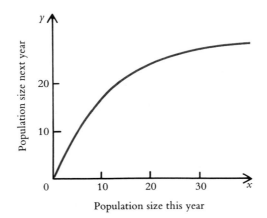

32. As in Exercise 31, but for the reproduction curve below.

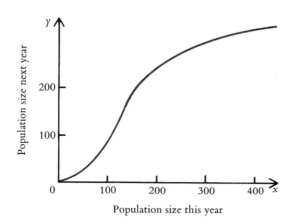

33. Suppose the cost of harvesting one individual from a population of size x is given by $20/(10 + 5x)$. Find the cost of harvesting five individuals from a population of size 20.

34. As in Exercise 33, but find the cost of harvesting five individuals from a population of size 80.

35. Suppose that the figure below gives the cost of harvesting an individual from a population of size x. Use this graph to estimate the cost of harvesting five individuals from a population of size 50.

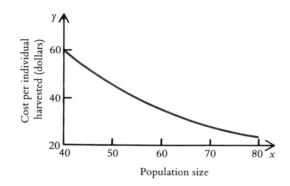

36. As in Exercise 35, but for a population of size 80.

37. Suppose that the reproduction curve for a population is as shown in graph (a) below and that the cost-of-harvesting function is as shown in graph (b). The price obtained for one harvested individual is $8. Find the net yearly revenue when a sustained-yield harvesting policy is used that maintains a population of size 20.

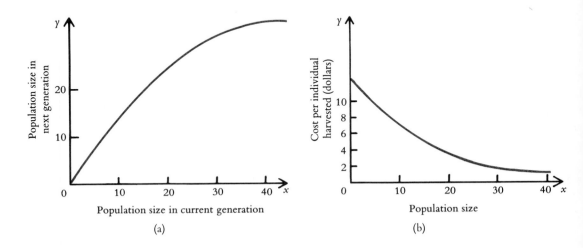

(a)   (b)

38. Solve the same problem as in Exercise 37 in the case where the price for one harvested individual is $6 and the population is maintained at size 15.

39. Discuss the question posed in Exercise 37 when the price for a single harvested individual is $4 and the population is maintained at size 15.

● **40.** (For this exercise you will need a calculator with EXP and LN buttons.) The accompanying table shows the U.S. population from 1790 to 1990.

Year	U.S. population (in thousands)	Year	U.S. population (in thousands)	Year	U.S. population (in thousands)
1790	3,929	1860	31,513	1930	123,077
1800	5,297	1870	39,905	1940	132,457
1810	7,224	1880	50,262	1950	152,271
1820	9,618	1890	63,056	1960	180,671
1830	12,901	1900	76,094	1970	205,052
1840	17,120	1910	92,407	1980	227,757
1850	23,261	1920	106,461	1990	250,000

Source: *Historical Statistics of the U.S.*, p. 8; 1930 through 1980 figures are from *Statistical Abstracts, 1988*, Bureau of the Census, p. 7, Table 2. The figure for 1990 is an estimate.

a. Plot the population versus year on ordinary graph paper. Then plot it on semilog paper, with the population on a logarithmic scale and the year on an ordinary linear scale (you'll need what's called "3-cycle" paper). On semilog scale, geometric growth appears as a straight line.

b. Use the calculator button marked either "ln" or "LN" (not "log," which stands for a different kind of logarithm) to calculate the logarithm of each population size. For example, for the United States in 1790, with a population of 3,929,000, we have $\ln(3,929,000) = 15.184$. Then plot the logarithm of population versus year on ordinary graph paper. What you have done is explicitly make the logarithmic transform that semilog paper does for you. Once again, geometric growth will appear as a straight line.

c. Over the early decades, the fit of the points to a straight line, in either of the graphs above, is excellent, showing that the population was in fact growing geometrically. On your graph from part **b,** draw in the straight line between the points [1790, $\ln(3,929,000)$ and 1860, $\ln(31,513,000)$], and extend it across the graph. This line has slope

$$m = \frac{\ln(31,513,000) - \ln(3,929,000)}{1860 - 1790} = \frac{17.266 - 15.184}{70} = 0.0297$$

This quantity is the average annual rate of growth r, 2.97% per year. Also, we can use this information to predict the population in year t after 1790 with the formula

$$P = 3,929,000 \times e^{0.0297(t - 1790)}$$

Use this formula and the EXP button (for powers of e) to make a table of predictions of the population in 1870 and subsequent decades.

d. In what year do you begin to notice a deviation of the population from the line? To what can you attribute this deviation?

● Optional exercise.

·15·

Measurement

Before 1600, people had nothing more than their eyes and sighting rods with which to see the universe. Nevertheless, they made precise measurements of the size of objects on the earth, the size of the earth itself, and the distances to the moon and sun.

The mathematics used to determine these distances is the geometry Euclid set down in his *Elements*, written sometime around 300 B.C. and consisting of 13 chapter-long "books." Although the *Elements* organized all the geometrical and arithmetical knowledge accumulated by the Greeks, thus incorporating earlier knowledge from Babylonia and Egypt as well, its greatest achievement was to show a natural sequence by which one result could be derived logically from another. For example, Thales (ca. 600 B.C.) is said to have discovered that every angle inscribed in a semi-circle is a right angle, or, as geometry texts sometimes say, has the same **measure** (in degrees or other units of measure) as a right angle

(see Spotlight 15.1, p. 412). We don't know how — or in fact whether — he proved that this is so, but in the *Elements* Euclid shows just how it can be derived from much simpler facts.

The basic ideas of *congruence* and *similarity* are developed in the *Elements*. Two triangles are **congruent** if one is an exact copy of the other; one congruent triangle can be made to fit exactly onto another. Two triangles are **similar** if they have the same shape but not necessarily the same size.

Book I of the *Elements* contains most of the standard facts about congruence for triangles. These facts are followed by the introduction of the famous parallel postulate (discussed in Chapter 16), which proved to be very important in the history of ideas and paved the way for Einstein's theory of relativity and other modern theories about the structure of the universe. Book I culminates with the **Pythagorean theorem**, which gives a way to calculate distances along a slant when the corresponding horizontal and vertical distances are known. This theorem is the basis of the standard ways of determining the distance between two points on a straight line or on a curve, and hence for all of analytic geometry. It is in turn the necessary preliminary for calculus, the tool that Newton invented for understanding the motions of the planets and of falling apples.

The fundamental role of the Pythagorean theorem is illustrated by the following straightforward problem (illustrated in Figure 15.1): In a city laid out on a grid of east-west and north-south streets, a house B is known to be four blocks south and three blocks east of another house, A. What is the straight-line distance AB and A to B?

Here is the answer: Because $AC = 4$ and $BC = 3$, by the Pythagorean theorem we know that $(AB)^2 = (AC)^2 + (BC)^2 = 4^2 + 3^2 = 16 + 9 = 25$. If $(AB)^2 = 25$, we know immediately that $AB = 5$.

In words, the Pythagorean theorem says that for any right triangle, the sum of the squares on the two short sides, or **legs**, is equal to the square on the long side, the **hypotenuse**.

How was the Pythagorean theorem discovered in the first place? We have no writings from Thales, and the presentation in Euclid is a highly systematic and organized presentation that hides the original roots. So, we don't know. For the particular case of the 3 by 4 by 5 right triangle, we have a figure (but not a written-out proof) in a manuscript dating to about 2000 years ago in China. Like Euclid's *Elements*, this manuscript compiles mathematical facts learned over the previous several hundred years.

Figure 15.2 shows the proof, which proceeds by subtraction from the area of the larger square. The askew square in the middle includes four large triangles, plus a tiny square in the center. Let's fasten on the the 3 by 4 by 5 right triangle in the upper-right-hand corner.

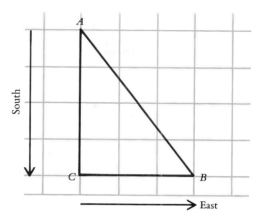

Figure 15.1 *Problem:* What is the distance, as the crow flies, from A ro B?

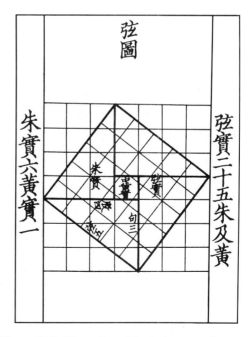

Figure 15.2 Illustration of the Pythagorean theorem for a 3 by 4 by 5 right triangle from the Chinese work *The Arithmetical Classic of the Gnomon and the Circular Paths of Heaven (Chou-pei suan-ching,* or *Zhōubì suànjīng* in the Pinyin system), the earliest Chinese writing on mathematics to survive to today. [From Frank J. Swetz and T. I. Kao, *Was Pythagoras Chinese?*, Pennsylvania State University Press, University Park, Pa., 1977.]

SPOTLIGHT 15.1 Thales

We know very little for certain about Thales, who lived 2600 years ago (ca. 600 B.C.). He is usually considered the founder of Greek philosophy and is one of the "seven sages" honored by the Greeks and Romans. He taught that in spite of the vast differences in the appearance of things, there is an underlying unity in the world. He also believed that everything originates in water. According to an often-repeated anecdote, Thales once fell into a well while looking at the stars, and a servant girl laughed at him, saying that he wanted to know what happens in the heavens but couldn't even keep track of his own feet.

However, if the ancient stories are true, he was also a versatile and practical man. He is said to have predicted the year in which a solar eclipse turned "day into night" during a battle of 585 B.C. One year he also cornered the market in olive oil presses, thereby making a great deal of money.

Thales was not the first to unravel the intricacies of mathematics and astronomy. For example, his prediction of an eclipse could not have been accomplished without the knowledge, observations, and geometric facts known before him in Egypt and Babylon. He probably began a systematic organization of these teachings just as Euclid later organized knowledge in the *Elements*.

We can see to its left a 3 by 3 square, and below it a 4 by 4 square; the two remaining blocks that make up the larger square are 3 by 4 rectangles. Meanwhile, its diagonal is a side of the 5 by 5 askew square; the remaining area of the larger square consists of four halves of 3 by 4 blocks that have been split down the diagonal. We can write an equation as follows:

Area of larger square
  = area of 3 by 3 square + area of 4 by 4 square + area of two 3 by 4 blocks
  = area of 5 by 5 square + area of four halves of 3 by 4 blocks

Equating the two 3 by 4 blocks with the four halves and subtracting this quantity from each of the last two expressions, we have the desired result.

A slightly different idea leads to a general proof that $a^2 + b^2 = c^2$ for a right triangle with hypotenuse c and legs a and b, using only the fact that the area of a right triangle is $\frac{1}{2}ab$. (See Spotlight 15.2, p. 418.)

Later books of the *Elements* go on to treat the standard theory of similarity, which is the basis for all map making and for most other methods of calculating and representing inaccessible distances, including the tools of trigonometry. In the remainder of this chapter we will show how these simple ancient tools, which are still routinely taught to all engineers, have shaped the way we view our universe.

ESTIMATING INACCESSIBLE DISTANCES

Our story concerns four men: Thales, Euclid, Aristarchus, and Eratosthenes, each of whom developed new techniques for measuring ever more distant objects. The last three lived about

300 B.C. and were probably born in the order they are named. Some 300 years earlier, Thales is supposed to have made two difficult measurements: (1) the distance of ships at sea, using congruence of triangles, and (2) the height of the Great Pyramid in Egypt, using similarity of triangles.

Suppose you want to find the distance of a ship at position *B* straight out at sea from your position *A* on the shore, using the method attributed to Thales (see Figures 15.3 and 15.4). Here is Thales's solution: Starting from *A*, walk along the shore any distance in a direction perpendicular to *AB*. Put a marker at *S*, making sure it is tall enough to see from a distance. Then walk the same distance to point *C*. Now turn at a right angle and walk away from the shore until you reach a point *E* from which your marker *S* is exactly lined up with the ship *B*. The distance *CE* you walked away from the shore is exactly the same as the distance *AB* of the ship from shore.

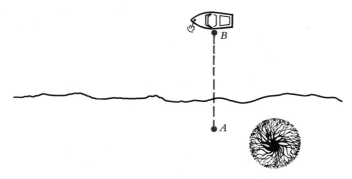

Figure 15.3 *Problem:* Find the distance of the boat from shore, using elementary geometry.

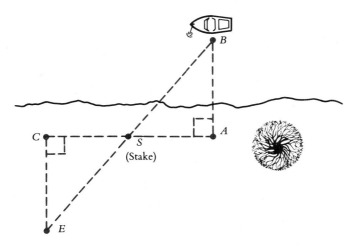

Figure 15.4 *Solution:* Thales found the distance of the boat from shore by putting a stake at *S* and then showing that triangles *SAB* and *SCE* are congruent.

Figure 15.4 shows why this works. The simple geometry that Thales used, long before Euclid, tells us that the two angles at S are congruent: vertical angles are congruent. (If the sides of an angle are extended through its vertex, then the angle of the triangle and the angle opposite it are called **vertical angles**.) The angles at A and C are equal (have the same measure) because both are right angles. The sides AS and SC are equal because they were paced off to be equal. By one of the congruence theorems for triangles, we know that triangle ABS is congruent to triangle CES and hence that the corresponding lengths AB and CE are equal, as we claimed. The congruence theorem that we use here says that two triangles are congruent if two angles and the included side of one are congruent to two angles and the included side of the other.

If a triangle is enlarged or reduced by a photocopy machine, then the resulting triangle is similar to the original triangle. Thus, corresponding angles of similar triangles have the same size, and the lengths of corresponding sides are in the same ratio. For example, a photograph and its enlargement are similar to each other and to the original (we will look into enlargements and similarity further in Chapter 17). Thales may have known that right triangles with corresponding angles are similar and hence have proportional sides. With this knowledge he could calculate the height of the Great Pyramid.

To take the height of the Great Pyramid (or any other vertical object such as a tall tree), hold an upright stick on the ground at the site of the object and measure its length and the length of its shadow. The right triangle whose legs are the stick and its shadow is similar to a right triangle whose legs are any other vertical object and its shadow (at the same place and time). Hence the ratio of these two lengths, stick and shadow, is the same as the ratio of the length (height) of the Great Pyramid to the length of its shadow (see Figure 15.5).

EXAMPLE: Finding the Height of the Great Pyramid. Suppose, to be specific, that in this case the length of the stick is 10 feet, it casts a 16-foot shadow, and Thales's measurement of the shadow of the Great Pyramid was 770 feet. From the equality of these ratios, $\frac{10}{16} = h/770$, he could calculate the height h of the Great Pyramid to be $\frac{10}{16} \times 770$, or 481 feet.

Notice that ratios (such as the $\frac{10}{16}$ of the height of the stick to the length of its shadow) are the key to finding all sorts of inaccessible heights. In fact, if the ratio of stick to shadow is a/b and the length of shadow of any object is s, then we can find the height of that object simply by multiplying the two: $h = (a/b) \times s$.

In later times extensive tables of such ratios showed, for any given angle of inclination of the sun, what the ratio of the length of the

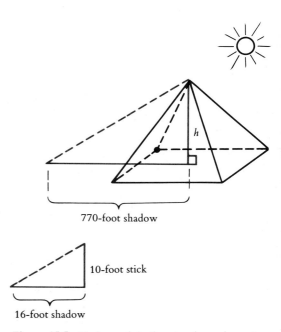

Figure 15.5 Thales used similar triangles to determine the height of the Great Pyramid. Since the shadow of the Great Pyramid is 48.1 times the shadow of the stick, then its height must be 48.1 times the 10-foot height of the stick, 481 feet.

stick to its shadow would be (see Table 15.1).

With such a table, the heights of inaccessible objects can be calculated without the use of the stick if a suitable instrument, such as a sextant, is available for measuring the angle of the sun above the horizon. Table 15.1 immediately tells us, for each angle of the sun, the ratio for

$$\frac{\text{Length of stick}}{\text{Length of its shadow}}$$

To summarize, the procedure for measuring height is

1. Measure the length s of the shadow of the object whose height you want to determine.

2. Measure the angle of the sun above the horizon, and look up the corresponding ratio r in the table.

3. Multiply $r \times s$ to find the height of the object.

This procedure was the beginning of trigonometry (triangle measurement) as we know it. The ratio we have tabulated is now called the **tangent** of the angle.

EXAMPLE: Measuring the Height of a Flag Pole. The Department of Physical Plant at Beloit College needed to know the height of the college flag pole in order to buy rope of the right length to raise the flag. The shadow of the flag pole was measured and found to be $s = 25.5$ feet. A Brunton compass from the Geology Department was used to measure the angle of the sun above the horizon, which was found to be 70°. From Table 15.1 (or from using the "tan" button on a calculator), we

TABLE 15.1 Table of tangents

Angle of the sun above the horizon	Tangent of the angle = $\dfrac{\text{length of stick}}{\text{length of its shadow}}$
5° (Sun nearly on the horizon)	.08749
10° (Sun somewhat higher)	.17633
20°	.36397
30°	.57735
40°	.83910
45° (Sun exactly halfway between horizon and directly overhead)	1.00000
50°	1.19175
60°	1.73205
70°	2.74748
80°	5.67128
87°	19.08114
89° (Sun almost directly overhead; higher than it ever gets in the United States)	57.28996

find the tangent of 70° is $r = 2.75$. So the height of the flag pole is $r \times s = 2.75 \times 25.5$ feet $= 70$ feet.

We can verify the entries in the tangent table by using a pocket scientific calculator — one with buttons for the "trig" functions sin, cos, and tan. Try this on the calculator: press the "tan" button for various angles and check that you get the same values given in Table 15.1. On some calculators there may be slight round-off errors, so you may get tan 45° = 0.99999 instead of tan 45° = 1.00000. (Be sure that the calculator is in *degree* mode. If you get 1.62 instead of 1.00, you are in *radian* mode instead. Radians are another unit for measuring the size of angles.) Essentially, however, you should be able to reproduce the table and even fill in the omitted entries if you wish.

DIGGING STRAIGHT TUNNELS

The famous Greek historian Herodotus, who lived some 100 years after Thales, described three engineering achievements on the Greek island of Samos. One was a tunnel that brought water through Mount Castro to the capital city, Samos (see Color Plate 13).

Nearly 2500 years later, in 1882, archeologists rediscovered the tunnel, exactly as Herodotus had described it. It was 1 kilometer (about 0.6 mile) in length and more than 2 meters (about 6 feet) high and wide. A deep ditch in its floor contained pipes, and the tunnel had vertical vents for changing the air and cleaning away rubble, and niches where workers placed their lamps. The ditch had a depth of some 2 meters at the upper end and 8 meters at the lower end, and was probably dug because the drop that had originally been planned turned out to be too small.

The remarkable thing about this tunnel was that the digging teams, proceeding from each end, met at the center with an error of only 10 meters (33 feet) horizontally and 3 meters (10 feet) vertically. We know this because at the center of the tunnel there is a jog of that size to make the two ends meet.

King Hezekiah of Judea was less successful. When he had a similar aqueduct constructed through the rocks near Jerusalem around 700 B.C., his workers had to check the direction of digging in a very primitive way, by means of vertical shafts from the top. The result was a zigzag tunnel twice as long as the distance between its ends.

How was the Samos tunnel dug without the benefit of guiding shafts? We do not know for sure, but a later writer, Heron, described a likely method, which we modify slightly here to bring out the essentials. In his view, the method used similar triangles in a considerably more complicated way than Thales had used them. We will describe this method in detail, following Figure 15.6.

Suppose the tunnel entrances are to be at A and B, on opposite sides of Mount Castro in Figure 15.6. Begin by marking off some convenient distance BE on any line at B. Following the figure, make a right turn at E and go to F. At F turn again and go to G, then turn again and go to H, which is chosen so that a right turn takes you straight to A. Suppose, to be specific, that the distances in this detour around the mountain are $BE = 750$ meters, $EF = 1000$ meters, $FG = 2000$ meters, $GH = 800$ meters, and $HA = 250$ meters. For the right triangle ABC, which is underneath the mountain and therefore not directly accessible, we know, by subtraction, that $AC = 200$ meters and $BC = 1000$ meters. Any right triangle with short sides in this same ratio (5 to 1) will be similar to ABC, thus its angles will be equal to those of ABC.

To find the direction to dig, we construct triangles BOP (say, $BO = 50$ meters, $OP = 10$

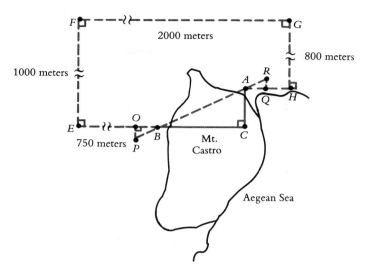

Figure 15.6 The plan for digging a tunnel through Mount Castro on the island of Samos. Using this plan, two digging teams starting at opposite ends A and B met at the center with only a small error.

meters) and AQR (say, $AQ = 50$ meters, $QR = 10$ meters) outside the proposed entrances A and B to the tunnel. (These are not drawn to scale in Figure 15.6.) The angle OBP is the same as the angle CBA, which tells us that PB is the direction in which to dig. Similarly, at the other end, RA is the direction to dig. Thus, the clever use of similar triangles over 2600 years ago helped to solve a major problem of civil engineering.

MEASURING THE EARTH

The modern use of similar triangles in engineering projects also depends on similarity principles. Our next example makes a big jump in the gradually increasing scale of distances: here we learn how to find the size of the earth itself. One of the truly spectacular achievements of ancient mathematical science was the determination, by a very simple method, of the **circumference** of the earth,

or the distance around the earth measured along a circle passing through the poles. The most accurate of these calculations was that of Eratosthenes in about 200 B.C. His method is illustrated in Figure 15.7.

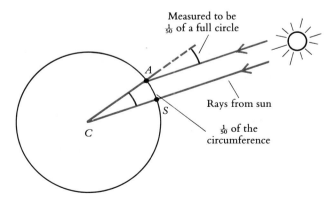

Figure 15.7 Eratosthenes noticed that the sun is about $\frac{1}{50}$ of a circle south of the zenith at Alexandria when it is directly overhead at Syene. Thus the distance between Alexandria and Syene must be about $\frac{1}{50}$ of the earth's circumference.

SPOTLIGHT 15.2 The Pythagorean Theorem

Many different proofs have been developed for the Pythagorean theorem—more, in fact, than for any other theorem of geometry. One proof is based on the two figures shown below.

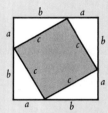 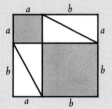

Each figure is a square with sides of length $a + b$. The first has been subdivided into four right triangles congruent to the original triangle and a square whose sides are equal to its hypotenuse. The second figure also contains four right triangles congruent to the original triangle. The theorem follows from the fact that the rest of the figure consists of two squares whose sides are equal to the legs of the triangle.

Expressing this algebraically in terms of the total areas of the two figures, we have

$$(a + b)^2 = 4\left(\frac{1}{2}ab\right) + c^2$$

and

$$(a + b)^2 = 4\left(\frac{1}{2}ab\right) + a^2 + b^2$$

Hence,

$$4\left(\frac{1}{2}ab\right) + c^2 = 4\left(\frac{1}{2}ab\right) + a^2 + b^2$$

and

$$c^2 = a^2 + b^2$$

One assumption that we made about the figure is not completely obvious. We have assumed that the four-sided figure in the center of the first figure is a square. This fact can be established with a little further argument.

[Adapted from Harold R. Jacobs, *Geometry*, 2d ed., Freeman, New York, 1987, pp. 329–330.]

It was known that at a certain time the sun was directly overhead at Syene (now Aswan), point *S*, in Egypt. At exactly the same time in Alexandria, lying straight north of Syene at point *A*, the position of the sun was measured to be $\frac{1}{50}$ of a full circle (that is, 7.2°) away from directly overhead. Because the sun is so far away from the earth, the two arrows in the figure that point to the sun are essentially parallel lines. Hence, the angle at *C*, the center of the earth, is also $\frac{1}{50}$ of a full circle because it is the **corresponding angle** when the two parallel lines are cut by the line *AC* (*AC* is traditionally called in this context a **transversal** of the parallel lines).

Then, because the angle at *C* is $\frac{1}{50}$ of 360°, the full circle, the distance *AS*, from Alexandria to Syene, is also $\frac{1}{50}$ of the complete circumference of the earth. It is only necessary to measure the distance from Alexandria to Syene (not a triviality in those days!) to have all the information needed. When the distance from Alexandria to Syene was found to be 5000 *stadia* (a Greek unit of measure; singular, *stadium*), this yielded 5000 × 50 = 250,000 stadia for the circumference of the earth.

Although we are not sure how large the stadium unit was, one estimate from Pliny is that a stadium was 157.5 meters. Using this value, we get 157.5 × 250,000 meters = 157.5 × 250 kilometers = 39,375 kilometers for the circumference of the earth. The kilometer was originally defined as $\frac{1}{10,000}$ of the distance from the North Pole to the Equator —one-quarter of the earth's circumference— so that the earth's total circumference is 40,000 kilometers or 24,800 miles. We see that Eratosthenes's result is nearly on the mark. We can be excused for thinking that some of Eratosthenes's numbers seem to be rounded off, and hence only accidentally accurate. Even so, we must admire his achievement. After all, in later years there was even some doubt that the earth was round!

MEASURING ASTRONOMICAL DISTANCES

Knowing the circumference of the earth (and hence its radius), the astronomer Aristarchus could consider even greater distances. His measurements of the distances of the moon and the sun from the earth were not as accurate as Eratosthenes' determination of the size of the earth, but his ingenious method is worth looking at. It shows how even a very simple understanding of triangle and circle geometry yielded information that completely revised his contemporaries' picture of the universe; they had imagined celestial distances to be much smaller than he showed them to be.

Using simple geometry, Aristarchus first determined how many times as far the sun is from the earth than the moon is from the earth (note that all our distances will be taken from the center of one object to the center of the other). He noticed that when the moon is exactly half full, that is, when we see exactly half the moon in shadow and half in the sun's light, the triangle *MES* formed by the moon, earth, and sun is a right triangle, with its 90° angle at the moon *M* (see Figure 15.8). (The two days in each month when the moon is exactly half full are marked on many modern calendars.)

By measuring the angle at *E*, we would be able to read off the ratio *MS/EM* from the

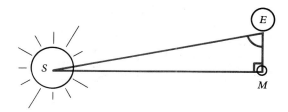

Figure 15.8 Aristarchus's method for estimating the ratio of the distances of the sun and moon from the earth.

tangents in Table 15.1. (We actually want the ratio ES/EM, which for such large distances is only slightly different. Another table, the "cosine" table, would give the ratio EM/ES if we wanted it.) Aristarchus estimates angle E as 3° less than a right angle, that is, $E = 87°$. From the table we see that tan 87° = 19; therefore, according to Aristarchus the sun is 19 times as far from the earth as the moon is.

Neither Thales nor Aristarchus actually had a table of tangents, but they were able to find these strictly geometric ratios in other, more complicated ways. Moreover, it is likely that Aristarchus realized that he had inaccurately estimated the crucial angle E, which could not easily be measured. A glance at the table shows that for large angles E, a small error makes a very large difference in the resulting ratio. The true value of E differs from 90° by less than one-sixth of a degree, so that E is more than $89\frac{5}{6}°$, and tan E is about 390, the true ratio of the distance of the sun to the distance of the moon from the earth. Even though Aristarchus's calculation was off by a factor of 20, his method was sound and his results revised upward the Greek estimates of the size of the universe by an enormous amount.

You will have noticed that Aristarchus's simple method gives only a ratio. To determine the actual distance to the sun, he needed to know the actual distance to the moon. By observing the time it takes for the shadow of the earth to cross the moon during a total eclipse of the moon, we can estimate this distance very accurately. Although Hipparchus, some 100 years after Aristarchus, used this method to come within 1% of the value we know today, the rougher estimates already available to Aristarchus were sufficiently accurate for his purposes.

Before describing this method, we need to remind ourselves of two simple facts from Euclidean geometry. In Euclid's treatise, facts about congruence appear at the very beginning. It is only after Euclid introduces proper-

ties of parallel lines that he is able to prove the fundamental result that the sum of the angles of a triangle is 180° (half of a complete circle). (This property of triangles—that all of them have the same angle sum, which is 180°—is one that dramatically distinguishes Euclid's geometry from *non-Euclidean geometries,* which are discussed in Chapter 16.)

A simple property of circles is also used in the derivation, namely, that the arc of a circle **subtended** by, or opposite, an angle at its center is proportional to the radius of the circle. In Figure 15.9, the arc subtended by angle A on the large circle is twice as long as the arc on the small circle because the radius of the large circle is twice that of the small one. When we measure the angle in degrees, the factor we have to multiply by to get the arc length is $\pi/180$ (about $\frac{3}{180}$, or $\frac{1}{60}$). Thus, for example, if a circle has a 10-foot radius, an angle of 90° at its center subtends (includes) an arc of $(\pi/180) \times 90 \times 10$, or approximately 16 feet.

Let's return to our main story: Aristarchus's measurement of the distance of the sun and the moon from the earth. Using the facts that the angle sum in a triangle is equal to a straight angle and that the arc subtended by a central angle of a circle is $\pi/180 \times$ radius $\times$ angle, we

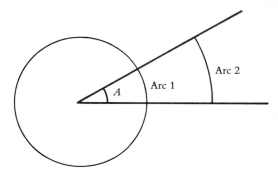

Figure 15.9 Arc 2 is twice as long as arc 1 because its circle has twice the radius of the other circle.

can understand the following calculation of the distance of the moon from the earth. We here combine Aristarchus's own method, a similar method used by Hipparchus a century later, and modern notation.

By observing the amount of time it takes the earth's shadow to cross the moon, Aristarchus knew that the diameter of this shadow was about two times the diameter of the moon, as shown in Figure 15.10.

Because the moon's and sun's discs are both about the same size in the sky and because both are about $\frac{1}{720}$ of the whole circumference of the circle they trace through the sky, the angles at C and D are easily found: $C = \frac{1}{1440}$ and $D = \frac{1}{720}$ of a complete circle. Hence $C + D = \frac{1}{1440} + \frac{1}{720} = \frac{1}{480}$ of a complete circle $= 0.75°$. Now, $A + B + E = C + D + E$ because $A + B + E$ is the angle sum in triangle AEB and hence is equal to the straight angle $C + D + E$. Because $A + B$ is equal to $C + D$, we know it is also $0.75°$. Moreover, angle A is very small compared to B (since the sun is much farther away than the moon), so angle B itself is approximately $0.75°$.

For angles as small as B, line segment TE is essentially equal to the arc TE of the circle with center at B and radius BT. Because the arc TE corresponding to angle B is proportional to the radius BT, we can write

$$TE = \frac{\pi}{180} \times BT \times B$$

Solving for angle B, we have

$$B = \frac{180}{\pi} \times \frac{TE}{BT}$$

Substituting $B \approx 0.75°$ ($\approx$ means "approximately equal to") and rearranging, we get

$$BT \approx \frac{180}{\pi} \times \frac{TE}{0.75} \approx 80 \times TE$$

That is, the distance from earth to moon, by this simple calculation, is about 80 earth radii.

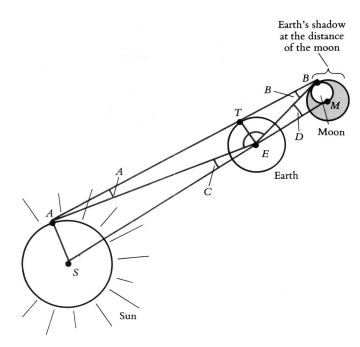

Figure 15.10 Aristarchus devised an ingenious method for determining the ratio between the moon's distance and the radius of the earth from the average duration of a lunar eclipse and the length of the month. His method resulted in an estimate for the moon's distance of 80 earth radii.

In fact, the distance is about 60 earth radii, a result that we get from only slightly more refined versions of this same calculation.

Once the distance to the moon has been calculated, the ratio of the sun's distance to the moon's distance (which Aristarchus thought to be about 19, but which is in fact nearly 400) yields the distance to the sun in earth radii. Combined with the still earlier calculation of the earth's radius, this gives the distance to the sun, a value which we now know to be 92,956,000 miles. Using the same elementary ideas, we can calculate the distances to and between other planets in the solar system.

In summary, a few simple tools of elementary geometry, combined with imaginative questions and insights about the possible structure of the universe, led to the calculation of inaccessible distances long before the development of sophisticated equipment. In more modern times, mathematicians have tried to understand the relation of Euclid's geometry to the physical world of light rays and the pull of gravity. These attempts led directly to the theory of relativity, which is based on a geometric model of the universe that includes time as one of the coordinates.

REVIEW VOCABULARY

Circumference The distance around a circle; for a sphere, the distance around a circle that passes through the poles.

Congruent Two geometric figures are congruent if they have the same shape and the same size. In effect, they are the same figure in different positions.

Corresponding angle When two parallel lines are cut by a third line (sometimes called a *transversal*), the angles formed match up; ones on the same side of the third line which are equal are called corresponding angles.

Elements Euclid's compilation and organization of the geometric and arithmetic knowledge of his time. Most high school geometry texts are strongly influenced by this work.

Hypotenuse of a right triangle The side opposite the right angle.

Legs of a right triangle The two sides adjacent to the right angle (in other words, the two sides that aren't "slanted").

Measure of an angle The size of an angle, as measured in degrees or other units of measure.

Pythagorean theorem "The sum of the squares on the two short sides of a right triangle is equal to the square on the long side." A fundamental tool for calculating distances, as in surveying.

Similar Two geometric figures are similar if they have the same shape, but not necessarily the same size — like the relationship between a photograph and an enlargement of the same photograph.

Subtended An angle at the center of a circle is said to subtend the arc of the circle that it includes, and the arc is said to be subtended by the angle.

Tangent of an angle The tangent of an angle in a right triangle is the ratio of the length of the side opposite the angle over the length of the side adjacent to the angle. (Two right triangles have to be similar if they have another pair of corresponding angles equal; so the tangent of an angle doesn't depend on what right triangle we pick to calculate it.)

Transversal A third line cutting across two parallel lines.

Vertical angles Two line segments that cross form two pairs of opposite angles, which are usually called vertical angles.

SUGGESTED READINGS

Heath, T.L.: Introduction to Books I and II, in *The Thirteen Books of Euclid's Elements,* vol. 1, Dover Publications, New York, 1956.

Jacobs, Harold: *Geometry,* 2d ed., W. H. Freeman, 1987, pp. 364–374, 390–398, 404–408. Similar triangles, the Pythagorean theorem, and the tangent ratio.

Layzer, David: *Constructing the Universe*, Scientific American Library, W. H. Freeman, New York, 1984.

van der Waerden, B.L.: *Science Awakening*, Science Editions, Wiley, New York, 1963.

EXERCISES

1. Triangles are *congruent* if they have the same shape *and* size. The corresponding parts (sides and angles) have the same measures. An *equiangular* triangle is one in which all three angles are the same size. Are all equiangular triangles congruent? Give an appropriate reason for your answer.

2. Does a diagonal of a square separate the square into two congruent triangles? Give an appropriate reason for your answer.

For Exercises 3 to 6, refer to the following: Two triangles are congruent if:

♦ The three sides of one triangle are congruent, respectively, to the three sides of the other triangle (*SSS*).

♦ Two sides and the included angle of one triangle are congruent, respectively, to two sides and the included angle of the other triangle (*SAS*).

♦ Two angles and the included side of one triangle are congruent, respectively, to two angles and the included side of the other triangle (*ASA*).

3. In the figure to the left below, ∠1 is congruent to ∠2, and line segment *AB* is congruent to line segment *AC*. Is △*ABD* congruent to △*ACD*? Give an appropriate reason for your answer.

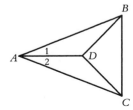
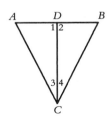

4. In the figure to the right above, ∠1 is congruent to ∠2, and ∠3 is congruent to ∠4. Is △*ACD* congruent to △*BCD*? Give an appropriate reason for your answer.

5. In the figure at the left below, △*ABC* is congruent to △*DCB*. List all of the pairs of corresponding parts that are congruent.

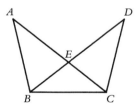
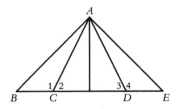

6. In the figure to the right above, ∠1 is congruent to ∠4, ∠2 is congruent to ∠3, and side *AB* is congruent to side *AE*. Is △*ABC* congruent to △*AED*? Give an appropriate reason for your answer.

7. A right triangle has legs of lengths 6 centimeters and 8 centimeters. Determine the length of its hypotenuse.

8. A right triangle has a leg of length 12 feet and a hypotenuse with length 13 feet. Determine the length of the other leg.

9. A 26-foot ladder is placed against a building, with the foot of the ladder 10 feet from the base of the building. How far above the ground does the top of the ladder lean against the building?

10. The Great Pyramid is no longer as high as it was in Thales's time. A modern measurement using the same 10-foot stick with a 16-foot shadow would find a pyramid shadow of only 720 feet. How high is the pyramid now?

11. Two smaller pyramids near the Great Pyramid were, in Thales's time, 471 feet and 215 feet tall, respectively. What lengths of shadows would they have cast when Thales's 10-foot stick was casting its 16-foot shadow?

12. What is the main practical difficulty in the way Thales measured the height of the Great Pyramid (short of getting a passport or the expense of traveling to Egypt)?

13. Suppose a 4-foot stick casts a 5-foot shadow at the same time that a pine tree casts a 50-foot shadow. How tall is the pine tree?

14. A 6-foot stick casts a 10-foot shadow at the same time that a tree casts a 120-foot shadow. How tall is the tree? What is the distance from the top of the tree to the tip of its shadow?

15. Suppose a woman notices that when her shadow is exactly the same length she is, the shadow cast by a neighboring building is 100 feet long. How tall is the building? (The earliest commentators say that this was the problem actually solved by Thales, not the slightly more complicated case requiring ratios.)

16. A man wishes to determine the height of the tree pictured in the figure below. He stands a yardstick vertically on the ground at *D*, 33 feet from *A*. From the ground, he then determines the point *B* on the ground such that the points *C*, *E*, and *B* are collinear. He next determines that the measure of the line segment *DB* is 6 feet. Determine the height of the tree.

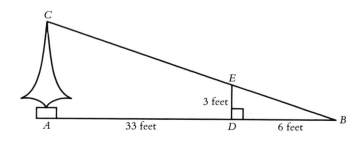

17. In the accompanying figure, △MNP is similar to △MRS. The lengths of some line segments are indicated. Determine the value of x.

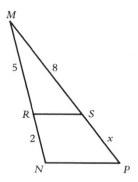

18. Two angles of △ABC have measures of 71° and 46°. Two angles of △DEF have measures of 46° and 63°. Determine if the two triangles are similar.

19. Consider △ABC in the accompanying figure. D is the midpoint of side AB. A line segment is drawn through D parallel to side AC and intersecting side AC at the point E. Is E the midpoint of AC? Give an appropriate reason for your answer. (Use the properties of similar figures that corresponding angles have the same measures and that the measures of corresponding sides are proportional.)

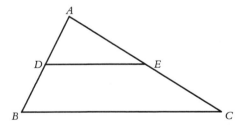

20. In the figure on the left below, determine the $m(\angle 1)$, the measure of $\angle 1$.

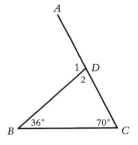
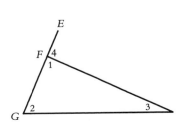

21. In the figure on the right above, determine $m(\angle 4)$, the measure of $\angle 4$, in terms of the measures of $\angle 2$ and $\angle 3$.

22. In the figure on the left below, determine the value of y.

   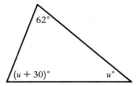

23. In the figure on the right above, determine the value of u.

24. In $\triangle ABC$, the measure of $\angle A$ is twice the measure of $\angle B$, and the measure of $\angle C$ is three times the measure of $\angle B$. Determine all three measures.

25. In $\triangle DEF$, the measure of $\angle E$ is equal to the measure of $\angle F$. The measure of $\angle D$ is 10° less than the measure of $\angle E$. Determine all three measures.

26. Explain why two right triangles that have two corresponding angles equal (apart from their right angles) are similar triangles.

27. From Table 15.1, what is the angle of inclination of the sun (to the nearest 10°) when a 10-foot stick has a 16-foot shadow? Use a calculator to refine your answer, by calculating several nearby values, and find the angle of inclination to the nearest 1°.

28. An earlier calculation of the circumference of the earth was made using the same technique Eratosthenes used. The only difference was that the angle was measured at Lysimachia [now near Gelibolu (formerly Gallipoli), Turkey] instead of at Alexandria. This angle was found to be $\frac{1}{15}$ of a complete circle, or 24°. It was thought that Lysimachia was 20,000 stadia straight north of Syene. Using these figures, what would be the circumference of the earth, in stadia? In miles?

29. The calculation in Exercise 28 leads to a crudely accurate estimate of the circumference of the earth, off by less than 20%. But there was a serious error: in fact, the two cities are only 1180 miles apart. Using 25,000 miles as the circumference of the earth, and assuming that Lysimachia was directly north of Syene, what should the angle have been instead of 24°?

30. Eratosthenes was also incorrect in assuming that Alexandria is 500 miles directly north of Syene (Aswan). Alexandria in fact is 490 miles north of Aswan (pretty close, Eratosthenes), but Aswan is 190 miles farther east and the cities are actually 520 miles apart. Regarding the situation of Exercises 28 and 29, Syene (Aswan) is 1130 miles south of Gelibolu but 390 miles farther east. How much does it matter that these cities are not in a straight north-south line?

31. *The American Heritage Dictionary*, Second College Edition (Houghton-Mifflin, Boston, 1982), gives a different estimate for the length of a Greek stadium: 185 meters. (The dictionary actually says "185 kilometers, or 607 feet"; as noted in an exercise in Chapter 13, "kilometers" should be "meters.") Based on this length for a stadium, what was Eratosthenes's estimate for the circumference of the earth? What was the percentage error compared to the true circumference?

32. The angle subtended by the sun is 0.5°. Using the fact that the sun is 92,956,000 miles from the earth, find the radius of the sun in miles.

Modern measurements of astronomical distances

Quantity	Symbol	Measurement (miles)
Radius of sun	s	432,000
Radius of earth	e	3,963.5
Radius of moon	l	1,080
Earth to sun (center to center average distance)	S	92,956,000
Earth to moon (center to center average distance)	L	238,857

33. The following figure shows a simple and practical method for finding the radius RM of the moon once the distance to the moon ER is known: Assume that $ER = $ 238,857 miles. The angle subtended by the whole moon, as seen by an observer at E on the earth, is 0.5°; hence, the angle E subtended by half the moon is 0.25°. Given that the tangent table says $\tan 0.25° = 0.00436$, find RM.

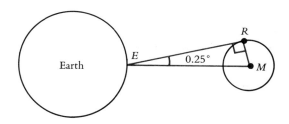

34. The sizes of the disks of the sun and moon as viewed from earth are nearly the same — a marvelous coincidence. The sun is 92,956,000 miles from earth and 432,000 miles in radius. The moon is (on average) 238,857 miles from earth; what is its radius?

35. We saw earlier how Eratosthenes was able to compute the circumference and hence the radius of the earth. From this fact, we can easily determine the distance from the earth to the moon:

a. In the following figure, A and B denote two points that measure 500 miles apart on the earth's surface. Compute the measure of angle AOB.

b. Compute the measure of angles OAB and OBA.

c. AH and BH represent the earth's horizontals at A and B; they are therefore tangent to the earth at these points. Use this fact to explain how to calculate the measures of angles ABH and BAH.

d. Simultaneous observations of the moon are made from points A and B so that angles MAH and MBH are determined. Given these angles, you can easily determine the measures of angles MAB and MBA. Now, explain how to calculate the measure of angle AMB.

e. Finally, explain how to calculate the distances from points A and B to the moon.

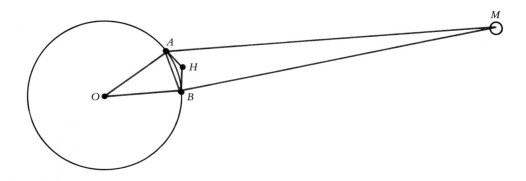

36. Assuming that the necessary measurements could actually be carried out, the following figure suggests a conceptually simple way of finding the distance of the moon from the earth. An observer at P sees the moon M directly overhead at exactly the same time as an observer at Q sees the moon right on the horizon. The two observers will be nearly a quarter of the way around the earth from each other. In fact, the central angle E is 89.07°. With a calculator you can find that tan 89.07° = 61.60295. Question: Assuming the earth's radius is 3963.5 miles, what is the distance PM (from the earth to moon) in miles? (Note that $PM = EM - 4000$ and that EM can be assumed to be equal to QM.)

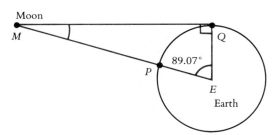

37. If we want to find the distance of the sun from Venus by the methods used to find the distance from the earth to the moon or the earth to the sun, we would have to be on Venus. Because this is not possible, we can use a more subtle and practical method. We can think of earth and Venus as points E and V moving in circular orbits around the sun. Notice in the following figure that the angle at E varies as the two planets travel around the sun and that it reaches its maximum value when the angle at V is a right angle.

Suppose we observe this angular separation of Venus and the sun throughout the year and find that the maximum value of angle E is $47°$. In right triangle EVS, we know $\tan 47° = VS/EV$. However, because we don't know either VS or EV, the tangent table is of no help. But another table, the *sine table*, would give the ratio VS/ES for various values of E. Using the value $\sin 47° = VS/ES = 0.73135$, find the distance of Venus from the sun to the nearest million miles.

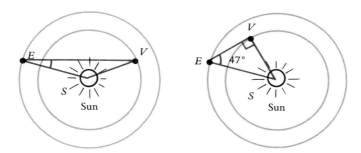

38. Hipparchus, who lived after Aristarchus, used an eclipse of the sun to estimate the distance to the moon. In this exercise we use a simpler calculation with his data to estimate the *radius* of the moon instead. As related later by Pappus, "Hipparchus starts from this observation: there was an eclipse of the sun which was exactly total in the region about the Hellespont [at Gelibolu in the Dardanelles, the strait that connects the Aegean Sea with the Sea of Marmara], no portion of the sun being seen, whereas at Alexandria in Egypt about four-fifths only of its diameter was obscured." The figure below shows the situation. Following the ancients, we let L be the distance from the earth to the moon, S the distance from the earth to the sun, and l and s the radii of the moon and sun, respectively. Note that angles A and P are right angles and that triangles AQH and KQP have equal vertical angles at Q, so these two triangles are similar. Write down an equation for proportional sides, and approximate $S - L$ by S; use the fact that the disk of the moon exactly covers the disk of the sun to replace L/S by its equal, l/s; and arrive at an equation for l in terms of AH. Finally, use the fact that the north-south distance AH between A and H is 640 miles to calculate l.

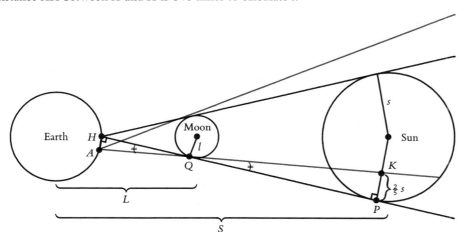

39. The situation of Exercise 38 required two observers a great distance apart. But if you by yourself observe a total eclipse of the sun, there is an easy way to estimate the distance to the moon. What you need to do is to *time* one of the next total eclipses that comes your way, from the first moment that the moon begins to move across the sun until the last instant before it is all the way across. In the United States, the eclipse of January 4, 1992, will be visible in part of California, and the one of May 10, 1994, will be visible in a band across the United States. Since you don't want to have to wait for one of those events to do this exercise, let's say that the elapsed time is 125 minutes. The figure below shows the situation. Although the earth is moving during the eclipse (in fact, it moves more than 125,000 miles!), the earth and sun stay in the same relative positions and the moon is carried along. In addition, the moon moves in its own orbit. How far? It moves two moon diameters in distance and about 1° of the 360° of its orbit.

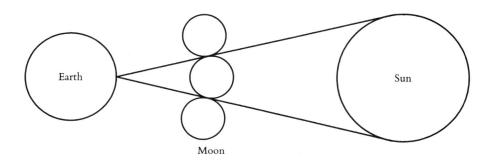

a. Write down two expressions for the circumference of the (assumed circular) orbit of the moon, one in terms of the distance from the earth to the moon and the other in terms of the radius of the moon. Use 1080 miles as the radius of the moon to arrive at an estimate of the distance to the moon.

b. You will notice that in part **a** we did not yet use any information from the eclipse! All we used is that the two moon diameters is about 1° of its full orbit, and it is this rough approximation that is the weak point in the calculation in part **a** and the explanation for the result being too large. We can use our timing of the eclipse to make a more refined estimate. We need to know one other easily measured quantity, the time it takes for the moon to complete one orbit (as seen from the earth); the ancients knew this quantity quite accurately. Modern measurements give 29 days, 12 hours, 44 minutes, and 2.8 seconds as the observed time from one new moon to the next. Convert this measurement to minutes, calculate what fraction of it 125 minutes is, arrive at a new estimate of the angular measure of the diameter of the moon, and make a new estimate of the distance to the moon.

40. A timing of a total eclipse of the moon also can be used to make a rough calculation of the distance to the moon. You may have observed the total eclipse of the moon of August 16, 1989, which was visible throughout most of the United States. In central Brazil, the moon was exactly overhead during this eclipse, which lasted 220 minutes there. The figure below shows an idealized situation: the observer is at *0*, on the side of the earth away from the sun, the moon is directly overhead and the sun is on the other side of the earth.

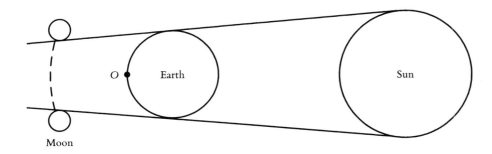

a. During the course of the eclipse, the moon moves approximately one earth diameter plus one moon diameter. Use this fact, the data in Exercise 39, and the fact that the radius of the earth is 3963.53 miles (as measured at the equator) to give an estimate of the distance to the moon.

b. In fact, the estimate in part **a** is way too large. The main reason is that the moon moves substantially *less* than one earth diameter plus one moon diameter, as the figure in fact suggests. For the sake of improving the model and method of part **a**, let's assume we know (by some other means) that the average distance from the sun to the earth is 92,956,000 miles and the sun's radius is 432,000 miles. Then the slant line through the top of the sun and the top of the earth dips 428,000 miles over a distance of 92.9 million miles, so it is dipping 1 mile for every 217 miles across. As it proceeds L miles farther to the moon, it dips another $L \times \frac{1}{217}$ miles. So in fact the moon during eclipse will travel one earth diameter ($2e$) plus one moon diameter ($2l$) minus $2(L \times \frac{1}{217})$. Use this new estimate of how far the moon travels to get a new estimate of the distance from the earth to the moon.

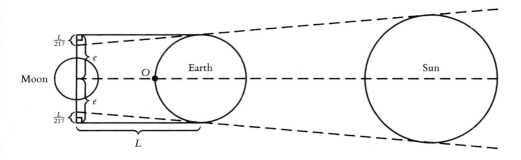

c. Even the new model gives a result that is too large. Can you think of defects in the new model?

d. If the moon were far enough away from the earth, it could never be completely eclipsed. The moon would have to be a distance, M far enough away that the disk of the earth just barely fails to shade the moon, so that a ray from the top of the sun would graze the earth and pass on to graze the moon. That ray and the line through the centers of the bodies intersect at a point z in space, forming several similar triangles. Use the idea from part **b** to estimate how far the moon would have to be.

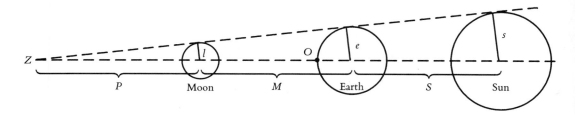

e. What are the observational difficulties of timing a lunar eclipse?

·16·

Measuring the Universe with Telescopes

Modern science arose in the seventeenth century with the work of Galileo Galilei (1564–1642), Johannes Kepler (1571–1630), and Isaac Newton (1642–1727). The first distinguishing characteristic of the new science was its experimental method. The second was its quantitative character. The physics of Aristotle (384–322 B.C.), which still held sway in the intellectual world of Galileo's time, gave only qualitative explanations for physical phenomena; for example, the old science tried to explain why an apple falls downward from a tree. In contrast, Galileo was more interested in how; he dismissed bare qualitative explanations as "fantasies" that are "not really worthwhile." He sought instead a mathematical description of such events as the motion of a freely falling body.

The third and crowning characteristic of modern science was its striving for a mathematical theory, which was highlighted in the work of Isaac Newton. A mathematical theory enables us to make predictions and—often incidentally—explains a wide variety of related phenomena. For example, Newton's work coordinated Kepler's observed laws of planetary motion with the laws of mechanics that appeared to govern terrestrial phenomena. His mechanics explained much about gravitation and ocean tides as well as planetary motion.

FROM GREECE TO GALILEO

Galileo began his assault on astronomy in 1609, after he learned that a Dutch lens maker had discovered how to achieve great magnification by arranging two lenses in a special way in a long tube. This, of course, was the invention of the telescope. Galileo then proceeded to build his own telescopes. He first achieved a threefold magnification. Then, after mastering the problems of grinding and polishing lenses and experimenting with the arrangement of the lenses in the tube, he was able to construct a telescope that magnified approximately 33 times (Figure 16.1). These instruments, although modest by today's standards, revealed some astonishing astronomical sights to Galileo.

Turning the telescope to the moon, he saw immediately that the surface of the moon had mountains and valleys and was not the "perfect" sphere of accepted Aristotelian theory. Later, Galileo discovered sunspots, showing that the sun, too, was not "perfect." These observations shocked his contemporaries by contradicting long-held beliefs about the nature of the universe.

The prevailing conception of the universe in Galileo's time derived primarily from the Greek philosopher Aristotle and the Alexandrian astronomer and geographer Claudius Ptolemy (second century A.D.). Briefly, the Aristotelian and Ptolemaic view maintained that the earth is the immovable center of the universe, which is a large celestial sphere that rotates about the earth and on which all the stars are fixed. Referred to as a *geocentric,* or earth-centered, theory, this view had prevailed for well over a thousand years; it had the support of almost all academicians and the official support of the Catholic Church, of Martin Luther, and of Jewish leaders.

An alternative to the geocentric theory had already been proposed by the ancient Greek astronomer Aristarchus of Samos in the third century B.C., but his work was largely ignored. Aristarchus held to a *heliocentric* theory, placing the sun at the center of the universe. This theory was revived in a modified form approximately 1800 years later by a young Polish student, Nicolaus Copernicus (1473–1543). Copernicus argued that all the planets, includ-

(a)

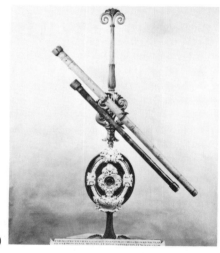

(b)

Figure 16.1 (a) After inventing the telescope, Galileo made some astonishing astronomical discoveries. (b) Two of Galileo's telescopes and the lens from another. [Photo (b): Scala/Art Resources.]

ing the earth, moved in concentric spheres, with only slight modification, about the sun.

Perhaps most devastating to proponents of the geocentric theory was Galileo's discovery of four moons revolving about the planet Jupiter (Figure 16.2). If Jupiter, a planet, possesses moons, then the earth, too, might also be a planet. Moreover, these newly discovered moons of Jupiter were not circling the earth,

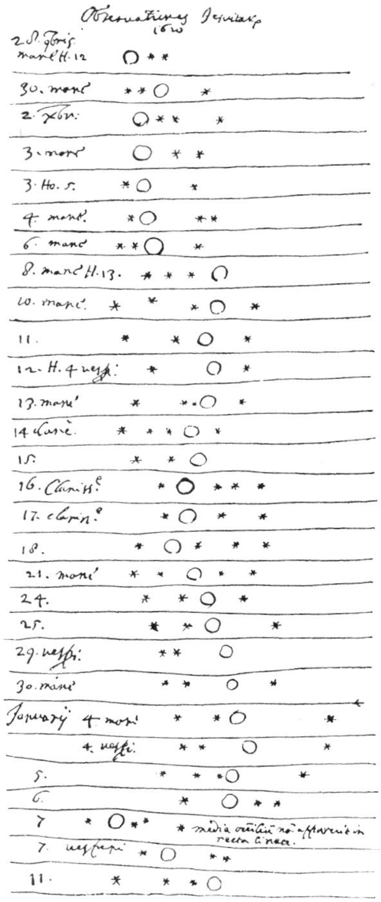

Figure 16.2 Galileo's drawings of Jupiter and its moons. He discovered the moons when noticing that four shining objects moved back and forth across Jupiter from one night to the next. [Yerkes Observatory.]

the presumed center of the universe around which all bodies should revolve.

IMPROVING THE TELESCOPE

A half-century after Galileo built his first telescope, Sir Isaac Newton turned his genius to

improving the instrument. Galileo's was a *refractor telescope,* one that bent light rays by means of lenses. Such instruments have two shortcomings: (1) the glass used for the lenses must be of high quality and free of flaws in order to minimize distortions, and (2) the bending of the light rays separates the colors contained in white light, introducing a distortion called *chromatic aberration.* The first of these problems is eliminated, and the second reduced, by using a mirror instead of a lens for the light-gathering work of the telescope. It was this idea that Newton exploited when he constructed the first *reflector telescope,* using a mirror to replace the light-gathering lens of the refractor telescope.

As a student and great admirer of Greek geometry, and himself one of the most profound geometers in all history, Newton knew that the best shape for a light-gathering mirror would be a parabola, a shape related to the well-known curve discovered 2000 years earlier. Parabolas possess a remarkable **reflection property,** illustrated in Figure 16.3, that makes them especially suitable. At the point V, where the parabola crosses its axis of sym-

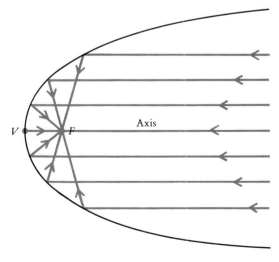

Figure 16.3 Parallel light rays reflect off a parabola and meet at its focus F.

metry, the curve is "sharpest." Point V is called the **vertex of the parabola** and the axis of symmetry simply the **axis of the parabola.** Lines parallel to the axis that come in from afar meet the parabola (on its concave side) at some acute angle and then "bounce off" the curve at the same angle. The parabola's remarkable feature is that all the bouncing-off lines pass through a single point F called the **focus of the parabola.** If the parabola is a reflecting surface, then the lines parallel to its axis can be regarded as light rays radiating from a distant heavenly body; these light rays reflect off the surface and accumulate, or focus, at the focus of the parabola, as in Figure 16.3.

Because of the difficulties of grinding a parabolic mirror, Newton compromised and constructed a spherical mirror instead. Such a mir-

ror, whose surface is a portion of a sphere, gathers light from afar and tends to accumulate it at the center, or focus, of the sphere. But such a light-gathering mirror presented another problem: the observer would have to be placed at the center of the sphere — directly in front of the mirror — thus blocking all the incoming light.

To make his telescope, Newton placed the spherical mirror at the bottom of a cylindrical tube so that the mirror would reflect the incoming rays of light onto one image point, the focus. To view the image, or focused light, from outside the tube, Newton placed a small plane mirror close to the focus in order to reflect the image to the side of the telescope, where he made a small hole in the cylindrical housing (see Figure 16.4 and Color Plate 15).

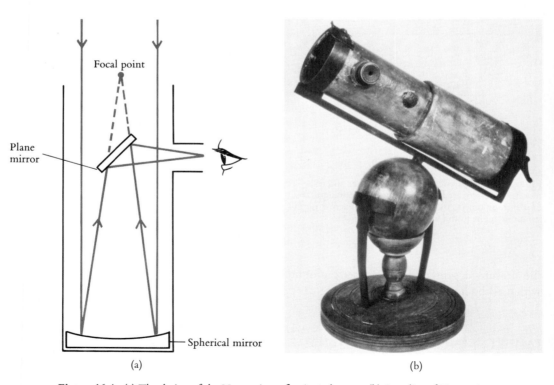

(a)   (b)

Figure 16.4 (a) The design of the Newtonian reflecting telescope. (b) A replica of Newton's reflecting telescope. [The Granger Collection.]

Less than four years after Newton built his first telescope, a report came to the French Academy that someone else, Guillaume Cassegrain, had invented still another reflecting telescope. Cassegrain had succeeded in grinding and polishing a large concave parabolic mirror to gather light, together with a smaller convex hyperbolic mirror to focus it (see Color Plate 15). Using both convex and concave mirrors makes distortions tend to cancel. The main advantage of the Cassegrain design is the short physical length of the telescope compared to its long focal length: a long focal length is needed to focus distant objects accurately, while the small size of the telescope is crucial in some applications. For example, the NASA Hubble Space Telescope launched into orbit in 1990 uses the Cassegrain system, with a parabolic mirror 94 inches in diameter. (See Spotlight 16.7, p. 463.)

lows a straight line to the earth; but the graph of distance fallen versus time forms half of a parabola (see Figure 16.5). The height y above the ground of an object t seconds after it is dropped is given by

$$y = h_0 - \frac{1}{2} g t^2$$

where h_0 is the height from which it is dropped, the minus sign indicates that the object gets lower with time (it falls down, not up), and g stands for the acceleration due to gravity. In the U.S. Customary System of units, $g = 32$ feet per second per second; in the metric system, $g = 9.8$ meters per second per second. We take $t = 0$ to be the instant when the object is dropped. At the end of the first second, the formula gives $y = h_0 - 16$ feet, indicating that the object has fallen 16 feet from

OTHER APPLICATIONS OF THE PARABOLA

Parabolas have been observed and used by humans from time immemorial. Projectiles, such as baseballs and bullets, follow parabolic paths (modified somewhat by air resistance), as does the water from a garden hose or a fire hose. When you cup your hands in front of your mouth to project your voice, or around your ears to help your hearing, the ideal shape is a parabola.

EXAMPLE: Galileo and the Leaning Tower of Pisa. Heavy objects fall with the same speed and acceleration as light ones (except for the effect of air resistance). A fanciful legend tells us that Galileo established this fact by dropping objects from the top of the Leaning Tower of Pisa. How are parabolas involved? An object dropped straight down follows

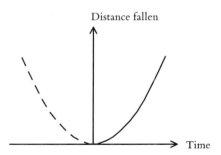

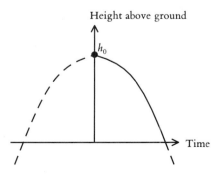

Figure 16.5 Parabolic graphs for an object falling from height h_0.

height h_0 since the drop. At the end of the second second, $y = h_0 - 64$, and the object is 64 feet below its original location. It has fallen 48 feet during the second second, compared to 16 feet during the first second; it is accelerating.

We can use the formula to determine how long it takes objects to fall from the Leaning Tower of Pisa. The Tower is 179 feet high, so $h_0 = 179$. The object hits the ground when $y = 0$; so the time it takes to reach the ground is the value of t that solves $0 = 179 - 16t^2$. Solving for t, we get $t = \sqrt{179/16} \approx 3.3$ seconds. Actually, we get $t = \pm\sqrt{179/16} \approx \pm 3.3$. The value $t = -3.3$ seconds corresponds mathematically to a point on the other half of the parabola and physically to throwing the object up from the ground—launched with the same speed as it later hits the ground—3.3 seconds before the moment of dropping.

EXAMPLE: Son of Galileo. According to an even less reliable legend, Galileo's son Claribel, who earned a mediocre living playing in a heavy-metal rock band, admired his father and wanted to pay him a tribute. (Claribel also needed more notoriety to help his own career along.) So for Galileo's birthday, Claribel booked a gig at the Leaning Tower and punctuated the act by throwing—not dropping, *throwing*—water balloons, guitars, and bathtubs from the top of the Tower. The anxious fans below were reassured once Galileo explained that all of the objects followed quite predictable parabolic paths, and that as long as Claribel threw the items in a somewhat upward direction, the fans had more than 3.3 seconds to get out of the way.

How much longer than 3.3 seconds? you ask. It's a good thing you asked. . . . The answer depends on the velocity and direction with which Claribel launched the object. We can analyze the problem by separating out a horizontal speed component s_h and a vertical speed

component s_v. Then the motion of the object is described by

$$y = -\frac{1}{2} gt^2 + s_v t + h_0 \qquad x = s_h t$$

where, as before, y is the height above the ground, and x is the distance away from the Tower.

To make these equations concrete, let's concentrate on the water balloons, which Claribel was launching at a modest $s_v = 24$ feet per second (about 16 mph) and $s_h = 30$ feet per second (about 20 mph) (the two combine according to the Pythagorean theorem, so that the total launch speed was $\sqrt{24^2 + 30^2} = 38$ feet per second = 26 mph). We can describe the motion of a balloon by $y = -16t^2 + 24t + 179$, $x = 30t$. To find out how long it takes to hit the ground, we have to find the t for which $y = 0$, that is, solve $0 = -16t^2 + 24t + 179$. The solutions can be found by the quadratic formula (which Claribel didn't remember) or by trial and error; they are $t = 4.2$ and $t = -2.7$. So the balloons took 4.2 seconds to reach the ground, by which time they were $x = 30(4.2) = 126$ feet from the Tower.

By solving the second equation for t, getting $t = x/s_h$, and substituting this expression into the first equation, we get

$$y = -\frac{1}{2}\left(\frac{g}{s_h}\right) x^2 + \left(\frac{s_v}{s_h}\right) x + h_0$$

which in our case specializes to

$$y = -\frac{16}{900} x^2 + \frac{24}{30} x + 179$$

the equation of the parabola that the balloons followed (see Figure 16.6).

Although the focal property of parabolas had been described by Apollonius (ca. 260–190 B.C.), the reflecting telescope appears to be

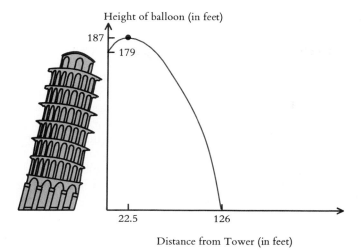

Height of balloon (in feet)

187

179

22.5   126

Distance from Tower (in feet)

Figure 16.6   Graph of the path of Claribel's water balloons.

its first technological application. The key ideas upon which the usefulness of the parabolic mirror rests are the focal property of the parabola and the fact that when light rays reflect off a smooth surface, the angle of incidence equals the angle of reflection (see Figure 16.7).

The conjunction of these two ideas has found a number of applications, including flashlights, searchlights, and the automobile headlight. These all reverse the job of the telescope. Instead of gathering incoming light and bringing it into focus at a point, they have the light source (a bulb) at the focus of a **paraboloid of revolution,** the surface made by rotating a parabola around its axis. The light is reflected outward along rays that are parallel to the axis (see the red rays in Figure 16.8a). Headlights with both high and low beams make double use of the parabola. The light

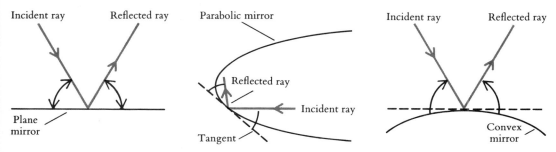

Figure 16.7   When light rays bounce off a smooth surface, the angle of incidence equals the angle of reflection.

source for the high beam is at the focus of the parabola, but the light source for the low beam is to the left of and above the focus. While the high beam reflects rays directly ahead, the low beam sends rays down and to the right, away from oncoming traffic (see the black rays in Fig. 16.8a).

Some fluorescent lamp tubes have housings above them with cross sections that are parabolas, so that the tube forms a line of foci for the different parabolas and the light is reflected straight down all across the length of the lamp. The shape of the surface of the housing is called a **parabolic cylinder.** Just as a normal (circular) cylinder can be thought of as lifting a circle straight up out of the plane, a parabolic cylinder can be thought of as lifting a

parabola straight up out of the plane. Another invention that uses a parabolic cylinder but focuses light for a different purpose is the solar cooker.

Another major application of the parabola's focusing property is the dishlike antenna used for radio telescopes, radar, microwave towers, and surveillance systems. Like the faint light rays caught and focused by the telescope, faint signals are caught by the antenna dish and reflected to the focus, where they are gathered and amplified into a strong signal (see Figure 16.8b).

In addition to its importance because of its focusing properties, the parabola plays a role in architecture. The main cables of a suspension bridge (such as the Golden Gate Bridge in Fig-

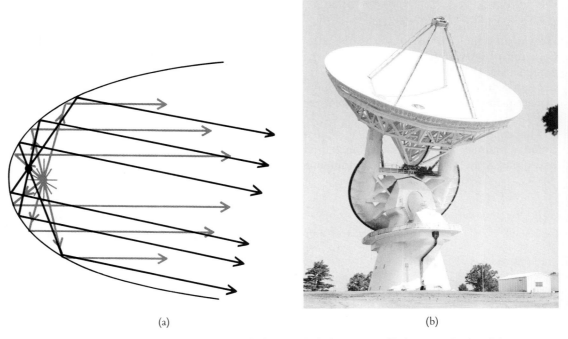

(a)   (b)

Figure 16.8 (a) In an automobile headlight, a parabola directs rays of light outward in parallel lines: straight out, for the high beam source located at the focus, and down and to the right, for the low beam source located above and to the left of the focus. (b) A 140-foot-wide radio telescope at Green Bank, West Virginia. [The National Radio Astronomy Observatory, operated by Associated Universities, Inc., under contract with the National Science Foundation.]

Figure 16.9   The main cables of the Golden Gate Bridge approximate a parabola. [Photo by Bob David, Golden Gate Bridge Highway and Transportation District Archives.]

ure 16.9) approximate a parabola, and many other bridges are supported from below by approximately parabolic arches. The key feature is that the parabolic shape allows support of a uniform horizontal load to be spread out in such a way that there is uniform stress. Did you ever wonder why the largest dinosaurs, such as *Stegosaurus, Brontosaurus,* and others, all seem to have big, humpy backs? They supported their enormous body weight with a spine that was approximately parabolic.

(As an aside, we mention that many common shapes we see that may appear to be parabolas in fact are *not* exact true parabolas. Some, such as telephone and power lines, empty clothes lines, the string going up to a kite, the cross section of a sail filled with wind, and Gateway Arch in St. Louis, are *catenaries*. A catenary is the curve formed by a freely hanging cable or rope supported at the ends. If a clothes line is loaded uniformly with clothes, the shape of the line deforms to a parabola, because of the loading. The ideal design for a roof that is to support only itself would have a cross section that is a catenary; the ideal cross section for one that would need to support a uniform load of snow is a parabola. Most roofs, of course, are either flat or have triangular cross sections, because those kinds of roof are easier to build. Steep A-frame roofs avoid the snow-loading problem by being so steep that buildups of snow cascade off.)

EXAMPLE: Finding the Focus of a Parabola. Given a parabola on paper, you can find the location of its focus approximately by drawing approximate tangents to the parabola and using the reflection property (more details follow in one of the exercises). But if you have the algebraic description of the parabola, in terms of its equation, you can easily find the focus exactly. With the appropriate choice of x- and y-axes so that the axis of the parabola lies along the y-axis, the parabola will be described by the equation $y = ax^2 + bx + c$. You can locate the focus on the axis at a distance $1/(4a)$ from the vertex, on the inside side of the parabola [if a is negative, the distance is $-1/(4a)$]. So, for example, the sample parabola $y = x^2$ has its vertex at the point $(0, 1/4)$, one-fourth unit above the vertex at $(0, 0)$.

CONIC SECTIONS

A parabola is one of several important curves that can be formed by cutting a (mathematical) cone. Imagine a circle drawn on a flat surface, such as a table top, with a line through the center of the circle perpendicular to the surface. Choose a point V on this line, above the table. The surface consisting of all the lines that simultaneously pass through both V and the circle is called a **right circular cone** with vertex V. A mathematical cone differs from a common cone in having two parts: the vertex separates the surface into two **nappes** (see Figure 16.10).

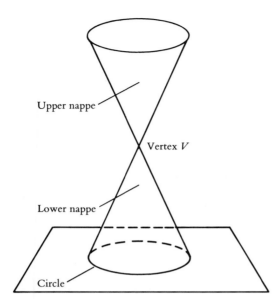

Upper nappe

Vertex *V*

Lower nappe

Circle

Figure 16.10   A cone.

If we slice (intersect) a mathematical cone with a plane, we get a curve called a **conic section.** By changing the angle of the slice, we can see the variety of possibilities for conic sections. The plane we started with — the table top — intersects the cone in a circle. When we tilt the plane a bit, the intersection becomes an **ellipse.** As we continue tilting the plane, the intersection remains an ellipse as long as the plane cuts the one nappe of the cone in a closed curve. However, there comes a point when the cutting plane, while still intersecting only one nappe, no longer intersects the nappe in a closed curve. If the slice is at the same angle as the side of the cone, the intersection is a **parabola.** Finally, by tilting the plane still further, it will cut both nappes of the cone, and the intersection curve is the twin branches of a **hyperbola** (see Figure 16.11). These four

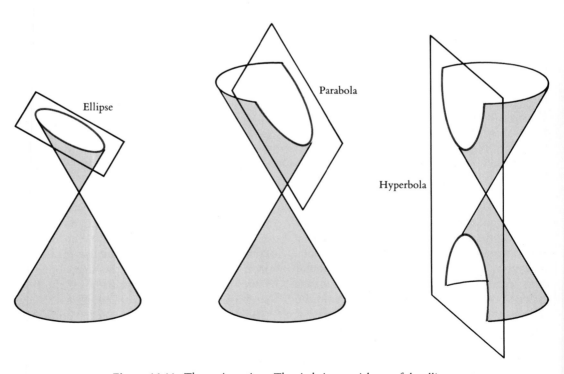

Ellipse

Parabola

Hyperbola

Figure 16.11   The conic sections. The circle is a special case of the ellipse.

curves, the circle, ellipse, parabola, and hyperbola, are the conic sections. As we will see, they *all* have focusing properties.

The ellipse may have first been discovered by the Greeks in connection with sundials, since the tip of the shadow of the *gnomon* (peg) of a sundial follows an elliptic path as the sun moves. The conic sections were probably investigated first by the Greek geometer Menaechmus in the fourth century B.C., and they were apparently studied by other Greek mathematicians, particularly Apollonius, whose studies were remarkably complete. It is no exaggeration to say that we now know only slightly more than he did about the properties of conic sections.

Figure 16.12   Johannes Kepler.

JOHANNES KEPLER

Galileo had discovered the importance of the parabola in making telescopes. But conic sections play an even more fundamental role in the operation of the solar system: planets indeed travel around the sun, but not in the circular orbits hypothesized by Galileo. The man who established that they in fact travel in *ellipses* was Johannes Kepler (1571–1620), a brilliant mathematician with a keen interest in geometry (see Figure 16.12).

Originally, however, Kepler had devised an elaborate mystical theory of the solar system, in which the six known planets were related to the five Platonic solids (see Spotlight 16.1, p. 444). In attempting to establish his mystical theory of celestial harmony, he had to use the ambiguous astronomical data available at the time. He realized that the construction of any theory would require more precise data. Those data, he knew, were in the possession of the Danish astronomer Tycho Brahe (1546–1601), who had spent 20 years making extremely accurate recordings of the planetary positions and the positions of 1000 stars.

Kepler became Brahe's mathematical assistant in February 1600 and was assigned a specific problem: to calculate an orbit that would describe the position of Mars at any time to within the accuracy allowed by observation, which at that time was 4 seconds of arc, or $\frac{1}{900}$ of a degree. Kepler boasted that he would have the solution *in eight days*.

Both the Copernican and the Ptolemaic theories held that the orbit should be circular, perhaps with slight modifications. Thus, Kepler sought the appropriate circular orbits for Earth and Mars. (The orbit for Earth, from which all the observations were made, had to be determined before one could satisfactorily use the data for the positions of the planets.) *After four years,* Kepler found a solution that seemed to fit Brahe's observations. However, on checking his orbits—by predicting the position of Mars and comparing it with more of Brahe's data—he found that one of his predictions was off by at least 8 minutes (= 480 seconds) of arc!

SPOTLIGHT 16.1 Kepler's Model of the Solar System

Kepler, in the *Mysterium Cosmographicum (The Cosmographic Mystery)*, published in 1596 a fantastic cosmological interpretation of the Platonic solids (here translated by Koyré, *The Astronomical Revolution*, p. 146):

Tetrahedron   Cube   Octahedron   Dodecahedron   Icosahedron

The five Platonic solids.

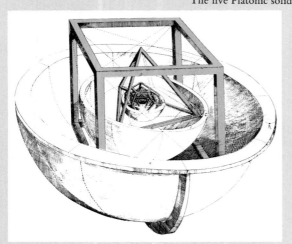

The Earth [the sphere of the Earth] is the measure for all the other spheres. Circumscribe a Dodecahedron about it, then the surrounding sphere will be that of Mars; circumscribe a Tetrahedron about the sphere of Mars, then the surrounding sphere will be that of Jupiter; circumscribe a Cube about the sphere of Jupiter, then the surrounding sphere will be that of Saturn. Now place an Icosahedron within the sphere of the Earth, then the sphere which is inscribed is that of Venus; place an Octahedron within the sphere of Venus, and the sphere which is inscribed is that of Mercury.

This shocking failure led to *two more years* of struggle, in which Kepler finally took the revolutionary step of discarding the long-held conviction that all heavenly bodies move in circular paths (or circular paths modified in some way by the imposition of smaller circles). This decision permitted him to find an accurate solution to the Mars problem and to put forth a new theory of planetary motion. The results of Kepler's six years of research were published in 1609 in his *Astronomia Nova*, in which he announced two of his three remarkable laws (see Figure 16.13 and Table 16.1):

1. *Law of elliptic paths.* The orbit of each planet is an ellipse with the sun at one focus.

2. *Law of areas.* During each time interval, the line segment joining the sun and planet sweeps out an equal area anywhere on its elliptic orbit. (A brief version is: equal areas are swept out in equal times.)

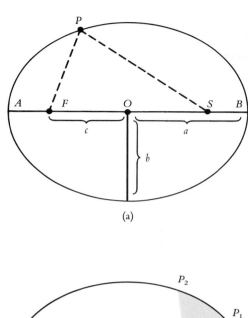

(a)

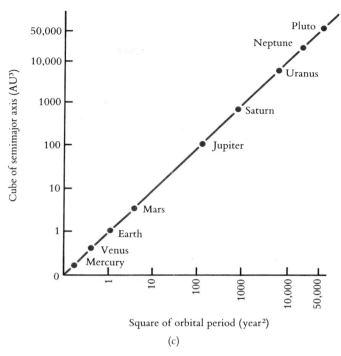

Square of orbital period (year²)

(c)

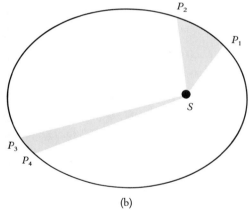

(b)

Figure 16.13 (a) The law of elliptic paths. The orbit of each planet is an ellipse with the sun at one focus. The sum $(PF + PS)$ of the distances from any point P of an ellipse to the two foci F, S is equal to the major diameter AB. (b) The law of areas. The shaded parts are of equal area; thus, the planet takes the same amount of time to move from P_1 to P_2 as to move from P_3 to P_4. (c) The law of times. The points in the graph fall along a straight line, verifying Kepler's discovery that the square of the orbital period equals the cube of the planet's average distance from the sun. (1 AU ≈ 93 million miles.) [(c) is from William J. Kaufmann, III, *Universe*, W. H. Freeman, 1985.]

Kepler's third law was published later and helped Isaac Newton formulate his law of gravity:

3. *Law of times.* The square of the time of revolution of a planet about the sun is proportional to the cube of that planet's average distance from the sun (where "average distance" means one-half the length of the ellipse's longer axis).

TABLE 16.1 A demonstration of Kepler's third law

Planet	Orbital period P (in years)	Average distance from sun (in AU)	P^2	a^3
Mercury	0.24	0.39	0.06	0.06
Venus	0.61	0.72	0.37	0.37
Earth	1.00	1.00	1.00	1.00
Mars	1.88	1.52	3.53	3.51
Jupiter	11.86	5.20	140.7	140.6
Saturn	29.46	9.54	867.9	868.3

Once again, conic sections played a crucial role in the development of science. We now recognize, more than 2000 years later, the incredible genius of the ancient Greeks in identifying and exploring this and other fundamental areas of knowledge. Their work with conic sections developed a subject that is now known to be fundamental to the study of physics, astronomy, architecture, and engineering.

EXAMPLE: The Orbits of the Planets.
Kepler's third law allows us to calculate how far from the sun other planets are. Take, for example, Jupiter, which takes 4332.4 earth days to complete its orbit. Let D_J and D_e be the average distances of Jupiter and the earth from the sun, and let T_J and T_e be the times it takes them to complete one orbit. Then the law says that $T_J^2 = KD_J^3$ and $T_e^2 = KD_e^3$, for the same K. In other terms,

$$\frac{T_J^2}{D_J^3} = K = \frac{T_e^2}{D_e^3}$$

Then substituting $T_J = 4332.4$ days, $T_e = 365.25636$ days, and $D_e = 92,956,000$ miles, we find $D_J^3 = 1.1300 \times 10^{26}$ and $D_J = 483.46$ million miles.

Alternatively, we could do the calculation in terms of earth years and astronomical units, where 1 **astronomical unit (AU)** equals the average distance of the earth from the sun, 92,956,000 miles. Jupiter completes its orbit in 11.861 earth years. The square of this time is $(11.861)^2 = 140.68$. By Kepler's law, the cube root of this number, $\sqrt[3]{140.68} = (140.68)^{1/3} = 5.20$, gives the average distance of Jupiter from the sun, in astronomical units. So Jupiter is (on average) 5.20 times as far from the sun as the earth is, and we can determine the distance in miles by multiplying by 92,956,000.

FURTHER APPLICATIONS OF THE ELLIPSE

The ellipse has many applications beyond the magnificent ones in Kepler's work. We observe an ellipse whenever we look at a circle that is tilted or deformed, such as a round plate, the top of a round glass, or the surface of liquid in a glass.

An unusual application of the idea that an ellipse can result from the flattening of a circle is in paleozoology. In a bed of fossils, the squeezing from the layers above can result in flattened fossils that mistakenly appear to be distinct and separate species; an example is *Ellipsolithe*, which turned out to be just an elliptic distortion of the circular chambered nautilus of Figure 13.8.

The ellipse has applications in mechanics. Elliptic gears in machinery can provide a quick-return mechanism and a slow power stroke (e.g., for heavy cutting) (see Figure 16.14a). Using the same idea, some racing bicycles have circular gears for the rear wheels (like ordinary 10-speed bikes) but an elliptic gear for one of the front gears, thereby allowing the gearing to match the natural cycle of available power in the rider's legs (Figure 16.14b). Ellipses even played a small role in the Battle of Britain in World War II. The British Spitfire fighter's excellent maneuverability and acceleration was due in part to the elliptic profile of its wings and tail (Figure 16.14c).

The most important geometric quality of an ellipse is its **reflection property:** A light ray passing through one focus of an elliptic mirror will reflect off the mirror and pass through the other focus. A visual illustration of the reflection property of the ellipse is provided by an elliptic pool table with a single pocket at one focus (see Figure 16.15): any shot without spin that passes over the other focus will bounce off

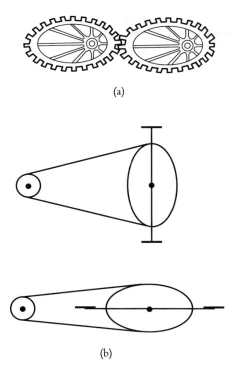

(a)

(b)

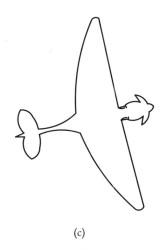

(c)

Figure 16.14 Applications of the ellipse. (a) Elliptic gears. (b) Bicycle with one elliptic gear. (c) British Spitfire airplane.

the cushion directly into the pocket. The reflection property of the ellipse has been used by acoustical engineers in designing whispering galleries, such as those in the Mormon Tabernacle in Salt Lake City and the Capitol building in Washington, D.C. If the shape of the cupola of a gallery or auditorium is elliptic, a weak whisper at one focus may be barely audible — even inaudible — in most of the room, except at the other focus, where the reflections of the whisper are brought together again.

A significant medical application of the reflection property of an ellipse is the use of the *lithotripter* in a noninvasive therapy to break up kidney stones. The patient lies in an elliptic tub of water, placed so that the kidney stone is

Figure 16.15 An elliptic pool table. [From *Inventing, Discovery, and Creativity*, by A. D. Moore. Copyright ©1969 Doubleday & Company, Inc. Reproduced by permission of the publisher.]

located at one focus of the ellipse. A high-energy shock is delivered at the other focus. The shock waves are reflected off the sides of the tub: at all other points, the waves cancel each other out (and the patient feels no pain or ill effects), while the waves concentrate at the other focus to blast the kidney stone. It may take as many as 2000 shocks over a half-hour period to break a stone into sand-sized particles that can pass through the urinary system; but the patient recovers in days, as opposed to up to six weeks if surgery is used instead.

EXAMPLE: Measuring How Skinny an Ellipse Is. As in the case of the parabola, you can use the reflection property and approximate constructions to locate approximately the foci of an ellipse. But it is easy to be exact. An ellipse is customarily described in terms of a, half of the length of its longer axis, and b, half of the length of its shorter axis (see Figure 16.13a). If you now place one end of a compass where the short axis and the ellipse intersect and mark off an arc of length a, the arc will intersect the long axis at the two foci. The distance of the foci from the center of the ellipse is customarily denoted by c, and the Pythagorean theorem gives $b^2 + c^2 = a^2$, or $c = \sqrt{a^2 - b^2}$.

If we think of the ellipse as the orbit of a planet around the sun at one focus (or a moon or satellite around a planet at one focus), we see that the maximum distance of the planet from the sun is $a + c$ and the minimum distance is $a - c$. The average of the two distances is just a. For example, consider the moon in orbit around the earth. The maximum distance of the moon from the earth (center to center) is 238,857 miles and the minimum distance is 221,463 miles. So we have $a = \frac{1}{2}(238,857 + 221,463) = 230,160$ miles, with $c = 8697$ miles, which locates the foci as being about one earth radius above the surface of the earth.

The length c is also involved in measuring how skinny the ellipse is. The technical term for skinniness is **eccentricity,** denoted by e, which is defined as $e = c/a$. Since c is always less than a and greater than or equal to zero, we have $0 \leq e < 1$. For the moon we have an eccentricity of $8697/230,160 = 0.038$.

For the special case of an ellipse that is a circle, we have $a = b$, $c = 0$, and $e = 0$. A definition of eccentricity in more general terms allows us to make it meaningful for parabolas (for all of which we have $e = 1$) and hyperbolas (for which $e > 1$).

THE HYPERBOLA AND ITS APPLICATIONS

It is likely that you most commonly see a hyperbola in the shadow cast by a shaded lamp, as Figure 16.16a. But hyperbolas find their most significant applications in navigation, particularly in the LORAN (LOng RAnge Navigation) and OMEGA navigational systems. LORAN systems use stations with known locations that broadcast a signal simultaneously. A ship observes the time interval between receiving signals from one pair of stations and determines its location as lying on one hyperbola, which is the curve of constant time difference. The ship then does the same with another pair of stations, thereby placing itself on a different hyperbola. Where the hyperbolas cross is the location of the ship (Figure 16.16b). (Some small correction is needed to account for the earth not being flat.)

Rotating a hyperbola around the axis between its two branches produces what mathematicians call a *hyperboloid of one sheet*. The hyperboloid is used as the design for cooling towers of nuclear power plants because it can be built from interlocking families of ordinary

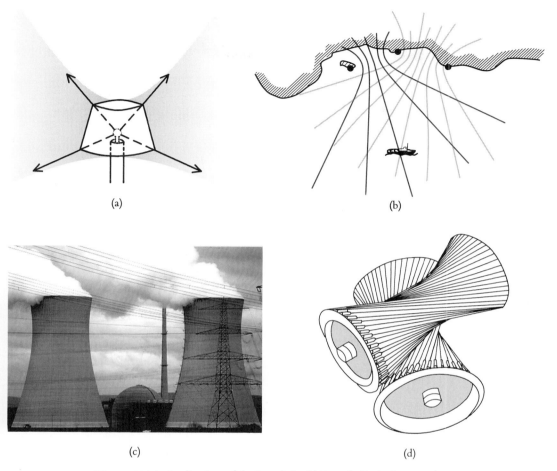

(a)

(b)

(c)

(d)

Figure 16.16 Applications of the hyperbola. (a) Hyperbolic shadows cast by a lamp. (b) Navigation by LORAN. (c) Nuclear cooling towers in the shape of a hyperboloid [Peter Arnold, Inc./Helga Lade.]. (d) Hyperbolic gears.

straight beams along the slant of the surface (Figure 16.16c).

A nonreturning comet, one that passes through our solar system only once, follows a hyperbolic path. Analogous to elliptic gears, there are hyperbolic gears, which can transmit rotation around one axis to rotation around another, as in the transmission of a car or truck,

where rotation of the wheels is transmitted to rotation of the drive shaft (Figure 16.16d).

Like its other conic section cousins, the hyperbola has a **reflection property.** A light ray proceeding in a direction toward (or away from) the focus of a hyperbolic mirror will reflect off the mirror in a direction toward (or away from) the focus of the other branch of the

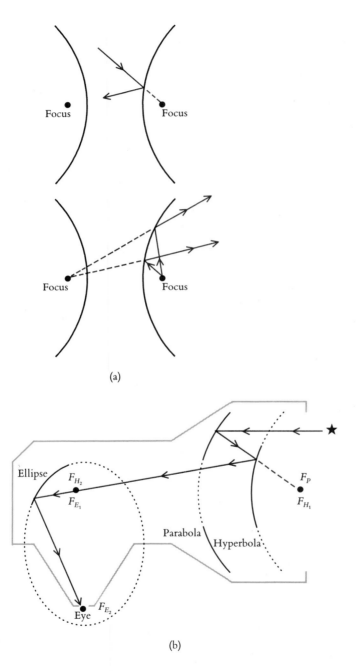

(a)

(b)

Figure 16.17 (a) Reflection property of the hyperbola. (b) Its use in the design of a telescope that uses the reflection properties of all three conics.

hyperbola (Figure 16.17). It is this property that Cassegrain telescopes use (see Color Plate 15).

NEWTON'S GREAT UNIFICATION

About 50 years after the death of Galileo, Sir Isaac Newton turned his attention to some of the same problems that had engaged Galileo and Kepler, particularly to the problems of terrestrial and celestial mechanics. In his famous *Principia,* whose full title is *Philosophiae Naturalis Principia Mathematica* (Mathematical Principles of Natural Philosophy), he unified terrestrial and celestial mechanics into one deductive mathematical science.

Writing in the spirit of Euclid, Newton began his *Principia* with definitions of terms such as *mass, force, inertia,* and *momentum.* He then presented three *laws of motion,* assumptions that constituted the starting point for his deductive system:

1. A body continues in a state of rest or in a state of constant unaccelerated motion in a straight line unless it is acted upon by an external force.

2. At any instant of time, the force acting on a body is equal to the product of its mass and acceleration.

3. To every action there is always opposed an equal reaction.

Using Kepler's third law (the law of times), Newton was led to the formulation of his *universal law of gravitation:* between any two bodies is a gravitational force of attraction that is proportional to the mass of each and inversely proportional to the square of the distance between them. The phrasing in words is not as easy to interpret as when the law is written as a formula:

$$F = \frac{Gm_1 m_2}{r^2}$$

where F is the force of attraction, m_1 and m_2 are the masses of the bodies, r is the distance between them, and G is the universal gravitational constant.

Firmly convinced of the validity of the universal law of gravitation, Newton used it as an assumption. Together with his three laws of motion, the law of gravitation enabled Newton to erect a masterpiece of mathematics in which he deduced the dynamics of Galileo, the statics that was developed by Archimedes and Galileo, the planetary laws of Kepler, and much more. Imagine — all this in one mathematical system that simultaneously vindicated the "heresies" of Copernicus, Kepler, and Galileo!

Newton's friend Sir Edmund Halley (1656–1742) persuaded him to publish his discoveries and financed the publication of the *Principia.* It contained a wealth of mathematical and physical discoveries even beyond those already mentioned: It explained the perturbations in the path of the moon, the motion of comets, the flattened shape of planets, and the phenomenon of tides. The *Principia* was published in 1687, but Newton had discovered many of its great ideas at a much earlier date. In fact, his law of gravitation must certainly have been known to him a decade earlier, for in 1679 he verified the law by calculations based on a new measurement of the earth's radius, together with observations of the moon's position.

Halley was inspired by Newton's work to check historical records of comets. Halley concluded that the comet he observed in 1682 was a returning comet in an elliptic orbit and predicted it would appear again at 76-year intervals; we know it as *Halley's comet.* Also on the basis of Newton's work, later astronomers were able to predict the existence and the positions of the planets Neptune and Pluto (the discoverer of Pluto in 1930 was still alive on the anniversary of the discovery in 1990!). Still further deviations of observations from the predictions of Newtonian theory suggest that there may be a very distant tenth planet in the solar system.

NEW GEOMETRIES TO MEASURE A NEW UNIVERSE

New kinds of geometry were conceived during the eighteenth century but were not to be born until the nineteenth century, when one of them became the basis for the next major revolution in physics and cosmology — the theory of relativity. We refer to **non-Euclidean geometries** — any of several sets of postulates, theorems, and corollaries that differ from Euclid's.

Ordinary geometry consists of statements, called theorems and corollaries, that are logical deductions from assumptions (other statements) called postulates. Euclid presented five postulates from which he developed a large body of theorems. We call this system **Euclidean geometry.**

From the fifth century B.C. until late in the nineteenth century, Euclidean geometry — including its extension to three dimensions — was thought to be the only science of space. Its theorems were thought to be statements of truth about the world we live in. No one could imagine a different geometry.

Euclid's five postulates, paraphrased somewhat, are as follows:

1. Two points determine a line.

2. A line segment can always be extended.

3. A circle can be drawn with any center and any radius.

4. All right angles are equal.

5. If *l* is any line and *P* any point not on *l*, then there exists exactly one line *m* through *P* that does not meet *l*.

These five statements were supposed to be absolute, self-evident truths. The first four are rather simple statements and are sufficiently unrelated that we may consider them **logically independent:** that is, none of them can be derived from the others by deduction. However, the postulate that *l* and *m* are parallel (i.e., they do not intersect) is another matter. Many early geometers thought this postulate was a logical consequence of the first four. In fact, Euclid himself may have thought his **parallel postulate** to be an unnecessary assumption for his geometry, for he derived nearly 30 theorems before using it.

The long history of these attempts to derive postulate 5 as the consequence of the first four postulates makes a fascinating story of failures; for whenever someone discovered a supposed proof, it was always found to be tacitly based on some assumption not contained in postulates 1 through 4. Attempts to prove the parallel postulate always failed because of an assumption that was **logically equivalent** to the parallel postulate; the reasoning was circular, hence invalid.

Among the hidden assumptions that were used in these purported proofs and that are logically equivalent to the parallel postulate are

1. The sum of the angles of a triangle equals 180°.

2. There is exactly one circle through any three points that are not on one line.

3. Parallel lines are equidistant.

HYPERBOLIC GEOMETRY

Nicolai Ivanovich Lobachevsky (1793–1856) and Janos Bolyai (1802–1860) are credited with independently inventing non-Euclidean geometry (see Spotlight 16.2). Both men considered Euclid's parallel postulate to be logically independent of the first four postulates and therefore reasoned that Euclid's postulate 5 could be replaced by a contradictory assumption. They both chose the following instead:

> **Postulate H:** If *l* is any line and *P* is any point not on the line, then there exists *more than one* line through *P* not meeting *l*.

The use of postulate H led to an entirely new system of theorems and corollaries, which we now call **hyperbolic geometry.** Some of the theorems in this system were exactly the same as those of the old, for those theorems that are derived only from postulates 1 through 4 must be valid in both systems. However, hyperbolic geometry provided some new and very surprising theorems. Color Plate 16 shows the *pseudosphere*, one model for hyperbolic geometry.

Figure 16.18 shows a line *l* and a point *P* not on *l*. We drop a perpendicular from *P* to *l*, calling *A* the foot of the perpendicular. Now, consider the line *PE*, which is perpendicular to

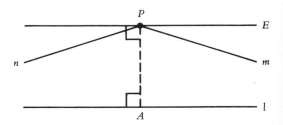

Figure 16.18 In hyperbolic geometry there is more than one parallel through a point *P* not on a given line *l*.

SPOTLIGHT 16.2 The Discovery of Non-Euclidean Geometry

Nikolai Ivanovich Lobachevsky.
[Novasti Press Agency (A.P.N.).]

Nikolai Ivanovich Lobachevsky (1793–1856) and Janos Bolyai (1802–1860) independently discovered non-Euclidean geometry. Lobachevsky was the first to publish an account of non-Euclidean geometry (1829), which he first called "imaginary geometry" and later "pangeometry." His work attracted little attention when it appeared, largely because it was written in Russian and the Russians who read it were very critical.

Bolyai published his work on non-Euclidean geometry as a 26-page appendix to a book (the *Tentamen,* 1831) by his mathematician father Wolfgang, who proudly sent the work by his son to Carl Friedrich Gauss, the leading mathematician of his day. Gauss replied to Wolfgang that he himself had earlier discovered non-Euclidean geometry!

There is no direct mathematical connection between hyperbolic geometry and the hyperbola, or between elliptic geometry (the other type of non-Euclidean geometry that we will examine shortly) and the ellipse; but there *is* a significance in the common origins of their names. The name "hypberbolic" geometry was given to the geometry of Lobachevsky and Bolyai by Felix Klein, a famous geometer later in the nineteenth century. "Hyperbolic" comes from the Greek word *hyperbole*, meaning to be excessive (we get the slang word "hyper" from the same root): in hyperbolic geometry there are *too many* parallels. Klein at the same time christened another non-Euclidean geometry, that of Riemann, in which there are *no* parallels, "elliptic" geometry, from a Greek word meaning "to fall short." Ordinary Euclidean geometry, with Euclid's parallel postulate, fits into Klein's scheme as "parabolic" geometry, from a Greek word meaning "to compare." Apollonius, the Greek who discovered most of the properties of the conic sections, had named them in similar fashion according to their *eccentricity,* or departure from circularity: an ellipse has an eccentricity between 0 (for a circle) and up to but not including 1 (for very long and thin ellipses); a hyperbola has an eccentricity from just above 1 (for a hyperbola that is almost a pair of parallel lines) to infinity (for a very thin hyperbola). All parabolas come in at an eccentricity of exactly 1, in terms of the mathematical definition of this concept.

PA. The line *PE* is parallel to *l*, and according to postulate 5 it is the only parallel to *l* through *P*. However, if we are using postulate H, then there is *another* line *m* that passes through *P* and is parallel to *l*. Let us assume that *m* makes an acute angle with *PA*, as shown in Figure 16.18. Then there must be another line *n* that makes the same acute angle with *PA* on the other side of *PA* and is therefore also parallel to *l*.

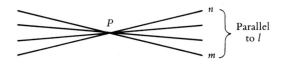

Figure 16.19 In hyperbolic geometry there are infinitely many parallels through a point P not on a given line l.

From this construction, it is easy to see the first astonishing conclusion: through P there are *infinitely many* parallels to line l. This is clear as soon as one considers the set of all lines through P, which are separated into two classes by m and n; one class contains PA and the other contains PE. The lines in the second class lie between n and m. All the lines in the second class are parallel to l (see Figure 16.19).

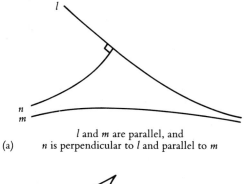

(a) l and m are parallel, and n is perpendicular to l and parallel to m

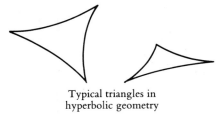

(b)   Typical triangles in hyperbolic geometry

Figure 16.20 In hyperbolic geometry, (a) given two parallel lines l and m, there exists a third line n perpendicular to one and parallel to the second; and (b) the sum of the angles of any triangle is less than 180°.

Using similar reasoning, Bolyai and Lobachevsky discovered many unusual theorems, of which we will list only three (Figure 16.20 illustrates theorems 1 and 3):

1. The sum of the angles in any triangle is less than 180°. (See Color Plate 17.)

2. Similar triangles are congruent—that is, triangles having the same shape also have the same size.

3. Given two parallel lines, there exists a third line perpendicular to one and parallel to the second.

Similar results were obtained by Carl Friedrich Gauss (see Spotlight 16.3), who investigated these unusual geometries before Lobachevsky and Bolyai but did not publish his results.

ELLIPTIC GEOMETRY

A generation after the discovery of hyperbolic geometry, G. F. Bernhard Riemann (1826–1866), a young German mathematician and disciple of Gauss, further scrutinized the basic assumptions of Euclidean geometry (see Figure 16.21). He analyzed postulate 2 and observed that "A line segment can always be extended" should be distinguished from "A line is infinite." That is, *unboundedness does not imply infinite extent.*

Think of the geometry on the surface of the earth. Going along what you would imagine was a line around the earth, you can travel another mile and another mile, and so on, and you would eventually return to your starting point. You have traveled on a finite path that is unbounded—that is, you can keep on traveling on and on. When Riemann investigated the consequences of a line coming back on

SPOTLIGHT 16.3 Carl Friedrich Gauss

Carl Friedrich Gauss. [Photo Deutsches Museum München.]

Nicolai Ivanovich Lobachevsky and Janos Bolyai are justly given the credit for the invention of non-Euclidean geometry because they had the courage to publish their revolutionary work. However, Carl Friedrich Gauss (1777–1855) is also given credit as a coinventor. From correspondence and private papers that became available after his death, we know that Gauss, too, long believed that Euclid's parallel postulate could not be proved from the first four postulates.

Like Lobachevsky and Bolyai, Gauss derived many theorems based on postulate H and in fact produced some of the most beautiful proofs in hyperbolic geometry. The one most familiar to all students of the subject is Gauss's proof that triangles cannot be arbitrarily large; the maximum area that a triangle may possess is the area contained by the trebly asymptotic triangle, the figure consisting of three lines, each of which is simultaneously parallel to the other two.

itself, he came to the conclusion that his rephrasing of Euclid's postulate 2 would also allow him to abandon Euclid's parallel postulate 5 in a strikingly novel way. Riemann replaced it with

Postulate E: *Every* two lines intersect.

Postulate E leads to **elliptic geometry.** We can see the plausibility of postulate E when we consider the geometry of the earth's surface, the **spherical geometry** used in navigation. What is a "straight line" in this geometry? The shortest distance between two points would be to tunnel directly through the earth, but to do so would be "out of bounds": we must stay on the surface of the sphere and consider our distances along it.

Figure 16.21 Georg Riemann. [Photo Deutsches Museum München.]

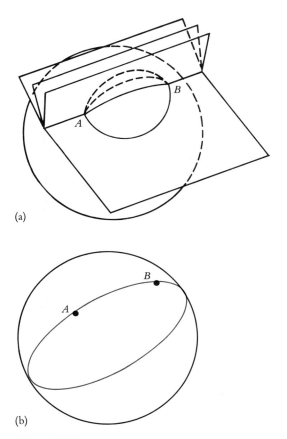

(a)

(b)

Figure 16.22 (a) The planes through A and B intersect the sphere in circles, the largest of which has the smallest curvature. (b) The great circle through A and B.

If we intersect the sphere in Figure 16.22 with a plane through A and B, the section is a circle passing through the given points, and the shortest arc, from A to B, of this circle is a candidate for the shortest path from A to B. Every plane section of the sphere is a circle, each with different curvature. *The larger the circle, the smaller the curvature; the smaller the curvature, the shorter the path between A and B.* Thus, the largest circle obtained as a section of the sphere gives rise to the shortest path. The largest circle is the **great circle**, obtained by hav-

ing the plane cut through A and B and the center of the sphere (see Color Plate 18).

This result of spherical geometry is familiar to pilots who fly long distances. If an airplane is on the equator and the pilot wishes to fly to another point on the equator, the pilot would simply fly along the equator, a great circle. However, if the airplane were at 10° latitude north and the pilot wished to fly to a destination at the same latitude, then the shortest distance would require going farther north. An airplane flying from New York to Naples, both at the same latitude, would actually have to travel quite far north in the Atlantic Ocean, while the shortest flight path from Washington, D.C., to Ho Chi Minh City, Vietnam, is over the North Pole (see Figure 16.23).

Returning to the idea of parallelism, we see that this concept simply doesn't exist in spherical geometry! Every pair of "lines" — that is, every pair of great circles — intersect. On the surface of the globe we note that triangles can have two or even three right angles: just put two vertices on the equator and one at a pole (see Color Plate 19). In fact, a general theorem of elliptic geometry states that *in an elliptic geometry, the sum of the angles of any triangle is greater than 180°.*

Figure 16.23 The shortest path between two cities is an arc of a great circle.

SPOTLIGHT 16.4 Angle Sums in a Triangle

The following simple proof shows that the angle sum in a triangle is equal to a straight angle. We will use triangle ABC as an example. Our goal will be to prove that $A + B + C =$ a straight angle ($= 180°$).

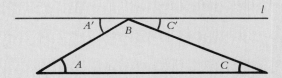

The Proof

At vertex B, draw a line l parallel to side AC. Then, by one of the first properties of parallels, which states that when parallel lines are cut by a transversal, the "alternate interior angles" are equal, angle $A =$ angle A'. Likewise, angle $C =$ angle C'. Now clearly $A' + B + C' =$ a straight angle. Hence, by substituting A for A' and C for C',

$$A + B + C = \text{a straight angle}$$

which is what we set out to prove.

We can't use this same proof on the sphere because it has no parallels at all. On the sphere every two straight-line paths, or great circle routes, eventually cross. In fact, on the sphere the sum of the angles of a spherical triangle is always *greater* than a straight angle.

On the other hand, in the plane of hyperbolic geometry there are "too many" parallels, and when the line l is drawn at B so that angle $A' =$ angle A, then angle C is always less than angle C'. Hence, the angle sum $A + B + C$ in triangle ABC is always less than the straight angle $A' + B + C'$.

Hence, in each of the three geometries we have investigated, we have a different sum for the angles in a triangle:

♦ $< 180°$ in hyperbolic geometry

♦ $= 180°$ in Euclidean geometry (sometimes called *parabolic geometry*)

♦ $> 180°$ in elliptic geometry

The property of ordinary planar Euclidean triangles—that all of them have the same angle sum, which is $180°$—is one that dramatically distinguishes Euclid's geometry from spherical geometry and hyperbolic geometry. In spherical geometry, the angle sum in a triangle is always greater than $180°$, and it is not the same for all triangles. We find triangles whose angle sum is $190°$ as well as triangles whose angle sum is $250°$. However, any two triangles having the *same area* do have the same angle sum (see Spotlight 16.4). In hyperbolic geometry (the geometry used by the artist M. C. Escher in his "Circle Limit" prints),

SPOTLIGHT 16.5 Angels and Devils

The Dutch artist M. C. Escher (1898–1972) was particularly interested in the metamorphosis of figures that change almost imperceptibly into other figures or into larger or smaller versions of themselves. He was able to discover a way to draw, inside a circle or square, figures that gradually get larger as they approach the outside of the enclosure. But it wasn't until he was shown a mathematician's representation of hyperbolic geometry (in which the sum of the angles of a triangle is always less than 180°) that he discovered how to make figures gradually get *smaller* toward the outside of a circle.

Douglas Dunham of the University of Minnesota, Duluth, has devised a computer program based on hyperbolic geometry that can produce an infinite variety of the type of drawings Escher so ingeniously drew. One of these, Dunham's "Circle Limit IV," is shown here. (For others, see Color Plate 14.)

The "lines" of this geometry are arcs of circles that are perpendicular to the outside circle. (The "lines" of spherical geometry are great-circle routes on the sphere.) This particular print is based on a regular tiling of the hyperbolic plane. The tiles shown here are regular quadrilaterals — their vertices are the points where the feet of three angels meet the feet of three devils. Six of these tiles meet at each vertex. Thus, the angle of each is 60° instead of 90°, which it would be for the corresponding tiling of the Euclidean plane, where four squares meet at each vertex.

The edges of some of the tiles (of the underlying tiling) have been drawn in so that they can be seen, and two are shaded. Although the tiles appear to get smaller (in the Euclidean sense) toward the edge of the outside circle, the hyperbolic geometry uses a different distance measure in which all of the tiles, including the two that are shaded, are congruent (the same size).

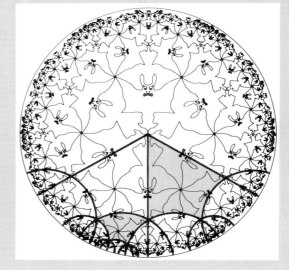

Dunham's "Circle Limit IV" plot. This computer-generated tiling in the hyperbolic plane creates an image very similar to that of M. C. Escher's "Angels and Devils." Note how the positions of feet and heads are related by radii and arcs that define the underlying tiling pattern. [Courtesy of Douglas Dunham.]

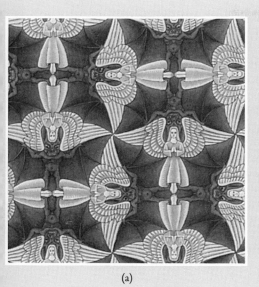

(a)

(b)

(c)

(a) M. C. Escher's *Heaven and Hell* (also
known as "Angels and Devils").
Photographed from one of Escher's
notebooks, this example demonstrates a
repeating pattern of the Euclidean plane.
Note the uniform size of the figures and
the central meeting of four angel and four
devil wing tips. [© M. C. Escher Heirs
c/o Cordon Art — Baarn — Holland.]
(b) Escher's angels and devils pattern
carved on an ivory sphere by Masatoshi.
This mapping of the pattern onto a
sphere shows the different effects of a
spherical geometry. Note that three angel
and three devil wing tips meet in this
version. [© M. C. Escher Heirs c/o
Cordon Art — Baarn — Holland.] (c) M.
C. Escher's *Circle Limit IV (Heaven and
Hell)*. This example shows the repeating
angels and devils pattern mapped onto a
hyperbolic geometry. At the center three
angel and three devil feet meet. As one
moves outward, the figures get smaller.
Note that four angel and four devil wing
tips meet. [© M. C. Escher Heirs c/o
Cordon Art — Baarn — Holland.]

the angle sum in a triangle is always less than 180°, but it is not the same for all triangles (see Spotlight 16.5, p. 458).

As we have seen, any geometry that differs from Euclidean geometry is a non-Euclidean geometry. The first departures from Euclid's postulates involved denying, in some way, his parallel postulate. This led to the development of hyperbolic geometry and elliptic geometry, which are now considered the *classical non-Euclidean geometries*. However, we now have many more geometries that differ from Euclid's in a variety of ways.

THE THEORY OF RELATIVITY

In 1905, Albert Einstein (Spotlight 16.6) put forth his *special theory of relativity*, a complicated theory that constituted the first step in the greatest revolution in physics since Newton's *Principia*.

Einstein proposed a new way of thinking about events in the history of the universe. An event takes place in our three-dimensional space at a specific time in history. Thus, an event is located in *space-time* by four coordinates: three determine its position in space, and the fourth determines its position in time. Of course, these coordinates locate the event relative to a specific coordinate system. Einstein observed that the location of an event in space-time therefore depends on the position of the observer—that is, on the origin and orientation of the coordinate system being used. Different observers may obtain very different views of events, especially if one observer is traveling very fast with respect to the other.

Let's consider these ideas geometrically. The *distance* between two events, usually in relativity theory called an *interval*, is split into two parts: a *space-part* and a *time-part*. The

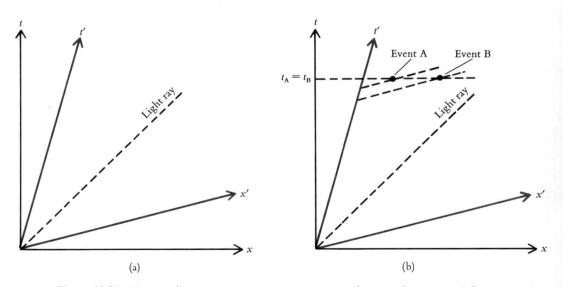

Figure 16.24 (a) A coordinate system representing space-time. The *t*-axes show time and the *x*-axes space. The black axes are a system at rest. Note that the red, moving system tilts toward the 45° light ray line. (b) Observers in the system at rest (black axes) will say the events *A* and *B* occur at the same time. Observers in the moving red system will say that event *B* occurs before event *A*.

SPOTLIGHT 16.6 Albert Einstein

Albert Einstein. [Yerkes Observatory.]

Einstein (1879–1955) was born in Ulm to a German Jewish family with liberal ideas. Although he showed early signs of brilliance, he did not do well in school. He especially disliked German teaching methods. In the mid-1890s he went to study in Switzerland, a country much more to his liking, where he went to work as a patent clerk. Einstein burst upon the scientific scene in 1905 with his theory of special relativity. In 1916 he published his theory of general relativity. General relativity was successfully tested in 1919, and his fame grew enormously. Nazism forced Einstein to leave Europe. He settled at the Institute for Advanced Study at Princeton, where he remained until his death at age 76.

space-part will be the part of the interval that comes from the position of the events in three-dimensional space, and the time-part will be the length of time that separates the events. This splitting up depends on the coordinate system and its orientation, so different results may be obtained by different observers (see Figure 16.24). However, the interval, being a line segment joining the two events in four-dimensional space-time, is absolute—in the sense that it is the same for an observer at rest and for all other observers who are traveling at a constant velocity with respect to the one considered at rest.

For example, let's imagine that the eruption of Mount St. Helens in Washington in 1980 took place at the very same time that someone on Mount Palomar in California observed an astronomical phenomenon 100 light-years away. (A **light-year** is the *distance* that light travels in a year.) For those of us on earth (at rest relative to the earth), the eruption and the astronomical phenomenon took place one century apart on our time scale: the interval between the two events has a space-part of 100 light-years and a time-part of 100 years.

For observers traveling at constant velocity with respect to the earth, say at 50 light-years away from earth, the space-part and time-part of the interval would be very different. One observer might determine that the two events took place 200 years apart, while another might conclude that the two events took place simultaneously. Their splitting of the interval into space-parts and time-parts would be very different from ours. The geometry of space-time is indeed strange: in its four-dimensional space, the distance between two points — now the interval between two events — remains invariant (in the sense we have described), but its respective parts vary.

Three years after Einstein published his first paper on the subject, the mathematician Hermann Minkowski (1864–1909) gave Einstein's work a geometric interpretation that accepted Einstein's strange calculation of intervals and greatly simplified the theory. The geometry that was used, justifiably called *Minkowskian geometry,* is certainly non-Euclidean. Further, it makes use of one of Riemann's far-reaching ideas — that the nature of a mathematical space is determined by the way distance is measured; the distance formula therefore determines the nature of the geometry.

If the coordinate representations of two events are given by

$$(x_1, y_1, z_1, t_1) \quad \text{and} \quad (x_2, y_2, z_2, t_2)$$

then the interval I between them, in Minkowskian space, is calculated by the formula

$$I = \sqrt{\begin{aligned} &(t_2 - t_1)^2 - (x_2 - x_1)^2 \\ &- (y_2 - y_1)^2 - (z_2 - z_1)^2 \end{aligned}}$$

whereas the distance d between them, if the points are in *Euclidean* four-dimensional space, would be computed by the formula

$$d = \sqrt{\begin{aligned} &(t_2 - t_1)^2 + (x_2 - x_1)^2 \\ &+ (y_2 - y_1)^2 + (z_2 - z_1)^2 \end{aligned}}$$

The second formula is a direct generalization of the Pythagorean theorem from Euclidean plane geometry.

Little more than a decade after introducing his special theory of relativity, Einstein came forth with his *general theory of relativity.* This work, too, astonished the scientific world. Among other revolutionary ideas was his contention that space was "curved." By this he meant that light rays, which are considered to travel on paths of shortest distance, don't actually follow "straight lines" but bend to follow shortest distance paths in the curved space. Light rays even bend to different degrees, depending on where in the universe they are; if they pass through a strong gravitational field, then they bend considerably.

A test of this contention was made in 1919 during a total eclipse of the sun, when the light rays from a distant star passed close to the sun and could be studied. Einstein was right; the rays did bend — and in an amount very close to his predictions. This observation showed that lines in the geometry of general relativity are not of the same character as Euclidean lines.

What sort of geometry was Einstein using? There are several answers to the question. First, the idea of "curved" space smacks of elliptic geometry, in the sense that a line through the universe comes back on itself. Second, Einstein used a variation on Minkowskian geometry in which the distance formula appropriate to the needs of physics varies from place to place in the universe, depending on the strength of the gravitational field. So, Einstein was using a form of Minkowskian geometry along with some considerably modified ideas of elliptic geometry. An appreciation of these non-Euclidean geometries very likely motivated his remark about the postulates of geometry in a famous 1921 lecture: "[They] are voluntary creations of the human mind. . . . To this interpretation of geometry I attach great importance, for should I not have been acquainted with it, I would never have been able to develop the theory of relativity."

SPOTLIGHT 16.7 The Hubble Space Telescope

In April, 1990, the Hubble Space Telescope was launched from the Space Shuttle *Discovery*. The Space Telescope was first proposed in 1946 by Lyman Spitzer, Jr., of Princeton University. Its name honors Edwin Hubble, an American astronomer who discovered in 1930 that the universe is expanding.

Dan Schroeder, Professor of Physics and Astronomy at Beloit College, Wisconsin, helped oversee its design and construction. He comments on the new potential for "exploring the universe through telescopes" that the Space Telescope represents:

> The great advantages of a telescope in orbit are that it is above the atmosphere, with its air that attenuates light and the air currents that distort it, and away from the light pollution from human activity.
>
> With the Space Telescope we should be able to see known objects much more clearly than before, and see many more objects than are now visible, including far-distant objects that may give us better clues to the age of the universe. To give you an idea of the telescope's resolution: it should enable us to read the inscriptions on a coin at a distance of ten miles.
>
> The instruments on board the craft include cameras, spectrographs (to gather data on the chemical composition, temperature, pressure, and density of objects), and a photometer (to measure the brightness of objects).

Like other telescopes of the Cassegrain design (see Color Plate 15b), the Space Telescope contains a hyperbolic mirror and a larger parabolic mirror (94.5 inches across). Early

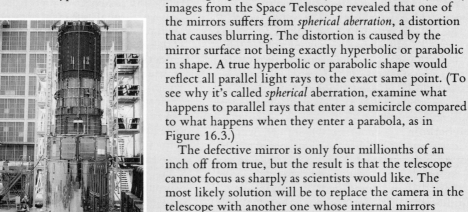

images from the Space Telescope revealed that one of the mirrors suffers from *spherical aberration*, a distortion that causes blurring. The distortion is caused by the mirror surface not being exactly hyperbolic or parabolic in shape. A true hyperbolic or parabolic shape would reflect all parallel light rays to the exact same point. (To see why it's called *spherical* aberration, examine what happens to parallel rays that enter a semicircle compared to what happens when they enter a parabola, as in Figure 16.3.)

The defective mirror is only four millionths of an inch off from true, but the result is that the telescope cannot focus as sharply as scientists would like. The most likely solution will be to replace the camera in the telescope with another one whose internal mirrors compensate precisely for the mirror imperfection.

The Hubble Space Telescope while being assembled and tested at Lockheed Missiles & Space Company in Sunnyvale, Calif.

REVIEW VOCABULARY

Astronomical unit (AU) The average distance from the earth to the sun, 92,950,000 miles.

Axis of a parabola The line dividing a parabola into two identical portions.

Circle The conic section formed when the cutting plane is parallel to the plane.

Cone The set of all lines, each of which passes through some point of a given circle and all of which pass through the same point not in the plane of the circle.

Conic section A curve formed when a plane intersects a cone.

Eccentricity of an ellipse A measure of the deviation of an ellipse from circularity.

Ellipse A conic section formed when a plane intersects one nappe of a cone in a closed curve.

Elliptic geometry A system of geometry in which there are no parallel lines.

Euclidean geometry The "ordinary" system of geometry based on the five postulates Euclid used, including the parallel postulate.

Focus (of a parabola) The single point at which rays parallel to the axis of a parabola come together after "bouncing off" a parabola. (An analogous definition can be given for the foci of ellipses and hyperbolas.)

Great circle The set of points that is the intersection of a sphere and a plane containing its center.

Hyperbola A conic section formed when a cutting plane intersects both nappes of a cone.

Hyperbolic geometry A system of geometry in which there exists more than one parallel line through a point P not on a given line l.

Light-year The distance that light travels in a year.

Logically equivalent Two statements are logically equivalent if each can be deduced from the other.

Logically independent Refers to a set of statements, no one of which can be logically deduced from the others.

Nappes The two surfaces of a cone separated by its vertex.

Non-Euclidean geometry Any geometry that differs from Euclidean geometry.

Parabola A conic section formed when a cutting plane is parallel to a generating line of the cone, thus cutting only one nappe to result in a curve that is not closed.

Parabolic cylinder A surface formed by translating a parabola in a direction perpendicular to the plane it lies in ("lifting it straight out of the plane").

Paraboloid of revolution A surface formed by rotating a parabola around its axis.

Parallel postulate A basic assumption of geometry that states whether through a point P, not on a given line l, there exists none, one, or more than one line parallel to given line l.

Postulate E *Every* two lines intersect.

Postulate H If l is any line and P is any point not on the line, then there exists *more than one* line through P not meeting l.

Reflection property of an ellipse A light ray passing through one focus of an elliptic mirror will reflect off the mirror and pass through the other.

Reflection property of a hyperbola A light ray proceeding in a direction toward (or away from) the focus of a hyperbolic mirror will reflect off the mirror in a direction toward (or away from) the focus of the other branch of the hyperbola.

Reflection property of a parabola A light ray entering a parabolic mirror parallel to its axis will reflect off the mirror and pass through the vertex; and vice versa, a light ray passing through the vertex will reflect off the mirror and leave the parabola on a line parallel to the axis.

Right circular cone A cone in which the line joining the vertex V to the center of the circle C is perpendicular to the plane P.

Spherical geometry The geometry of a sphere or the earth's surface.

Vertex of a parabola The point where a parabola crosses its axis of symmetry.

SUGGESTED READINGS

Cohen, I. Bernard: *The Birth of a New Physics,* revised and updated, Norton, New York, 1985.

———— : *Revolution in Science,* Belknap Press, Cambridge, England, 1985.

Crowe, Donald W.: Some Exotic Geometries, in Anatole Beck, Michael Bleicher, and Donald W. Crowe (eds.), *Excursions into Mathematics,* Worth, 1969, Chapter 4, pp. 211–247, 308–314. A comparison of spherical, Euclidean, and hyperbolic geometries.

Greenberg, M. J.: *Euclidean and Non-Euclidean Geometries,* W. H. Freeman, New York, 1980.

Jacobs, Harold R.: *Geometry,* 2d ed., W. H. Freeman, 1987. Chapters 6 and 16 on parallel lines and on non-Euclidean geometries.

Taylor, E. F., and J. A. Wheeler: *Spacetime Physics,* W. H. Freeman, New York, 1966.

Whitt, Lee: "The Standup Conic Presents: The Parabola and Applications; The Ellipse and Applications; The Hyperbola and Applications" *The UMAP Journal* 3(3)(1982) 285–313, 4(2)(1983) 157–183; 5(1)(1984) 9–21. Source for many of the applications of conics mentioned in this chapter; includes references to works with more details.

EXERCISES

1. A stone is thrown upward with motion described by the equation

$$y = -16t^2 + 48t + 32$$

where y is the height (in feet) of the stone above the ground at time t (seconds).

a. The person throwing the stone at time $t = 0$ is on top of a building. How high above the ground is her hand when she lets go of the stone?

b. When will the stone hit the ground?

c. How high does the stone go? (Hint: What kind of curve is the graph of the equation of motion, and what kind of symmetry does it have?)

2. As in Exercise 1, but for $y = -16t^2 - 56t + 32$. What is the significance of the minus sign in front of the 56?

3. Here is an alternative definition of parabola in terms of solely geometric considerations: a parabola is the set of points in a plane equidistant from a fixed point (called the *focus*) and a fixed line (called the *directrix*) in that plane. Carry out the following steps in drawing a parabola according to this new definition.

a. On an ordinary-size sheet of paper, draw a line d and mark a point F 1 or 2 inches away from d.

b. Locate a point V that is halfway between F and d. Why is V a point of the parabola?

c. Using compasses or by trial and error, locate a point P (different from V) whose distance from F is equal to the perpendicular distance from P to d.

d. Repeat part **c** several times, locating five or six such points P on the parabola.

e. How does symmetry help you locate still more points on the parabola?

f. Now connect the points of the parabola to obtain a smooth curve.

4. The point V constructed in the previous exercise is the vertex of the parabola, and line VF is the axis of the parabola.

   a. Explain why the axis of the parabola is the axis of symmetry of the parabola.

   b. On a fairly accurate drawing of a parabola, draw its axis.

   c. Draw FP, where P is a point of the parabola, and then draw a line through P that is parallel to the axis.

   d. Draw a tangent (by the eyeball method) to the parabola at P, and measure the angles that this tangent makes with FP and with the line through P parallel to the axis. Does this verify the focal property of the parabola?

   e. Repeat parts **c** and **d** with another point of the parabola that is not the reflection of P in the axis.

   f. What are the measures of the angles of incidence and reflection if $P = V$?

5. In this exercise we derive the equation for a parabola from the definition in Exercise 3. To implement the definition of Exercise 3, we need a formula for the distance between two points. If the two points are $P_1(x_1, y_1)$ and $P_2(x_2, y_2)$, then the distance between them is

$$d(P_1, P_2) = \sqrt{(x_1 - x_2)^2 + (y_1 - y_2)^2}$$

So, for example, the distance between the points $(2, 4)$ and $(5, -3)$ is

$$\sqrt{(2 - 5)^2 + [4 - (-3)]^2} = \sqrt{9 + 49} = \sqrt{58} \approx 7.6$$

One of the pleasures of coordinate geometry is that we can place the coordinate axes to suit our purposes. So, we will assume that our parabola is nicely centered with the y-axis as its axis. Suppose the parabola has its focus at the point $(0, p)$ on the y-axis and has as directrix the line $y = -p$, the horizontal line p units below the x-axis. What we do now is consider an arbitrary point $P(x, y)$ and see what must be true of it if it is to be on

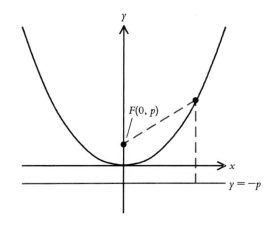

our parabola. The parabola consists of all points that are equally far from the focus point and the directrix line. From our distance formula, the first distance is

$$\sqrt{(x-0)^2 + (y-p)^2} = \sqrt{x^2 + (y-p)^2}$$

The distance to the directrix is along the vertical line from $P(x, y)$ to the directrix; this vertical line intersects the directrix at $(x, -p)$. So the distance of point $P(x, y)$ to the directrix is

$$\sqrt{(x-x)^2 + [y-(-p)]^2} = \sqrt{(y+p)^2}$$

Equating the two distances, we have that the point $P(x, y)$ will be on the parabola exactly when the coordinates x and y satisfy

$$\sqrt{x^2 + (y-p)^2} = \sqrt{(y+p)^2}$$

After we square both sides and multiply out, we have

$$x^2 + y^2 - 2py + p^2 = y^2 + 2py + p^2$$

which simplifies to

$$y = \frac{1}{4p}x^2$$

We've drawn the figure assuming that $p > 0$, in which case the parabola opens upward; if instead, $p < 0$, then it opens downward.

For the parabola described by the equation $y = \frac{1}{6}x^2$:

a. Determine the coordinates of its vertex.

b. Determine the coordinates of its focus.

c. Determine the equation of its axis.

d. If the point $(c, 6)$ lies on the parabola, what is the value of c?

e. If the point $(-2, b)$ lies on the parabola, what is the value of b?

6. Consider the parabola described by the equation $y = x^2$.

a. Starting with the graph of that parabola, indicate how to get the graph of $y = x^2 + 3$.

b. Again starting with the given parabola, indicate how to get the graph of $y = -(x-2)^2 + 3$.

7. Here's a geometric definition of an ellipse: an ellipse is the set of all points P having the property that the sum of the distances from P to two fixed points F_1 and F_2 (the foci) is constant.

a. Draw a horizontal line on a sheet of paper and mark two points F_1 and F_2 approximately 4 inches apart. Construct the perpendicular bisector of segment F_1F_2.

b. On the perpendicular bisector you constructed in part **a**, mark a point Q approximately 2 inches from the line F_1F_2. Off to the side, draw a segment whose length is the sum of the lengths of segments QF_1 and QF_2. Point Q will be a point of the ellipse you are constructing, so the length $QF_1 + QF_2$ will serve as the constant referred to in the definition.

c. Using compasses, determine several other points X, such that $XF_1 + XF_2 = QF_1 + QF_2$, thus giving you more points on the ellipse.

d. Explain why, in general, determining one point of the ellipse gives you three others almost immediately.

e. Now connect the points of the ellipse that you've constructed to form a smooth curve.

8. Pick a general point P on an ellipse and draw a tangent at P. Construct line segments from the two foci to P and measure the angles of incidence and reflection in a manner analogous to the one you used with the parabola. Do you have an analogous result?

9. A circle is a special kind of ellipse in which the two foci coincide; it is the set of points in a plane that are a given distance from a fixed point. The fixed point is the *center* and the given distance is the *radius*. In terms of coordinates, a circle in the xy-plane with center at the point (h, k) and radius equal to r is described by the equation $(x - h)^2 + (y - k)^2 = r^2$.

Determine the equation of a circle whose center is at the point $(-4, 3)$ and whose radius is 3.

10. A *diameter* of a circle is a line segment through the center of the circle with its endpoints on the circle. Determine the equation of a circle that has a diameter whose endpoints are $(2, 3)$ and $(-6, 5)$.

11. As in Exercise 9, but for the circle whose center is at the point $(5, -6)$ and which passes through the point $(1, 2)$.

12. In this and the next few exercises, we use the definition of an ellipse in Exercise 7 to derive the equations of ellipses, together with the distance formula from Exercise 5. We choose our coordinate system so as to locate the two foci on the x-axis at $(c, 0)$ and $(-c, 0)$. An ellipse is the set of points $P(x, y)$ with the property that the sum of the distances from P to the foci is constant. For reasons that will become clear shortly, let that constant be denoted by $2a$. So a point $P(x, y)$ will be on the ellipse if x and y satisfy the equation

$$\sqrt{[x - (-c)]^2 + (y - 0)^2} + \sqrt{(x - c)^2 + (y - 0)^2} = 2a$$

or

$$\sqrt{(x + c)^2 + y^2} + \sqrt{(x - c)^2 + y^2} = 2a$$

To simplify this equation, we subtract the second square root from both sides and then square both sides; simplifying that produces

$$a - \frac{c}{a}x = \sqrt{(x - c)^2 + y^2}$$

Well, there's nothing to do but square again and simplify some more, which produces the result of

$$\frac{x^2}{a^2} + \frac{y^2}{a^2 - c^2} = 1$$

Usually we make the definition $b = \sqrt{a^2 - c^2}$, so that the final equation is

$$\frac{x^2}{a^2} + \frac{y^2}{b^2} = 1$$

Ellipses are sometimes referred to as "flattened" circles. The equation $x^2 + y^2 = 1$ represents a circle of radius 1 centered at the origin. Solving for x gives $x = \pm\sqrt{1 - y^2}$.

a. Let's stretch the circle in the horizontal direction by multiplying the x coordinates of its points by 2, so that $x = \pm 2\sqrt{1 - y^2}$. Show that the result is an ellipse, and identify the coordinates of its foci.

b. As in part **a** but stretch the circle in the vertical direction instead. Caution: In this situation, a and b change roles.

13. A point moves in the xy-plane such that the sum of its distances from (3, 0) and $(-3, 0)$ is 10. What is the equation of the resulting ellipse?

14. As Kepler realized, the earth follows a circular orbit with the sun at one focus. Books of astronomical data usually give the *mean* distance of the earth from the sun — by which they mean one-half of the longer axis — as 92,956,000 miles. The eccentricity of the earth's orbit is 0.0167. Put these facts together to determine a, b, and c, and write an equation that describes the earth's orbit.

15. As in Exercise 14, but for the orbit of the moon around the earth, using the facts that the moon's greatest distance from the earth is 252,710 miles and its least distance is 221,643 miles.

16. The planet Neptune has an orbit with a mean distance (see Exercise 14) from the sun of 2.793 billion miles and an eccentricity of 0.0082. The planet Pluto has an orbit with average distance from the sun of 3.666 billion miles and an eccentricity of 0.2481. Show that Pluto is sometimes closer to the sun than Neptune is (such is the case for the years 1969–2009).

17. Kepler's third law, the law of times, applies not only to the planets of the sun but also to the moons and artificial satellites of any planet (but with a different constant K). In this exercise we use what we know about the moon to calculate the height of an artificial satellite from its period of rotation; for simplicity, we assume circular orbits. The moon, which has as its true period of revolution 27 days, 7 hours, 43 minutes, and 11.5 seconds (the time from one new moon to the next is a couple days longer because of motion of the earth), is at an average distance of 238,857 miles from the center of the earth. The first artificial satellite, Sputnik I, launched in 1957, took only 88 minutes to orbit the earth. How high was it? (Hint: Don't forget to take into account the earth's radius, 3963.5 miles.)

18. As in Exercise 17, apply Kepler's third law again, this time to calculate the period of rotation of the Hubble Space Telescope, launched in April 1990 into an orbit 381 miles above the surface of the earth.

19. We can apply the ideas of Exercise 17 to an application we take for granted today. If a satellite is at just the right height, it will orbit the earth in exactly 24 hours, so that it is always over the same spot on earth. Communications satellites, including those for television, are at this height. How high are they?

20. The orbits of most comets are parabolic, which means that the comet appears only once in our solar system and then departs forever. A few comets have hyperbolic orbits. Most, however, including Halley's comet, have elliptic orbits with very high eccentricity. From the period of Halley's comet, 76.1 years, and assuming an eccentricity of 0.999, calculate how far away from the sun it goes.

21. Here is a geometric definition of a hyperbola: the set of all points P having the property that the difference of the distances from P to two fixed points F_1 and F_2 (called the foci) is constant. Using this definition, construct a hyperbola in the same way you constructed the parabola and ellipse.

22. A hyperbola (with two branches) divides the plane into three mutually exclusive regions: one containing focus F_1, one containing the other focus F_2, and the other containing no focus.

   a. Let P be a general point on that branch of the hyperbola that isolates F_1. Draw a tangent to the hyperbola at this point. Now, draw a segment starting at F_1 and meeting the hyperbola at P. If you imagine F_1 to be a source of light and the hyperbola to be a mirror, then you have an angle of incidence. Finally, draw the line r that represents the ray of light reflected off the hyperbolic mirror at P, according to the usual law: the angle of incidence equals the angle of reflection.

   b. Extend line r "backward" and see how close it comes to passing through F_2. If your construction of the hyperbola is fairly accurate and if you have guessed right in constructing the tangent line at P, then your line r should pass very close — if not right through — F_2. (It can be proved, theoretically, that line r does pass through F_2.)

   c. Let Q be a point outside the region containing F_1. The line segment joining G to F_1 meets the hyperbola at a point that we call P. As before, draw a tangent at P and imagine the hyperbola to be mirrored on its convex side. A ray of light emanates from Q, hits the hyperbolic mirror at P, and is then reflected. Draw a line that represents the reflected ray and examine how close this line comes to passing through F_2.

   d. Summarize the discoveries you made in parts **b** and **c** to give a full and clear statement of the focal properties of the hyperbola.

23. Imitate the analysis of Exercise 12 to derive the equation that describes a hyperbola with foci at $(c, 0)$ and $(-c, 0)$:

$$\frac{x^2}{a^2} - \frac{y^2}{b^2} = 1$$

with $b = \sqrt{c^2 - a^2}$.

24. A point moves in the xy-plane such that the difference of its distances from the two points $(-4, 0)$ and $(4, 0)$ is 6. Use the result of Exercise 23 to find the equation that describes the point's motion.

25. Examine the diagram of the Cassegrain telescope in Figure 16.17b and explain why it works.

26. Conic sections that we haven't mentioned in this chapter occur when the plane that sections the cone passes through the vertex of the cone. These sections are called *degenerate*. Can you find three distinct types of degenerate conic sections?

27. Suppose a planet is moving at a constant velocity along a straight-line path. Show that the line from the sun to the planet sweeps out equal areas in equal time.

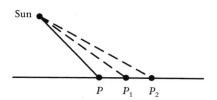

28. Newton's law of gravitation tells us that the gravitational attraction of a body diminishes with the square of the distance. Use Newton's formula for the force of gravitational attraction to locate a point between the earth and the moon where the attractions of the two are exactly equal and opposite. (The mass of the earth is 5.979×10^{24} kg, the mass of the moon is 7.35×10^{22} kg, and the two are on average 383,403 km apart.)

29. As in Exercise 28, but for the earth and the sun. The sun has 330,000 times the mass of the earth, and it is on average 92.956 million miles away.

30. The star (besides the sun) that is closest to the earth is Alpha Centauri, which is 4.3 light-years away. How far is that in miles? In astronomical units? (Light travels 186,292 miles per second.)

31. The *parsec* is a unit for distances between stars; an object 1 astronomical unit across subtends an angle of 1 second of arc $(= \frac{1}{3600}°)$ at the distance of 1 parsec. How large is a parsec in miles? In light-years?

32. By producing a specific example, show that there is a triangle in elliptic geometry in which all three angles are right angles, so that the sum of the angles of the triangle is 270°. (Hint: Spherical geometry is an elliptic geometry.)

33. When Galileo observed the moon, he noted "lofty mountains and deep valleys." He proceeded to measure the shadows cast by the mountains in order to compute the approximate height of the mountains. Galileo concluded that the moon's mountains were 4 miles high (which he thought to be higher than any mountain on the earth).

    a. Galileo determined that the ratio of the diameter of the earth to that of the moon was 7/2, and he believed that the earth's diameter was 7000 miles. (Today we know that the earth's diameter is closer to 7900 miles.) Using his data, compute the diameter of the moon, the circumference of the earth, and the circumference of the moon.

    b. The following figure represents a view of the moon that shows one-quarter of its surface illuminated. Point T represents the top of a specific lunar mountain whose height Galileo was calculating. Line *TB*, which is tangent to the moon at *B*, represents a ray of sunlight. Thus, the surface of the moon immediately below segment *TB* was in the shadow of the mountain.

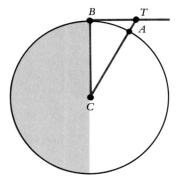

Draw segment *CT*, calling *A* the intersection of *CT* with the surface of the moon. Galileo measured the small arc *AB* to be $\frac{1}{20}$ of the diameter of the moon. He reasoned that this length was approximately equal to the length *TB*. What number did Galileo use for the length *TB*?

c. What theorem about circles enabled Galileo to conclude that triangle *BCT* was a right triangle? You can now do as he did and calculate the length *CT*, namely, the distance from the center of the moon to the top of the lunar mountain.

d. Finally, you can determine the length of segment *AT*, the height of the lunar mountain. How close have you come to Galileo's calculation for the height of the mountain?

34. Since the earliest time that a celestial sphere was conceived, it has been known by observers in the northern hemisphere that the North Star was directly — or very close to directly — over the North Pole; hence, the name Polaris for the North Star. In the accompanying diagram, an observer at *O* is sighting Polaris in the direction of *P*.

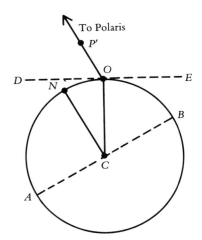

a. Which line in the diagram represents the equator and which line represents the horizon line for the observer?

b. The angle between the horizontal and the sighting of a star is called the *altitude* of the star. What angle is the altitude of Polaris for the observer at O?

c. What angle in the diagram represents the latitude of the observer?

d. What is the relationship between the latitude of the observer and the altitude of Polaris?

35. Let ABC be any triangle. Euclid's fifth postulate implies that there is exactly one line l through C that is parallel to AB. Draw line l and use it to prove that the sum of the measures of the interior angles of the triangle ABC is $180°$.

36. Prove the converse of the previous exercise: if the sum of the angles of a triangle is $180°$, then Euclid's parallel postulate holds. (This is considerably harder than Exercise 35.)

37. The quadrilateral $ABCD$ in the following figure is drawn to suggest the situation in hyperbolic geometry. The angles at A and B are right angles, whereas the angles at C and D are acute.

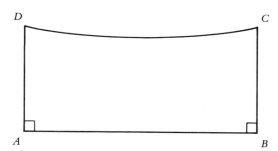

a. Prove that if $AD = BC$, then angles C and D are of equal measure. (You may use the Euclidean theorems on the congruence of triangles, for they do not depend on the parallel postulate.)

b. Prove that the sum of the angles of triangle ABD or of triangle BCD is less than $180°$. (In fact, the sum of the angles of each of the triangles is less than $180°$, but that is harder to prove.)

38. The surface of a solid figure whose faces are polygons is called a *polyhedron* (meaning "many planes"). The polygonal faces have vertices and edges that are the vertices and edges of the polyhedron. We want to explore the numerical relationship between the vertices, edges, and faces of an ordinary polyhedron (or solid) in which there is nothing peculiar, such as a hole.

a. If V is the number of vertices, E the number of edges, and F the number of faces, find V, E, and F for the cube and the triangular prism.

b. Notice that $V + F$ is larger than E. How much larger in the case of the cube? And how much larger in the case of the cube with a square pyramid pasted on a face? Write your discovery as a formula: $V - E + F = \_\_\_$.

c. Check this formula on all the polyhedrons that follow in Exercise 39.

39. A *regular polyhedron* (or solid) is one in which all the faces are alike and all the vertices are alike. To be more precise, we require that all faces be congruent regular polygons and that each vertex be surrounded by the same number of faces. Let p denote the number of edges on each face and q denote the number of faces that meet at a vertex. (A cube, for example, yields $p = 4$ and $q = 3$.) The climaxing theorem in Euclid's work on geometry is a proof that there are only five regular solids, now known as the Platonic solids (see Spotlight 16.1).

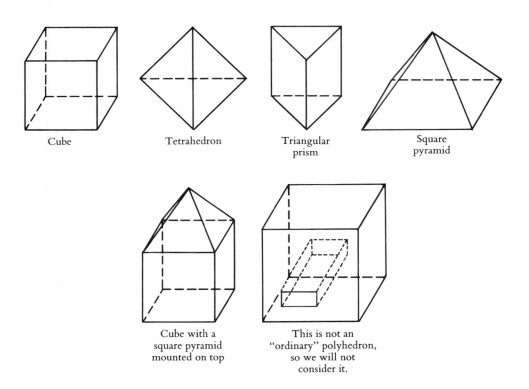

Cube

Tetrahedron

Triangular prism

Square pyramid

Cube with a square pyramid mounted on top

This is not an "ordinary" polyhedron, so we will not consider it.

a. Determine p and q for the tetrahedron and the octahedron.

b. Compare the values of pF and qV for the cube, tetrahedron, and octahedron.

c. What is the meaning of pF?

d. Show that $qV = 2E$.

e. The prefix *dodeca-* derives from the Greek word for "twelve"; hence, the regular solid with 12 faces is called a dodecahedron. Determine the number of vertices and edges of a regular dodecahedron.

f. The prefix *icosa-* derives from the Greek word for "twenty." Determine the number of vertices and the number of edges of a regular icosahedron.

40. Newton's proof of Kepler's second law: Newton knew, from the work of Galileo and the Dutch scientist Simon Stevin (1548–1620), that the combined effect, or *resultant,* of two given forces acting on a body is found by the so-called parallelogram of forces. In the following figure, the two forces F_1 and F_2 are represented by arrows whose directions represent the direction of the forces and whose lengths represent the magnitude of the forces. The resultant force is then represented by the arrow along the diagonal of the parallelogram, as indicated in the figure.

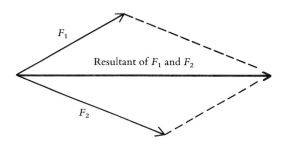

a. Imagine a heavenly body moving at a rate of v feet per second, with no external force acting on it. Let S be any point in space not on the line of motion of the planet and let the planet be at P_0 at the outset, at P_1 after 1 second, at P_2 after 2 seconds, and at P_3 after 3 seconds [see diagram (a) below]. Compare the areas of triangles SP_0P_1, and SP_2P_3.

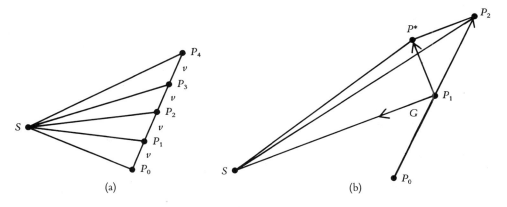

b. Now, imagine the sun at S exerting a gravitational force of attraction G. At time $t = 1$, for example, G pulls the planet toward S along the line P_1S while the inertial force is pulling the planet from P_1 to P_2. In diagram (b) below, which directed segment represents the resultant of the gravitational and inertial forces on the planet in the time interval $t = 1$ to $t = 2$?

c. Compare the area of triangle SP_1P^* with that of SP_1P_2. (Hint: The two triangles have the same base.) Then explain why Kepler's second law actually follows from this result.

·17·

Patterns

"The senses delight in things duly proportional." So said the famous philosopher-theologian Thomas Aquinas more than 700 years ago, in noting human aesthetic appreciation. In this chapter we will examine some of the elements of that aesthetic appreciation, particularly what we call *symmetry*.

Symmetry, like beauty, is very difficult to define. Dictionary definitions talk about "correspondence, equivalence, or identity among constituents of a system," "correspondence of form and arrangement of parts," and "beauty as a result of balance or harmonious arrangement" (*The American Heritage Dictionary,* 2nd ed.).

In the narrowest sense, symmetry refers to "mirror-image" correspondence between parts of an object. Crystals, in both their appearance and their atomic structure, provide examples of symmetry in this sense. Taken in a wider sense, though, symmetry includes notions of *balance, similarity,* and *repetition.*

It is our sense of symmetry that leads us to appreciate patterns. As we noted in the introduction to this part of the book, mathematics is the study of patterns, and we will see that mathematics gives important insights into symmetry.

Patterns abound in nature. The successive sections of the beautiful chambered nautilus grow according to a very strict and specific spiral pattern, a broader kind of symmetry. This spiral has the property that it has the same shape at any size: a photographic enlargement of it would fit exactly on the original spiral. This feature of self-similarity at a change in scale is a fundamental aspect of the fractals discussed in Chapter 20 and is no doubt associated with their aesthetic appeal.

Botanists have long appreciated other spirals, for instance, the intricate interlocking spirals of *phyllotaxis,* a phenomenon of plant growth from a central stem. The seeds of a sunflower are arranged in spirals, as are the

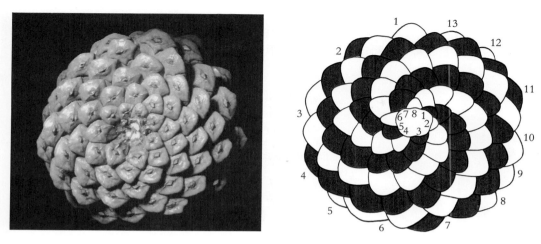

Figure 17.1 Spirals of scales on a pine cone: 8 right, 13 left. [Photo from The National Audubon Society Collection/PR.] Illustration after Verner E. Hoggatt, Jr., *Fibonacci and Lucas Numbers,* Houghton Mifflin, New York, 1969, p. 81.]

scales on a pineapple or a pine cone (Figure 17.1), and the petals on a daisy. Like the chambers of the nautilus in Figure 17.2b, the spirals on these plants are similar one to another, and they are arranged in a regular way, with balance and "proportion." These plants have a kind of symmetry we would naturally call **rotational.**

Associated with the geometric symmetry of phyllotaxis, curiously, there is also a kind of

numeric symmetry, with a "proportion" in the sense of a ratio of numbers. Strangely, the number of spirals in these plants is never just any whole number but always comes from a particular sequence of numbers, called the **Fibonacci numbers:** 1, 1, 2, 3, 5, 8, 13, 21, 34, 55, 89, 144, 233, 377, ... This sequence begins with the numbers 1 and 1 again and each other number is obtained by adding the two preceding numbers.

(a)

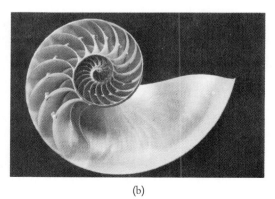

(b)

Figure 17.2 (a) This sunflower has 55 spirals in one direction and 89 spirals in the other direction. [© Dwight Kuhn.] (b) A chambered nautilus shell. [Photo by Nancy Rodger.]

SPOTLIGHT 17.1 Leonardo Pisano Bigollo ("Fibonacci")

A "portrait" of Leonardo Pisano ("Fibonacci") of unlikely authenticity. [David Eugene Smith Collection, Rare Book and Manuscript Library, Columbia University.]

Leonardo Pisano Bigollo was born in Pisa in 1170. He has been known more popularly for the past century and a half as "Fibonacci." This nickname refers to his descent from an ancestor named Bonaccio, but the nickname is of modern origin, and there is no evidence he was ever known as Fibonacci in his own time.

Leonardo was the greatest mathematician of the Middle Ages. His stated purpose in his book *Liber abbaci* (1202) was to introduce Hindu-Arabic numerals and calculation with them into Italy, to replace the Roman numerals then in use. Other books of his treated topics in geometry, algebra, and number theory.

We know little of Leonardo's life apart from a short autobiographical sketch in the *Liber abbaci:*

I joined my father after his assignment by his homeland Pisa as an officer in the customhouse located at Bugia [Algeria] for the Pisan merchants who were often there. He had me marvelously instructed in the Arabic-Hindu numerals and calculation. I enjoyed so much the instruction that I later continued to study mathematics while on business trips to Egypt, Syria, Greece, Sicily, and Provence and there enjoyed discussions and disputations with the scholars of those places.

(Quoted from L. E. Sigler, *Leonardo Pisano Fibonacci, The Book of Squares: An Annotated Translation into Modern English,* Academic Press, New York, 1987.)

The *Liber abbaci* contains a famous problem about rabbits, whose solution is the sequence now called the Fibonacci sequence. Leonardo did not write further about it.

Look at the sunflower in Figure 17.2a. You see a set of spirals running in a counterclockwise direction and another set in the clockwise direction. It is (just barely) possible to count the number of spirals in both directions; in the sunflower there are 55 in one and 89 in the other direction, two consecutive Fibonacci numbers. In the case of the pineapple, there are three sets of spirals, one each along the three directions through each hexagonally shaped scale. For the common grocery pineapple *(Ananas comosus),* there are always 8 spirals to the right, 13 to the left, and 21 vertically—again, consecutive Fibonacci numbers.

Why are the numbers of spirals in plants the same numbers that appear next to each other in a purely mathematical sequence? The question has been the subject of extensive research, and there is no easy answer; there are several intricate theories of the dynamics that must be involved in the plant's growth.

THE GOLDEN RATIO

The ancient Greeks fastened on a specific numerical proportion as being essential to their ideas of beauty. Known variously as the **golden ratio, golden mean,** or **divine proportion,** its geometric origin and mathematical properties were investigated by Euclid in Book II of the *Elements.*

The geometric origin was in the desire to construct a square with the same area as a given rectangle. For a rectangle with sides w and l, the square would need to have a side of length s satisfying

$$s^2 = lw$$

The length s is said to be the **mean proportional,** or **geometric mean,** between the two extremes of l and w, since s will always be between the two in size, and the equation above can be expressed in terms of the proportions

$$\frac{w}{s} = \frac{s}{l}$$

Given the lengths l and w, the Greeks devised a very simple geometric construction with compass and straightedge to produce a segment of length s and then a square with sides of length s.

A variation is to start with a *single* line segment and find a point on it that divides it into two pieces so that the larger piece is the mean proportional between the whole segment and

Figure 17.3 Point C divides segment AB in the golden ratio: $AC/CB = AB/AC$.

the smaller piece. In Figure 17.3 point C divides segment AB "in mean and extreme ratio" — in the golden ratio — provided

$$\frac{AC}{CB} = \frac{AB}{AC}$$

Let us follow tradition and denote this common ratio by the Greek letter ϕ (phi). Also, we note that substituting $AB = AC + CB$ gives

$$\phi = \frac{AC}{CB} = \frac{AC + CB}{AC} = \frac{AC}{AC} + \frac{CB}{AC}$$

so

$$\phi = \frac{AC}{CB} = 1 + \frac{CB}{AC}$$

The fraction on the right is just $1/\phi$, so we have

$$\phi = 1 + \frac{1}{\phi}$$

This is an equation in ϕ; we solve it for ϕ. Multiplying through by ϕ gives

$$\phi^2 = \phi + 1$$

or

$$\phi^2 - \phi - 1 = 0$$

This is a quadratic equation of the form

$$ax^2 + bx + c = 0$$

with variable ϕ in place of x and $a = 1$, $b = -1$, and $c = -1$. To find the solutions of the

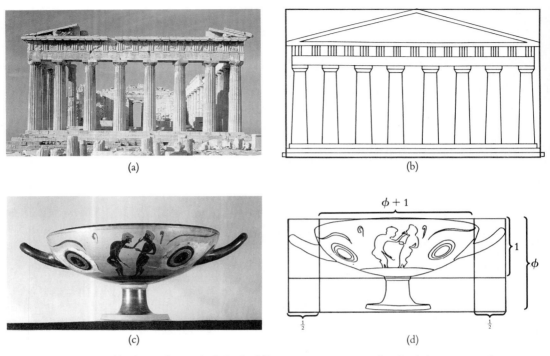

Figure 17.4 (a) The Parthenon, built in the fifth century B.C., as it is today. [Scala/Art Resource.] (b) When intact, the Parthenon fitted almost exactly into a golden rectangle. [After H. E. Huntley, *The Divine Proportion,* Dover Publications, New York, 1970, p. 63.] (c) A Greek drinking cup with proportions in the golden ratio. [Musée Vivenel, Compiègne. Giraudon/Art Resource.] (d) The grid shows the proportions for the drinking cup in (c).

equation, we apply the famous quadratic formula

$$x = \frac{-b \pm \sqrt{b^2 - 4ac}}{2a}$$

to get the two solutions

$$\phi = \frac{1 \pm \sqrt{5}}{2} = \frac{1 + \sqrt{5}}{2} \approx 1.618034$$

and

$$\frac{1 - \sqrt{5}}{2} \approx -0.618034$$

Not being a positive number, this second solution does not give a solution to our original problem about ratios of line segments, so we discard it. It is the first solution that is called

the golden ratio (though some authors give the same name to $1/\phi \approx 0.618$). It occurs often in other contexts in geometry — for example, in a pentagon in which all of the sides are equal, ϕ is the ratio of a diagonal to a side.

The Greeks felt that rectangles whose sides are in the golden ratio — **golden rectangles** — prove especially beautiful. The facade of the Parthenon is framed by such a rectangle, the Greek sculptor Phidias used proportions based on golden rectangles, and other Greek artifacts use combinations of golden rectangles (see Figure 17.4).

In more modern times the noted impressionist G. Caillebotte (1848–1894) used the

(a)

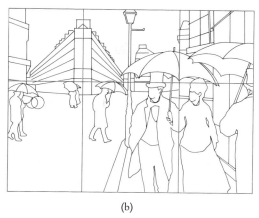

(b)

Figure 17.5 Gustave Caillebotte used golden ratio proportions in laying out his famous painting *Paris, A Rainy Day*, shown in (a). [Oil on canvas, 1877, 212 × 276.2 cm, Charles H. and Mary F. S. Worcester Collection, 1964. 336. © 1990 The Art Institute of Chicago, all rights reserved.] In (b) we show a grid over a schematic of the painting. The horizontal line divides the height of the painting in the golden ratio, and the vertical lines divide the right and left halves in the golden ratio. [After Kirk Varnedoe, *Gustave Caillebotte*, Yale University Press, 1987, pp. 34–39.]

golden ratio in such paintings as *Paris, A Rainy Day* (1877) (Figure 17.5), as did the post-impressionist Georges Seurat (1859–1891) in *The Circus* (1891), *Invitation to the Sideshow (La Parade)* (1889), and the famous pointillistic *Sunday Afternoon on the Island of the Grande Jatte* (1884–1886). Psychological research indicates that we do in fact prefer rectangles with shapes close to that of the golden rectangle, and this empirical fact has been exploited in product marketing by shaping products and product containers as golden rectangles.

The human body exhibits examples of ratios close to the golden ratio, as Leonardo da Vinci (1452–1519) noted in his drawings (Figure 17.6), and as you can see by comparing the height of your navel to your overall height. The twentieth-century Swiss-born architect Le Corbusier (Charles-Édouard Jenneret, 1887–1965) used the golden ratio (including the navel-height feature) as the basis for his "Modulor" scale of proportions.

There is an intriguing connection between the spiral of the nautilus and the spirals of the sunflower, between the golden ratio and the Fibonacci sequence. The nautilus shape follows what is known as an *equiangular* or *logarithmic* spiral, which in its turning determines a

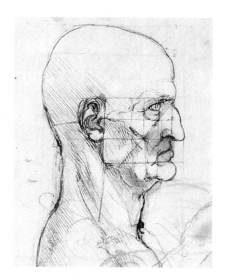

Figure 17.6 Leonardo used the golden ratio in his drawing *Head of an Old Man*. Some of the rectangles inside the square grid over the face approximate the golden ratio. [Leonardo, Studies of Physiognomy, Venice, Academy. Scala/Art Resource.]

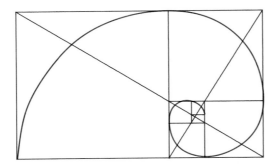

Figure 17.7 A logarithmic spiral determines a sequence of golden rectangles and corresponding squares.

sequence of golden rectangles (see Figure 17.7). The spirals of the sunflower are in fact approximations to a logarithmic spiral. The mathematical reason for this connection is that the ratios of consecutive Fibonacci numbers

$\frac{1}{1}$	1.0	$\frac{8}{5}$	1.6
$\frac{2}{1}$	2.0	$\frac{13}{8}$	1.625
$\frac{3}{2}$	1.5	$\frac{21}{13}$	1.615 . . .
$\frac{5}{3}$	1.666 . . .	$\vdots$	$\vdots$

provide alternately under- and overapproximations to $\phi = 1.618034$ . . . .

SYMMETRY

The spiral distribution of the seeds in a sunflower head and the spiraling of leaves around a plant stem highlight the aspects of symmetry of *similarity* and *repetition*. Another aspect of symmetry is *balance,* which refers to how the repetitions are arranged. A *single* spiral, although having a regularity, is not balanced, since there is no repetition; indeed, it may seem to be perpetually rotating, or just ready to rotate. Balance in symmetry is more common in such human-made objects as pottery, wallpaper, and buildings.

In considering patterns with repetition, we will distinguish the individual element or figure of the design (sometimes called the *motif*) from the pattern of the design — *how the copies of the motif are arranged.*

Mathematicians describe a variety of kinds of balance by using the geometric notion of **rigid motion,** also known as an **isometry** (which means "of the same measure"). A rigid motion is a specific kind of variation on the original pattern: we pick it up and move it, perhaps rotate it, possibly flip it over — but we don't change its size or shape. (To connect this concept with the language of Chapter 13, the original figure and its image are *congruent;* but we won't need to use that terminology here.)

Figure 17.8 shows the results of various motions applied to the rectangle in Figure 17.8a. Figure 17.8b shows the result of shrinking each side by 50%: not a rigid motion, because the size of the rectangle changes. For Figure 17.8c, we have imagined that the rectangle has rigid sides but hinges at the corner; like an unbraced bookshelf, it has sagged: again, this is not a rigid motion. In Figure 17.8d we have rotated the rectangle 90° (a quarter turn) clockwise around the center of the rectangle: this is a rigid motion. Similarly, in Figure 17.8e, rotating by 180° (a half turn) is a rigid motion; in fact, rotating by any angle, around any point as center, is a rigid motion. In Figure 17.8f we have reflected the rectangle along a vertical mirror down the middle: could you tell? The right and left halves have exchanged places. A reflection across *any* line is a rigid motion; Figure 17.8g shows the result of reflecting across a diagonal of the rectangle.

All reflections and all rotations are rigid motions. So are all **translations** — translation is just the mathematical word for a transformation that moves everything a certain distance in one direction.

The only remaining kind of rigid motion in the plane is a hybrid of reflection and translation. Known as a **glide reflection,** it is the kind of pattern your footprints make as you walk along: each successive element of the design (footprint) is a reflection of the previous one (see Figure 17.9). The motion combines a translation ("glide") with reflection. However, if you were to follow a translation with a reflection, or vice versa, you wouldn't get the same effect as a glide reflection. A glide reflection really is a separate entity from either a translation or a reflection.

Any rigid motion (in the plane) must be one of the four types we have described:

1. Reflection (across a line)

2. Rotation (around a point)

3. Translation (in a particular direction)

4. Glide reflection (across a line)

In terms of symmetry, we will be especially interested in rigid motions like those of Figure 17.8e and 17.8f that **preserve the pattern:** that is, ones for which the pattern looks exactly the same, *with all the parts appearing in the same places,* after the motion is applied.

You may enjoy thinking of applying these motions as a game, "The Pattern Game": *You turn your back, I apply a transformation, then you turn back and see if you can tell if anything is changed.*

The 90° rotation of Figure 17.8d does not preserve the pattern: the transformed pattern doesn't look just like the original. The moved

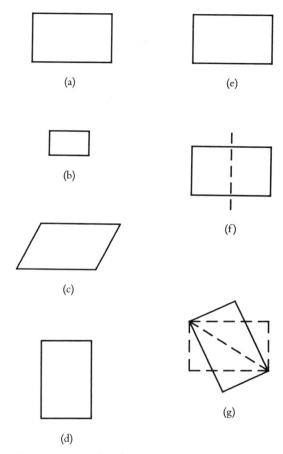

(a)

(b)

(c)

(d)

(e)

(f)

(g)

Figure 17.8 Results of various motions applied to a rectangle. (a) The original rectangle. (b) 50% reduction (not a rigid motion). (c) Sagging (not a rigid motion). (d) Quarter turn. (e) Half turn. (f) Reflection along the vertical line down the middle. (g) Reflection along a diagonal line.

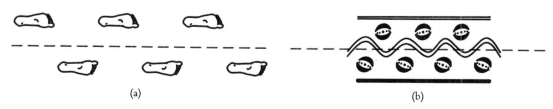

(a)

(b)

Figure 17.9 Glide reflection of (a) footprints and (b) design elements on a pot from San Ildefonso Pueblo. [From Dorothy K. Washburn and Donald W. Crowe, *Symmetries of Culture: Theory and Practice of Plane Pattern Analysis,* University of Washington Press, Seattle, Wash., 1988, p. 51.]

SPOTLIGHT 17.2 Symmetry in Modern Design: Scott Kim, an Artist in Symmetric Forms

Scott Kim—whom author Isaac Asimov has called "the Escher-of-the-Alphabet"—created in the 1980s a new art form with words. His calligraphy is playful, surprising, elegant, and fun. He calls the results *inversions:* words that can be read right side up, upside down, and every which way. His inversions use symmetries, distorting letters a little here or there, reflecting them as in mirrors, or rotating them around central points. The two illustrations shown here, taken from his book *Inversions* (W. H. Freeman, 1989), illustrate rotation *(MAN)* and left-right reflection *(WOMAN)*. He comments:

[Copyright Scott Kim, 1989.]

Deceptively simple constructions involving the letter *M*. Adding a single crossbar to a symmetric zigzag is enough to distinguish three asymmetrically placed letters: *M, A,* and *N*. A lowercase *a* is used in *WOMAN*. Notice that the two words, although closely related, have different symmetries: If you look at this design in a mirror, *WOMAN* will look the same but *MAN* will not. If you turn this design 180°, *MAN* will look the same but *WOMAN* will not.

rectangle doesn't fit exactly over the original rectangle (or over a copy of it that was present in the original pattern).

On the other hand, the 180° rotation of Figure 17.8e does preserve the pattern. It's true that the top of the original rectangle is now on the bottom of the transformed version; but you can't tell that that has happened, because you can't distinguish the two from appearance. If you had turned your back while the motion was applied, you wouldn't be able to tell that

anything had been done. A rotation by any multiple of 180° would also preserve the pattern.

Similarly, the mirror reflection along the vertical line in Figure 17.8f preserves the pattern, while the one in Figure 17.8g, where the mirror is along a diagonal, does not. (For an illustration of rotation and left-right reflection in calligraphy, see Spotlight 17.2.)

The pattern of footsteps in Figure 17.9a is not preserved under just reflection along the direction of walking — there is not a left footprint directly across from a right footprint. The pattern is preserved under a glide reflection along the direction of walking, as well as by a translation of two steps, or one of four steps, etc. — but not by a translation of one step.

Given a pattern, we will analyze it by *determining which rigid motions preserve the pattern.* These are often referred to as the **symmetry operations (or symmetries) of the pattern.** We will then be able to classify the pattern by which rigid motions preserve it.

We may think of a pattern as a recipe for repeating a figure (motif) indefinitely. Of course, any pattern we see in nature or art has only finitely many copies of the figure; but if the recipe for repetition is clear, we may imagine that we are looking at just a part of a pattern that extends indefinitely far.

Patterns in the plane can be divided into those that have indefinitely many repetitions in

* no direction — the **rosette patterns**

* exactly one direction (and its reverse) — the **strip patterns**

* more than one direction — the **wallpaper patterns**

A rosette pattern describes the possible symmetry operations for a flower. There is just one flower in the pattern; the repetition aspect of symmetry consists of the repetition of the petals around the stem. Translations and glide reflections do not come into play. The pattern is preserved under a rotation by certain angles, corresponding to the number of petals. There may or may not be reflections that preserve it, depending on whether the petal is symmetric. Most flowers have symmetric petals (Figure 17.10a), but some do not. An everyday example of rosette pattern — a human-made one — that does not have reflection symmetry is a

(a)

(b)

Figure 17.10 (a) Alpine aster with symmetric petals. [Photo by John Shaw.] (b) Pinwheel. [Photo by Travis Amos.]

pinwheel (Figure 17.10b). A rosette with *n* petals is said to have *n***-fold symmetry.** If there is no reflection symmetry, the motif of the pattern (the element that is repeated) is an entire petal; if there is reflection symmetry, the motif is just half a petal, because the entire pattern can be generated by rotation and re-

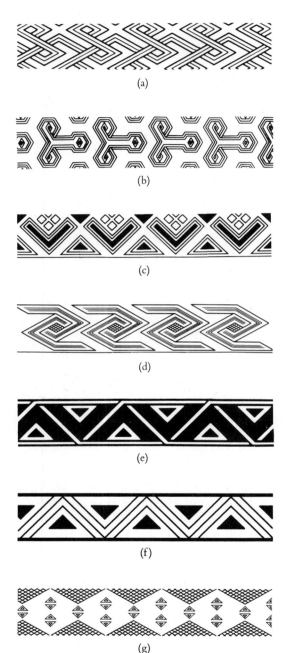

(a)

(b)

(c)

(d)

(e)

(f)

(g)

Figure 17.11 Bakuba patterns. (a) Carved stool. (b) Pile cloth. (c) Pile cloth. (d) Embroidered cloth. (e) Embroidered cloth. (f) Carved back of wooden mask. (g) Carved box.

flection of a half petal. The fact that these are the only possibilities is sometimes called *Leonardo's theorem,* after Leonardo da Vinci, who, in the course of planning the design of churches, needed to decide if chapels and niches could be added without destroying the symmetry of the central design.

We will illustrate the different kinds of strip patterns, and their "ingredient" symmetry operations, with patterns in art of the Bakuba people of Zaire, who are noted for their fascination with pattern and symmetry (see Spotlight 17.3).

All of the strip patterns offer repetition and translation symmetry along the direction of the strip. For our purposes, we will always position the pattern so that its repetition runs horizontally.

It may be that the pattern has no other rigid motions that preserve it apart from translation, as in Figure 17.11a.

The simplest other rigid motion to check for preservation of the pattern is reflection in a line, often called **bilateral symmetry** or *mirror symmetry:* the figure looks the same on both sides of a line, except that the two sides are mirror images of each other.

For a strip pattern, the center line of the strip may be a mirror line; if so, as in Figure 17.11b, we say that the pattern has symmetry across a horizontal line.

There may instead be mirror reflection across a *vertical* axis, such as the vertical lines through or between the V's in Figure 17.11c.

What kind of rotational symmetry can a strip pattern have? The only possibility for a strip pattern is a rotation by 180° (a half turn), since any other angle won't even bring the strip back into itself. (We don't count rotations of 360° or integer multiples (full turns), since any pattern at all is preserved under these.) Figure 17.11d shows a strip pattern that is unchanged by a 180° rotation about any point at the center of the small crosshatched regions.

SPOTLIGHT 17.3 Patterns Created by the Bakuba People

Among the Bakuba people of Zaire (shaded area of map), it is considered an achievement to invent a new pattern, and every Bakuba king had to create a new pattern at the outset of his reign. The pattern was displayed on the king's drum throughout his reign and, in the case of some kings, on his dynastic statue.

When missionaries first showed a motorcycle to a Bakuba king in the 1920s, he showed little interest in it. But the king was so enthralled by the novel pattern the tire tracks made in the sand that he had it copied and gave it his name.

(Adapted from Jan Vansina, *The Children of Woot,* University of Wisconsin Press, 1978, p. 221.)

The pattern made by tire tracks fascinated the Bakuba people. [Travis Amos.]

Two women with raffia cloths from the Bakuba village of Mbelo, July 1985; Mpidi Muya with embroidered raffia cloth (left) and Muema Kenye with plush and embroidered raffia cloth (right). [Photo by Dorothy K. Washburn.]

What about glide reflections? A row of alternating p's and b's has glide reflection:

Glide   p p p p p p p p p

Reflection   p p p p p p p p p
                         b b b b b b b b b

Glide reflection   p b p b p b p b p

For glide reflection, a p is translated as far as the next b and is then reflected upside down. Figure 17.11e shows a Bakuba pattern whose only symmetry (except for translation) is glide reflection.

Having examined symmetry operations on strip patterns, we can ask: what *combinations* of the four are possible? It turns out that apart from the five kinds of patterns we have already seen, there are only two other possibilities: we can have vertical line reflection and half turns, with either glide reflection but not horizontal line reflection (Figure 17.11f), or else with both glide reflection and horizontal line reflection (Figure 17.11g).

By analyzing cases, it is possible to show that these are the only seven possibilities; that is, mathematical analysis reveals that *there are only seven ways to repeat a pattern along a strip.* That this number is so small is quite surprising, since there is a myriad of different design elements (motifs); two designs may look entirely different yet share the same pattern of reproducing their design elements.

The standard notation of crystallographers for the strip patterns uses four symbols:

1. The first symbol is always a *p* (patterns other than the strip patterns also use the four-symbol system, and their notation does not necessarily begin with a *p*).

2. The second symbol is *m* if there is a vertical line of reflection; *1* otherwise.

3. The third symbol is *m* if there is a horizontal line of reflection, *a* if there is a glide reflection but no horizontal line of reflection, and *1* otherwise.

4. The fourth symbol is *2* if there is half-turn rotational symmetry, and *1* otherwise.

Figure 17.12 gives a flowchart for identifying patterns, together with the notations for them. The four-symbol notation is sometimes abbreviated by dropping the first and fourth symbols and replacing *a* with the more natural *g* (for "glide"). There is one exception: *p112* becomes *p12*.

So far we have classified the patterns with no translation repetition (the rosette patterns) and those with repetition in one direction (the strip patterns). What about those that have repetition in more than one direction—the wallpaper patterns? It turns out that there are exactly 17 of those. Illustrations, notation, and a flowchart are given in Spotlight 17.4 (p. 492).

The method we have developed here for classifying patterns, by the combinations of symmetry elements present, was originally developed by crystallographers in the nineteenth century. They wanted to classify and recognize the three-dimensional patterns associated with crystal structure. They proved—after several years of different crystallographers coming up with different totals!—that there are exactly 230 *crystal patterns*.

In applying these classification schemes to patterns on real objects, we need to take into account that the pattern itself may not be perfectly rendered. Also, patterns that are not on flat surfaces—for example, the pattern around the rim of a bowl, or around the body of a jar—require some latitude in our interpretation.

EXAMPLES: Patterns on Pueblo Pottery.
The pitchers in Figure 17.13 are from a thousand-year-old Pueblo site at Starkweather

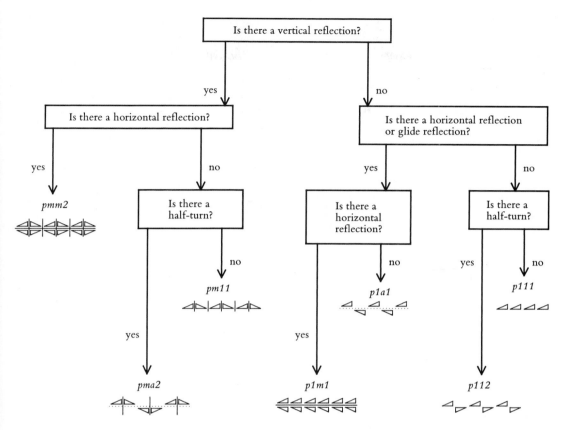

Figure 17.12 Flowchart for the seven strip patterns. [From Dorothy K. Washburn and Donald W. Crowe, *Symmetries of Culture: Theory and Practice of Plane Pattern Analysis,* University of Washington Press, Seattle, Wash., 1988, p. 83.]

Ruin near Reserve, New Mexico. We consider the patterns on the main bodies of the pitchers, which continue on the back sides. We suppose that they could be unwrapped and continued as strip patterns, and we consider them as such. We disregard the patterns on the spouts and handles.

We immediately come up against the question of the perfectness of the patterns. In Figure 17.13a the "teeth" on the left design element on the main body are "sharper" than those on the right. Is this lack of pattern, or just

lack of perfection in executing one? For our analysis, we opt for the latter.

Similarly, what are we to make of the diagonal lines on the pitcher in Figure 17.13b? In the narrowest interpretation, these lines are part of the pattern and any rigid motion that is to qualify as a symmetry of the pattern must preserve them. More liberally, we may consider the lines as a kind of shading, a way to make the region appear gray; indeed, to an observer at a distance, that is the effect of the lines.

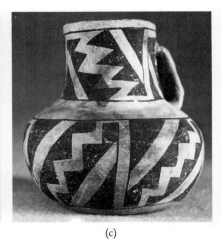

(a)   (b)   (c)

Figure 17.13 Reserve black-on-white pitchers from the Pueblo II horizon (900–1100 A.D.), excavated in 1935–1936 from Starkweather Ruin by Professor Paul H. Nesbitt and students from Beloit College. [Courtesy of Logan Museum of Anthropology, Beloit College. Photos by Paul J. Campbell.]

For the pattern on the body of the pitcher in Figure 17.13c, we notice that the jagged white line in the design element on the left has three "steps," while that in the one on the right has four. If we were really strict, we would decide that the two are different design elements. But we do detect a similarity of the two that we do not want to deny totally; we attribute the variations in the jagged lines to artistic license and for our purposes consider the two jagged lines to be the same.

If we follow the flowchart in Figure 17.12, we get the following:

◆ Figure 17.13a: Is there a vertical reflection? *No.* Is there a horizontal reflection or glide reflection? *No.* Is there a half turn? *No.* Hence the pattern is *p111.*

◆ Figure 17.13b (narrow interpretation of the diagonal lines): Is there a vertical reflection?

No. Is there a horizontal reflection or glide reflection? *No.* Is there a half turn? *Yes* (e.g., around the center of each cross). The pattern is *p112.*

◆ Figure 17.13b (liberal interpretation — diagonal lines as shading, their direction doesn't have to be preserved): Is there a vertical reflection? *Yes* (e.g., on a vertical line through the center of a cross). Is there a horizontal reflection? *Yes* (e.g., through the center of a cross). The pattern is *pmm2.*

◆ Figure 17.13c: Is there a vertical reflection? *No.* Is there a horizontal reflection or glide reflection? *No.* Is there a half turn? *Yes* (e.g., around the center of each jagged white line). The pattern is *p112.* (This pitcher has the interesting feature that the patterns on the neck and the body are mirror images of each other.)

The women who made the pots at Stark-weather Ruin strongly preferred the symmetry of half turns; very few of the pots have any reflection symmetry, either mirror or glide. The avoidance of mirror symmetry was a consistent feature of pottery of the indigenous peoples of the Western hemisphere. Spotlight 17.5 (p. 497) discusses the significance of pattern classification for anthropologists.

We emphasize again that our analysis of patterns does not refer to the *design* in the pattern, but to how its repetition is structured across the plane. There is an infinite variety of possible designs that artists can devise. You should imagine that the artist has created one copy of the design and is then contemplating how to place equal-sized copies of it in other parts of the (infinite) plane, in a way that is symmetric. It is those strategies for placement of which there are very few.

A slightly more involved analysis allows mathematicians to refine the classification of patterns to take into account colors that are repeated in a symmetric way.

Figure 17.14 Mosaics from the Hakim Bey Mosque, Kenya. [Bildarchiv Foto Marburg/Art Resource.]

PATTERNS AND TILINGS

When our ancestors used stones to cover the floors and walls of their houses, they selected shapes and colors to form pleasing designs. We can see the artistic impulse at work in mosaics, from Roman dwellings to Moslem religious buildings (see Figure 17.14). The same intricacy and complexity arise in other decorative arts — on carpets, fabrics, baskets, and even linoleum.

Such patterns have one feature in common: they use repeated shapes to cover a flat surface, without gaps or overlaps. If we think of the shapes as tiles, we can call the pattern a **tiling,** or *tessellation*. Even when efficiency is more important than aesthetics, designers value clever tiling patterns. In manufacturing, for example, stamping the components from a sheet of metal is most economical if the shapes of the components fit together without gaps — in other words, if the shapes form a tiling.

REGULAR POLYGONS

The simplest tilings are those using only one size and shape of tile, known as **monohedral** tilings.

We are interested especially in tiles that are **regular polygons,** figures all of whose sides

SPOTLIGHT 17.4 The 17 Wallpaper Patterns

There are exactly 17 wallpaper patterns. We give an example of each, together with a flowchart for identifying the patterns.

The International Crystallographic Union has established a standard notation for the wallpaper patterns. The full notation consists of four symbols:

- The first symbol is *c* (for "centered") if all rotation centers lie on reflection lines, and *p* (for "primitive") otherwise.
- The second symbol is either *1*, *2*, *3*, *4*, or *6*, whichever is the highest order of rotational symmetry present.
- The third symbol is either *m*, *g*, or *1*, corresponding to the presence of mirror, glide, or no reflection symmetry.
- The fourth symbol (*m*, *g*, or *1*) is for describing symmetry relative to an axis at an angle to the symmetry axis of the third symbol.

(Note: The patterns *p31m* and *p3m1* provide an exception to the notation.)

Below each pattern illustration we give both the standard abbreviation (on top) and the full notation (below).

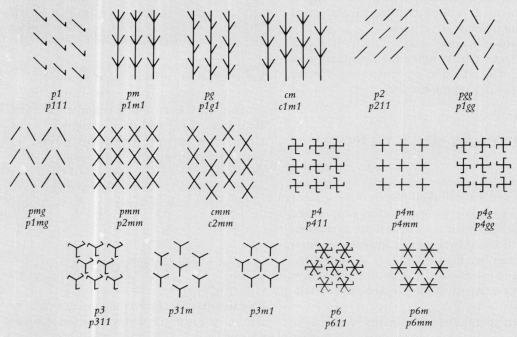

The 17 wallpaper patterns, with both the abbreviations and the full notation used by crystallographers.

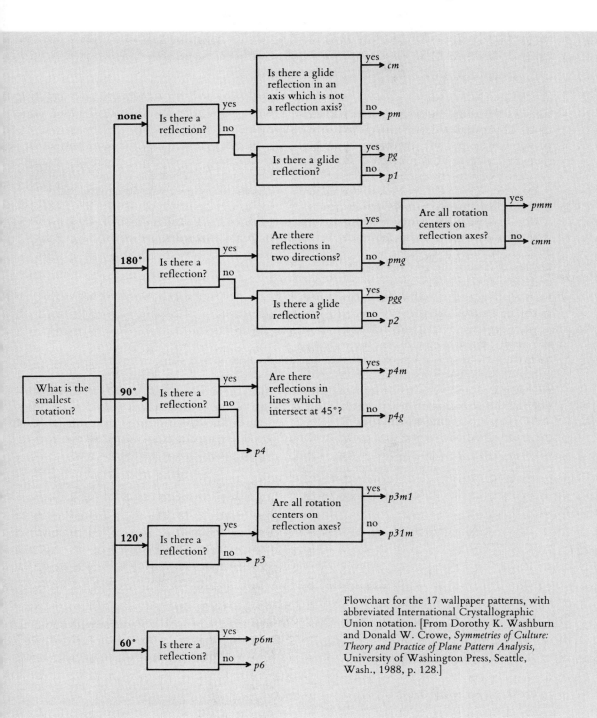

Flowchart for the 17 wallpaper patterns, with abbreviated International Crystallographic Union notation. [From Dorothy K. Washburn and Donald W. Crowe, *Symmetries of Culture: Theory and Practice of Plane Pattern Analysis,* University of Washington Press, Seattle, Wash., 1988, p. 128.]

are the same length and all of whose angles are equal. A square is a regular polygon with four sides and four equal interior angles; a triangle with all sides equal (an *equilateral* triangle) is also a regular polygon. A polygon with five sides is a pentagon, one with six sides is a hexagon, and one with *n* sides is an **n-gon.** Regular polygons are especially interesting because of their high degree of symmetry; each has the symmetry of a rosette pattern.

By a convention dating back to the ancient Babylonians, angles are measured in degrees, with the measure of an angle that goes all the way around a point being arbitrarily set to 360°. As we note in Chapter 16, the **interior angles** of a triangle add up to 180°; in an **equilateral triangle,** each angle equals 60°. Each of the interior angles of a square is one-fourth of a circle, or 90°.

Since we have deliberately referred to interior angles, you can be sure that there are also such things as exterior angles, and they too will come into play in our discussion. An **exterior angle** of a polygon is one formed by one side and the extension of an adjacent side (Figure 17.15). Proceeding around the polygon in the same direction, we see that each interior angle is paired with an exterior angle. If we bring all the exterior angles together at a single point, they will add up to 360° (see Figure 17.15). If the polygon has *n* sides, then each

exterior angle must measure 360/*n* degrees. For example, a square with *n* = 4 sides has 4 exterior angles, each measuring 90°; a pentagon with *n* = 5 sides has 5 exterior angles, each measuring 72°; while a regular hexagon with *n* = 6 sides has 6 exterior angles, each measuring 60°.

Notice that each exterior angle plus its corresponding interior angle make up a straight line, or 180°. Thus the interior angles for a regular triangle must be 60°; for a square, 90°; for a regular pentagon, 108°; for a regular hexagon, 120°; and for regular polygons with more than six sides, between 120° and 180°. This last consideration will prove crucial shortly.

REGULAR TILINGS

A monohedral tiling whose tile is a regular polygon is called a **regular tiling.** A square tile is the simplest case. Apart from varying the size of the square, which would change the scale but not the pattern of the tiling, we can get different tilings by offsetting one row of squares some distance from the next.

However, there is only one tiling that is **edge-to-edge,** that is, the edge of a tile coincides entirely with the edge of a bordering tile (see Figure 17.16 for a tiling that is not edge-to-edge and another that is). For simplicity, from now on we will consider only edge-to-edge tilings. For edge-to-edge tilings (even ones with tiles of different shapes and sizes), edges of different tiles meet at points that are surrounded by tiles and their edges; the particular arrangement of polygons around a point is its **vertex figure.**

Any tiling by squares can be refined to one by triangles by drawing a diagonal of each square; but these triangles are not regular (equilateral). Equilateral triangles can be arranged in rows by alternately inverting trian-

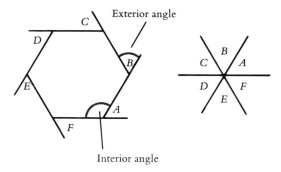

Figure 17.15 The exterior angles of a regular hexagon, like those of any regular polygon, add up to 360°. Each interior angle measures 60°.

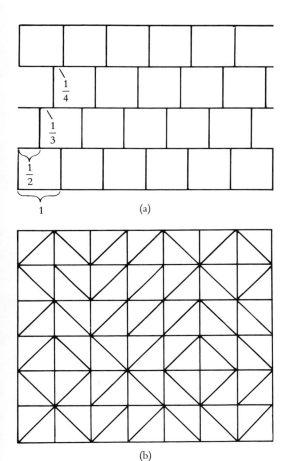

Figure 17.16 (a) A tiling that is not edge-to-edge; the horizontal edges of two adjoining squares do not exactly coincide. (b) A tiling by right triangles that is edge-to-edge.

gles; as with squares, there is only one pattern of equilateral triangles that forms an edge-to-edge tiling.

What about tiles with more than four sides? An edge-to-edge tiling with regular hexagons is easy to construct (see the hexagonal tiling of a moth's eye in the photograph on p. 342).

However, if we look for a tiling with regular pentagons, we won't be able to find one. How do we know whether we're just not being clever enough or there really isn't one to

be found? This is the kind of question mathematics is uniquely equipped to answer. In the other sciences, phenomena may exist even though we have not observed them; such was the case for bacteria before the invention of the microscope. In the case of an edge-to-edge tiling with regular pentagons, we can conclude with certainty that there isn't one.

The proof is very easy. As we calculated earlier, the interior angles of a pentagon are each 108°. At a point where several hexagons meet, how many can meet there? The total of all of the angles around a point must be 360°. Four pentagons at a point would be too many (they'd have to overlap), and three would be too few (some of the area wouldn't be covered). Since 108 does not evenly divide 360, regular pentagons can't tile the plane.

With this argument, we can do something that is a favorite with mathematicians: we can generalize it. Its main idea is a criterion for when a regular pentagon can tile the plane: when the size of its interior angles divides 360 evenly. We can apply this criterion to determine exactly which other regular polygons can tile the plane.

EXAMPLE: Identifying the Edge-to-Edge Regular Tilings. A regular hexagon has interior angles of 120°; 120 divides 360 evenly, and 3 regular hexagons fit together exactly around a point. A regular 7-gon — or any regular polygon with more than six sides — will have interior angles that are larger than 120° but smaller than 180°. Now 360 divided by 120 gives 3, and 360 divided by 180 gives 2 — and there aren't any other possibilities in between. Angles between 180° and 120° divided into 360° will give a result *between* 2 and 3, and consequently not an integer. So there are no edge-to-edge regular tilings of the plane with polygons of more than 6 sides; *the only edge-to-edge regular tilings are the ones with equilateral triangles, with squares, and with regular hexagons.*

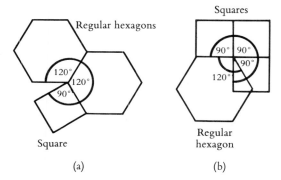

Figure 17.17 Polygons that come together at a vertex in a tiling must have interior angles that add up to 360° — no less, no more.

The follow-up question, of course, is which *combinations* of regular polygons of different numbers of sides can tile the plane edge-to-edge? Recall that the particular arrangement of polygons around a vertex is called the *vertex figure*. A systematic tiling that uses a mix of regular polygons with different numbers of sides but in which *all vertex figures are alike*—the same polygons in the same order—is called a **semiregular tiling.**

As before, our technique of adding up angles at a vertex (to be 360°) eliminates some impossible combinations, such as "square, hexagon, hexagon" (Figure 17.17). Having found the arrangements that are not numerically impossible, we must confirm the actual existence of each tiling by constructing it (i.e., show that it is geometrically possible). For example, even though a possible arrangement of regular polygons around a point is "triangle, square, square, hexagon," it is not possible to construct a tiling with that vertex figure at every vertex.

The result of this investigation is that in a semiregular tiling no polygon can have more than 12 sides. In fact, polygons with 5, 7, 9, 10, or 11 sides do not occur either. Figure 17.18 exhibits all of the eight semiregular tilings.

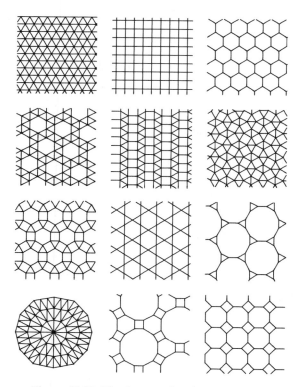

Figure 17.18 The three regular tilings and the eight semiregular tilings, plus one tiling that does not belong to either group. Can you identify it? [Courtesy of Darrah P. Chavey, Beloit College.]

NONPERIODIC TILINGS

All the patterns we've exhibited and discussed so far have been **periodic tilings.** If we transfer a periodic tiling to a transparency, it is possible to slide the transparency a certain distance horizontally, without rotating it, until the transparency exactly matches the tiling everywhere; we can achieve the same result by moving the transparency some second direction (possibly vertically) a certain (possibly different) distance.

Another way to think of a periodic tiling is that in it you can identify a *fundamental region* that you can use to tile the plane by transla-

SPOTLIGHT 17.5 Symmetry in Ancient Design

Archeologist Dorothy Washburn comments on the significance of design as an indicator of a culture's history and how design can signal change in a culture:

Human beings do things in a very consistent fashion; and over the years, within a given cultural group, their activities are nonrandom. In an archeological context we can see repetition of these behavior patterns in their material culture. But what we did not see before the mathematics of symmetry was that the structure of a culture's decorative designs are consistent over time and through space within a given cultural group.

That is, although some people do produce random patterns, by far the largest number of decorative designs that cultural groups produce are based on symmetry. The point is that any time you repeat a motif in a systematic fashion, you're using one of the four rigid motions to make that repetition.

Early descriptions of design were largely idiosyncratic and dealt just with the individual types of material — types of textiles or types of pottery, and the features that typified these types. But now we can study and compare the way the motifs in design are arranged in material from all different cultures throughout the world — regardless of whether they're contemporary cultures or past cultures — and see how these designs are put together and how they change through time and space.

One of the most interesting things we've found is that a given cultural group, a tribe or a band unit, will choose just a few symmetries to structure its designs. I've tested this observation through studies of California Indian baskets, with the work of Bakuba cloth weavers, and I've even found a consistency in the archeological record among the Anasazi, one of the prehistoric traditions of the American Southwest.

Let's take an example from material found on Crete. The site of Knossos had 3,000 years of uninterrupted prehistory. For 1,500 of those years, only two of the 7 one-dimensional symmetries were used. Then suddenly 5 more symmetries came into use. The design motifs were the same, but it was the rearrangement of these motifs into patterns based on different structures that suggested to us that something really interesting was happening.

That something interesting was the beginning of trade in the Aegean. We could see that simply by noting the introduction of new symmetry patterns from cultures outside that island. The increase in the number and variety of symmetries indicated that trade was coming in; the change in design structure was an incredibly sensitive marker of change.

tions. In the terms we used earlier in the chapter, the periodic tilings are ones that are preserved under translations in more than two directions. (What we have called wallpaper patterns are sometimes called *periodic plane patterns.* In this part of the chapter we are concerned with how to assemble such patterns from polygons; that is, we are concerned with the design elements more than with the patterns.)

There are infinitely many periodic tilings and infinitely many nonperiodic ones. As an example of a nonperiodic tiling, we can consider again a non-edge-to-edge tiling by squares, where the second row from the bottom is offset half a square to the right from the bottom row, the third row a third of a square more, and so forth (see Figure 17.16a). An elegant theorem in number theory assures us that $\frac{1}{2} + \frac{1}{3} + \frac{1}{4} + \cdots + 1/n$ never adds up to a whole number, so this type of tiling never repeats: there is no direction in which we can move the entire tiling and have it coincide with itself. (You should try horizontal, vertical, and diagonal shifts to convince yourself.)

For an edge-to-edge example (after all, we said we would stick to edge-to-edge tilings), take the usual edge-to-edge square tiling and divide each square into two right triangles by adding at random either a rising or a falling diagonal. The chance of this procedure resulting in a periodic tiling by right triangles is 0 (see Figure 17.16b). If you aren't willing to leave matters to chance, then just consider the tiling of Figure 17.19.

As far as mathematicians know, for monohedral tilings, every shape that can be used to construct a nonperiodic tiling of the plane can also be used to assemble a periodic one. It is an open question whether this assertion is true for every shape of tile whatsoever. To settle the question in the affirmative, mathematicians would have to give a proof that worked for any shape of tile. Alternatively, to settle the matter in the negative, they would have to either devise a specific tile, use it in a nonperiodic tiling, and prove that it cannot form a periodic tiling; or else prove that there is such a tile, even if they can't describe it precisely.

THE PENROSE TILES

For a long time, mathematicians believed that if you can construct a nonperiodic tiling with a set of tiles, you can construct a periodic one. But in 1964 the first set of tiles was found that permits only nonperiodic tiling. No wonder it took so long to discover that set — it contains 20,000 different shapes! Over the next several years smaller sets of this type were discovered, with as few as 100 shapes. But it was still amazing when in the 1970s Roger Penrose, a mathematical physicist at Oxford, found a new set that would do the job — with just 2 tiles! (See Figure 17.20.)

Penrose called his tiles "darts" and "kites." It is easy to specify their construction, as one of each can be obtained from a single rhombus. (A *rhombus* is a figure in the plane with four equal sides and equal opposite interior angles.) The particular rhombus from which the Penrose tiles are constructed has interior angles of 72° and 108°. If we divide the longer diagonal according to the golden ratio — the proportion $(1 + \sqrt{5})/2 \approx 1.618/1$, so dear to the Greeks — and connect the dividing point to the remaining corners, we split the rhombus into a dart and a kite (Figure 17.20).

Since the two Penrose pieces come from a rhombus and that rhombus can be replicated to tile the plane periodically, you must have

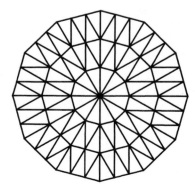

Figure 17.19 An edge-to-edge nonperiodic tiling.

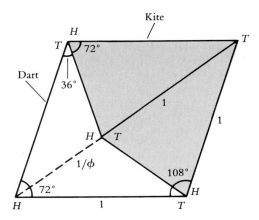

Figure 17.20 Construction of Penroses's "dart" (white area) and "kite" (colored area). The length $1/\phi \approx 0.618$; ϕ is the golden ratio.

guessed that the rules for fitting the Penrose pieces together do not allow the rhombus arrangement. The actual plastic pieces (produced in a limited edition in the 1970s) have little nicks and corresponding bumps that enforce this prohibition. Instead, we may label the front and back vertices of the dart with H (for head) and its two wing tips with T (for tail), and do the reverse for the kite. Then our rule is that only vertices with the same letter may meet: heads must go to heads, and tails to tails.

A prettier method of enforcing the rules, proposed by John Conway of Cambridge University, is to draw circular arcs of different colors on the pieces and require that adjacent edges must join arcs of the same color. The result is the pretty patterns of Color Plate 12. In fact, Conway thinks of the darts as children, each with two hands. The rule for fitting the pieces together is that children are forced to hold hands. Penrose patterns become dancing circles of children.

Color Plate 11a shows a tiling by a different pair of pieces, both rhombuses, that tile the plane only nonperiodically. Color Plate 11b shows a modification of the Penrose pieces into two bird shapes. Color Plate 11c shows a

coloring of one particular tiling with the Penrose pieces so that no two adjacent pieces have the same color.

Although tilings with Penrose's pieces cannot be periodic, the tilings possess unexpected symmetry. As you recall, we have explored our intuitions of symmetry in terms of *balance, similarity,* and *repetition.* Patterns made with the Penrose pieces certainly involve repetition, but it is the balance in the arrangement that we seek. What balance can there be in a nonperiodic pattern? It turns out that some Penrose patterns have a single line of reflection. But most surprising of all, every Penrose pattern has arbitrarily large regions with fivefold and tenfold rotational symmetry!

Consider, for example, any one of the 10 red pieces of Figure 17.21. If we rotate the pattern around its center through one-fifth of a turn, the region surrounded by the red pieces looks exactly the same as before (but parts of the

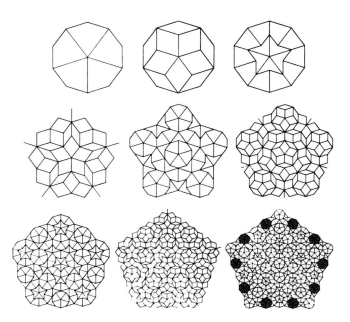

Figure 17.21 Successive deflation (i.e., the systematic cutting up of large tiles into smaller ones) of patches of tiles of a Penrose nonperiodic tiling. [Courtesy of Roger Penrose.]

pattern farther away may not exactly match). Note that rotation through one-tenth of a turn would not preserve all of that region. In Conway's metaphor, whenever a chain of children closes, the region inside has fivefold symmetry.

There are, in fact, two—and only two—Penrose tilings for which the *entire* pattern has the fivefold rotation symmetry of a rosette. Figure 17.21 shows how to construct both of these, beginning with five kites meeting at a vertex. We cut the darts and kites up into smaller darts and kites and then enlarge the new ones to the same size as the old, that way covering more area each time (the figure doesn't show the enlargement). Conway calls this operation *deflation*. As we proceed with successive steps, we get partial tilings alternately by two different patterns, and each has a fivefold rotational center: one has five kites at the center, the other has five darts.

Where does this rotational symmetry come from? The original rhombus we split up has the angles shown in Figure 17.20. Except in the recess of the dart and matching part of the kite, all the internal angles of the kite and dart are either 72° or 36°. Now, 72° goes into 360° five times, and 36° goes 10 times. If we recall that it is the interior angles that matter in arranging polygons around a point, we see that fivefold or tenfold symmetry could conceivably result from using such tiles.

The reverse of Conway's deflation, *inflation,* is the key idea in a simple argument to show that a Penrose pattern must be nonperiodic. For the inflation process, cut each dart down its middle and put glue on the short edges of the resulting triangles (but not on the cut itself). The result is a pattern of larger kites and darts!

We show that a Penrose pattern is nonperiodic by proceeding by contradiction. Suppose (contrary to what we want to establish) that some Penrose pattern is periodic, that is, it has translation symmetry. Let *d* be the distance along the translation direction to the first repetition. Performing inflation does the same thing to each repetition, so the inflated pattern *must* still have translation symmetry and a distance *d* along the translation direction to the first repetition. Keep on performing inflation, time after time, until the darts and kites are so large that they are more than *d* across. The pattern, as we have just argued, must still have translation symmetry at a distance *d*; but it can't, because there's no repetition inside a single tile! We reach a contradiction. So what's wrong? Our initial supposition, that the pattern was periodic in the first place, must have been erroneous. We conclude that all Penrose tilings are nonperiodic.

Despite their being nonperiodic, all Penrose patterns are somewhat alike, in the following sense: Any finite region in one pattern is contained somewhere inside every other pattern; in fact, it occurs infinitely many times in every pattern.

Penrose tilings have another feature that allows us to characterize them as "quasiperiodic." Robert Amman introduced onto the two rhombic Penrose pieces used in Color Plate 11a lines that are now known as *Ammann bars.* In any Penrose tiling, these bars line up into five sets of parallel lines, each set rotated 72° from the next, forming a pentagonal grid (Figure 17.22). The distance between two adjacent parallel bars is one of only two values, either *A* or *B*. Do you want to guess what the ratio of the longer *A* is to the shorter *B*? You don't think it could possibly be anything but the golden ratio, do you? And so it is.

EXAMPLE: Musical Sequences. What about the order in which the *A*'s and *B*'s occur, as we move from left to right in Figure 17.22? Is there any pattern to that? From the limited part of the pattern we can observe, we see the sequence as

A B A A B A B A A B A B A . . .

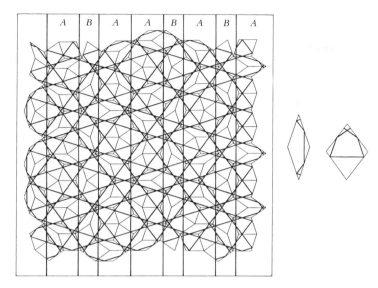

Figure 17.22 Penrose tilings with Amman bars. Specially placed lines on the tiles produce five sets of parallel bars in different directions. [Courtesy of Paul J. Steinhardt, University of Pennsylvania.]

You might think from the figure that the pattern continues repeating the group

$$A \; B \; A \; A \; B \; . \; . \; .$$

indefinitely. After all, there are five symbols in this group. But such is not the case. Known as a *musical sequence,* the sequence of intervals between Ammann bars is nonperiodic — it cannot be produced by repeating any finite group of symbols. We can think of it as a one-dimensional analogue of a Penrose tiling.

There is some regularity in musical sequences. Two *B*'s can never be next to each other, nor can we have three *A*'s in a row. Just as any finite part of any Penrose tiling occurs infinitely often in any other Penrose tiling, any finite part of any musical sequence appears infinitely often in any other one. The order of the symbols is neither periodic nor random, but between the two — quasiperiodic.

The ratio of darts to kites in an infinite Penrose tiling, or of *A*'s to *B*'s in a musical sequence, is exactly the golden ratio, approximately 1.618. So if you are going to play with sets of Penrose pieces and see what kinds of patterns you can create, you will need about 1.6 times as many kites as darts.

As pointed out by geometers Marjorie Senechal (Smith College) and Jean Taylor (Rutgers University), Penrose tilings have three important properties:

1. They are constructed according to rules that force nonperiodicity.

2. They can be obtained from a substitution process (inflation and deflation) that features self-similarity (self-similarity at a change in scale is one of the hallmarks of fractals, which are discussed in Chapter 20).

3. They are quasiperiodic.

Research of the late 1980s indicates that these properties are somewhat independent, meaning that one or two may be true of a tiling without all three being true.

SPOTLIGHT 17.6 Quasicrystals

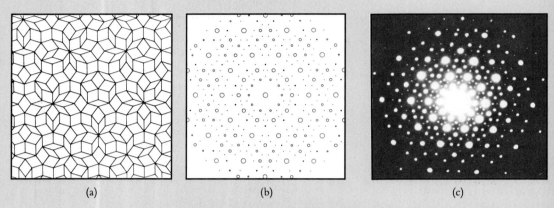

(a)   (b)   (c)

(a) A nonperiodic Penrose tiling. (b) A three-dimensional crystallike structure based on this tiling. (c) The crystal pattern observed by chemist Daniel Shechtman in a special manganese-aluminum alloy. Note the similarity to the pattern in (b). (d) Scanning tunneling microscope image of a perfect Penrose tiling formed by four layers of atoms on the surface of an aluminum-cobalt-copper alloy. (e) The tiling corresponding to the image in (d); the aluminum atoms (open centers) tend to settle into pentagonal rings around copper or cobalt atoms (filled centers). [(a), (b), and (c) from D. R. Nelson and B. I. Halperin. *Science* 229:233–238 (1985); (d) and (e) courtesy of A. Refik Kortan, AT&T Bell Laboratories.]

In 1984 Paul Steinhardt, a physicist at the University of Pennsylvania, and one of his graduate students, Don Levine, calculated the diffraction patterns that their three-dimensional Penrose patterns would produce if the building blocks were real atoms instead of imaginary tiles.

Diffraction patterns are the windows physicists use to peer inside materials. When a beam of electrons or x-rays pass through a solid material, they are diffracted, or scattered, by the atoms inside. The diffracted beams can be photographed head-on, and the images they form reflect the atomic architecture of the solid.

The most distinct diffraction patterns contain sharp, isolated dots. These patterns are the portraits of crystal, and they owe their clearly defined spots to the periodicity of the underlying structure. In a few preferred directions, depending on the arrangement of the atoms, the diffracted beams reinforce one another, producing bright spots on the film. A crystal is a little like an orchard planted in a rigid geometric grid. Most lines of sight are blocked by trees, but you can see right through to the other side in a few directions.

In another class of diffraction patterns the dots are either spread out into fuzzy rings or altogether absent. These are the images formed by glassy materials. Glasses, in contrast to crystals, are made up of atoms or molecules stuck together randomly; they're more like random forests than well-planned orchards. Because they offer no preferred directions for diffraction, the patterns they produce contain no sharp dots.

The computed diffraction pattern for Levine and Steinhardt's imaginary solid contained a surprise: unmistakable sharp points. Since the atomic arrangement of their solid was nonperiodic, it

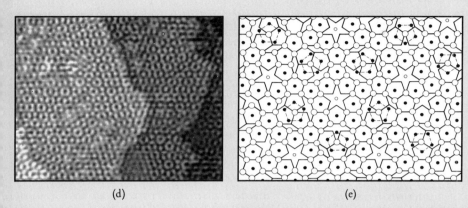

(d)   (e)

should have produced the fuzzy diffraction pattern characteristic of glassy substances. Since the dots in the pattern were arranged with fivefold symmetry, the solid wasn't a crystal either. Steinhardt decided to call it a *quasicrystal*.

In the fall of 1984 a colleague of Steinhardt's showed him a diffraction image made from a real substance, Shechtman's alloy of aluminum and manganese. The picture looked amazingly similar to Steinhardt and Levine's computer simulation.

In short order more than a hundred alloys with fivefold symmetry were discovered; and sevenfold, ninefold, and other symmetries proved to be possible. But no one could think of a mechanism by which millions upon millions of real atoms could arrange themselves spontaneously in those intricate patterns.

Anyone who tries to assemble Penrose pieces into tilings quickly realizes that it's not easy. You have to think ahead and keep the whole pattern in mind when adding a tile; otherwise, there is trouble. *Local rules*, or instructions for fitting a tile into a particular niche, don't seem sufficient to build the entire pattern without *global rules* that force you to plan ahead and check the configuration of tiles at distant points.

Local rules for adding tiles are analogous to forces that attract and hold new atoms to the surface of a growing quasicrystal. The atoms on a growing surface do not plan ahead. If quasiperiodic patterns could be constructed only with the help of global rules, they could not be assembled by real atoms in real alloys, and quasicrystals could not exist in nature.

In 1988 playfulness paid off once more. George Onoda, an IBM ceramics expert, started toying with about 200 Penrose tiles. Unconvinced that he wasn't supposed to be able to do it, he learned how to assemble flawless tilings of any size using only local rules.

For a complete theory of quasicrystals, the local rules will have to be generalized to three dimensions, and they must be shown to correspond to actual atomic forces. In the meantime, experimentalists continue to report bigger, more perfect quasicrystals.

(Adapted from Hans C. von Baeyer, "Impossible Crystals," in *Discover* 11(2):69–78, 84 (February 1990).)

SHECHTMAN'S CRYSTALS AND BARLOW'S LAW

Although Penrose's discovery was a big hit among geometers and in recreational mathematics circles in the mid-1970s, few people thought that his work might have practical significance. In the early 1980s some mathematicians even generalized Penrose tilings to three dimensions, using solid polyhedrons to fill space nonperiodically. Like the two-dimensional Penrose patterns, these have orderly fivefold symmetry but are nonperiodic.

Yet in 1982 scientists at the U.S. National Bureau of Standards discovered unexpected fivefold symmetry while looking for new ultrastrong alloys of aluminum (mixtures of aluminum with other metals).

Manganese doesn't ordinarily alloy with aluminum, but the experimenters were able to produce small crystals of alloy by cooling mixtures of the two metals at a rate of millions of degrees per second. Following routine procedures, chemist Daniel Shechtman began a series of tests to determine the atomic structure of the special crystals. But there was nothing routine about what he found: the atomic structures of the manganese-aluminum crystals were so startling that it took Shechtman 3 years to convince his colleagues they were real.

Why did he encounter such resistance? His patterns — and the crystals that produced them — defied one of the fundamental laws of crystallography. Like our discovery that the plane cannot be tiled by regular pentagons, **Barlow's law,** also called the **crystallographic restriction,** says that no crystal can have more than one center of fivefold symmetry.

Peter Barlow was a nineteenth-century British mathematician whose name survives today in the name of a book of mathematical tables. His argument was a very simple proof by contradiction, similar to Conway's proof in which we saw earlier that Penrose patterns are

not periodic. Suppose (contrary to what we intend to show) that there is more than one fivefold rotation center. Let A and B be two of these that are closest together (see Figure 17.23). Rotate the pattern of Figure 17.23 by one-fifth of a turn clockwise around B, which carries A to some point A'. Since the pattern has fivefold symmetry around B, the point A', which is the image of the fivefold center A, must itself be a fivefold center. Now we use A as a center and rotate the pattern by one-fifth of a turn counterclockwise, which carries B to some point B'; as we just argued for A', B' must also be a fivefold center. But A' and B' are closer together than A and B, which is a contradiction. Hence our original supposition must be false, and a pattern can have at most one fivefold rotation center (as the patterns in Figure 17.21 in fact do).

Barlow's law, as a mathematical theorem, shows that fivefold symmetry is impossible in a periodic tiling of the plane or of space. Chemists, for good theoretical and experimental reasons, believe that crystals are modeled well by three-dimensional tilings. An array of atoms with no symmetry whatever would not be considered a crystal. Yet until Penrose's discovery, no one realized that non-

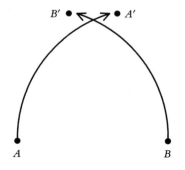

Figure 17.23 Barlow's proof that no pattern can have two centers of fivefold symmetry. [From Martin Gardner, *Penrose Tiles to Trapdoor Ciphers,* W. H. Freeman, New York, 1989, p. 27.]

periodic tilings—or arrays of atoms—can have the regularity of fivefold symmetry.

Chemists could simply say that Shecht-man's alloys aren't crystals. In the classical sense they aren't, but in other respects they do resemble crystals. It is scientifically more fruitful to extend the concept of crystal to include them rather than rule them out; they are now known as *quasicrystals* (see Spotlight 17.6, p. 502.)

Once again, as so often in history, pure mathematical research anticipated scientific applications. Penrose's discovery, once just a delightful piece of recreational mathematics, is now prompting a major reexamination of the theory of crystals.

Not so surprising, either, is where we finish this chapter compared to where we started. We started to investigate how mathematics can contribute to our understanding of beauty through analyzing symmetry in patterns. We encountered the unexpected Fibonacci sequence in nature, and its companion—the golden ratio—in both natural and human-made forms. We then did some pure mathematics in asking about the existence of nonperiodic patterns, only to find ourselves returning once again to the golden ratio, discovering more subtle patterning in apparent chaos, and finding subtlety—and beauty—unexpectedly in still more natural forms.

REVIEW VOCABULARY

Barlow's law, or the **crystallographic restriction** A law of crystallography which states that a crystal may have only rotational symmetries that are twofold, threefold, fourfold, or sixfold.

Bilateral symmetry Ordinary mirror (that is, reflection) symmetry, as seen in the letter A.

Divine proportion Another term for the **golden ratio.**

Edge-to-edge tiling A tiling in which bordering tiles meet only along full edges of each.

Equilateral triangle A triangle with all three sides equal.

Exterior angle The angle outside a polygon formed by one side and the extension of an adjacent side.

Fibonacci numbers The numbers in the sequence 1, 1, 2, 3, 5, 8, 13, 21, 34, . . . (each number after the second is obtained by adding the two preceding numbers).

Geometric mean The number s is the geometric mean between numbers l and w if $l/s = s/w$, or $s^2 = lw$.

Glide reflection A combination of translation

(= glide) and reflection in a line parallel to the translation direction. Example:

. . . p b p b p b . . .

Golden ratio, golden mean The ratio of ϕ to 1, where $\phi = (1 + \sqrt{5})/2$.

Golden rectangle A rectangle the lengths of whose sides are in the golden ratio.

Interior angle The angle inside a polygon formed by two adjacent sides.

Isometry Another word for rigid motion. Angles and distances, and consequently shape and size, remain unchanged by a rigid motion. (For plane figures there are only four possible isometries: reflection, rotation, translation, and glide reflection.)

Mean proportional Same as **geometric mean.**

Monohedral tiling A tiling with only one size and shape of tile (the tile is allowed to occur also in "turned-over," or mirror-image, form).

n-fold symmetry A pattern has n-fold symmetry if a rotation of $360°/n$ preserves the pattern.

n-gon A polygon with n sides.

Periodic tiling   A tiling that repeats at fixed intervals in two different directions, possibly horizontal and vertical.

Preserves the pattern   A transformation preserves a pattern if all parts of the pattern look exactly the same after the transformation has been performed.

Regular polygon   A polygon all of whose sides and angles are equal.

Regular tiling   A tiling by regular polygons, all of which have the same number of sides and are the same size.

Rigid motion   A motion that preserves the size and shape of figures; in particular, any pair of points is the same distance apart after the motion as before.

Rosette pattern   A pattern that has no repetitions by translation.

Rotational symmetry   A figure has rotational symmetry if a rotation about its "center" leaves it looking the same, like the letter S.

Semiregular tiling   A tiling by regular poly-

gons; all polygons with the same number of sides must be the same size.

Strip pattern   A pattern that has indefinitely many repetitions in one direction.

Symmetry (operation) of the pattern   A transformation of a pattern is a symmetry operation (or symmetry) of the pattern if it preserves the pattern.

Tiling   A covering of the plane without gaps or overlaps.

Translation   A rigid motion that moves everything a certain distance in one direction.

Translation symmetry   An infinite figure has translation symmetry if it can be translated (slid, without turning) along itself without appearing to have changed. Example:

. . . A A A A A A . . .

Vertex figure   The pattern of polygons surrounding a vertex in a tiling.

Wallpaper pattern   A pattern in the plane which has indefinitely many repetitions in more than one direction.

SUGGESTED READINGS

Ascher, Marcia: "Patterned Strip Decorations," Chapter 7 in *Ethnomathematics: Mathematical Ideas in Other Cultures,* Wadsworth & Brooks/ Cole, Pacific Grove, Calif., 1990.

Barber, Frederick, et al.: *Tiling the Plane,* Faculty Advancement in Mathematics Module. COMAP, Arlington, Mass., 1989.

Gallian, Joseph A.: "Symmetry in Logos and Hubcaps," *American Mathematical Monthly* 97(3):235–238 (March 1990).

Gardner, Martin: "Mathematical Games: Extraordinary Nonperiodic Tiling That Enriches the Theory of Tiles," *Scientific American* 236:1, 110–121, 132 and front cover (January 1977). Reprinted with additional material in *Penrose Tiles to Trapdoor Ciphers,* by Martin Gardner, W. H. Freeman, New York, 1989, pp. 1–29.

Hoggatt, Verner E., Jr.: *Fibonacci and Lucas Numbers,* Houghton Mifflin, New York, 1969.

Huntley, H. E.: *The Divine Proportion,* Dover Publications, New York, 1970.

O'Daffer, Phares G., and Stanley R. Clemens: *Geometry: An Investigative Approach,* Addison-Wesley, Reading, Mass., 1976, Chapters 1–5. A gentle introduction to the geometry of symmetry, with lots of examples and illustrations. Chapter 4 gives an elementary proof that there are only four kinds of rigid motions in the plane.

Penrose, Roger: *The Emperor's New Mind,* Oxford University Press, 1989.

Runion, Garth E.: *The Golden Section and Related Curiosa,* Scott Foresman and Company, Glenview, Ill., 1972.

Senechal, Marjorie, and Jean Taylor: "Quasicrystals: The View from Les Houches," *Mathematical Intelligencer* 12(2):54–64, (Spring 1990).

Sibley, Thomas Q.: *Geometric Patterns: A Study in*

Symmetry, Saint John's University, College-ville, Minn., 1989.

Steinhardt, Paul Joseph: "Quasicrystals," *American Scientist* 74(6):586–597 plus cover (November–December 1986). Includes illustrations of nonperiodic tilings with sevenfold and ninefold symmetry.

von Baeyer, Hans C.: "Impossible Crystals," *Discover* 11(2):69–78, 84 (February 1990). Tells how the playfulness of mathematicians and physicists led to the discovery of quasi-crystals.

Washburn, Dorothy K., and Donald W. Crowe: *Symmetries of Culture: Theory and Practice of Plane Pattern Analysis,* University of Washington Press, Seattle, Wash., 1988. An introduction to the mathematics of symmetry, splendidly illustrated with photographs of patterns from cultures all over the world. Includes a complete analysis of patterns with two colors. Appendices contain proofs of the facts that there are only four rigid motions in the plane and that there are exactly seven strip patterns.

EXERCISES

1. A pair of newborn male and female rabbits is placed in an enclosure to breed. The rabbits start to bear young 2 months after their own birth. At the end of each month, they have another male-female pair, which in turn matures and starts to bear young 2 months later. Assuming that none of the rabbits dies, how many pairs of rabbits will there be at the end of a year (just *before* any births)? (Hint: Draw a month-by-month chart of the situation at the end of the month, just before any births, and check the standings after 1, 2, 3, . . . , etc., months.)

2. Sometimes a sequence of numbers is specified by stating the value of the first term or couple terms and then giving an equation to calculate succeeding terms from preceding ones. This is called a *recursive* rule. Let's denote the nth Fibonacci number by F_n; then the Fibonacci sequence can be defined by

$$F_1 = 1, F_2 = 1, \text{ and } F_{n+1} = F_n + F_{n-1} \quad \text{for} \quad n > 2$$

The recursive rule just expresses in algebraic form that the next Fibonacci number is the sum of the previous two.

Another sequence closely related to the Fibonacci sequence is the *Lucas sequence,* which is formed using the same recursive rule but different starting numbers. The nth Lucas number L_n is given by

$$L_1 = 1, L_2 = 3, \text{ and } L_{n+1} = L_n + L_{n-1} \quad \text{for} \quad n > 2$$

a. Calculate L_3 through L_{10}.

b. Calculate the ratio of successive terms of the Lucas sequence:

$$\frac{L_2}{L_1}, \frac{L_3}{L_2}, \cdot \cdot \cdot , \frac{L_{10}}{L_9}$$

What do you notice?

3. Believe it or not, there is a society, the Fibonacci Association, devoted to fostering interest in Fibonacci and related numbers. (We would love to be able to tell you that Leonardo Pisano founded the society 800 years ago, but in fact it is only 30 years old.) The November 1988 issue of the society's publication, *The Fibonacci Quarterly,* contains the article "Suppose More Rabbits Are Born" (pp. 306–311), by Shari Lynn Levine (who was a high school student when she wrote it). The article begins:

> How would Fibonacci's age-old sequence be redefined if, instead of bearing one pair of baby rabbits per month, the mature rabbits bear two pairs of baby rabbits per month?

The article goes on to discuss properties of the resulting "Beta-nacci" sequence and the sequences that result from even greater rabbit fertility. Here we ask you to rediscover some of Shari's results about the Beta-nacci sequence:

a. How many rabbits will there be each month for the first 12 months?

b. What is the recursive rule for the nth Beta-nacci number B_n?

c. For the terms of the sequence in part **a**, calculate the ratios B_{n+1}/B_n of successive terms. (Motivating hint: It's not the golden ratio this time.)

d. Suppose that the ratio of successive terms approaches a number x. We show how to find x exactly. For very large n, $B_{n+1} \approx xB_n \approx x^2B_{n-1}$. Substituting these values into the recursive rule for the sequence and dividing by B_{n-1} gives us the equation $x^2 = x + 2$. Solve this equation for x (you can use the quadratic formula).

e. Make a table of values of $3B_n$ versus 2^n. From the evidence, can you suggest a formula for B_n?

4. Generalize Exercise 3, parts **a** through **d**

a. to the case of each pair of rabbits having three pairs of rabbits (the "Gamma-nacci" sequence).

b. to the case of each pair of rabbits having q pairs of rabbits.

5. For a sequence specified by a recursive rule, finding an explicit expression for the nth term is not easy, nor is the form necessarily simple. An exact expression for the nth term of the Fibonacci sequence is given by the *Binet formula:*

$$F_n = \frac{1}{\sqrt{5}} \left(\frac{1 + \sqrt{5}}{2} \right)^n - \frac{1}{\sqrt{5}} \left(\frac{1 - \sqrt{5}}{2} \right)^n$$

a. Verify the formula for $n = 1$ and $n = 2$ (by multiplying out, not by using a calculator).

b. Use the Binet formula and your calculator to find F_5.

c. In fact, the second term on the right of the equation gets closer and closer to 0 as n gets large. Since we know that the Fibonacci numbers are integers, we can just round off the result of calculating the first term. Find F_{13} by calculating the first term with your calculator and rounding.

6. The Fibonacci sequence and the Lucas sequence of Exercise 2 are intimately related.

  a. Make a table of values for $L_n - F_{n-1}$ for n from 2 through 10. What do you notice?

  b. Make a table of values for $F_n + L_n$ for n from 1 through 10. What do you notice?

  c. Make a table of values for $F_n L_n$ for n from 1 through 5. What do you notice?

7. We can translate the geometric construction problems of the Greeks into problems of algebra:

  a. Find the mean proportional between 3 and 27.

  b. Find the length of a side of a square that has the same area as a rectangle that is 4 by 64.

8. Here are more algebraic versions of Greek construction problems:

  a. Find the geometric mean of 4 and 9.

  b. You divide a line segment of length 6 in mean and extreme ratio. How long are the two pieces?

9. Although the Fibonacci numbers get bigger and bigger, their units digit (the rightmost, or last, one) just keeps going through digits between 0 and 9. In fact, we can show that the sequence of units digits has to repeat.

  Our idea is that if we ever come to two consecutive Fibonacci numbers that have the same units digits as two previous consecutive Fibonacci numbers, then the whole sequence of units digits from that previous point on has to repeat. The reason we need to look at two consecutive Fibonacci numbers is that their units digits completely determine the units digit of the next Fibonacci number: if one Fibonacci number ends in 5 and the next one in 9, then the following one has to end in 4. If we ever come to another 5 followed by a 9, the next has to be a 4 again, and so on.

  So how do we know that we will ever get such a repetition? We make use of a simple but effective tool that mathematicians call the *pigeonhole principle:* if you have more pigeons than holes, then some hole has more than one pigeon. For our situation, the pigeons are the pairs of consecutive Fibonacci numbers (of which we may have as many as we please) and the pigeonholes are the possible pairs of units digits (of which there are only 100: 0, 0 through 9, 9.) The units digits must have begun repeating by the time we get to the 101st Fibonacci number. In fact, the repetition starts a bit sooner, but not before

$$F_{49} = 7,778,742,049 \qquad F_{50} = 12,586,269,025$$

When does the repetition start?

10. We investigate repetitions in generalizations of the Fibonacci sequence:

  a. When does the units digit of the Beta-nacci sequence of Exercise 3 repeat?

  b. When does the units digit of the Gamma-nacci sequence of Exercise 4 repeat?

  c. When do the last *two* digits of the Beta-nacci sequence of Exercise 3 repeat?

11. The Greeks treated lengths geometrically; for them it was important to construct lengths using straightedge and compass. In the accompanying figure, we construct a golden rectangle that is 1 by ϕ. We start with a straight line with an interval of length 1 marked off on it. Placing one point of the compass at one end of the interval, we use the other end to mark off another interval of length 1. Using in turn each of the two

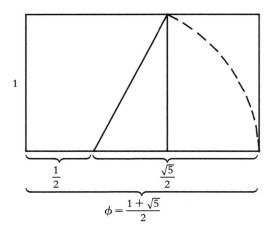

$$\phi = \frac{1 + \sqrt{5}}{2}$$

opposite ends of these intervals as a center for the compass and an opening of length 2, we make arcs above and below the line. Their points of intersection determine a perpendicular bisector of the line, giving us the first right angle of a square of side 1. Continuing, we can go on to construct the remaining sides. If we bisect the original segment, we get a new point that divides it into two pieces of length one-half each. Using this new point and a compass opening equal to the distance from it to a far corner of the square (see the figure above), we can cut off another interval on the original line.

Show that this last interval, added to the adjoining interval of length one-half, gives an interval of length ϕ.

12. In this exercise we construct geometrically the geometric mean between two segments of length a and b. We do it in three steps, constructing at each stage a right triangle. At each stage we employ the Pythagorean theorem (see Chapter 15). Exercise 11 describes how to construct a right angle at a point; the only other fact we will need is that a triangle inscribed in a semicircle is a right triangle.

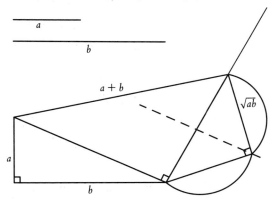

STEP 1. Construct a right triangle with legs of lengths a and b.

STEP 2. At one end of the hypotenuse of that first right triangle, construct a perpendicular. Center the compass at the other end of the hypotenuse and mark an arc of length $a + b$ that intersects the perpendicular. This forms a new right triangle.

STEP 3. Draw the perpendicular bisector of the vertical leg of the new right triangle to find the center of the leg. Draw a semicircle using the center of the leg as center and half its length as radius. The two ends of the leg and the point where the semicircle intersects the perpendicular bisector form a triangle inscribed in a semicircle: our third right triangle.

Perform the construction and use the Pythagorean theorem (three times!) to show that a leg of the final right triangle has length $\sqrt{ab}$.

13. In this exercise we show that the length of a diagonal of a regular pentagon is the golden ratio times the length of a side. For simplicity we assume that the length of the side is 1, so we show that the length of the diagonal, which we will denote by x, is in fact ϕ. Consider the accompanying figure. From the facts that the measure of an interior angle of a regular pentagon is $108°$, that the measures of the angles of a triangle sum to $180°$, and that angles opposite equal sides of a triangle must have equal measure, conclude that triangles FAE and BDE are congruent, and that triangles FBD and BDE are similar. From the fact that similar triangles have proportional sides, arrive at an equation that you can solve to find $x = \phi$.

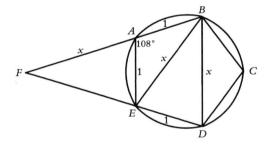

14. Consider a regular decagon (10-sided polygon) inscribed in a circle. Show that the ratio of the radius to the length of one of the sides is ϕ.

15. For each of the shapes in parts (a) through (e) of the accompanying figure, determine all lines of symmetry.

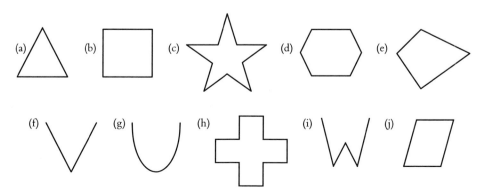

16. As in Exercise 15, but for the shapes in parts (f) through (j).

17. Determine whether each of the following statements is always true or sometimes false. (Drawing some sketches may be helpful.)

a. A line reflection preserves *collinearity* of points. That is, if the points *A*, *B*, and *C* are in a straight line *(collinear)*, then their images reflected in some other line also lie in a straight line.

b. A line reflection preserves betweenness. That is, if the collinear points *A*, *B*, and *C* (with *B* between *A* and *C*) are reflected about a line, then the image of *B* is between the images of *A* and *C*.

c. The image of a line segment under a line reflection is a line segment of the same length.

d. The image of an angle under a line reflection is an angle of the same measure.

e. The image of a pair of parallel lines under a line reflection is a pair of parallel lines.

18. Determine whether each of the following statements is always true or sometimes false. (Drawing some sketches may be helpful.)

a. The image of a pair of perpendicular lines under a line reflection is a pair of perpendicular lines.

b. The image of a square under a line reflection is a square.

c. Label the vertices of a square *A*, *B*, *C*, and *D* in a clockwise direction. Then their images *A′*, *B′*, *C′*, and *D′* under a line reflection also follow a clockwise direction.

d. The perimeter of a geometric figure is equal to the perimeter of its image under a line reflection.

e. The image of a vertical line under a line reflection is always a vertical line.

19. Which of the 26 capital letters of the alphabet have

a. a horizontal line of reflection symmetry?

b. a vertical line of reflection symmetry?

c. rotational symmetry?

d. combinations of the above?

(Assume that each letter is drawn in the most symmetric way. For example, the upper and lower loops of "B" should be the same size.)

20. As in Exercise 19, but for the lowercase letters.

21. In *The Complete Walker III*, 3rd ed., Knopf, 1984, p. 505, Colin Fletcher's answer to "What games should I take on a backpacking trip?" is the game he calls "Colinvert": "You strive to find words with meaningful mirror (or half-turn) images." Some of the words he found are

MOM   WOW   pod   MUd   bUM

a. Which of his words reflect into themselves?

b. Which of his words rotate into themselves?

c. Find some more words or phrases of these various types — the longer, the better.

22. As in Exercise 21, but for words written vertically instead of horizontally.

23. For each of the following strip patterns, identify the rigid motions that preserve the pattern:

a. A A A A A A A A A A   **c.** X X X X X X X X X X

b. B B B B B B B B B B   **d.** F F F F F F F F F F

24. As in Exercise 23, but for

a. N N N N N N N N N   **c.** d b p q d b p q d b p q

b. b d b d b d b d b d

25. Use the flowchart in Figure 17.12 to identify (by International Crystallographic Union notation) the types of the strip patterns from San Ildefonso Pueblo, New Mexico, shown in the accompanying illustration.

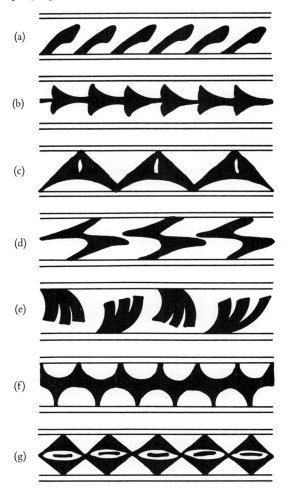

(a)

(b)

(c)

(d)

(e)

(f)

(g)

26. As in Exercise 25, for the accompanying patterns from Hungarian needlework.

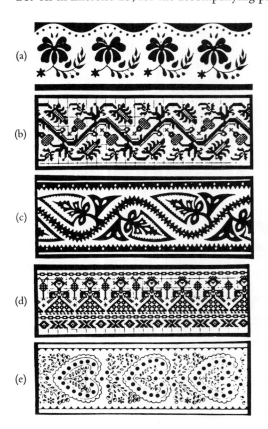

Hungarian needlework designs. (a) Edge decoration of table cover from Kalocsa, southern Hungary. (b) Pillow end decoration from Tolna County, southwest Hungary. (c) Decoration patched onto a long embroidered felt coat of Hungarian shepherds in Bihar County, eastern Hungary. (d) Embroidered edge decoration of bed sheet from the eighteenth century. (Note the deviations from symmetry in the lower stripes of the pattern.) (e) Shirt from Karád, southwest Hungary. (f) Pillow decoration pattern from Torockó (Rimetea), Transylvania, Romania. (g) Grape leaf pattern from the territory east of the river Tisza. [Courtesy of István Hargittai and Györgyi Lengyel, *Journal of Chemical Education* 61(12): 1033–1034 (December 1984).]

27. As in Exercise 25, for the accompanying eight strip patterns, all of which appear on the brass straps for a single lamp from nineteenth-century Benin in West Africa. [From H. Ling Roth, *In Great Benin.*]

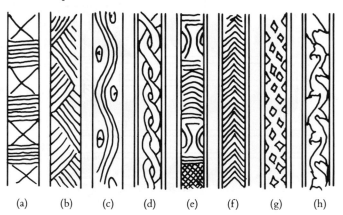

28. In each of the four accompanying examples, two adjacent triangles of an infinite strip are shown. For each example:

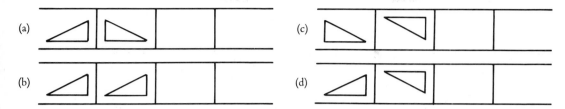

(a)   (c)

(b)   (d)

    a. Determine a motion (translation, reflection, rotation, or glide reflection) that takes the first (=left) triangle to the second (=right) one.

    b. Draw the next four triangles of the infinite strip that would result if the second triangle is moved to the next space by another motion of the same kind, and so on.

    c. Identify (by notation) the resulting strip as one of the seven possible strip patterns.

● **29.** For each of the Bakuba cloths shown in the accompanying illustration, use the flowchart in Figure 17.12 to identify (by notation) the type of wallpaper pattern.

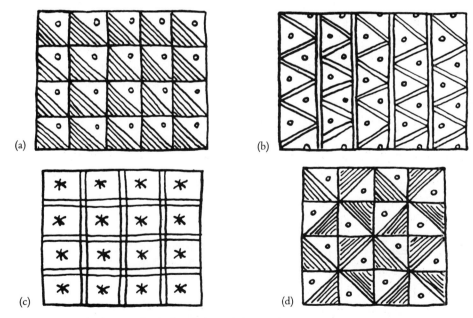

(a)   (b)

(c)   (d)

Patterns on Yoruba (West Africa) *adire* cloth, made by starching a pattern onto white cloth, then dyeing the cloth blue before rinsing out the starch, so that the starched portion remains as a white design against a blue background. [Courtesy of Donald W. Crowe, in Claudia Zaslavsky, *Africa Counts: Number and Pattern in African Culture,* Prindle, Weber, & Schmidt, Boston, Mass., 1973, p. 195.]

● Optional exercise.

30. The following table shows comparative data about the frequency of occurrence of strip designs of various types on pottery (Mesa Verde, United States) and smoking pipes (Begho, Ghana, Africa) from two different continents.

Frequency of strip designs on Mesa Verde pottery and Begho smoking pipes

| Strip type | Mesa Verde | | Begho | |
	Number of examples	Percentage of total	Number of examples	Percentage of total
p111	7	4	4	2
p1m1	5	3	9	4
pm11	12	7	22	10
p112	93	53	19	8
p1g1	11	6	2	1
pmg1	27	16	9	4
pmm1	19	11	165	72
Totals	174		230	

a. Which types of motions appear to be preferred for designs from each of the two localities?

b. What other conclusions do you draw from the data of this table?

c. On the evidence of the table alone, in which locality is each of the strip patterns in the accompanying figure most likely to have been found?

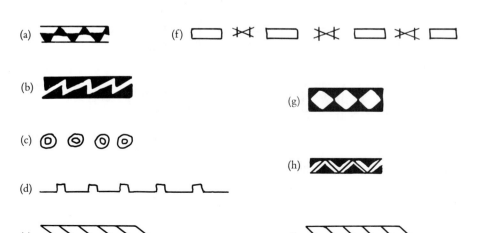

31. Determine the measure of an exterior angle and of an interior angle of a regular octagon (eight sides).

32. Discover a formula for the measure of an interior angle of a regular *n*-gon.

33. For each of the tiles below, show how it can be used to tile the plane.

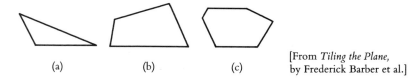

(a)   (b)   (c)   [From *Tiling the Plane*, by Frederick Barber *et al.*]

34. Can every triangle tile the plane?

35. Can every quadrilateral (four-sided polygon) tile the plane?

36. You know already that the plane cannot be tiled with regular pentagons. Monohedral tilings are possible with *some* irregular convex pentagons; *convex* means that any line segment joining two points of the figure does not contain any points outside the figure. We do not require the tiling to be edge-to-edge. Five types of such pentagons were discovered in 1918; by 1968 a mathematician who had worked for 35 years on the problem of finding all such pentagons announced that there were only three more types. For example, any pentagon having a pair of parallel sides will tile; this is one of the types. When an account of the "complete" classification into eight types appeared in *Scientific American,* the article provoked an amateur mathematician to discover a ninth type! A second amateur, Marjorie Rice, a housewife with no formal education in mathematics beyond high school "general mathematics" 36 years earlier, then found *four more* over the next 2 years. Thirteen more years have passed, and no one knows if the classification is complete.

   a. Show how an arbitrary pentagon with two parallel sides can tile the plane.

   b. Shown below is a tile of type 13, the latest type to have been discovered (by Marjorie Rice in December 1977). Show how it can tile the plane. (Its parts satisfy the following relations: $A = C = D = 120°$, $B = E = 90°$, $2A + D = 360°$, $2C + D = 360°$, $a = e$, $a + e = d$.)

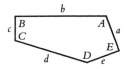

   [You can read more about this problem in "In Praise of Amateurs," by Doris Schattschneider, in *The Mathematical Gardner,* edited by David A. Klarner, pp. 140–166, plus Color Plates I–V opposite p. 166 (Wadsworth, Belmont, California, 1981).]

   To complete the classification of monohedral tilings by convex polygons that we have begun in Exercises 34, 35, and 36, we note that there are exactly three types of convex hexagons that will tile the plane, and that *no* convex polygon with seven or more sides will tile the plane.

● **37.** One way to arrive at all of the regular and semiregular tilings of the plane is to enumerate all the possibilities for regular polygons meeting at a point in terms of the numbers of each kind.

a. Calculate the interior angles for regular polygons of 3, 4, . . . , 12 sides.

b. Write a computer program using nested loops that tries all combinations of interior angles for polygons of 3 through 12 sides and identifies those that add up to 360°. (In an interpreted language like BASIC, such a program may involve several hours of run time.) You should arrive at 12 combinations that are feasible.

c. Analyze the results of your program. One of the 12 combinations that you should have arrived at in part **b** in fact won't work because the polygons won't fit properly at other vertices; another combination gives rise to two different vertex figures, depending on the order of the polygons around the point; and a tiling using one of the combinations may require for its completion one of the others as well.

d. What about polygons with more than 12 sides? The search can be shortened by noting that using such a polygon means that we cannot have in the vertex figure with it more than 3 triangles, 2 squares, or 1 polygon with 6 through 11 sides; and we cannot have any polygons with it that have more than 11 sides. Write a program that searches for vertex companions for polygons of 13 through 42 sides. (This program should run much more quickly than the previous one; you should arrive at 5 possibilities.)

e. Argue that since the limited possibilities for companions of a polygon of more than 42 sides all have interior angles with whole numbers of degrees, "the big one" must too. Write a program that tries all the integer possibilities between 172° through 179° for this polygon of more than 42 sides.

f. From among the possibilities that your programs have generated, identify all the 21 feasible vertex figures (there are 21 of them). Only 11 of these have the same vertex figure at each vertex. These are the regular and semiregular tilings (there will be 3 regular ones and 8 semiregular ones). The others make nice tilings too; they just aren't regular or semiregular.

(Note: This is a long project. The details of the geometry can be found in *Geometry: An Investigative Approach* by P. G. O'Daffer and S. R. Clemens, pp. 91–101. The specific ideas for the computer programs are from John Lamb, Jr., of the Mathematics Department of East Texas State University and appear here through his courtesy.)

● **38.** Here is another extensive project: What tilings with regular polygons are possible on a sphere? (To see an example of a semiregular tessellation of the sphere, find a soccer ball.)

39. The rabbit problem of Exercise 1 can lead us directly into nonperiodic patterns and musical sequences. Let *A* denote an adult pair of rabbits and *B* denote a baby pair. We will record the population at the end of each month, just before any births, in a particular systematic way—as a string of *A*'s and *B*'s. At the end of their second month of life, a rabbit pair will be considered to be adult. At the end of the first month our sequence is just *A*; and the same is true at the end of the second month. When an adult pair *A* has a baby pair *B*, we write the new *B* immediately to the right of the *A*. So at the end of the

● Optional exercise.

third month, our sequence is *AB*; at the end of the fourth, it is *ABA,* since the first baby pair is now adult; at the end of the fifth month we have *ABAAB.*

Mathematicians and computer scientists call this manner of generating a sequence a *replacement system.* At each stage we replace each *A* by *AB* and each *B* by *A.*

   a. What is the sequence at the end of the sixth month?

   b. Why can't we ever have two *B*'s next to each other?

   c. Why can't we ever have three *A*'s in a row?

   d. Show that from the fourth month on, the sequence for the current month consists of the sequence for last month followed by the sequence for 2 months ago.

40. Just as for Penrose patterns, we can define an inflation and a deflation for any sequence of *A*'s and *B*'s. Both of these operations preserve musicality: if we inflate or deflate a musical sequence, we get a musical sequence.

   a. Inflation can be used to generate musical sequences. Start at the first stage with just a *B*. Inflation consists of replacing each *A* with *AB* and each *B* with *A.* Show that at the nth stage there are F_n symbols in the sequence.

   b. Deflation can be used to check whether a finite block of *A*'s and *B*'s can belong to a musical sequence or not. Each deflation stage has two parts: first replace each *A* with $(A/2)B(A/2)$ and each *B* with $(A/2)(A/2)$, then combine pairs of adjacent $(A/2)$'s into a single *A* so that no fractional *A*'s remain. Another way to get the same result is to proceed from left to right, replacing *B* with *A*, *AA* with *B*, and deleting single *A*'s. The deflated block will be shorter. If at any stage we have a block with two or more *B*'s in a row, or three or more *A*'s in a row, then the original block could not be part of a musical sequence; otherwise, the original block will eventually deflate to a single symbol, at which point we conclude that the original block is a part of a musical sequence (in fact, infinitely often, a part of every musical sequence). Check the two sequences *ABAABABAAB* and *ABAABABABA.*

·V·

Computers

In Part V, we examine the role of mathematics in the development and workings of modern computational machines. To many people, computers and mathematics appear to be one subject, but the relationship is more subtle than that. To get a handle on that relationship, we need to look back and ask a fundamental question: What is mathematics? What lies at the foundation of the subject?

At heart, mathematics works with the notions of proof and truth, and we use proof to discover truth. We manipulate symbols, without regard to their meaning, according to fixed logical rules. The conclusions we draw reveal true statements in the real world, where our assumptions and symbols make sense. We begin our investigation with this relationship between mathematics and the world we live in—between proof and truth.

This study became truly focused in 1900 when David Hilbert, the most influential mathematician of his time, set forth his famous research program in a speech to the International Congress of Mathematics. Two major goals of the Hilbert program were to prove that mathematics is consistent (gives rise to no contradictions) and complete (what is true can be proved). In the 1930s, after years of intense work on Hilbert's program, Austrian mathematician Kurt Gödel proved a remarkable result: he showed that mathematical theories powerful enough to do arithmetic contain true statements that cannot be proved—so-called undecidable statements.

A computer-generated view of a microscopic detail of the Mandelbrot set. The boundaries of the view rectangle are minimum x = − 0.746987, *maximum* x = − 0.745921; *minimum* y = − 0.114763, *maximum* y = − 0.115563. *[Rollo Silver, San Cristobal, N. Mex.]*

This result shocked the mathematical world. It meant that mathematics has inherent limitations. British mathematician Alan Turing tried another tack. Gödel's result guaranteed the existence of "undecidable" propositions. Could we perhaps decide in advance which ones they were? To look for some automatic procedure to detect the undecidable statements, Turing first had to formalize the notion of procedure. He came up with the concept of a computing machine. Now, this idea of a machine was simply a mental construct that enabled Turing to describe what he meant by a computable procedure, or function; in fact, he discovered that it is not possible to determine in advance which statements are undecidable. But the door was opened. What began as an idea for defining a computational procedure quickly became a practical reality.

The story involves many characters, but perhaps the most important is John von Neumann, one of the finest mathematicians of this century. A pure mathematician, von Neumann was fascinated by practical applications. One of von Neumann's many interests was fluid dynamics, a branch of physics with applications from weather forecasting to wing design. But the equations that represent moving fluids are so complicated that even a single problem could take a roomful of clerks with desk calculators weeks to solve. Although this was acceptable in peacetime, World War II brought urgent demands for faster results. Enormous fluid-dynamics calculations were needed for technological developments, including the atom bomb.

This type of time constraint led von Neumann to begin work on computer development. He not only theoretically designed a machine but actually directed the construction of a computer embodying his ideas. This machine was built in the late 1940s at Princeton University and became the prototype of the modern computer.

Since that time, improvements and innovations have occurred at an incredible pace. New developments in computer graphics have led to new mathematical discoveries. Computer verifications have even been incorporated into proofs of major new results. So, in a sense, we've come full circle. The computer began as an idea to help us understand the meaning of mathematical proof. But mathematics is always growing and expanding. Today, the computer not only helps us to do our calculations and draw our pictures, but it is even changing our notion of proof and the very image of mathematics itself.

·18·

Computer Algorithms

Contemporary computing devices have had a significant effect on our lives for two reasons. First, the modern computer is significantly faster than any of its predecessors. Personal computers are capable of performing several hundred thousand computations per second, and the fastest supercomputers exceed a billion computations per second. Equally important is the ability of modern computers to store programs and data. With these features, we can program modern computers to handle tasks ranging from the lofty to the mundane, from space exploration to printing junk mail.

If we define "computer" as a nonhuman device that computes, we may include as examples both early devices, such as the sun dial and the Chinese abacus, and modern ones, such as the Apple Macintosh and the IBM-PC personal computers and the CRAY-XMP and NASA's MPP supercomputers. To make a computer do what we want it to do, we need to provide it with a **program,** or detailed sequence of instructions describing the task to be performed. By controlling a computer's behavior through programs, we can vary the task of the computer without changing the equipment itself, making the computer an immensely versatile device. In this sense, computers and programs can be likened to record players and records: the output from a computer is determined by the program it is executing in much the same way as the output of a record player is determined by the particular record being played.

ONE STEP AT A TIME

To see the advantages of storing programs in computers, consider the following problem:

Mary intends to open a bank account on the first day of the month with an initial deposit of $100. She intends to deposit an additional $100 into this account on the first day of each of the next 19 months, for a total of 20 deposits (including the initial deposit). The account pays interest at the rate of 5% per annum, compounded monthly. Mary would like to know what the balance in her account will be at the end of each of the 20 months in which she will be making a deposit.

In order to solve this problem, we need to know how much interest is earned each month. Because the annual interest rate is 5%, the monthly interest rate is $\frac{5}{12}$%. Consequently, the balance at the end of the first month is

$100.00 + interest on $100.00

$$= \$100.00 + \frac{5}{12}\% \text{ of } \$100.00$$

$$= \$100.00 + \left(\frac{5}{1200} \times \$100.00 \right)$$

$$= \$100.00 \times \left(1 + \frac{5}{1200} \right)$$

$$= \$100.00 \times \frac{241}{240}$$

More generally, we can compute the balance at the end of any month from the balance at the beginning of the month using the equation

End-of-month balance
$$= \text{initial balance} \times \frac{241}{240}$$

With this analysis out of the way, we can use the following steps to compute the balance at the end of each month:

STEP 1. Let balance denote the current balance. The starting balance is $100, so we set balance = 100.

STEP 2. The balance at the end of the month is $241/240 \times$ balance at the beginning of the month.

STEP 3. If 20 months have not elapsed, then add 100 to the balance to reflect the deposit for the next month and go to step 2; otherwise, we are done.

Suppose that we have to compute the monthly balances using a computing device that cannot store the computational steps and associated data. A nonprogrammable calculator is one such device. The above steps translate into the following process:

STEP 1. Turn the calculator on.

STEP 2. Enter the initial balance as the number 100.

STEP 3. Multiply by 241 and then divide by 240.

STEP 4. Record the result as a monthly balance.

STEP 5. If the number of monthly balances recorded is 20, then stop.

STEP 6. Otherwise, add 100 to the previous result.

STEP 7. Go to step 3.

If you try this process out on any electronic calculator, you will notice that the total time you spend is not determined by the speed of the calculator but by how fast you can enter the required numbers and operators (add, multiply, and so forth) and how fast you can record the monthly balances. Even if your calculator could perform a billion computations per second, you would not be able to solve Mary's problem any faster.

When we use a stored-program computing device, we need to enter the instructions into the computer just once. The computer can then execute these instructions at its own speed. Because the instructions are entered just once (rather than 20 times), we get almost a twentyfold speedup in the computation. If the balance for 1000 months is required, the speedup increases almost by a factor of 1000. We have achieved this speedup without making our computing device any faster. We have simply cut down on the input work required of the slow human.

Instruction sequences are provided to a computer through a programming language. Over one thousand programming languages are in use today. Some of the more popular ones are BASIC, COBOL, FORTRAN, and Pascal. Our seven-step computational process translates into the BASIC program shown in Figure 18.1. In Pascal, it takes the form shown in Figure 18.2. These two programs are not only written in different languages but also represent different programming styles. The Pascal program has been written to permit us to make changes with ease. The number of months, interest rate, initial balance, and monthly additions are more easily changed in the Pascal program.

```
line    program account (input, output);
 1      {compute the account balance at the end of each month}
 2      const   InitialBalance = 100;
 3              MonthlyDeposit = 100; {additional deposit per month}
 4              TotalMonths = 20;
 5              AnnualInterestRate = 5; {percent rate}
 6      var balance, interest, MonthlyRate : real; month : integer;
 7      begin
 8          MonthlyRate := AnnualInterestRate /1200; {rate per $}
 9          balance := InitialBalance ;
10          writeln('   Month     Balance');
11          for month := 1 to TotalMonths do
12          begin
13              interest := balance * MonthlyRate ;
14              balance := balance + interest ;
15              writeln(month :10,'    ', balance :10:2);
16              balance := balance + MonthlyDeposit ;
17          end;
18          writeln;
19          writeln('Balance is balance at end of month');
20      end
```

Figure 18.2 A Pascal program for Mary's problem.

ALGORITHMS

The three methods we have just described for solving Mary's problem — using the calculator, using the BASIC program, and using the Pascal program — are all **computational procedures.** A computational procedure consists of a finite number of steps, each of which is definite and effective. By **definite,** we mean that the outcome of each is well defined. The step "set x to $1/0$" is not definite because $1/0$ is not a well-defined quantity. By **effective,** we mean that the step can be completed in a finite amount of time, using a finite amount of computational resource. It is easy to see that each of the steps in our calculator, BASIC, and Pascal examples is both definite and effective.

A computational procedure that terminates on every possible input after executing a finite

```
10    balance = 100
20    month = 1
30    balance = 241 * balance/240
40    print month, "$", balance
50    if month = 20 then stop
60    month = month + 1
70    balance = balance + 100
80    goto 30
```

Figure 18.1 A BASIC program for Mary's problem.

number of steps and produces some output is called an **algorithm.** It is important to note that a procedure whose description consists of a finite number of lines does not necessarily stop after executing the commands in each of the lines. For example, the following algorithm written as a five-line BASIC program describes the execution of not five but nine steps. The leftmost numbers 10, 20, etc., are line numbers.

$$10 \quad A = 3$$

$$20 \quad A = A + 1$$

$$30 \quad \text{if } A < 6 \text{ go to } 20$$

$$40 \quad \text{print } A$$

$$50 \quad \text{end}$$

Here are the steps executed by this program:

STEP 1. Line 10 gives A the value 3.

STEP 2. Line 20 changes the value of A from 3 to 4.

STEP 3. Line 30 determines that A (current value 4) < 6 and directs the program to line 20.

STEP 4. Line 20 changes the value of A from 4 to 5.

STEP 5. Line 30 determines that A (current value 5) < 6 and directs the program to line 20.

STEP 6. Line 20 changes the value of A from 5 to 6.

STEP 7. Line 30 determines that A (current value 6) is not < 6 (hence the program is directed to line 40).

STEP 8. Line 40 prints 6 (the current value of A).

STEP 9. Line 50 stops the program execution.

When counting the total number of steps executed by an algorithm we count each repetition of a line as an additional step. Thus a single line executed 7 times is equivalent to 7 different instructions each executed once.

The number of steps executed by an algorithm could also depend on the input into the algorithm. Consider an algorithm that inputs a positive integer N and proceeds to add the first N positive integers and prints out their sum. If the input into the algorithm is 3, that is, $N = 3$, the algorithm will compute $1 + 2 + 3$, performing two additions. If the input into the algorithm is 10, $1 + 2 + 3 + 4 + 5 + 6 + 7 + 8 + 9 + 10$ will be calculated, requiring 9 additions. An algorithm must terminate after executing a finite number of steps for any input into the algorithm.

The computational procedures we have seen so far are all algorithms. Each terminates after a finite number of steps and each produces output, for example, the 20 monthly balances in Mary's problem. For an example of a computational procedure that is not an algorithm, consider the following sequence of BASIC steps:

$$10 \quad x = 1$$

$$20 \quad x = x + 1$$

$$30 \quad \text{go to } 20$$

$$40 \quad \text{print } x$$

$$50 \quad \text{end}$$

This sequence of commands produces the following steps:

STEP 1. At line 10 the value of x is set to 1.

STEP 2. At line 20 the value of x is changed from 1 to 2.

STEP 3. At line 30 the program is directed to line 20.

STEP 4. At line 20 the value of x is changed from 2 to 3.

STEP 5. At line 30 the program is directed to line 20.

Steps at lines 20 and 30 repeat indefinitely, making it impossible for the procedure to stop in a finite number of steps. Notice that although there are steps at lines 40 and 50 of the procedure, these steps are never executed.

For an example in which the input determines whether a procedure terminates, consider

10	input x
20	$x = x + 1$
30	if $x > 0$ go to 20
40	print x
50	end

Through the execution of line 10, the user specifies (inputs) an initial value of x. If this value happens to be ≤ -1, at line 20 it is incremented by 1 but still remains ≤ 0. Execution is then directed to line 40 and the procedure terminates after printing the modified value of x. However, if the user specifies an initial value of x that is greater than -1, say 3, lines 20 and 30 will repeat indefinitely and the procedure will not terminate. Therefore, this is not an algorithm since it does not terminate for every possible input.

Although there are certain real situations, such as nuclear-reactor monitoring, in which a computational procedure should never terminate, we generally want our computational procedures to be algorithms. Moreover, we want our algorithms to terminate in a reasonable period of time; after all, an algorithm that does not produce an answer in time for us to use it is not worth much. Take chess, for example, in which a kind of brute force algorithm

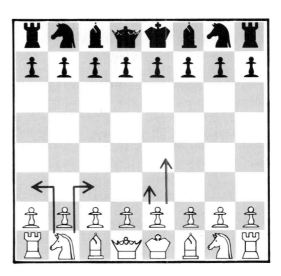

Figure 18.3 White has 20 possible opening moves.

will determine whether white can always win. We have 20 possible opening moves for white (Figure 18.3) and 20 responses for black. An algorithm to analyze chess could examine each of the 20×20, or 400, board configurations that are possible after the first pair of moves. The algorithm would figure out all the moves white could make next, then what black could do against all those configurations, and so on. In a typical game of chess, 40 pairs of moves add up roughly to 10^{95} board configurations. Even with a superfast computer, this algorithm would require fantastic amounts of time. Suppose we examine 1 billion configurations per second: it would still take 10^{77} centuries to determine all the possible game outcomes. (By comparison, the age of the universe is only 10^9 centuries.) Even though the algorithm terminates in theory, it takes so long that it is useless in practice. Actual computer chess systems use a completely different method, modeled more on the way human chess masters play the game.

It should be noted that an algorithm is a method for obtaining a solution to a problem, not the solution itself; an algorithm for baking a cake is a recipe, not the cake. An algorithm for computing the taxes we owe is embedded in the IRS 1040 form shown in Figure 18.4. This form should lead us through all the steps necessary to calculate the tax we owe. Although some people might quibble about how well defined some of the instructions are, let's agree that if we follow them precisely, we will get an answer. This algorithm tells us the method, but the actual solution comes only by performing the computations according to the algorithm.

It is a common misconception that in order to solve a difficult problem, you have to understand it thoroughly. In truth, we can solve problems without understanding them if we have methods that yield solutions to such problems. For example, we don't need to learn the entire U.S. tax code to figure out our tax obligation. As long as we have a good algorithm available, we don't have to understand the subject itself at all. We just follow the steps of the algorithm and arrive at the answer. Computers do the same thing; they do not "understand" problems but are able to follow algorithms through to a solution. Of course, it's not quite that easy; the algorithm has to be written so the computer can interpret it, which means specifying all the steps in a language the

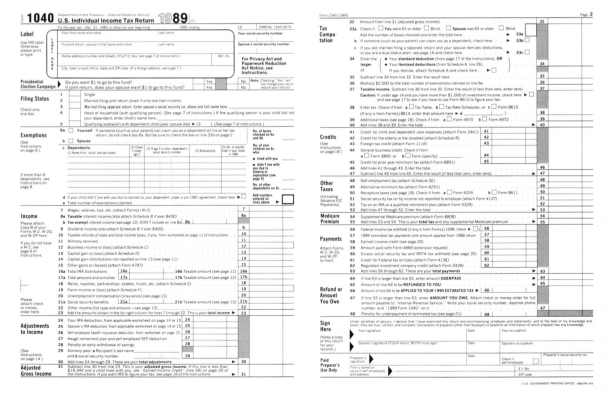

Figure 18.4 The IRS 1040 form contains a familiar algorithm.

computer can understand. And that's what most computer programs actually are: algorithms translated into a precisely defined language the computer can interpret.

SORTING THINGS OUT

Underlying the amazing variety of tasks computers can handle are some fundamental procedures that they do over and over. For example, computers do a lot of **sorting,** putting in order items such as numbers, names and addresses, or pictures from Saturn. So computer scientists have given a lot of attention to developing algorithms that tell a computer how to sort things.

Without getting too technical, let's say that sorting means putting a list of similar things into a sequence determined by some specified attribute. Numbers might be sorted according to size, for example, from smallest to largest, whereas names and addresses would probably be sorted alphabetically. Suppose you want to sort the following 10 numbers according to their size, from smallest to largest, on a computer that can do essentially two things: it can compare two numbers, and it can move a number from one position to another.

42 12 17 98 56 63 34 72 25 83

Here is an algorithm for sorting these randomly arranged numbers, using only comparisons and movements.

EXAMPLE: Insertion Sort Algorithm. Start with the second number and move it to a temporary holding place. Compare it with the first number. If the first number is larger, move it to the right and insert the one from the holding position in its place. (This sequence of data moves is shown in Figure 18.5.) This guarantees that the first two numbers will be in increasing order. Next, move the third

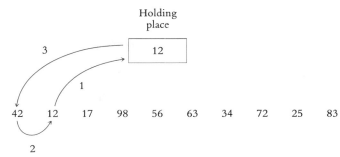

Figure 18.5 The initial data movements for insertion sort.

number into the holding place. Move each number that is to the left of the third number (the first and second) to the right if it happens to be larger than the number in the holding place. (In our particular example, the second number, 42, will move to the right, but the first number will remain in its current location.) Now insert the number from the holding position into the vacant spot. As shown in Figure 18.6, these moves guarantee that the first three numbers on our list are in ascending order.

Continuing this way, we can sort all 10 numbers. Notice that at each step, we move one of the numbers to the holding place. As we do that, we know that all the numbers to its left are already sorted. Now we move these sorted numbers to the right, one at a time, until we find the correct spot, and we insert the number from the holding place at that spot.

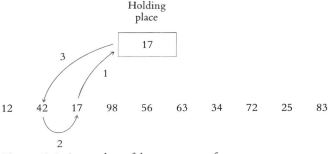

Figure 18.6 A second set of data movements for insertion sort.

```
line   procedure insertsort        (var k              : datarray; {array to be sorted}
  1                                 N                   : integer); {# of items in the array}
  2
  3    {this procedure sorts the array k (of size N) using the insertion sort
  4    method. insertsort only traverses the array; it calls the subprocedure
  5    insert to insert specific items into the array.}
  6
  7                 var      ctr  : integer;              {used to step through the array}
  8    {.......................................................................}
  9
 10                 procedure insert(var sortarray       : datarray;
 11                        loc                            : integer); {of next item in file}
 12
 13                 {this procedure is a subprocedure to insertsort for inserting the
 14                 locth entry of the array sortarray into its proper position among
 15                 sortarray [1], sortarray [2], ... , sortarray [loc − 1].}
 16
 17                        var cnt   : integer;
 18                            recerd : integer;          {variable for array entry}
 19
 20                 begin {of insert}
 21
 22                    recerd := sortarray [loc];         {hold the value at the loc}
 23                    cnt := loc − 1;                    {start at previous item}
 24
 25                    while recerd < sortarray [cnt] do  {compare to find place}
 26                        begin {of while}
 27                            sortarray [cnt + 1] := sortarray [cnt]; {move next item up}
 28                            cnt := cnt − 1              {so to previous item}
 29                        end; {of while}
 30
 31                    sortarray [cnt + 1] := recerd      {put item in correct place}
 32
 33                 end; {of insert}
 34    {.......................................................................}
 35
 36
 37    begin {of insertsort}
 38
 39                 k [0] := − maxint;                    {initialize to small value}
 40
 41                 for ctr := 2 to N do                  {call on insert to locate}
 42                     insert (k, ctr);                  {k[2], . . . , k[N] into k}
 43
 44    end; {of insertsort}
```

Figure 18.7 A Pascal program for insertion sort.

Insertion sort is a reasonable algorithm for solving our problem. Of course, in order to be executed by a computer it would have to be translated into a programming language that the computer can interpret. Figure 18.7 shows a Pascal program segment for the insertion sort. Instead of giving a detailed explanation of each line in Figure 18.7 (to readers, some of whom may not have any familiarity with Pascal), we provide a general description of the major features of the program segment. First of all, any material enclosed by braces "{"and"}" is strictly for human consumption and will be ignored by the computer.

The actual insertion of a number into its correct position is done in the section between the two rows of dotted lines. Here, the first, second, etc., numbers have the names "sortarray[1]," "sortarray[2]," etc. In the portion between lines 25 and 29, the number to be inserted, "recerd," is successively compared to the first, second, etc., of the numbers until its proper position is determined, with insertion actually taking place at line 31.

Insertion sort is not the only algorithm that can be used for sorting. Let's look at another algorithm for sorting the same list of numbers.

EXAMPLE: Merge Sort Algorithm.
Start with a list of numbers, but this time imagine that each number constitutes a separate sublist. Because each sublist contains only one number, all initial sublists are already sorted. Now we merge each adjacent pair of sorted sublists, producing half as many new sorted lists of two numbers each. The algorithm simply tells us to take successive passes in which adjacent sublists that are already sorted are merged to form fewer sublists with a larger number of entries, until all the numbers appear in a single sorted list. Figure 18.8 shows the structure of the sublists in successive passes of merge sort over the 10 numbers.

When we have two or more items in each list, we need to be more explicit about how to

Initial	{42}	{12}	{17}	{98}	{56}	{63}	{34}	{72}	{25}	{83}
Pass 1	{12	42}	{17	98}	{56	63}	{34	72}	{25	83}
Pass 2	{12	17	42	98}	{34	56	63	72}	{25	83}
Pass 3	{12	17	34	42	56	63	72	98}	{25	83}
Pass 4	{12	17	25	34	42	56	63	72	83	98}

Figure 18.8 Successive passes of merge sort.

merge adjacent lists. We start with markers positioned on the first number in each list, compare these numbers, and determine which number is smaller. Move that number to the first place in a new list, and advance the marker to the next item in the old list. Now repeat the process, comparing the two numbers now opposite the markers.

For example, at the end of Pass 1, to merge the first two lists {12 42} and {17 98} into {12 17 42 98}, we start with markers positioned at the first numbers of {12 42} and {17 98}:

{12 42}   {17 98}
  ↑   ↑
 marker   marker

The smaller of the two marked numbers, 12, is now copied into the merge list and that marker is advanced while the other marker remains in its present position. We now have

{12 42}   {17 98}
     ↑   ↑
 marker   marker

The smaller of the two marked numbers, 17, is now copied into the merge list (making it {12 17} at this point), and this marker, and only this marker, is advanced. At this time we have

{12 42}   {17 98}
     ↑   ↑
 marker   marker

The smaller of the marked numbers, 42, is now placed into the merge list (making it {12 17 42}), and the marker is removed since it cannot advance beyond the end of the list. The remaining value, 98, is now moved to the merge list, completing the merger.

When the number of lists is uneven — five, for example — pairing leaves out one list. In this example, the right-hand list of two numbers doesn't get paired until the fourth round, when we have to merge one list of eight with this list of two. At the in-between stages, we keep copying that pair intact. (There is a shortcut that will speed up the process: when a marker moves all the way through one of the lists, the algorithm instructs the computer to just copy the remaining numbers of the other list.)

COMPARISON OF ALGORITHMS

Because we can have more than one algorithm for a single problem, our concern will be to determine which is the best algorithm. Before we confront this issue seriously, we need to clarify what we mean by "best": we could mean the most simple and direct, we could mean the algorithm that uses the least amount of computer memory, or we could mean the fastest one. To compare the two sorting algorithms, insertion sort and merge sort, let's agree that by better we mean faster.

One way to determine which algorithm is faster is to take performance measurements, wherein we monitor the actual time taken by a computer to execute the algorithms. To do this, we need to refine our algorithms into computer programs written in a specific programming language. When this is done, the two programs can be given worst-case data (initial arrangement of numbers in a list of a given length that requires the longest time to

sort) or average-case data and the actual time taken to sort is measured for lists of different sizes. For the insertion sort, a list initially arranged in descending order represents the worst case and a list with randomly arranged values represents the average case. The generation of worst- and average-case data is itself a challenge. When it becomes difficult to generate the worst-case or average data, we resort to simulations.

Suppose we wish to measure the average performance of our two sorting algorithms using the programming language Pascal on a VAX-11/780, a minicomputer widely used in universities and in business. For this experiment, 2000 random numbers are generated in groups of 100. Table 18.1 gives the compara-

TABLE 18.1 Time (in seconds) of insertion and merge sorts on a VAX-11/780

N	Insertion sort	Merge sort
100	0.03	0.02
200	0.14	0.04
300	0.29	0.07
400	0.57	0.09
500	0.84	0.11
600	1.19	0.15
700	1.68	0.16
800	2.15	0.20
900	2.73	0.22
1000	3.38	0.25
1100	4.12	0.29
1200	4.82	0.33
1300	5.77	0.35
1400	6.63	0.39
1500	7.59	0.41
1600	8.71	0.43
1700	9.84	0.46
1800	11.19	0.50
1900	12.19	0.52
2000	13.48	0.57

tive sorting times for insertion sort and merge sort and Figure 18.9 makes a graphic comparison of these sorting times; in both, N represents the length of the list being sorted. As you can see, as N grows larger, so does the amount of time saved by using the merge sort. The poor performance of insertion sort cannot be overcome simply by using a faster computer. We get experimental evidence for this by contrasting the times required for sorting random numbers by insertion sort on a VAX-11/780 computer and the time required for sorting the same numbers by merge sort on the much slower AT&T PC-6300 Plus microcomputer. Table 18.2 shows the results. Although the performance of insertion sort does at first reflect the advantage of being run on a faster

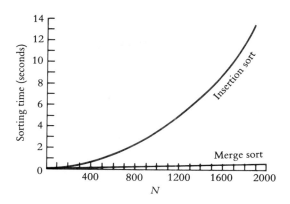

Figure 18.9 As N gets large, merge sort becomes more efficient compared to insertion sort.

computer, merge sort manages to overtake it at $N = 2400$. In other words, for large values of N, the choice of algorithm is more critical than computer speed.

The number of steps executed by an algorithm is called the **time complexity** of that algorithm. Obviously, the time complexities of our sorting algorithms depend on N, the number of numbers that are being sorted. (The actual time complexities of the insertion sort and merge sort are proportional to N^2 and $N \log N$, respectively.)

To develop insight about the execution times of algorithms with different time complexities, consider Table 18.3, which gives

TABLE 18.2 Time (in seconds) of insertion and merge sorts

N	Insertion sort VAX-11/780	Merge sort AT&T PC-6300+
200	0.14	0.96
400	0.57	2.36
600	1.19	3.27
800	2.15	4.83
1000	3.38	6.74
1200	4.82	7.60
1400	6.63	9.96
1600	8.71	10.57
1800	11.10	12.89
2000	13.48	13.53
2200	14.82	15.50
2400	18.81	16.98
2600	19.48	18.87
2800	22.59	20.08
3000	29.50	21.51
3200	33.45	23.24
3400	35.40	24.99
3600	37.35	26.14
3800	43.22	28.31
4000	49.03	29.98

TABLE 18.3 Comparative growth of some functions

N	$N \log N$	N^2	N^3	2^N
5	8.0	25	125	32
10	23.0	100	1,000	1,024
15	40.6	225	3,375	32,768
20	60.0	400	8,000	1,048,576
25	80.4	625	15,625	33,554,432
30	102.0	900	27,000	1,073,741,824
35	124.4	1225	42,875	34,359,738,368

TABLE 18.4. **Time to compute $f(N)$ instructions on a 1 billion instructions per second computer**

N	$f(N) = N$	$f(N) = N \log N$	$f(N) = N^2$	$f(N) = N^3$	$f(N) = N^{10}$	$f(N) = 2^N$
10	0.01 μs	0.03 μs	0.1 μs	1 μs	10 s	1 μs
20	0.02 μs	0.09 μs	0.4 μs	8 μs	2.84 hr	1 ms
30	0.03 μs	0.15 μs	0.9 μs	27 μs	6.83 d	1 s
40	0.04 μs	0.21 μs	1.6 μs	64 μs	121.36 d	18.3 min
50	0.05 μs	0.28 μs	2.5 μs	125 μs	3.1 yr	13 d
100	0.10 μs	0.66 μs	10 μs	1 ms	3171 yr	4×10^{13} yr
1,000	1.00 μs	9.96 μs	1 ms	1 s	3.17×10^{13} yr	32×10^{283} yr
10,000	10.00 μs	130.3 μs	100 ms	16.67 min	3.17×10^{23} yr	
100,000	100.00 μs	1.66 ms	10 s	11.57 d	3.17×10^{33} yr	
1,000,000	1.00 ms	19.92 ms	16.67 min	31.71 yr	3.17×10^{43} yr	

μs = microsecond = 10^{-6} seconds
ms = millisecond = 10^{-3} seconds
s = seconds
min = minutes
hr = hours
d = days
yr = years
Source: Sartaj Sahni, *Software Development in Pascal*, Camelot Publishing, Fridley, Minn., 1985, p. 415.

values of N, $\log N$, N^2, N^3, and 2^N for $N = 5$, 10, . . . , 35. The function 2^N grows very rapidly with N. In fact, if a program needs 2^N steps for execution, when $N = 40$ the number of steps needed is approximately 1.1×10^{12}. On a computer executing 1 billion instructions per second, this would require 18.3 minutes. If $N = 50$, the same program would run for about 13 days on this computer; when $N = 60$, about 311 years will be required; and when $N = 100$, more than 10^{11} centuries will be needed. If a program needs N^{10} steps, then using our 1 billion instructions per second computer, we will need 10 seconds when $N = 10$; 3171 years when $N = 100$; and 3.17×10^{13} years when $N = 1000$. Had the program's time complexity been N^3 instead, then we would have needed 1 second of computation time for $N = 1000$, 111 minutes when $N = 10,000$, and 11.57 days when $N = 100,000$.

Table 18.4 gives the time needed by a 1 billion instructions per second computer to execute a program of time complexity $f(N)$ instructions. We should note that currently only the fastest computers can execute 1 billion instructions per second. From a practical standpoint, for reasonably large N (say $N > 100$) only programs of small time complexity (such as N, N^2, N^3, and so forth) are feasible. Furthermore, this would be the case even if we could build computers capable of executing 10^{12} instructions per second, because this would only decrease the computing times of Table 18.4 by a factor of 1000. The economic implications of more modest improvements of linear-programming algorithms are discussed in Spotlight 18.1.

Although we have clear evidence of the superiority of merge sort over insertion sort, that does not necessarily mean we should always use the merge sort algorithm. We have other factors to consider. For instance, sorting is a data-sensitive task: which method is most efficient depends to some degree on the original arrangement of the input data. Imagine that

SPOTLIGHT 18.1 The Economic Value of Algorithms

Algorithms that we can implement on computers have significant economic value. We can see a good example of their potential value in the case of Karmarkar's algorithm. In 1984, Narendra Karmarkar, a researcher at AT&T Bell Laboratories, announced his discovery of a revolutionary new linear-programming algorithm. As discussed in Chapter 4, linear-programming methods are used to solve complex real-world problems—especially in the transportation and communications industries—of scheduling, routing, and planning.

Following the announcement of its discovery, Karmarkar claimed that his algorithm would solve problems many times faster than the currently used simplex method, the standard tool for solving linear-programming problems for over 40 years. Because breakthroughs such as Karmarkar's are rare, his claims of superior performance in solving complicated problems met some initial skepticism. By 1986, however, growing experimental evidence showed that implementations of Karmarkar's algorithm in specialized computer programs did solve certain types of linear-programming problems between 1.7 and 4.5 times faster than the standard computer implementation of the simplex method. It is particularly noteworthy that the computer program employing Karmarkar's algorithm improved the relative speed of solving larger problems (those with 1000 constraints and between 2000 and 3000 variables) more than it improved the speed of solving problems with fewer variables. These tests were performed on the same large mainframe computer so that comparisons could be made.

Although current programs are experimental, the potential cost-saving implications for Karmarkar's algorithm are enormous. When successfully implemented in commercial software programs, Karmarkar's method could save airlines, railroads, trucking firms, federal and state agencies (such as NASA), communications companies, and others millions of dollars annually by solving complex problems more quickly and efficiently. Because the flow of goods, services, and information through intricate networks in the United States and around the world is essential to global economic health and growth, the application of Karmarkar's algorithm would affect everyone. Furthermore, such a fundamental discovery in applied mathematics can have extraordinary effects on companies that develop products based on this new mathematics and on companies that use those products to serve large populations.

your boss gives you a randomly arranged set of 500 index cards containing names and addresses and tells you to sort them alphabetically. You're probably going to start making individual piles for each letter of the alphabet, or maybe for groups of letters. When you have sorted all the cards into piles, you alphabetize each pile and then stack up the piles in alphabetical order.

On the other hand, imagine that the telephone company in a small town has hired you to put together next year's directory. You take the current directory and have to incorporate 20 new names at the proper alphabetic places.

Are you going to put the new names at the end and then merge sort the whole list? Of course not. You're going to find the right place to insert each new name. The point is that for a common task like sorting, you have many algorithms from which to choose. Which one you should pick depends on your particular problem, the nature of the data you are working with, and your goals in terms of time, money, and simplicity.

COMPLEX ALGORITHMS

When comparing merge sort and insertion sort, we concluded that improved performance was more a matter of choosing the faster algorithm than using the more powerful computer. But sorting, even if applied to a very long list, is relatively simple and can be tackled by any computer. For more complex problems, the search for better algorithms is bound up with the development of faster computers.

Everyone relies on weather forecasts, and everyone would like to have reliable predictions. At the National Weather Service (NWS) in Maryland, mathematicians and computer scientists are hard at work on improving forecasting techniques. The starting point for local weather forecasters is information transmitted by the NWS. The NWS collects observations all over the world—from human observers, weather balloons, ships and planes, and satellites. Twice a day computers turn all that data into national and regional forecasts. The NWS stays ahead of nature partly by running its numerical weather prediction algorithm on one of the world's fastest supercomputers. This supercomputer is designed to repeat a sequence of calculations at maximum speed, exactly what the weather prediction algorithm calls for.

The algorithm divides the atmosphere into a grid of 256,000 points in three-dimensional space (see Color Plate 20). Observations give four basic values for each of these points: temperature, pressure, moisture content, and wind velocity. Taken together, these values yield a model of what the atmosphere looked like when the observations were taken. The next step is to apply a set of equations based on physics to the values at each point. The equations make use of what we know about surrounding points to predict new values for a given point 10 minutes in the future. This same process is applied to all points on the grid; the process is then repeated for successive 10-minute intervals until we can view a model of what the atmosphere will be like in 24 hours.

The NWS research and development team is always trying to make the forecast more accurate. But accuracy is not allowed to interfere with deadlines. Changes in the forecasting program are not adopted until it has been proved that they do not slow the process down.

SOME UNSOLVED PROBLEMS

Because algorithms are so useful, it seems reasonable to ask if there is an algorithm to solve any given problem. Unfortunately, the answer to this question is no. Consider the equations

$$x^2 + y^2 - 2 = 0$$

and

$$x^3 + y^3 = z^3$$

Do these equations have solutions that are whole numbers? With a few guesses, we can see that $x = 1$ and $y = 1$ are whole-number solutions for the first equation. But no matter how long we keep making guesses for the second equation, all we can say is that we haven't found a solution yet. For centuries, mathematicians looked for an algorithm that would determine whether any such equation has a whole-number solution. Finally, in 1970 Yuri Matijasevic, a 22-year-old Russian mathematician, proved that such an algorithm does not exist.

For many practical problems, we simply do not know whether an algorithm exists. Suppose we wish to make leather goods, for example, coin purses and shoes. The parts for these products are leather shapes like the ones you see in Figure 18.10. To cut each of these parts from large pieces of leather, without any waste, we would like to fit the shapes together in a pattern with no gaps or overlaps. Mathematicians call this type of pattern a *tiling* (see Chapter 17).

Although it is easy enough to discover a tiling for rectangular pieces, it is not so clear whether we can arrange shoe shapes in a pat-

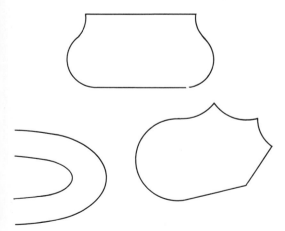

Figure 18.10 We do not know if an algorithm exists that will tell us if these leather shapes can be fit together to form a tiling.

tern without any wasted leather. It would be nice to have an algorithm that would tell us whether a particular shape will form a tiling. So far, such an algorithm has not been discovered, and we do not know if one exists; yet mathematicians are also unable to prove that it does not exist.

CONCLUDING COMMENTS

As we have seen, algorithms are useful in solving a variety of problems. Because many problems have multiple algorithmic solutions, we often need to choose a specific algorithm from among those available to us. In such instances, a comparative analysis of algorithms can be very beneficial.

We have four possibilities for the existence of algorithms for a given problem.

1. For relatively simple problems such as sorting, many good algorithms are known.

2. For highly complex problems such as weather forecasting, better algorithms and better computer techniques are leading to better solutions.

3. For problems such as whole-number solutions to equations, we can prove there are no algorithms.

4. For problems such as the tiling of shapes, we do not know if an algorithm exists.

What should keep mathematicians and computer scientists properly humble is that there is no algorithm that can determine which category a given unsolved problem belongs to. In other words, there is not, and cannot be, a comprehensive algorithm that proves or disproves the existence of an algorithmic solution for a particular problem. For this reason, the application of mathematics to computers will continue to be an experimental science.

REVIEW VOCABULARY

Algorithm A computational procedure that terminates on every input and produces some output.

Computational procedure A finite sequence of definite and effective steps for arriving at a solution to a problem.

Definite (step) A step in a computational procedure whose outcome is well defined.

Effective (step) A step in a computational pro-cedure that can be computed in a finite amount of time.

Program A sequence of instructions in a language that a computer can interpret.

Sorting The arrangement of data into a prespe-cified order.

Time complexity (of an algorithm) The exe-cution time (up to a proportionality constant) of an algorithm.

SUGGESTED READINGS

Dewdney, A. K.: *The Turing Omnibus,* Computer Science Press, New York, 1989.

Harel, D.: *Algorithms: The Spirit of Computing,* Addison-Wesley, Reading, Mass., 1987.

Horowitz, E., and S. Sahni, : *Fundamentals of Computer Algorithms,* Computer Science Press, New York, 1978.

Knuth, D.: "Ancient Babylonian Algorithms," *Comm. ACM,* 15(7):671–677 (July 1972).

——— : *The Art of Computer Programming: Sorting and Searching,* vol. 3, Addison-Wesley, Reading, Mass., 1973.

Sahni, S.: *Software Development in Pascal,* Camelot Publishing, Fridley, Minn., 1985.

EXERCISES

1. The following instructions are printed on a shampoo container:

   wet hair
   apply shampoo
   lather
   rinse
   repeat

Is this an algorithm for using the shampoo? Explain.

2. Is the following procedure an algorithm? Explain.

10	$x = 0$
20	$x = x + 1$
30	if $x > 0$ go to 20
40	print x
50	end

3. The following algorithm, analyzed in detail in the text, prints the number 6 for its output. How would the output of the algorithm change if the statements at lines 30 and 40 are interchanged?

10	$A = 3$
20	$A = A + 1$
30	if $A < 6$ go to 20
40	print A
50	end

4. What will be printed as a result of the execution of the following algorithm? (The * indicates a multiplication.)

10	$N = 1$
20	print $N, N * N, N * N * N$
30	$N = N + 1$
40	print $N, N * N, N * N * N$
50	$N = N + 1$
60	print $N, N * N, N * N * N$
70	end

5. What will be printed as a result of the execution of the following algorithm? (The * indicates a multiplication.)

10	$N = 1$
20	print $N, N * N, N * N * N$
30	$N = N + 1$
40	if $N < 21$ go to 20
50	end

6. How many times is each statement of the algorithm given in Exercise 5 executed?

7. What will be printed as a result of the execution of the following algorithm? (The * indicates a multiplication.)

10	$N = 1$
20	print N
30	$N = 2 * N$
40	print N
50	$N = 2 * N$
60	print N
70	end

8. What will be printed as a result of the following algorithm? (The * indicates a multiplication.)

10	$N = 1$
20	print N
30	$N = 2 * N$
40	print N
50	if $N < 200$ go to 30
60	end

9. How many times is each statement in the algorithm of Exercise 8 executed?

10. Give an algorithm that inputs four numbers and prints the sum of the four numbers.

11. Modify your algorithm of Exercise 10 so that, in addition to the sum, each of the four numbers is also printed.

12. Give an algorithm that inputs a positive integer N and prints the sum of the first N positive integers.

13. Give an algorithm that inputs an integer N and then inputs N numbers and prints their sum.

14. How many times is each statement of your algorithm of Exercise 13 executed?

15. Modify your algorithm of Exercise 10 to obtain an algorithm for computing the average of the four numbers that the algorithm inputs.

16. Modify your algorithm of Exercise 13 to obtain an algorithm for computing the average of the N numbers that the algorithm inputs.

17. How many times is each statement of your algorithm of Exercise 16 executed?

18. Give an algorithm that inputs two numbers and prints out the largest of the two numbers regardless of the order in which they were entered.

19. Give an algorithm that inputs three numbers and prints out the largest of the three numbers regardless of the order in which they were entered.

20. Give an algorithm that inputs an integer N and then inputs N numbers and prints out the largest of the N numbers. A particularly inefficient way of doing the exercise is to start by sorting the N numbers using the insertion sort or the merge sort algorithm and then choosing the last number in the list. Give a more efficient solution than this approach.

21. How many times is each statement of your algorithm of Exercise 20 executed?

22. John puts $1000 in a savings account that gives 6% per annum interest, compounded monthly. Give an algorithm that prints out the value of John's investment at the end of the first and second months following the initial deposit.

23. Modify the algorithm of Exercise 22 so that John could find the value of his initial $1000 investment at the end of each month for a period of 2 years.

24. Figures 18.5 and 18.6, respectively, illustrate the data movements necessary to insert the second and third data elements into their correct positions, relative to their predecessors. With similar figures, illustrate the data movements required for the insertion of the fourth through the tenth numbers.

25. By reviewing the figures from Exercise 24, determine the number of data movements and comparisons used by insertion sort when sorting the 10 numbers that were considered in Exercise 24.

26. Sort the four numbers 6, 37, 12, 9 by using the insertion sort. Illustrate your intermediate results in a manner similar to that given in Figures 18.5 and 18.6.

27. Determine the number of data movements and comparisons used in Exercise 26.

28. Sort the four numbers given in Exercise 26 by using the merge sort. Give the sorted sublists at each pass and indicate the position of the markers during the second pass.

·19·

Computer Codes and Data Storage

Thirty years ago, during the computer's early stages of development, computers were used almost exclusively to do numerical calculations. Although we continue to use computers this way, most computers today are also used to perform manipulations on nonnumeric data. We can view modern computing systems as systems that process information. Using such a system, we can enter data into computers and extract data from computers; we can store information and manage it to produce new results organized in useful ways. Computers can process many different kinds of information, and different computers often deal with a given type of data in different ways. In this chapter, we look at some of the coding schemes associated with the storage of certain types of information in modern computing systems.

If we are to view computers as information processors, we must interpret *data* somewhat more broadly than simple numeric information. Here are some examples of data commonly managed by computers:

1. Numeric data such as integers and real numbers.

2. Textual data consisting of the letters of the alphabet, punctuation marks, and special symbols.

3. Information that computers use to generate visual patterns on display units.

4. Information that can be used to generate sounds through such devices as music synthesizers.

5. Complex information structures such as records of employees of a corporation. Such records might contain subsections that could be numeric or textual.

6. Information that identifies where other data is stored. This type of data is called address, or pointer, data.

7. Information that tells the computer what it should do, for example, programs in any programming language.

In order to store, reproduce, and manipulate such diverse types of information, computers use **codes** to represent the specific type of information being processed; codes are schemes for representing information. For example, our alphabet is a code of 26 symbols that we use to represent our speech in written form; a biological coding scheme, based on chemical patterns that make up DNA molecules, is used to transmit biological information from one generation to the next.

One powerful insight from mathematics is that codes can be used to represent information. In a sense, all mathematics uses this idea. For example, we can represent things in the real world, such as the motion of an object, with an equation—a code of numbers and symbols—and then manipulate the equation to find results that we can apply to the real world.

In mathematics, a code consists of two elements: a group of symbols and a set of rules for interpreting those symbols. In music, those same elements make up a code called a score. The musical score helps us to understand the codes that are used by computers.

The basic musical symbol is the note as shown in Figure 19.1. Of course, there are other symbols, such as the one shown in Figure 19.2 which signifies a longer duration of the sound. But what about interpretation?

Figure 19.2 Musical symbol signifying a longer duration of sound.

How does the musician know which notes to play? In other words, how can a code with just one symbol encode so many different sounds? The answer is that the information encoded by the note varies according to its location on the staff. Mathematicians refer to such encoding systems as **place-value systems** because the value of the symbol is determined in part by its position. Notes at different locations on the staff represent different sounds.

We use 10 different symbols to represent numbers; musicians use a few more to represent all kinds of music. When we use our place-value system to look at numbers, there's no mystery to the difference between 20 and 200. We recognize without thinking about it that the symbol "2" in 20 represents two tens and that the same symbol in 200 represents two hundreds. In music, the same principle allows a trained performer to quickly recognize the values of the notes.

BINARY CODES

Because computers are built from two-state electronic components, it is natural to use two-state codes to represent information in computing systems. Floppy disks (for a graphic view of such a disk, see Color Plate 22) and magnetic tapes, the common storage media of computers, are covered with a thin layer of ferromagnetic material that can be magnetized by an electric current into one of two states: positive or negative. (If you have used a computer, you are probably familiar with floppy disks. The drives into which these disks fit can read data stored on the disk by

Figure 19.1 Basic musical symbol.

sensing magnetic states, or they can write data onto the disk by altering existing magnetic states. The reading and writing are done through the oval window in the disk, which exposes the disk to the disk drive.) Each cell on a floppy disk can be polarized by electric current into positive or negative states. The magnetic patterns created by these two possible orientations determine the information contained on the disk. The key to this system is a scheme for representing data with only two symbols: the **binary code.** In binary, we customarily use the symbols 0 and 1 to represent the two states.

REPRESENTATION OF INTEGERS WITHIN COMPUTERS

To see how binary codes are actually used by modern computers, we look at how computers store and manipulate integers, one of the simplest forms of data. To appreciate how integers are stored in computers, it helps to first review how we humans store integers, that is, how we write them on a piece of paper. When we write 5472, we use the sequence of decimal digits, 5, 4, 7, and 2 to represent the integer that is equal to

$$5 \times 10^3 + 4 \times 10^2 + 7 \times 10^1 + 2 \times 10^0$$

(Recall that in exponential notation, $10^0 = 1$.) The significance of each digit in our representation of 5472 is determined by the digit itself and by the position of that digit within the sequence that comprises 5472. The two 7s in the representation of 7174 have different interpretations: the leftmost digit has the factor of 10^3 associated with it, whereas the rightmost 7 is associated with the factor 10^1. The ordinary notation that we use to write integers is called the **decimal positional notation** because we associate powers of 10 with the positions of the digits within the sequence of digits.

If we had a computer that could store items such as digits (for example, if its storage cells could assume any one of 10 different states), we could store integers in a computer in much the same way we record them on paper. The 10 states then could be used to represent the digits 0, 1, . . . , 9; numbers such as 3438 could be represented as a sequence of states within the computer that represents the digits 3, 4, 3, and 8, in that order. However, because of the electromagnetic materials used to build them, computers are generally capable of storing things only in 1 of 2 states, not in 1 of 10 states.

To store integers on a medium whose storage cells can be in only one of two states, we can associate the digit 1 with one of the states, say the positive state, and the digit 0 with the negative state. In this two-state storage scheme, each of the two digits, 0 and 1, is called a **bit,** short for binary digit.

If we choose to write integers in binary (base 2 positional) notation, instead of base 10, then we can represent all positive integers by sequences of 0s and 1s. In this notation, the rightmost digit is associated with 2^0 (instead of 10^0 in the decimal case); the next digit to the left is associated with 2^1, and so on. Analogously to the decimal positional notation, the significance of each binary digit is determined by the digit itself and the position of the digit within the sequence. The rightmost 1 in 101 represents $1 \times 2^0 = 1$, the 0 represents $0 \times 2^1 = 0$, and the leftmost 1 represents $1 \times 2^2 = 4$. Thus, 101 in base 2 represents 5 in decimal, or base 10. For the remainder of this chapter we will use subscripts to designate the base of a number; sometimes we will even indicate a base of 10 for emphasis such as in $101_2 = 5_{10}$. When subscripts are not used, base 10 should be assumed.

EXAMPLE 1: Find the Decimal Representation of 1101101_2. Moving right-to-left, and using successive powers of 2 (starting with 2^0), we get

$$1 \times 2^0 + 0 \times 2^1 + 1 \times 2^2 + 1 \times 2^3$$
$$+ 0 \times 2^4 + 1 \times 2^5 + 1 \times 2^6$$
$$= 1 \times 1 + 0 \times 2 + 1 \times 4 + 1 \times 8$$
$$+ 0 \times 16 + 1 \times 32 + 1 \times 64$$
$$= 1 + 0 + 4 + 8 + 0 + 32 + 64$$
$$= 109$$

thus, $1101101_2 = 109_{10}$.

From this example, we can see that 7 binary digits (bits) are needed to store 109_{10}. In actual computer implementation a fixed number of contiguous, or adjacent, bits are grouped to form a **word,** and all integers are stored within the bits of a single word. In a computer with 16-bit word size, 109_{10} would be stored as

$$0000 \quad 0000 \quad 0110 \quad 1101$$

Most small microcomputers—for example, the Apple II, TRS-80, Commodore, Attari, or IBM-PC—use 16 bits for integer storage. Some of the newer microcomputers as well as most medium-size computers use 32 bits for integer storage. It should be clear from the following examples that as we deal with larger numbers, higher powers of 2 will be needed. For convenience, we list the powers of 2 up to 2^{16} in Table 19.1.

TABLE 19.1 Powers of 2

$2^0 = 1$	$2^6 = 64$	$2^{12} = 4{,}096$
$2^1 = 2$	$2^7 = 128$	$2^{13} = 8{,}192$
$2^2 = 4$	$2^8 = 256$	$2^{14} = 16{,}384$
$2^3 = 8$	$2^9 = 512$	$2^{15} = 32{,}768$
$2^4 = 16$	$2^{10} = 1{,}024$	$2^{16} = 65{,}536$
$2^5 = 32$	$2^{11} = 2{,}048$	

EXAMPLE 2: If a 16-Bit Word Consisting of 0001 0011 1010 1101 Represents an Integer, What Is the Integer? The quick answer to this problem is 0001 0011 1010 1101_2. To represent the answer in a form we are more accustomed to, we convert 0001 0011 1010 1101_2 to decimal notation:

$$0001\ 0011\ 1010\ 1101_2$$
$$= 1 \times 2^0 + 0 \times 2^1 + 1 \times 2^2 + 1 \times 2^3$$
$$+ 0 \times 2^4 + 1 \times 2^5 + 0 \times 2^6 + 1 \times 2^7$$
$$+ 1 \times 2^8 + 1 \times 2^9 + 0 \times 2^{10} + 0 \times 2^{11}$$
$$+ 1 \times 2^{12} + 0 \times 2^{13} + 0 \times 2^{14} + 0 \times 2^{15}$$
$$= 1 + 0 + 4 + 8$$
$$+ 0 + 32 + 0 + 128$$
$$+ 256 + 512 + 0 + 0$$
$$+ 4096 + 0 + 0 + 0 = 5037_{10}$$

Given a particular integer, say 21_{10}, by reversing the calculations of the previous examples, we can determine the bit pattern that will be used to store the integer. Let us again assume that our computer uses 16-bit words. The bit pattern required for storing 21_{10} is given by the base 2 representation of 21_{10}, with the necessary 0s added on the left to fill all 16 positions. The base 2 positional representation of 21_{10} can be obtained by successively extracting the largest possible powers of 2 from 21_{10} until 21_{10} can be written as a sum of powers of 2. Thus,

$$21 = 16 + 5$$
$$= 2^4 + 5$$
$$= 2^4 + 4 + 1$$
$$= 2^4 + 2^2 + 2^0$$
$$= 1 \times 2^4 + 0 \times 2^3 + 1 \times 2^2$$
$$+ 0 \times 2^1 + 1 \times 2^0$$

and $21_{10} = 10101_2$. Therefore, 21_{10} is stored as 0000 0000 0001 0101.

EXAMPLE 3: How Would 353 Be Stored in a 12-Bit Word Size Computer? First, we write 353_{10} in base 2:

$$353 = 256 + 97$$
$$= 2^8 + 97$$
$$= 2^8 + 64 + 33$$
$$= 2^8 + 2^6 + 33$$
$$= 2^8 + 2^6 + 32 + 1$$
$$= 2^8 + 2^6 + 2^5 + 2^0$$

We see that $353_{10} = 1\ 0110\ 0001_2$. Within the computer, 353_{10} would be stored as 0001 0110 0001.

EXAMPLE 4: How Would 1814_{10} Be Represented in a Computer That Uses 16-Bit Words? First convert 1814_{10} to base 2. Using the powers of 2 from Table 19.1,

$$1814 = 1024 + 790$$
$$= 2^{10} + 790$$
$$= 2^{10} + 512 + 278$$
$$= 2^{10} + 2^9 + 278$$
$$= 2^{10} + 2^9 + 256 + 22$$
$$= 2^{10} + 2^9 + 2^8 + 22$$
$$= 2^{10} + 2^9 + 2^8 + 16 + 6$$
$$= 2^{10} + 2^9 + 2^8 + 2^4 + 4 + 2$$
$$= 2^{10} + 2^9 + 2^8 + 2^4 + 2^2 + 2^1$$

Hence, $1814_{10} = 111\ 0001\ 0110_2$ and its representation within the computer is 0000 0111 0001 0110.

Once stored in words inside the computer, the patterns of bits that represent numbers can be added in a manner very similar to the method we use with a pencil and paper to add ordinary decimal numbers. The details are illustrated in the following example.

EXAMPLE 5: Show the Steps in the Binary Addition of 4502_{10} and 1234_{10} When 16 Bits Are Used to Store Integers:

$$4502 \longrightarrow 0001\ 0001\ 1001\ 0110$$

and

$$1234 \longrightarrow 0000\ 0100\ 1101\ 0010$$

(The symbol $\rightarrow$ signifies "is stored as.")

To compute the sum of 4502 and 1234, the computer can work directly on the bits representing 4502 and 1234 and perform the logical equivalent of bit-by-bit addition in the same way that we do ordinary addition: add columns from right to left, carrying when necessary:

	0 0000 0011 0010 110	carry bits
4502 $\longrightarrow$	0001 0001 1001 0110	
1234 $\longrightarrow$	0000 0100 1101 0010	
6736 $\longrightarrow$	0001 0110 0110 1000	

Here are the steps in detail:

1. Add the bits 0 and 0 in the rightmost column, producing a 0 and a carry bit of 0.

2. Next, add 1 and 1 and the 0 carry bit from the previous addition, obtaining a sum of 10, which is recorded as 0 with a carry bit of 1.

3. Next, add the bits 1 and 0 and the 1 carry bit from the previous addition, getting a sum of 10: record the 0 and carry the 1.

4. Continue the process from right to left.

Computers contain special circuitry to do addition in binary form, just as we have illustrated here.

To keep matters simple, we have restricted this section to a discussion of the storage of positive integers. There are several schemes for storing negative numbers, the simplest of which is called the **sign-magnitude** method. In this case, all but the leftmost bit in a word are used to store the absolute value (magnitude) of the integer, and the leftmost bit is adjusted to 0 or 1, depending on whether the integer is positive or negative. This effectively reduces the number of bits that can be used to store the actual number by 1. For example, we have seen that 109_{10} is represented by 0000 0000 0110 1101 in a computer with 16-bit words. If sign magnitude is used, -109_{10} will be represented by 1000 0000 0110 1101. Two additional, more complex but also more useful, methods for storing negative integers are discussed in the following optional section.

Optional REPRESENTATION OF NEGATIVE INTEGERS

The main disadvantage of the sign-magnitude method of representing negative integers is that it complicates computer implementation of arithmetic. To add integers represented in this format, different logical steps must be followed, depending on the signs of the integers. If both integers are positive, that is, if both have 0 leading bits, then the sum of the two integers would have to be computed in two steps: first, the computer-stored binary numbers are added; second, the leading bit of the sum is set to 0 to indicate that the sum is positive. Similarly, if both the integers are negative, then the leading bit of the sum must be set to 1 to indicate a negative sum. A more complicated algorithm has to be followed to determine the sum of integers of different signs, because the sign of the sum would be determined by the magnitudes of the two integers, not just by their signs. Because of these complications, the sign-magnitude method is not often used to store integers.

A better strategy for storing negative integers is to use the **one's complement** of the binary number that represents the positive integer: the sequence of bits that are the exact opposite of the original bits. In this scheme, when the 12-bit representation of 109 is 0000 0110 1101, the negative integer -109 would be coded as 1111 1001 0010.

To illustrate the use of negative integers in one's-complement notation, consider the addition of 4502 and -1234:

$$
\begin{array}{rl}
& 1\ 1110\ 0110\ 0111\ 100 \quad \text{carry bits} \\
4502 \longrightarrow & 0001\ 0001\ 1001\ 0110 \\
-1234 \longrightarrow & \underline{1111\ 1011\ 0010\ 1101} \\
3268 & 0000\ 1100\ 1100\ 0011
\end{array}
$$

A quick look at the result should indicate to you that something is wrong. The computed sum, 0000 1100 1100 0011, represents 3267, not 3268. (Note the 1 in the rightmost position — sure sign of an odd number.) What has gone wrong here is that the last carry bit is not being captured. To correct this error, the last carry bit is always added to the results of the bit-by-bit addition. This additional 1 will correct the rightmost bit and force a corresponding change in the rest of the numbers:

$$
\begin{array}{r}
0000\ 1100\ 1100\ 0011 \\
\underline{+\quad 1} \\
0000\ 1100\ 1100\ 0100
\end{array}
$$

(For sums of two positive integers, as in Example 5, this carry bit is zero, so it can be ignored.)

The necessity of adding the last carry bit to the bit-by-bit sum complicates the logic of integer addition. However, the ability to add integers without concern for their sign, positive or negative, makes this type of addition considerably simpler than addition in sign-magnitude format.

To further simplify integer arithmetic, computer scientists developed yet another method of storing integers. In this scheme, called **two's complement,** negative integers are coded by first complementing all the bits of the original (positive) integer, and then adding 1 to the complemented bits. For example, in a 12-bit configuration, the representation of -109 would be obtained by taking the complement of the representation of 109 and then adding 1, all in binary:

$$
\begin{array}{ll}
109 \longrightarrow & 0000\ 0110\ 1101 \\
-109 \longrightarrow & \\
& \text{complement}\ (0000\ 0110\ 1101) + 1 \\
& = 1111\ 1001\ 0010\ + 1 \\
& = 1111\ 1001\ 0011
\end{array}
$$

The 16-bit two's-complement representation of -4502 and -1234 would be

$$
\begin{aligned}
4502 &\longrightarrow & 0001\ 0001\ 1001\ 0110 \\
-4502 &\longrightarrow & 1110\ 1110\ 0110\ 1001 + 1 \\
& = & 1110\ 1110\ 0110\ 1010 \\
1234 &\longrightarrow & 0000\ 0100\ 1101\ 0010 \\
-1234 &\longrightarrow & 1111\ 1011\ 0010\ 1101 + 1 \\
& = & 1111\ 1011\ 0010\ 1110
\end{aligned}
$$

Notice that the convention of having a leading 0 bit for positive integers and a leading 1 bit for negative integers is continued. Also—and very important—the representation of x can be obtained by taking the two's complement of the representation of $-x$:

$$
\begin{aligned}
-4502 &\longrightarrow & 1110\ 1110\ 0110\ 1010 \\
-(-4502) &\longrightarrow & 0001\ 0001\ 1001\ 0101 + 1 \\
& = & 0001\ 0001\ 1001\ 0110
\end{aligned}
$$

To find the decimal value of the integer x represented in two's-complement notation by 1110 1101 0010 0010, we note first that x must be negative because its leading bit is 1. Hence, the absolute value of x must be represented by $TC(x)$, the two's complement of x:

$$
\begin{aligned}
TC(x) &= TC(1110\ 1101\ 0010\ 0010) \\
&= 0001\ 0010\ 1101\ 1101 + 1 \\
&= 0001\ 0010\ 1101\ 1110
\end{aligned}
$$

Converting $TC(x)$ from binary to decimal, we get (reading from right to left)

$$
\begin{aligned}
2 + 4 + 8 + 16 + 64 + 128 + 512 + 4096 \\
= 4830
\end{aligned}
$$

hence $x = -4830$.

We can still do integer addition by bit-by-bit additions, but in the two's-complement case, we can ignore the last carry bit (see Spotlight 19.1). Although two's-complement representation is somewhat more complicated,

integer addition is now simplified because we can ignore the carry bit at the end of the bit-by-bit addition without jeopardizing the result. •

WORD SIZE

One of the decisions that must be made in designing a computer is the number of bits to be allocated to integer representation. If relatively few bits are used, then only integers with small magnitudes can be stored and manipulated directly by the computer. On the other hand, if a large number of bits are used, fewer integers can be stored in a fixed portion of a memory or storage device. Suppose a computer uses 8 bits to store integers. Regardless of the precise storage scheme, only 7 bits would be available to represent positive integers. Thus, 0111 1111, or 127, would be the largest positive integer that could be accommodated. For almost any application involving integer arithmetic, this would be a severe limitation.

In general, if n bits are used for integer storage, then the largest integer that could be accommodated will be 011 . . . 1, representing

$$
2^{n-2} + 2^{n-3} + \cdots + 2^1 + 2^0 = 2^{n-1} - 1
$$

Most microcomputers use 16 bits for integer storage. The largest integer that can be directly manipulated by such machines is therefore

$$
2^{15} - 1 = 32{,}767
$$

Most medium-size computers and some microcomputers use 32 bits to store integers. In these cases, integers up to and including

$$
2^{31} - 1 = 2{,}147{,}483{,}647
$$

can be handled.

SPOTLIGHT 19.1 Two's-Complement Arithmetic

To illustrate the way two's-complement notation facilitates arithmetic in computers, we will do two arithmetic problems:

$$\begin{array}{cc} 4502 & 1234 \\ -\underline{1234} \quad \text{and} & -\underline{4502} \end{array}$$

First, we need to find the binary representation of 4502 and 1234:

$$4502 = 4096 + 216 + 128 + 16 + 4 + 2 \longrightarrow 0001\ 0001\ 1001\ 0110$$

$$1234 = 1024 + 128 + 64 + 16 + 2 \longrightarrow 0000\ 0100\ 1101\ 0010$$

Now we can compute the two's complement of each of these binary numbers to get the representations of -4502 and -1234:

$$\begin{aligned} -1234 \longrightarrow\ &TC(0000\ 0100\ 1101\ 0010) \\ =\ &1111\ 1011\ 0010\ 1101 + 1 \\ =\ &1111\ 1011\ 0010\ 1110 \end{aligned}$$

$$\begin{aligned} -4502 \longrightarrow\ &TC(0001\ 0001\ 1001\ 0110) \\ =\ &1110\ 1110\ 0110\ 1001 + 1 \\ =\ &1110\ 1110\ 0110\ 1010 \end{aligned}$$

The bit-by-bit addition for $4502 - 1234$ will be

$$\begin{array}{lll} & 1\ 1110\ 0110\ 0111\ 110 & \text{carry bits} \\ 4502 \longrightarrow & 0001\ 0001\ 1001\ 0110 & \\ -1234 \longrightarrow & \underline{1111\ 1011\ 0010\ 1110} & \\ & 0000\ 1100\ 1100\ 0100 & \end{array}$$

The answer, $2048 + 1024 + 128 + 64 + 4 = 3628$, is the sum of 4502 and -1234.

We handle the other problem in much the same way:

$$\begin{array}{lll} & 0\ 0001\ 1001\ 1000\ 010 & \text{carry bits} \\ -4502 \longrightarrow & 1110\ 1110\ 0110\ 1010 & \\ 1234 \longrightarrow & \underline{0000\ 0100\ 1101\ 0010} & \\ & 1111\ 0011\ 0011\ 1100 & \end{array}$$

However, because the answer now begins with a 1, it must represent a negative number, which we can find by taking the negative of the integer represented by its two's-complement:

$$\begin{aligned} TC(1111\ 0011\ 0011\ 1100) &= 0000\ 1100\ 1100\ 0011 + 1 \\ &= 0000\ 1100\ 1100\ 0100 \end{aligned}$$

This yields $4 + 64 + 128 + 1024 + 2048 = 3268$; therefore, 1111 0011 0011 1100 represents -3268.

BINARY CODES AND LOGIC

As often happens, the mathematics needed to take advantage of electromagnetic storage schemes used by computers was developed long before modern computers were a practical possibility. In the 1850s the English logician George Boole (see Spotlight 19.2) had an ambitious goal: to codify the laws of thought by studying the ways we draw logical conclusions from different combinations of statements.

Suppose that statement A is "The balloon is red," and statement B is "The balloon has polka dots." A is true for the balloon in Figure 19.3a but is false for the one in Figure 19.3b. Boole's innovation was to create a mathematical approach to the logical relationships of statements. He assigned 0 to represent false and 1 to represent true. Thus, statement B, "The balloon has polka dots," would have value 0 in Figure 19.3a and value 1 for the one in Figure 19.3b. The two statements together can be described by four possible combinations:

	Statements	
	A	B
Truth values	0	0
	0	1
	1	0
	1	1

We can now form a compound statement: "A and B." In terms of our example, this says, "The balloon is red and has polka dots." We want "A and B" to be true (to have the value 1) only when both A is true and B is true. We can summarize the information about this compound statement by expanding our original table into the following **truth table:**

	Statements		
	A	B	A and B
Truth values	0	0	0
	0	1	0
	1	0	0
	1	1	1

This truth table transforms our intuitive notion of "and" into a mathematically precise definition: "A and B" is true if and only if the balloon is red and has dots.

More than 80 years after Boole's time, in the late 1930s, a similar inspiration occurred to a number of persons working toward automated calculating machines. In the United States, that insight struck both George Stibitz (see Spotlight 19.3, p. 552), a mathematician at Bell Laboratories, and Claude Shannon, a master's degree candidate at the Massachusetts Institute of Technology. These men and others realized that Boole's algebra of logic could be physically embodied in electrical circuits. The circuits that engineers used in the 1930s were made of relay switches, which ba-

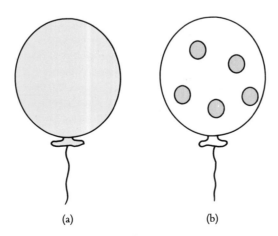

(a)   (b)

Figure 19.3 We can use logical statements and truth values to describe these two balloons.

SPOTLIGHT 19.2 George Boole

George Boole. [The Granger Collection.]

George Boole (1815–1864) was an English mathematician and logician. According to no less an authority than Bertrand Russell, "Pure mathematics was discovered by Boole, in a work which he called *The Laws of Thought.*" Boole's singular contribution was the development of a two-valued algebra, which serves as a workable method for interpreting logical truth and falsity. He set up a system of logical interpretation that could be treated as a numerical algebra in which the symbols are restricted to 0 and 1. As Boole said, "We may in fact lay aside the logical interpretation of the symbols in the given equation; convert them into quantitative symbols, susceptible only of the values 0 and 1; perform upon them as such all the requisite processes of solution; and finally restore them to their logical interpretation." Although Russell's statement is an exaggeration, it is nevertheless fair to say that Boole's work was essential to the development of modern logical thought.

sically resemble light switches. They have two positions, on or off, which correspond to current flowing or not flowing. We can represent on and off with the values 1 and 0, respectively. Stibitz and Shannon's breakthrough was the recognition that relay switches could represent logical statements and circuits could represent relationships between statements.

Let's return to our statements *A* and *B*, which are either true or false: that is, have truth value 1 or 0). We can think of *A* and *B* each representing a relay switch in a circuit. Then we can represent the logical function "*A* and *B*" by a circuit. If current is flowing in both *A* and *B*, then the "and" circuit permits current to flow; otherwise, it does not. You can see that this circuit is like two switches wired in series, the way Christmas tree lights often used to be. In a modern computer, this circuit, called an AND gate, is represented by a tiny intersection on a microchip and is designated in circuit diagrams by the symbol given in Figure 19.4.

Other logical functions are fundamental both to Boolean algebra and to the design of

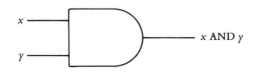

Figure 19.4 The circuit diagram for an AND gate.

SPOTLIGHT 19.3 George Robert Stibitz

George R. Stibitz
[Photo courtesy of Denison
University Archives.]

George Stibitz (1904–) received a B.S. degree from Denison University in 1926 and went on to earn a Ph.D. in mathematical physics from Cornell University in 1930. From 1930 to 1941 he was a research mathematician with the Bell Telephone Laboratories. A search for mathematics that would define the state (open or closed) of switching circuits led him to use telephone relays both to calculate in binary notation and to control calculation steps in a computer that would replace a desk calculator. The Complex Calculator, first in a series of relay computers, was designed for calculation with complex numbers and time-sharing operation from remote consoles. It was placed in service in 1939 and was demonstrated in 1940, with a terminal in Hanover, New Hampshire, and the computer in New York.

During World War II, Stibitz was a technical aide with the National Defense Research Committee, later reorganized as the Office of Scientific Research and Development, from 1941 to 1945. His model 2, the Relay Interpolator, used for ballistic calculations for gun directors, was placed in service in 1943. It utilized changeable programs and "self-checking." Later models in this series of relay computers included such refinements as memory indexing, a logic calculator, and automatic selection of programs.

From 1946 to 1964 Stibitz consulted in applied mathematics with clients in private industry and the government. During these years he became involved in a variety of projects. Among the many subjects he investigated were propeller vibration computations, stability of discontinuous servos, an electronic computer for commercial application, small-tool design, blood flow in an elastic artery, a mechanical change computer for a machine vender, an electronic organ and tone synthesizer, a heart-beat rate alarm, an arterial pump, and a ballistocardiograph.

Because of his interest in applying mathematics to physiological problems, he became a professor in the physiology department at Dartmouth College in 1966 and was granted emeritus status in 1970. At Dartmouth he has engaged in applying mathematics to such biomedical problems as biochemical binding to plasma, radiation dosimetry, random-walk transmission of nutrients through capillary walls, and the electrical properties of cell membranes. The author of numerous scientific articles and two books, he received the Harry Goode Memorial Award in 1966, the Emanuel R. Piore Award in 1977, the Computer Pioneer Award in 1982, and was elected to the National Inventors Hall of Fame in 1983. He holds 34 patents and has received three honorary Doctor of Science degrees.

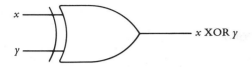

Figure 19.5 The circuit diagram for an XOR gate.

computers. A particularly useful relationship, called the *exclusive or,* abbreviated as XOR, has the following truth table:

	Statements		
	A	*B*	*A* XOR *B*
Truth values	0	0	0
	0	1	1
	1	0	1
	1	1	0

It is easy to confuse the XOR function with the OR function, which assigns a truth value of 1 to A OR B if A, B, or *both* A and B are true. A XOR B is true if either A is true or B is true, but false if *both* A and B are true. Again, we can build a circuit to physically implement this function. An XOR circuit is designated by the engineering symbol given in Figure 19.5.

COMPUTER ADDITION

Modern computers derive their power from their ability not only to store information but also to perform operations on the stored data. To illustrate this latter capability of computers, we consider the simple problem of adding two binary integers. There are only four possible pairs of single-digit binary integers. These pairs of numbers and their corresponding sums are given by

$$
\begin{array}{cccc}
0 & 0 & 1 & 1 \\
+0 & +1 & +0 & +\,1 \\
\hline
0 & 1 & 1 & 10
\end{array}
$$

Note that in the last case, when we add 1 to 1, we get a sum of zero and a *carry* of 1. Remember that 10 is the way we represent decimal 2 in binary.

The simplest circuitry that performs additions, called the *half adder,* takes two single-digit inputs and produces a sum and a carry digit. The addition pattern we have just observed can be implemented by circuits that take the inputs and produce the outputs summarized below.

Input 1 *x*	Input 2 *y*	Output 1 sum	Output 2 carry
0	0	0	0
0	1	1	0
1	0	1	0
1	1	0	1

If you look at the truth tables for the AND and XOR functions, you will notice that

$$\text{sum} = x \text{ XOR } y$$

and

$$\text{carry} = x \text{ AND } y$$

A circuit with the diagram given in Figure 19.6 will perform this addition.

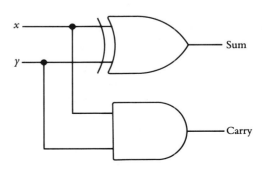

Figure 19.6 The circuit diagram for the half adder.

Figure 19.7   A replica of the model *K* computer built by George Stibitz [Photo courtesy of Denison University.]

Figure 19.7 shows the first adder, called the model *K*, which was built by Stibitz on his kitchen (that is where the *K* comes from) table in 1937. In the 50 years since Stibitz and Shannon, computers have evolved from automatic calculators into the most versatile machines ever designed. But they still work in essentially the same way, by acting on two-valued functions. We can now appreciate the secret of their flexibility: the 0s and 1s in these functions can represent numbers; but they can also represent other kinds of information, such as text for word processing, elements of a picture, and even musical notes.

TEXT DATA

A large number of computing applications, such as automated letter writing, billing, and interoffice communications, deal mostly, if not exclusively, with text data. Formally, **text data** signifies sequences of characters from the character set consisting of 26 uppercase letters, 26 lowercase letters, 10 digits, about 8 punctuation marks, and a number of special charac-

ters, including "@," "%," and "$." In addition to these 90 or so printable characters representing symbols that appear in written text are a number of nonprintable characters whose effects are visible on the computer screen but that do not represent written text. These are called **control characters:** they govern how data is transmitted and the way the computer screen behaves. Characters that cause carriage returns, line feeds, and page breaks are examples of control characters.

To store text data or, more simply, a single character such as "A," "7", or "?," we need to devise a code to store these characters as sequences of binary bits. If such a code consisted of 2 binary bits, then only 4 characters, represented by 00, 01, 10, and 11, could be accommodated. If 3 bits are used, 8 different characters can be represented by 000, 001, 010, 011, 100, 101, 110, and 111. If n bits are used, then 2^n characters can be represented.

Unfortunately, no single character coding is accepted as an industry standard. Currently systems employ character sets of size 64, 128, and 256. The two most widely accepted character codes are **EBCDIC (Extended Binary Coded Decimal Interchange Code)** and **ASCII (American Standard Code for Information Interchange)**. Because ASCII is the preferred coding scheme on microcomputers, we will confine our discussion to it.

In ASCII format, a **byte** (8 consecutive bits) is used to store each character. As a result, ASCII can accommodate up to 2^8, or 256, printable or control characters. The full set of printable characters, their binary representations, and their decimal equivalents are shown in Table 19.2. Notice that uppercase letters start with a code of 65 for *A* and progress sequentially to 90 for *Z*. The lowercase letters have codes ranging from 97 for *a* to 122 for *z*. The blank character has code 32, and the null character (it is convenient to have a character that does not produce any effect on the computer screen) has code 0. Notice that if alphabetic characters were sorted in ascending order according to their ASCII code, the characters would appear in their correct alphabetic order.

Readers who know the rudiments of the BASIC or Pascal programming languages will see that the execution of the simple BASIC program in Figure 19.8 or the equivalent Pascal program in Figure 19.9 will produce the code table used on a particular computer.

```
10   rem Program to Generate Character Code Table
20   print "CODE","CHARACTER"
30   for I = 0 to 127
40       print I, chr$(I)
50   next I
60   end
```

Figure 19.8 A BASIC program for generating ASCII characters and their codes.

TABLE 19.2 **Printable characters and their ASCII codes**

Character	ASCII code (decimal)	ASCII code (binary)	Character	ASCII code (decimal)	ASCII code (binary)
(space)	32	0010 0000	8	56	0011 1000
!	33	0010 0001	9	57	0011 1001
''	34	0010 0010	:	58	0011 1010
#	35	0010 0011	;	59	0011 1011
$	36	0010 0100	<	60	0011 1100
%	37	0010 0101	=	61	0011 1101
&	38	0010 0110	>	62	0011 1110
'	39	0010 0111	?	63	0011 1111
(	40	0010 1000	@	64	0100 0000
)	41	0010 1001	A	65	0100 0001
*	42	0010 1010	B	66	0100 0010
+	43	0010 1011	C	67	0100 0011
,	44	0010 1100	D	68	0100 0100
-	45	0010 1101	E	69	0100 0101
.	46	0010 1110	F	70	0100 0110
/	47	0010 1111	G	71	0100 0111
0	48	0011 0000	H	72	0100 1000
1	49	0011 0001	I	73	0100 1001
2	50	0011 0010	J	74	0100 1010
3	51	0011 0011	K	75	0100 1011
4	52	0011 0100	L	76	0100 1100
5	53	0011 0101	M	77	0100 1101
6	54	0011 0110	N	78	0100 1110
7	55	0011 0111	O	79	0110 1111

Continued on next page

```
 10    program info1 (output);
 20    {This program produces a table of ASCII for the characters with codes from 0 through 127}
 30
 40    var code : integer;
 50    begin
 60        writeln; writeln;
 70        writeln(' CODE     CHARACTER');
 80        for code := 0 to 127 do
 90            writeln(code:4, chr(code):12)
100    end
```

Figure 19.9 A Pascal program for generating ASCII characters and their codes.

TABLE 19.2 Printable characters and their ASCII codes *(continued)*

Character	ASCII code (decimal)	ASCII code (binary)	Character	ASCII code (decimal)	ASCII code (binary)
P	80	0101 0000	h	104	0110 1000
Q	81	0101 0001	i	105	0110 1001
R	82	0101 0010	j	106	0110 1010
S	83	0101 0011	k	107	0110 1011
T	84	0101 0100	l	108	0110 1100
U	85	0101 0101	m	109	0110 1101
V	86	0101 0110	n	110	0110 1110
W	87	0101 0111	o	111	0110 1111
X	88	0101 1000	p	112	0111 0000
Y	89	0101 1001	q	113	0111 0001
Z	90	0101 1010	r	114	0111 0010
[	91	0101 1011	s	115	0111 0011
\	92	0101 1100	t	116	0111 0100
]	93	0101 1101	u	117	0111 0101
^	94	0101 1110	v	118	0111 0110
_	95	0101 1111	w	119	0111 0111
'	96	0110 0000	x	120	0111 1000
a	97	0110 0001	y	121	0111 1001
b	98	0110 0010	z	122	0111 1010
c	99	0110 0011	{	123	0111 1011
d	100	0110 0100	|	124	0111 1100
e	101	0110 0101	}	125	0111 1101
f	102	0110 0110	~	126	0111 1110
g	103	0110 0111			

These programs will transmit the printable as well as nonprintable characters to a computer terminal or printer. You can observe the effect of these nonprintable control characters without any specific character actually being exhibited. If a listing of only the printable characters is desired, the loop counters in both programs should start at 32.

Optional POINTER DATA

In many computing applications, including those found in such diverse areas as banking, inventory control, and payroll management, the structure of each item of data is composed of many distinct subportions. For example, a student record may consist of the following: the name of the student (up to 25 characters long); the student's address (up to 100 characters long); the number of credit hours the student has accumulated (an integer); the student's grade point average (a real number); and the student's mailbox number (an integer). Figure 19.10 shows the layout of a student record in graphic form, with an indication of the number of bytes dedicated to each subsection of the record. In this example, each record would be 137 bytes long.

One way of organizing the data for, say, 10,000 students would be to take a large enough area (1,370,000 bytes) on some storage medium and alphabetically locate the 10,000 records in contiguous positions in that area. This type of structure, called an **array** of records, is illustrated in Figure 19.11.

Although structurally simple, arrays can have serious disadvantages. Suppose that in the illustration given in Figure 19.11, the second student, Daniel Bonar, drops out of school. To delete Bonar's record, records numbered 3 through 10,000 (9998 in all) would have to be moved up one spot. This means moving

Full name	Address	Credits	GPA	Mailbox
25 bytes	100 bytes	4 bytes	4 bytes	4 bytes

Figure 19.10   A sample student record.

$9998 \times 137 = 1,369,726$ bytes of data, which would require a significant amount of computing resources. Of course, if the record deleted comes from the lower portion of the array, fewer records would have to be moved. On the average, 5000 record or 685,000 byte movements would be required for each record deletion or addition.

A more efficient but somewhat more complicated storage organization will result if we attach to each record an additional field containing the address of the record alphabetically next in line. This type of data structure is called **address,** or **pointer,** data. Because we can now move from one record to the next by following pointers, we may store records anywhere and need not locate them contiguously.

Bates David A	. . .	. . .	. . .	. . .
Bonar Daniel D	. . .	. . .	. . .	. . .
Cameron James S	. . .	. . .	. . .	. . .
.	.	.	.	.
.	.	.	.	.
Snyder Rita E	. . .	. . .	. . .	. . .
Sterrett Andrew	. . .	. . .	. . .	. . .
Stout Elliott R	. . .	. . .	. . .	. . .
.	.	.	.	.
.	.	.	.	.
Wetzel Marion D	. . .	. . .	. . .	. . .

Figure 19.11   An array of student records.

Full name \\	Address \\	Credits	GPA	Mailbox	Pointer
25 bytes	100 bytes	4 bytes	4 bytes	4 bytes	4 bytes

Figure 19.12 A single record with a 4-byte pointer subfield.

Figure 19.12 gives the layout of a single record with a 4-byte pointer subfield. Figure 19.13 illustrates the **linked list** structure for the entire data set.

To delete student Andrew Sterrett's record, we follow the pointers to Rita Snyder's record and redefine the pointer embedded there, so that it points to Elliot Stout's record (see Figure 19.14). To insert a new record for Emmett

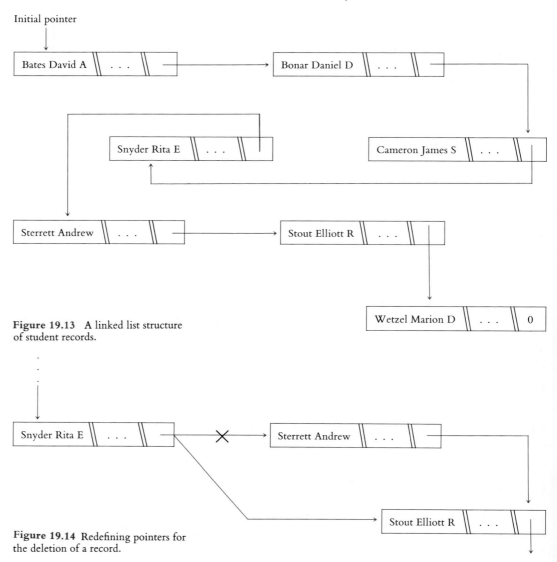

Figure 19.13 A linked list structure of student records.

Figure 19.14 Redefining pointers for the deletion of a record.

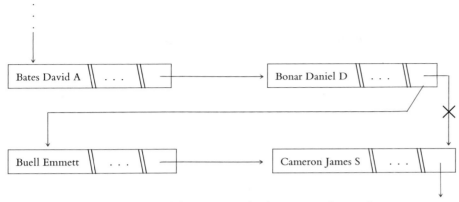

Figure 19.15 Redefining pointers for the insertion of a record.

Buell, we traverse the linked list until we determine the proper location for this record (between Daniel Bonar and James Cameron, in this case) and redefine the two pointers in the manner indicated in Figure 19.15.

To manage storage efficiently when records are added and deleted, the computer must have the ability (1) to handle storage that is released once records have been deleted and (2) to locate new storage areas to accommodate records that are to be inserted. Provisions for pointer data as well as mechanisms for access and release of storage areas are built into certain higher-level programming languages such as Pascal and PL/1. •

DATA INTERPRETATION

Most computers are **byte-addressable,** in the sense that a unique location number, or address, identifies each byte of memory. The overall memory size is usually given by the total number of bytes that make up the memory. A memory size of 16K bytes (or 16 kilobytes) refers to $16 \times 2^{10} = 16 \times 1024 = 16,384$ bytes of memory addressed by 0, 1, . . . , 16,383.

As we have seen, the byte is a convenient unit for storing characters. Larger numbers of bits generally are required for storing integers and other, more complex types of data. This is accomplished by using 2, 3, or 4 contiguous bytes for storing 16-, 24-, or 32-bit integers. Suppose a computer uses the 8-bit ASCII code for storing characters and a 16-bit format for storing integers. If two contiguous bytes within the memory of this computer contain the binary configuration

$$0100 \quad 0100 \quad 0100 \quad 0001$$

then the content of these two bytes could be the characters A and D (stored in right-to-left order), or the 16 bits could represent the integer

$$2^{14} + 2^{10} + 2^6 + 2^0$$
$$= 16,384 + 1024 + 64 + 1 = 17,473$$

Just by looking at the content of these two bytes, we cannot determine whether the data located there is the integer 17,473, the two characters A and D, or possibly some other form of encoded data.

Sometimes the information to be stored is a mixture of text and numeric data, as in the case of addresses that include zip codes and street numbers. The problem of interpreting the street number as numeric versus text can be sidestepped in this situation by treating everything as text data. After all, we generally would not consider performing arithmetic operations on zip codes and street numbers.

However, there are potentially more confusing situations. An employer may want to store an employee's name and address as text data and her salary as numeric data, with the anticipation of performing arithmetic operations on the salary to compute such things as tax obligations and fringe benefits. We need to make sure that we (more importantly, computers) do not get confused in such situations. The potential confusion is avoided by performing only the allowed operations on a specified data type. For example, text data representing a zip code will not be confused with integer data as long as we do not initiate a computer instruction to add (addition is not permissible for text data) the contents of memory locations holding zip codes.

Knowing the *context,* or how the data is to be used, removes the ambiguity in interpretation. If an integer addition is being executed and the addition instruction specifies a context for 0100 0100 0100 0001, then the data is interpreted as the integer 17,473. If, however, our task was to print out two characters and we encountered 0100 0100 0100 0001, then in this context, this bit configuration would be interpreted as the two characters *A* and *D*.

ERRORS AND ERROR CORRECTION

Although errors by programmers and computer operators are typically called "computer errors," true computer errors occur infrequently. Errors that do occur could be caused by equipment failure or even by background

radiation, which we cannot control. In this section, we consider a method from Richard Hamming, a mathematician at Bell Laboratories, for detecting and correcting certain types of errors in binary data.

In 1948 Hamming proposed an efficient method of detecting and correcting errors in binary coded data. Suppose we need to store 4 bits (single binary digits) in memory in such a way that we can correct any single error in memory. We can arrange the 4 bits (0 1 1 0, in our case) in a diagram as shown in Figure 19.16a. Next, we fill the three empty spaces in the diagram with a 1 or a 0, so that each circle contains an even number of 1s, as illustrated in Figure 19.16b. We can think of these 7 bits as a representation, a redundant one, of the central 4 bits that we want to preserve in memory. The 7 bits are called an **error-correcting code** because through a simple algorithm, we can correct any single error. To understand how the algorithm works, suppose there is an error in one of the bits, the bit that is under-

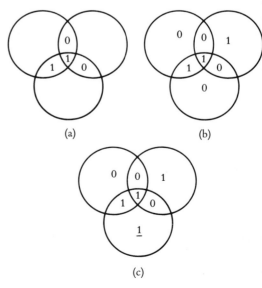

Figure 19.16 (a) We want to store the 4 bits, 0, 1, 1, and 0. (b) To complete the Hamming code, we add error correcting bits. (c) The underlined bit is an error.

lined in Figure 19.16c. When we count the number of 1s in each of the three circles, we find that two of them have an even number of 1s. We label these circles "good" and label the one circle with an odd number of 1s "bad." The bad circle tells us there is an error somewhere in the diagram. Once we have detected an error, we try to correct it.

Obviously we should try changing a bit's value to make all the circles good. Because changing any of the bits outside the bad circle is not going to help, we consider the 4 bits inside the bad circle. If we change the leftmost bit (from 1 to 0) or the rightmost bit (from 0 to 1), we fix the bad circle but ruin the one that was good. If we change the "innermost" 1 to 0, things get worse, because we end up fixing the bad circle but ruining both good circles. We have only one bit left to try: the one that has the error. When we change it, all the circles become good and the error is corrected.

The power of the **Hamming code** comes from the fact that there is just one way to fix a single error: the correct way. Note that in the Hamming code, each code "word" is 7 bits long (4 message bits plus 3 error-correcting bits), and every possible sequence of 7 bits is either a correct message or a message with a single correctable error. Thus, every code bit makes a contribution. Such codes are called *perfect codes*.

The Hamming code has its limitations, however. Suppose, owing to two errors, the code depicted in Figure 19.17a is transformed into the one in Figure 19.17b. If we count up the number of 1s, we find two good circles and one bad one. The error-correcting procedure can make all the circles good by changing one of the 1s to a 0 (Figure 19.17c). But this makes matters worse by introducing a third error.

To appreciate the value of the Hamming code, consider a modest-size memory bank that has about a million 8-bit words. If no error-correcting mechanisms are in place, there will be a mistake due to radiation about

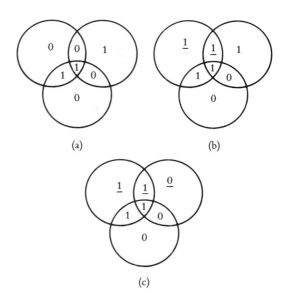

Figure 19.17 (a) A Hamming code, with error correcting bits. (b) The code now has two errors (underlined). (c) A "corrected code" with the errors.

every 43 days on the average. To use a single error-correcting code similar to the one we just considered, we would have to enlarge the memory by a little over 20%. That should not be surprising because we cannot expect to get "error insurance" for free. This larger memory will be prone to even more frequent errors, about one every 36 days. (Each memory location has a certain chance of being struck by radiation, and the more memory we have, the greater the chance that some location will be hit.) It might seem as if we have made the problem worse. But remember that a single error can be fixed, and we have a problem only if there are two or more errors in the same code word. We can calculate that this will happen about once in 63 years. Thus, by designing a clever code, Hamming was able to reduce the frequency of errors from once every 43 days to once every 63 years.

Some computer applications, such as flight-control management, need more protection than can be provided by single-error correction. In some such cases, redundant computing units can be used to reduce the chance of error. In less-critical instances, more complex error-correcting codes can be used. A tiny flaw on a compact disk can affect a large number of bits. Therefore, a code that can correct up to 14,000 consecutive errors is used by many compact disk players. This code requires only 1 correcting bit for every 3 "sound bits," an overhead of only 33%.

CONCLUDING COMMENTS

Any type of data that can be coded in binary form can be stored in modern computing systems. Specific binary coding schemes are used for representing integers and text data. Obviously, coding methods exist for representing other commonly used data types such as real numbers, or, as computer scientists sometimes call them, floating point numbers. It is also possible, through clever codes, to store more exotic data types such as visual images and musical scores.

As we saw in this chapter, to store a large number of records, it is better to include a pointer field within each record. The use of pointers allows records to be added and deleted more efficiently. In general, as coding schemes become more complex, the issues of storage efficiency and computational efficiency become more significant.

REVIEW VOCABULARY

Address A numeric value representing the location of data in memory.

Array A collection of similar data objects stored sequentially in memory.

ASCII (American Standard Code for Information Interchange) A binary coding scheme for representing character data.

Binary code A coding scheme that uses two symbols.

Bit The smallest storage unit in computing systems, generally represented by a binary digit.

Byte A group of 8 consecutive bits in a computer-storage medium such as memory or floppy disk.

Byte-addressable computer A computer in which each byte of memory is uniquely addressed by successive integers starting with 0.

Code A mechanism used to represent information. A code consists of a group of symbols and a set of rules for interpreting the symbols.

Control character A nonprintable character that, when transmitted to an output device such as a console or a printer, governs how the device behaves.

Decimal positional notation The usual (base 10) notation for writing integers.

EBCDIC (Extended Binary Coded Decimal Interchange Code) A binary coding scheme for representing character data.

Error-correcting code A code in which certain, but not necessarily all, types of errors can be identified and corrected.

Hamming code An error-correcting code, proposed by Richard Hamming, that can correct all single errors.

Linked list A collection of data elements in which all elements except the last one have a pointer to their successor.

One's complement A binary representation for storing integers, in which nonnegative integers are coded by their usual binary digits. Negative integers are represented by the complements of the bits used by their positive counterparts.

Place-value system A coding scheme in which the value of a symbol is determined in part by its position.

Pointer data   A data element whose value represents the memory address of other data.

Sign-magnitude   A coding scheme for storing integers, in which the leading bit represents the sign of the integer and the remaining bits represent its magnitude (absolute value).

Text data   Data comprised of the characters consisting of the letters of the alphabet, decimal digits, punctuation marks, and special characters such as "@", "%", and "$."

Truth table   A tabular representation of the truth values of a statement.

Two's complement   A binary representation for storing integers in which nonnegative integers are coded by their usual binary digits. Negative integers are represented by the bits obtained by complementing the bits of the corresponding positive integer and adding 1 to the result.

Word   A group of contiguous bits. The number of bits within the group varies among computers.

SUGGESTED READINGS

Dewdney, A. K.: *The Turing Omnibus,* Computer Science Press, New York, 1989.

Hamming, R. W.: *Coding and Information Theory,* Prentice-Hall, Englewood Cliffs, N.J., 1980.

McEliece, Robert J.: "The Probability of Computer Memories," *Scientific American,* 252(1) (January 1985).

MacKenzie, Charles E.: *Coded Character Sets: History and Development,* Addison-Wesley, Reading, Mass., 1980.

Pohl, Ira, and Alan Shaw: *The Nature of Computation: An Introduction to Computer Science,* Computer Science Press, New York, 1981.

EXERCISES

1. Write 86 in base 2 positional notation. If a computer uses 8 bits to store integers in sign-magnitude format, how would 86 be stored inside the computer?

2. Do Exercise 1 using -109 instead of 86.

3. What integer (in decimal notation) is specified by 100111_2? How would this number be stored in a computer that uses 8 bits to store integers in sign-magnitude format?

4. Do Exercise 3 using -100111_2 instead of 100111_2.

5. Assume that a computer uses 16 bits to store integers in sign-magnitude form. What bit patterns will represent the integers 3419, -3419, 17842, and -17842?

6. In the computer described in Exercise 5, what integers are represented by the bit patterns 0000 0011 1010 0110 and 1111 0101 1100 0010?

● **7.** Assume that a computer uses 16 bits to store integers in one's-complement form. What bit patterns will represent the integers 3419, -3419, 17842, and -17842?

● Optional exercise.

● **8.** In the computer described in Exercise 7, what integers are represented by the bit patterns 0000 0011 1010 0110 and 1111 0101 1100 0010?

● **9.** Assume that a computer uses 16 bits to store integers in two's-complement form. What bit patterns will represent the integers 3419, −3419, 17842, and −17842?

● **10.** In the computer described in Exercise 9, what integers are represented by the bit patterns 0000 0011 1010 0110 and 1111 0101 1100 0010?

11. Complete the second column of the following table, which gives 4-bit integer representations of the sign-magnitude storage method.

Bit configuration	Value represented in sign-magnitude
0000	0
0001	
0010	
0011	
0100	
0101	
0110	
0111	
1000	
1001	
1010	
1011	
1100	
1101	
1110	
1111	−7

● **12.** Add columns for one's complement and two's complement to the table constructed in Exercise 11.

13. In the 4-bit sign-magnitude integer representation of Exercise 11, what is the largest integer that can be accommodated? What is the smallest?

14. What is the largest integer that can be stored by a computer using 8-bit sign-magnitude format? What if 12 bits are used? What if 20 bits are used?

15. Show the steps in the binary addition of 734 and 1081 when 12 bits are used to store integers. See Example 5.

16. In decimal positional notation, the presence of a 0 as the least significant digit is a simple test for divisibility by 10. Similarly, the presence of a 0 as the least significant digit is a test for divisibility by 2 when binary positional notation is used. Is the presence

● Optional exercise.

of a 0 in the rightmost position a test for divisibility by 2 when the sign-magnitude storage scheme is used? (Look at the table you constructed in Exercise 11 and consider positive and negative integers.)

17. Exercise 16 deals with a test for divisibility by 2. Propose a more general rule for testing divisibility by 4, 8, 16, and other powers of 2.

18. Write down your first name and then use Table 19.2 to write the sequence of ASCII codes (in decimal) that would represent your name inside a computer.

19. Continue your work from Exercise 18 and determine the sequence of bits that would be required to store your name.

20. Suppose that a computer uses the 8-bit ASCII code for storing characters (see Table 19.2) and 8-bit sign-magnitude for storing integers. If an 8-bit memory location that holds the character "I" is somehow misinterpreted as an integer, what would the decimal value of that integer be?

21. If the character "3" instead of "I" were misinterpreted in Exercise 20, what would be the decimal value of the integer?

22. Suppose a computer uses the 8-bit ASCII code for storing characters and 12-bit sign-magnitude format for storing integers. The three consecutive memory locations where the characters "$a = 3$" are stored is somehow misinterpreted as two successive integers. What are the decimal values of these integers?

23. (For students with programming backgrounds.) Use one of the programs given in Figure 19.8 or 19.9 to determine whether the computer you are using stores characters by their ASCII codes. By making a suitable adjustment to these programs, you can find out if the computer you are using has a 256-character implementation.

● **24.** (For students with programming backgrounds.) Most BASIC compilers and interpreters allow the user to perform bit-by-bit AND operations on integers. A single-bit AND operation is given by 1 AND 1 = 1, 0 AND 1 = 0, 1 AND 0 = 0, and 0 AND 0 = 0. The result of an AND operation on two integers is obtained by ANDing the corresponding bits of the integers. Write a BASIC program that will input an integer A and use the AND operation to compute and print out the bit configuration of A. Using negative values for A, find out if the system you are using stores integers in sign-magnitude, one's-complement, or two's-complement format.

25. The negation of the statement A, NOT A, is defined to have the opposite truth value of A. Give a truth table for NOT A.

26. The logical inclusive OR function of statements A and B, A OR B, yields a true value if and only if A is true or B is true or both A and B are true. Give a truth table for A OR B.

27. Use truth tables to show that for all possible truth values of A and B, NOT (A OR B) is identical to (NOT A) AND (NOT B).

28. Use truth tables to show that for all possible truth values of A and B, A XOR B is identical to ((NOT A) AND B) OR (A AND (NOT B)).

● Optional exercise.

●**29.** In this chapter we discussed the advantages of using pointers when storing 10,000 student records (see Figures 19.12 and 19.13). One disadvantage of using pointers is the storage space they require. How many bytes of additional storage will be necessary to store the 10,000 student records in the format described in Figure 19.12? What percentage increase of storage space does this represent over the array method of storage?

●**30.** In the array structure, an average of $N/2$ record movements required to insert a record in an N-record array. If we wanted initially to build the entire array by successive record insertions, N would grow as we made insertions. On the average, how many record movements will be required to insert the first record? How many record movements will be required to insert the second record, etc., the kth record?

31. Starting with 1010, construct a diagram similar to that shown in Figure 19.16b. Consider the four possible single-bit data errors that may result (0010, 1110, 1000, and 1011), and show in each case that the error can be detected and corrected.

32. Assume that the 4-bit pattern, 1010, of Exercise 31 is erroneously recorded as 1100. Can this error be detected? If so, can it also be corrected?

● Optional exercise.

·20·

Computer Graphics

In the last decade, innovations in computer technology have transformed the computer into an important tool for people in many areas of our society. The way in which the individual interacts with the computer has become an important concern for computer hardware and software system designers. Computer-generated graphics is an extremely important part of this computer/human interface. The more people understand about graphics, the more creatively they can use the computer, and the more effectively they can assimilate the information that the computer presents to them.

In addition to contributing to the user interface, computer graphics has become commonplace as a design and analysis tool in its own right in such diverse activities as automotive design, publishing and commercial art layout, and drafting architectural plans. Computer graphics is used by artists in the development of animated cartoons, the generation of geometric patterns for use in textile design, and

the production of special effects for the feature film industry.

In the scientific community, computer graphics is used by chemists to create models of complex chemical molecules for the analysis of chemical reactions. Astronomers generate graphic models of galaxies to better understand the interaction of celestial bodies. These models are created using data sets that often contain millions of values, and the graphic presentation of these data sets allows the user to more efficiently interpret these values and their interrelationships.

Computer graphics contributes to the speed, the productivity, and, in some cases, the creativity with which people do their work. By becoming aware of the fundamentals of graphics algorithms and operations, the user gains a sense of the mechanics involved in presenting the computer-generated information; he or she can then use the new insight to work with the computer more easily.

Computer graphics can be defined as the synthesis of pictures by a computer. This synthesis may take place by the computer's presenting a predefined image, such as the "icons" in the user menus of many personal computers or the "sprites" that comprise the on-screen players in a video game. Alternately, the computer may reproduce the picture on a display device by executing a program that uses an algorithm based on a mathematical formulation of an object. This algorithm is created to transform the usually three-dimensional data about the object into an internal format that the computer can process and convert into a two-dimensional picture. The picture should have enough detail so that we can easily recognize it. However, the physical constraints of display devices make it impossible to produce perfect representations of the mathematical formulation.

To reduce the difficulties involved in constructing a complex figure, we place simple objects together in a suitable combination to form the complex object. Through arrangements of line segments, for example, we can produce pictures of considerable complexity. In actual practice, special graphics packages are used to generate and manipulate a collection of primitive, or simple, objects, thereby enabling the user to design a large variety of figures made up from the primitive forms.

With the current state of computer technology a person can create and manipulate pictures on a computer screen by simply pressing a few keys. The intent of this chapter is to take the mystery out of this process by exploring some of what happens inside the computer when simple commands are issued by the user. To do this, we first consider the physical characteristics of display devices and the way such devices produce pictures. This is followed by a discussion of some of the basic geometric transformations that are used in the production and manipulation of some simple graphic objects.

HOW PICTURES APPEAR AT A DISPLAY DEVICE

A computer graphics system consists of the display device on which the picture is presented and the hardware and software necessary to formulate the data required to construct the picture. In later sections, we describe the computer and its component software; here, we consider the device used to display the picture generated by the computer. A wide range of devices are available for presenting the computer-created picture to the user. In the simplest case, devices such as line printers or plotters render the picture in a "hard-copy" format. Other devices use light-emitting diodes (LEDs) or gas-discharge technologies, such as plasma panels.

The device most commonly used — from home computer systems to the most sophisticated computer graphics environments — is the **cathode ray tube,** or CRT. We can roughly classify CRTs as belonging to either the stroke (vector) family or to the raster family. Although these types function differently, they both use the same basic technology to display a picture. The front of the display surface of a vacuum tube (depicted in Figure 20.1a) is coated with a material called phosphor. If bombarded by electrons, phosphor becomes "excited" and emits a glow as it returns to its normal preexcitation state. The key to the operation of a CRT is to control this bombardment so that the phosphor glow is predictable and can represent a picture. Control is attained by using an electron "gun" in the narrow end of the tube, which emits the electrons, together with a set of deflection magnets, which control the direction of the flow of a finely focused beam of electrons to the phosphor surface.

The electron gun in Figure 20.1b consists of a heat-generating filament surrounded by a cathode. When the filament is heated, the cathode gives off electrons in all directions. A

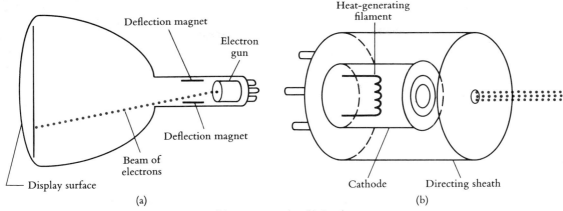

(a)   (b)

Figure 20.1 (a) A vacuum tube. (b) An electron gun.

directing sheath surrounds the cathode, allowing these electrons to escape only through a hole in the sheath. Once the electrons have left the sheath, they are focused into a precise beam by a series of electrostatic fields set up by several plates of different electric potential. A similar process accelerates this ray at different rates in order to control the intensity of the glowing phosphor.

Once the electron ray exits the focusing acceleration structure, the deflection system directs it to the phosphor field on the screen. Deflection is controlled by the display controller, which has its values either hard-wired (built in as part of the display equipment), as in

the raster system, or supplied by the computer, as in the stroke system. This difference in deflection methodology separates CRTs into the stroke and raster families.

As shown in Figure 20.2, the display controller of the stroke system receives a starting and ending value as well as an intensity value from the host computer. It then deflects the ray to the starting value and moves it in a straight line to the ending value, exciting all the phosphors it encounters on the way at the intensity level determined by the acceleration of the beam. By contrast, the display controller of the raster system (Figure 20.3) receives intensity, or color, values from the computer for

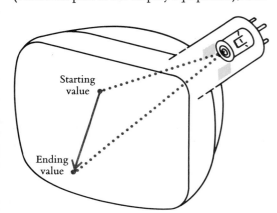

Figure 20.2 A stroke device.

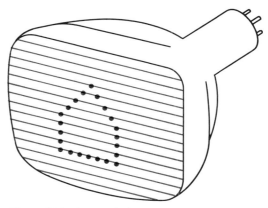

Figure 20.3 A raster device.

the points on a horizontal line. The picture is then formed by sequentially scanning the horizontal lines of points.

In both cases, the display controller must perform its tasks in rapid succession, because the phosphor glows for only a very short time. A typical rate is once every $\frac{1}{50}$ of a second for stroke display units and once every $\frac{1}{30}$ of a second for raster display devices. The precise speed depends on the type of phosphor used. If the repetition is too slow, the picture goes away for a period, and we see a "flicker."

We get color by using phosphors with different characteristics. A CRT uses phosphors that can produce red, green, and blue. A wider spectrum of colors is achieved by exciting the different types of phosphors singly or in various combinations. The $2^3 = 8$ different colors that can be obtained from the three primary colors are shown in Figure 20.4. Yellow is derived from red and green; magenta from red and blue; cyan from green and blue; white from red, green, and blue; and black simply from the absence of all three primary colors. The intensity controls the amount of color, giving a smooth range of color transitions.

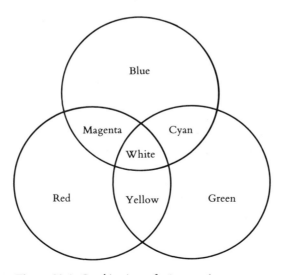

Figure 20.4 Combinations of primary colors.

There are currently many kinds of computers on the market that can be used for computer graphics: business computers, sophisticated engineering workstations, and special-purpose graphics supercomputers. The most widely available computers capable of performing the operations necessary to do graphics are the personal computers such as the Macintosh, the Amiga, and the IBM-PC family.

Even the simplest graphics requires a special combination of a computer chip, computer memory, software, and especially designed hardware necessary for converting a standard display into one that can be used for creating a picture. A display screen consists of "dots" or locations that can be turned on (illuminated) or off. These dots, called **pixels,** are the smallest renditions of a point on the graphics screen. To produce a picture of reasonable quality, about a quarter of a million pixels will need to be placed in the proper "on" or "off" combination in order to create the patterns that comprise the graphics image. In the next few sections we will see how dozens, even hundreds, of operations may be required to determine the on-off state of each pixel on the screen. Because of the limited memory capacity and relative slowness of personal computers, special graphics circuit boards are typically used to provide the interface necessary for graphics.

IBM-PCs use one of several graphics boards to provide this interface between the computer and the display monitor. They use a style of graphics called **bit-mapped graphics.** In this scheme, a bit (or series of bits) in memory is associated with each pixel on the screen. In order to create a picture, these bits are manipulated by the software to take on appropriate values, and they are then sent to the graphics board to drive the display monitor. The number of pixels determines the resolution of the display. For example, the most common board used for graphics on the PC can accommodate 640×350 pixels. The number of bits associated with each pixel determines the number of colors or intensities associated with each location of the display. If a single bit is used, the display can only be black and white (0 or 1). If the board allows 2 bits, each of which can be 0 or 1, then it is possible to display 4 colors. In general, if n bits are available, 2^n colors can be represented. The most common values of n are 4 (16 colors) and 8 (256 colors).

The simplest graphics board for the PC is called the *color-graphics adapter* (CGA). Because of its limitations, it is usually used only to provide color text and very primitive graphics. The *extended graphics adapter* (EGA) offers a major improvement over the CGA, including more colors, more video modes, programming features, and higher resolutions. The *video graphics adapter* (VGA) is the next step up from the EGA board. It allows resolutions up to 640×480. At this resolution, 16 colors can be used.

More recently, the SVGA, or super-VGA, has been added to the IBM family of graphics boards. In addition, there are a significant number of non-IBM boards that can be used, ranging from the very popular monochrome Hercules board to some very high resolution (1024×1024 or greater) adapters. Still other boards offer not only higher resolution displays, but also incorporate some more common graphics operations. For example, the Texas Instruments Graphics Architecture (TIGA) interface uses the TI 34010 graphics processor chip to give some outstanding graphics capability to personal computers.

SIMPLE GEOMETRIC OBJECTS

The simplest geometric objects we study in school are points, lines, and polygonal shapes. Although the notion of a point as a dimensionless object is appropriate in the study of geometry, a point on a display screen (repre-

sented by a pixel) will necessarily have a definite size. Objects such as straight lines or curves will be composed of a set of pixels. It is convenient to associate an (x, y) coordinate pair with each pixel on the screen in the same manner as we do in analytic geometry. If pixel sizes are small and their density great, a large number of pixels could be used in the representation of a figure to make it appear smooth.

The simplest figure we can consider is a straight-line segment. To display the line segment joining the two points $A = (x_1, y_1)$ and $B = (x_2, y_2)$, we need to "turn on" the pixels lying on the path from A to B. To do this we need to give a mathematical characterization of straight lines that is a little different from the $y = mx + b$ form. You know that, on a very dark night, you can move a flashlight about very rapidly and produce the illusion of figure 8s or circles. In a similar way, the CRT displays a straight line by exciting phosphor dots in rapid succession to produce the illusion of a line. So a line is produced on the CRT by exciting dots up and down the line in rapid succession. To see how this is done, think about the two points that define a straight line; in Figure 20.5 they are A and B with coordinates (x_1, y_1) and (x_2, y_2), respectively. When

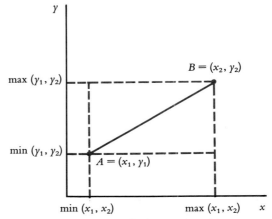

Figure 20.5 To display the line segment AB, we turn on the pixels lying in the path from A to B.

we specify the coordinates for the points A and B, the computer can easily excite the pixel locations. That produces only the two endpoints of the line and not the line in between. How does it get the points in between? Let's focus our attention on the x coordinates of these "in-between" points (the description of the in-between y coordinates in analogous). If we want the point x that is one-half of the way from x_1 to x_2, we add one-half of the distance $x_2 - x_1$ to x_1; mathematically stated, this is

$$x = x_1 + \tfrac{1}{2}(x_2 - x_1)$$
$$= (1 - \tfrac{1}{2})x_1 + \tfrac{1}{2}x_2$$

Suppose instead of one-half we wanted to go some other fraction of the way from x_1 toward x_2. Let's call the fraction u; then we would write the x coordinate as

$$x = ux_2 + (1 - u)x_1$$

You can see that if $u = 0$, then $x = x_1$ and if $u = 1$, $x = x_2$. For all intermediate values of u between 0 and 1, we get intermediate values of x between x_1 and x_2, as illustrated in Figure 20.6.

In a similar fashion we can get the y coordinate of the point a fraction u of the distance from y_1 to y_2 to be

$$y = uy_2 + (1 - u)y_1$$

Now if $u = 0$, then $x = x_1$ and $y = y_1$, which are coordinates for the point A, and if $u = 1$, $x = x_2$ and $y = y_2$, which are the coordinates for point B. For all intermediate values of u between 0 and 1, we get intermediate points between A and B, as illustrated in Figure 20.7.

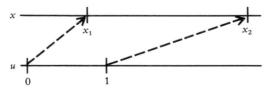

Figure 20.6 Conversion from the u scale to the x scale.

We can now give a simple algorithm for displaying the line segment from A to B:

1. For each u with $0 \le u \le 1$, compute

$$x = ux_2 + (1 - u)x_1$$

and

$$y = uy_2 + (1 - u)y_1$$

2. "Turn on" the pixel at location (x, y).

The representation of a line segment with the equations

$$x = ux_2 + (1 - u)x_1$$
$$y = uy_2 + (1 - u)y_1$$

with $0 \le u \le 1$ is called the *parametric form* of the line segment from (x_1, y_1) to (x_2, y_2). Its value to us here is that it lends itself to computations that produce the line segment from (x_1, y_1) to (x_2, y_2). The next step is to design an algorithm that uses this mathematical formulation to display the line segment at a graphics terminal.

Before stating the line-segment-drawing algorithm, we should note that because there are

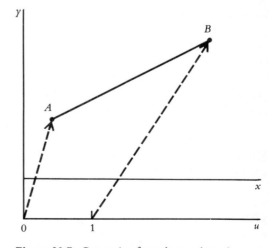

Figure 20.7 Conversion from the u scale to the segment AB.

infinitely many points on the line segment, infinitely many points will satisfy the condition stipulated by the parametric form of the line. We will therefore need infinitely many pixels for an "exact" rendition of the line segment. A computational procedure based on this approach will not be effective, that is, will not terminate in finite time. The following algorithm, through the variable E, controls the number of pixels that will be used in the display of the line segment.

1. Input (x_1, y_1) and (x_2, y_2).

2. $u = 0$.

3. $E = 0.05$.

4. Repeat the following as long as $u \leq 1$:

5. $\quad x = ux_2 + (1 - u)x_1$.

6. $\quad y = uy_2 + (1 - u)y_1$.

7. $\quad$ "Turn on" the pixel that is closest to the location (x, y).

8. $\quad$ Increment u by E.

9. End the algorithm.

As u is incremented 20 times, each time by $E = 0.05$, 21 pixels are "turned on" near the ideal locations of the line segment. If E were given value 0.01 or 0.005, the quality of the graphic display would significantly improve.

Once we have the ability to produce line segments, we can easily produce a polygon by specifying the coordinates of its vertices and connecting these vertices with line segments. To produce a polygonal figure that looks solid, we can fill the polygon by turning on all the pixels that lie within the polygon. The brute force method of filling a polygon would be to test each pixel on the screen and to turn on the pixel if it lies inside the polygon. Because this approach would be extremely inefficient, we now consider an alternative method for filling polygons.

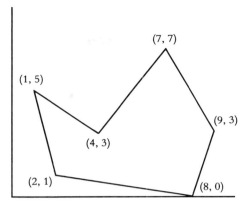

Figure 20.8 We want the computer to shade the inside of the polygon.

Consider the polygon shown in Figure 20.8, with vertices $(1, 5)$, $(2, 1)$, $(4, 3)$, $(7, 7)$, $(8, 0)$, and $(9, 3)$. We could fill it by scanning the figure vertically (from top to bottom) and drawing horizontal line segments that lie within the polygon; clearly, we need not consider line segments above the line $y = 7$ or below $y = 0$. Our scan would be limited to the range $y = 7$ to $y = 0$.

A crude algorithm for filling this polygon would be

1. $T = 7$.

2. $E = 0.05$.

3. Repeat the following as long as $T \geq 0$:

4. $\quad$ Draw horizontal line segment(s) at $y = T$ that lie in the polygon.

5. $\quad$ $T = T - E$.

6. End the algorithm.

Except for step 4, this algorithm is simple. To draw the horizontal line segments, we must decide which sides of the polygon will be "active," or involved in the determination of the endpoints of the line segment(s). Suppose the sides of the polygon are labeled as shown in

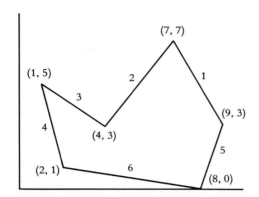

Figure 20.9 The same polygon with its sides labeled.

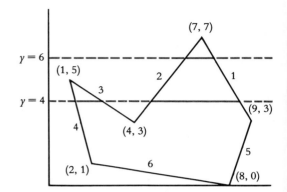

Figure 20.10 The active sides of the polygon for $y = 4$ and $y = 6$.

Figure 20.9. For $y = 6$, Figure 20.10 shows that only sides 1 and 2 need to be active; whereas for $y = 4$, sides 1, 2, 3, and 4 would be active. Information regarding the sides of the polygon and their activity status could be stored in the memory of the computer in tabular form, as shown in Table 20.1.

Notice that the table is constructed so that the y coordinates of the vertices associated with a line are arranged in nonincreasing order. Since the filling algorithm starts with $y = 7$, we find two sides (sides 1 and 2) coming out of vertex (7, 7). The activity status of each of these sides is changed to "on." As y is decremented down to 5, the two x coordinates on sides 1 and 2 are computed and arranged in increasing order — x_1, x_2 — and a line segment is drawn from x_1 to x_2. This is illustrated for $y = 6$ in Figure 20.11a.

When y assumes value 5, sides 3 and 4 also become active, and for values of y between 5 and 3, the four intersection points with the active sides (x_1, x_2, x_3, and x_4) are arranged in nondecreasing order. Horizontal line segments are drawn from x_1 to x_2 and from x_3 to x_4. Figure 20.11b illustrates this for $y = 4$. In general, if the intersection with the active sides gives $x_1, x_2, \ldots, x_n$ when the sides are arranged in increasing order, then segments from x_1 to x_2, from x_3 to x_4, and so forth are drawn.

In Figure 20.11c, when y assumes value 3, two x values, both equal to 4, are generated because sides 2 and 3 are active. When y becomes less than 3, sides 1, 2, and 3 are changed to "off" and side 5 is changed to "on." This process will continue until y assumes a negative value and all sides have "off" status.

TRANSFORMATIONS

We can do a lot even with simple polygonal shapes if we can reconstruct a given picture through movement, magnification, or reloca-

TABLE 20.1 **How a computer might store data about the sides of the polygon**

Side no.	Endpoint	Endpoint	Activity status
1	(7, 7)	(9, 3)	Off
2	(7, 7)	(4, 3)	Off
3	(1, 5)	(4, 3)	Off
4	(1, 5)	(2, 1)	Off
5	(9, 3)	(8, 0)	Off
6	(2, 1)	(8, 0)	Off

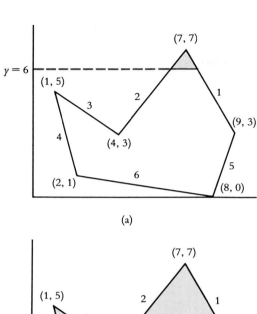

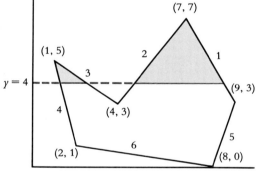

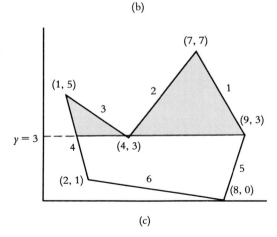

Figure 20.11 (a) When $y = 6$, two sides of the polygon are active. (b) When $y = 4$, four sides are active. (c) When $y = 3$, four sides are active but two of them generate the same value, 4.

tion. For example, we could produce animation (motion of an object on a graphics screen) with a rapid succession of movements of the object. This section considers translation, reflection, scaling, and rotation, four basic forms of motion that can be produced on a graphics screen. You have already seen examples of translations, reflections, and rotations as part of the discussion of symmetry in Chapter 13.

Translation is the movement of a figure from one location to another without altering its shape, size, or orientation. To move a picture, say, 2 units to the right and 5 units above its current location, we can look at the coordinates (x, y) of all pixel locations that are "on" and compute the coordinates (X, Y) of the points 2 units to the right and 5 units above (x, y). You can see from Figure 20.12 that this makes $X = x + 2$ and $Y = y + 5$. Now we turn on the pixel at location (X, Y). If we want

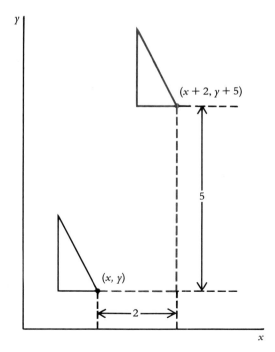

Figure 20.12 The image is translated 2 units to the right and 5 units up.

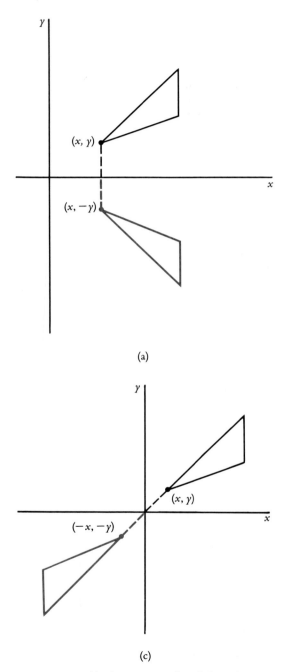

(a)

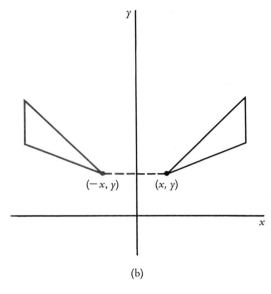

(b)

the original figure to disappear, we can turn off the pixel at (x, y). In general, to re-create a figure h units to the right and k units above its original location, we execute the following algorithm:

1. Input h, k.

2. For all pixels that are "on":

3.   Determine the coordinates (x, y).

4. $X = x + h$.

5. $Y = y + k$.

6.   Turn on pixel at location (X, Y).

7.   Turn off pixel at location (x, y).

8. End the algorithm.

If h or k, or perhaps both, are negative, the movement will be in the opposite direction. A sense of motion or animation can be created by repeatedly applying a translation with h and k values that are of small magnitude (absolute value).

(c)

Figure 20.13 (a) The image is reflected about the x-axis. (b) The image is reflected about the y-axis. (c) The image is reflected about the origin.

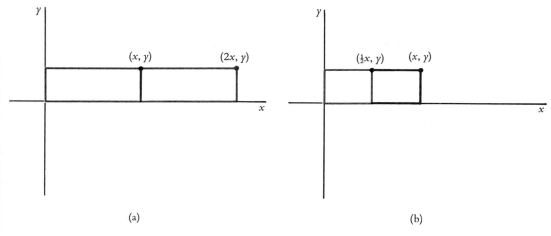

Figure 20.14 (a) The image is stretched in the horizontal direction, with $r = 2$.
(b) The image is shrunk in the horizontal direction, with $r = \frac{1}{2}$.

Another motion the computer can generate is **reflection,** which is the movement of a figure to its mirror image in respect to an axis or the origin.

1. To reflect a figure along the x-axis, we compute (X, Y) with $X = x$ and $Y = -y$ in lines 4 and 5 of the above algorithm (Figure 20.13a).

2. Reflection along the y-axis could be achieved with $X = -x$ and $Y = y$ at lines 4 and 5 (Figure 20.13b).

3. Reflections about the origin are done with $X = -x$ and $Y = -y$ at lines 4 and 5 (Figure 20.13c).

We can also easily **scale**—stretch or shrink —a given object in the horizontal or vertical direction. For scaling in the horizontal direction, we take a point on the object as the origin, then for each point (x, y) on the object, we compute the point (rx, y) and turn on the pixel at this point. In Figure 20.14a, the rectangular object is stretched in the horizontal direction; in Figure 20.14b, it is shrunk in the horizontal

direction. In general, this type of transformation will have the effect of stretching the figure in the horizontal direction if $r > 1$ and shrinking it if $r < 1$.

We can get similar results in the vertical direction by transforming points (x, y) to points (x, sy) as in Figure 20.15. If we desire

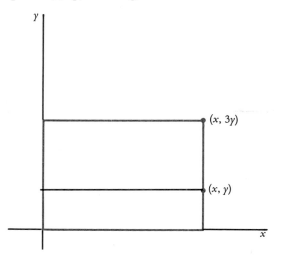

Figure 20.15 The image is stretched in the vertical direction with $s = 3$.

scaling in both directions, (x, y) could be transformed to (rx, sy), as in Figure 20.16.

We need a few results from trigonometry to describe how to **rotate** a figure about a given point. Suppose on a coordinate system we have

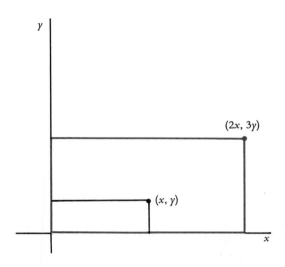

Figure 20.16 The image is scaled in the horizontal and vertical directions, with $r = 2$ and $s = 3$.

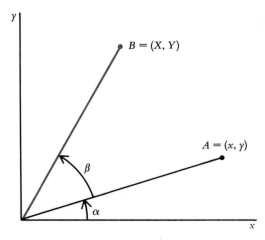

Figure 20.17 The point A is rotated counterclockwise by an angle β.

a point $A = (x, y)$, at distance R from the origin, on a line making an angle α with the x-axis. As shown in Figure 20.17, to rotate point A counterclockwise about the origin by an angle β, we need to determine the coordinates of $B = (X, Y)$, which has the same distance R from the origin as A but makes an angle $\alpha + \beta$ with the x-axis. The rotation algorithm is developed in the next section.

To rotate a complex figure, we could, of course, rotate each point on the figure, but this process is generally computationally expensive; typically we use more sophisticated methods.

Optional ROTATIONS

This section uses some basic trigonometric relationships to describe how a point (consequently a figure) can be rotated about the origin. From trigonometry we know that for the points A and B at distance R from the origin, depicted in Figure 20.17,

$$\cos \alpha = \frac{x}{R} \qquad \text{or} \qquad x = R \cos \alpha$$

and

$$\sin \alpha = \frac{y}{R} \qquad \text{or} \qquad y = R \sin \alpha$$

Similarly

$$X = R \cos (\alpha + \beta)$$
$$Y = R \sin (\alpha + \beta)$$

Using the sum-of-angles identities from trigonometry, we have

$X = R \cos(\alpha + \beta)$
$\quad = R(\cos \alpha \cos \beta - \sin \alpha \sin \beta)$
$\quad = (R \cos \alpha)\cos \beta - (R \sin \alpha)\sin \beta$
$\quad = x \cos \beta - y \sin \beta$

$Y = R \sin(\alpha + \beta)$
$\quad = R(\cos \alpha \sin \beta + \sin \alpha \cos \beta)$
$\quad = (R \cos \alpha)\sin \beta + (R \sin \alpha)\cos \beta$
$\quad = x \sin \beta + y \cos \beta$

Thus, to rotate point (x, y) counterclockwise β degrees about the origin, compute:

$$X = x \cos \beta - y \sin \beta$$
$$Y = x \sin \beta + y \cos \beta$$

and turn on the pixel closest to the point (X, Y).

EXAMPLE: Perform a 45° Counterclockwise Rotation of the Point $A = (2, 1)$. If we let B be the point that is obtained from this rotation, the situation can be envisioned as in Figure 20.18. Our task is to determine the coordinates X and Y of B from our knowledge of $x = 2$, $y = 1$, and $\beta = 45°$. From basic trigonometry we know that $\cos \beta = \sin \beta = \sqrt{2}/2$. Therefore,

$$X = x \cos \beta - y \sin \beta$$
$$= 2\frac{\sqrt{2}}{2} - 1\frac{\sqrt{2}}{2}$$
$$= \frac{\sqrt{2}}{2} \cong 0.707$$

$$Y = x \sin \beta + y \cos \beta$$
$$= 2\frac{\sqrt{2}}{2} + 1\frac{\sqrt{2}}{2}$$
$$= 3\frac{\sqrt{2}}{2} \cong 2.121 \qquad \bullet$$

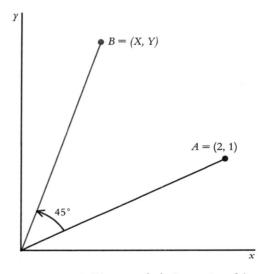

Figure 20.18 A 45° counterclockwise rotation of the point (2, 1).

FRACTALS

As we saw in earlier sections, it is easy to describe points, lines, and polygonal shapes so that computer programs can "draw" them on display screens. It is also easy to combine these primitive shapes to form three-dimensional objects of prescribed forms. However, a large class of objects is not so easy to describe. Natural objects such as mountains, clouds, subatomic structures, and planets have forms that are either amorphous or very complex.

The concept of **fractal dimension,** or **fractal,** was developed in order to describe the shapes of natural objects. In geometry, we consider a line a one-dimensional object and a plane a two-dimensional object. What dimension should a jagged line have? Within the concept of fractals, a jagged line is given a dimension between 1 and 2, with the exact value determined by the line's "jaggedness." An interesting property of fractal objects is that as we magnify a figure, more details appear but the basic shape of the figure remains intact. Spotlight 20.1 gives an overview of the origin of fractal geometry.

The first phase for the construction of a simple fractal object with dimension between 1 and 2 can be described as follows:

1. Start with a line segment *AB*.

2. Divide the line segment *AB* into three equal segments *AC, CD,* and *DB*.

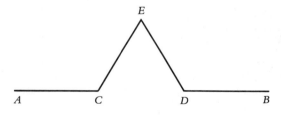

3. Construct an equilateral triangle *CED* on the middle segment and then remove the middle segment *CD*.

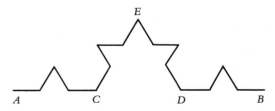

In the second phase, we repeat steps 1, 2, and 3 on each of the four segments *AC, CE, ED,* and *DB,* which yields

In the third phase, the same three steps are repeated on the 16 segments produced by the previous phase, giving

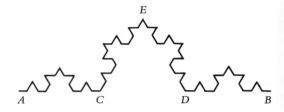

This process can be repeated until we get the desired level of complexity.

One variation of the construction we have given here is to start with a square and successively remove one-ninth of the square from the middle of the square(s). The first three stages of this process would yield

[From B. B. Mandelbrot: *The Fractal Geometry of Nature,* W. H. Freeman, New York, 1982, p. 144.]

A more complicated variation can be obtained by moving to a higher dimension and starting with a rectangular box and successively removing central portions from it. An illustration of this type of construction is given in Figure 20.19. Spotlight 20.2 describes a problem that produces some interesting and complex fractal figures.

SPOTLIGHT 20.1 Fractal Geometry and Benoit Mandelbrot

Fractal geometry as a serious mathematical endeavor began about 15 years ago with the pioneering work of Benoit B. Mandelbrot, a Fellow of the Thomas J. Watson Research Center, IBM Corporation. Fractal geometry is a theory of geometric forms so complex they defy analysis and classification by traditional Euclidean means. Yet fractal shapes occur universally in the natural world. Mandelbrot has recognized them not only in coastlines, landscapes, lungs, and turbulent water flow but also in the chaotic fluctuation of prices on the Chicago commodity exchange.

Though Mandelbrot's first comprehensive publication of fractal theory took place in 1975, mathematicians were aware of some of the basic elements during the period from 1875 to 1925. However, because mathematicians at that time thought such knowledge of "fractal dimension" deserved little attention, their discoveries were left as unrelated odd and ends. Also, the creation of fractal illustrations — a laborious and nearly impossible task at the turn of the twentieth century — can now be done quickly and precisely using computer graphics. (Even personal computers can now be used to generate fractal patterns with relative ease.)

The word *fractal* — coined by Mandelbrot — is related to the Latin verb *frangere,* which means "to break." The ancient Romans who used *frangere* may have been thinking about the breaking of a stone, since the adjective, derived from this action combines the two most obvious properties of broken stones — irregularity and fragmentation. The adjectival form is *fractus,* which Mandelbrot says led him to fractal.

Mandelbrot also has stated that fractals and fractions are related, especially if one views a "fraction" as a number that lies between integers (whole numbers). Similarly, a fractal set, a precise mathematical term defined by Mandelbrot, can be seen as lying between classical Euclidean shapes.

When dealing with these classical Euclidean shapes (circle, square, sphere, cube) we need only two parameters — position and length — to describe them. Fractals, however, require three parameters. The first is length; the second is "fractal dimension," denoted by D; the third is called "random seed" or "chance." The most interesting and important of the three is parameter two.

In Euclidean geometry a line segment and a square can be described as self-similar shapes; that is, all line segments or squares are simply enlarged or reduced replicas of each other. From these self-similar Euclidean forms, Mandelbrot saw that the ratio $D = \log N/\log (1/r)$ is the same as the form's dimension. (N is any integer; L is the sum of N straight segments of length $r = L/N$.) From this definition Mandelbrot defined a fractal set (the entire family of fractal shapes) as a mathematical set such that D is greater than the topological dimension D_T, or $D > D_T$.

(Modified from "Mandelbrot Gets Franklin Medal," *SIAM News,* vol. 19, no. 3, May 1986, p. 12. Reproduced with permission of The Society for Industrial and Applied Mathematics.)

SPOTLIGHT 20.2 Newton's Method and Computer Graphics

Much of the work of modern scientists and engineers depends on solving equations. One of the oldest and best known methods for solving equations is Newton's method. This method first requires us to estimate an approximate root to an equation. Then we insert that estimate into a formula and calculate a new, more accurate estimate of the root. The formula is then applied to this new number and a further refinement is obtained. The process is continued until we are satisfied with the accuracy of the answer. The formula for Newton's method can be transformed into a computer algorithm that can efficiently find one or more roots for most polynomials. The trick, however, is to select an appropriate starting point or first estimate so that the process converges quickly on a specific root. Not all first guesses converge quickly, and some will cause Newton's method to fail under certain conditions. These conditions arise when the chosen point falls on a boundary between regions associated with different roots or when the process gets stuck oscillating between two numbers, neither of which is a root.

Knowing when and why methods of solution such as Newton's will fail or produce strange results can be enormously important to scientists. Because of the complexity of effects involved, computer graphics can provide a clearer picture of what is occurring to equations near their boundary regions. For a given equation and starting value, the computer runs the algorithm implementing Newton's method and calculates the root toward which that value converges. If the computer then assigns a position and a color to the resulting point, a graphical picture can be created. Calculating the results for 1 million initial values and plotting the points in color can build a detailed picture of the mathematical activity (see Color Plates 27, 28, and 29). Shades of color indicate how fast a point converges on a root.

The large areas of color are known as *basins of attraction,* and starting points selected from those regions will converge on a root after a reasonable number of repeated calculations, or iterations. For mathematicians the interesting areas are those near or at a boundary. Graphically these border regions appear as complicated swirling patterns. Moving a starting point a slight amount in a boundary area can cause widely divergent results that would adversely affect the accuracy of an equation's solution.

If a portion of the boundary region is magnified, its complexity does not diminish and in fact looks much like the original image of the boundary. Further magnification only shows a repetition of the same pattern. This similarity of an intricate pattern regardless of scale is explained by fractals (see Spotlight 20.1).

Of particular interest to current researchers is the notion of "chaos." Depending on the initial conditions, Newton's method may indicate that there is no basin of attraction for a root; that is, no choice of starting points lead to a root. As the suspected roots become mathematically unstable, chaos occurs.

Because engineers use Newton's method to solve problems—such as designing supersonic aircraft—it is vital to understand under what conditions these methods will demonstrate chaotic behavior. At high speeds and high temperatures, stresses on materials become acute, and seemingly small design faults can prove disastrous. Thus not only can knowledge of a solution method's quirks determine the success or failure of an engineering project, but it can be crucial in safeguarding those humans touched by such projects. Research on chaos is still developing, and it will be some time before the important equations are fully analyzed and their chaotic behavior mapped.

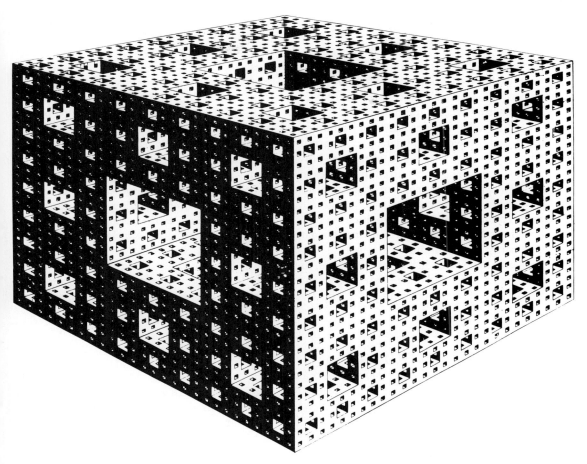

Figure 20.19 A three-dimensional fractal construction. [From B. B. Mandelbrot: *The Fractal Geometry of Nature*, W. H. Freeman, New York, 1982, p. 145.]

CONCLUDING COMMENTS

This chapter has focused on some of the mathematical ideas that are needed for an understanding of computer graphics. The theory as well as the practical applications of computer graphics is a rapidly evolving branch of computer science. Ever-increasing applications of computer graphics have led to the development of better algorithms for generating pictures and better devices for displaying pictures.

Increasing computation speed has dictated a change in the approach to making pictures. At the same time, decreasing costs have given rise to more diversified techniques for the development of image-making programs. These advances and innovations will no doubt continue to take place.

As a result of intensified research in the discipline of human/computer interfaces, traditional methods of communicating with the computer and interacting with the internal structures of the computer are being replaced with new methods. One method that is increasing in popularity relies completely on graphics. The standard computer terminal is being replaced by a graphics terminal, whose user selects commands and receives feedback by interacting with a graphics menu.

Traditionally the user has typed on a keyboard, and the words and data were transmitted as a series of letters and numerals. Now information can be transmitted in pictures instead. Even computer programs can be written by organizing series of graphic images. In short, computer-generated graphics are playing a much larger role in the information, data processing, and communications fields.

Thanks to research in graphics software, the computer is now able to simulate images that approximate reality, creating more effective presentations of simulated situations that either are not possible to capture in film or real-life scenarios, or are too expensive or dangerous to test. Aerospace simulations, for example, use more realistic visual feedback and thus train pilots and astronauts more effectively. Expensive movie sets are being replaced with computer-generated sets, with the actors composited over them later.

As computer-generated graphics continues to enter the realm of practical situations, it becomes increasingly important for people to gain a further understanding and appreciation of its basic mathematical foundations. Through research, the efficiency of computer graphics software will continue to increase while the price of graphics systems decreases. This will make computer graphics a standard tool in almost every area of human activity.

REVIEW VOCABULARY

Bit-mapped graphics   A graphics scheme in which bits in memory are associated with each pixel on the screen.

Cathode ray tube   Commonly used video display device for computer graphics.

Fractal dimension, or **fractal**   A measurement of the jaggedness of an object.

Pixel   The smallest rendition of a point on a computer graphics device.

Reflection   (of an object about an axis) The mirror image of the object along the axis.

Rotation   (of an object) The circular movement of an object about a given point.

Scaling   (of an object) The stretching or shrinking of an object in a horizontal or vertical direction.

Translation   (of an object) The movement of an object from one location to another without altering its shape, size, or orientation.

SUGGESTED READINGS

Artwick, B. A.: *Microcomputer Displays, Graphics, and Animation.* Prentice-Hall, Englewood Cliffs, N.J., 1985.

Demel, John T., and Michael J. Miller: *Introduction to Computer Graphics,* Brooks/Cole, Monterey, Calif., 1984.

Hearn, Donald, and M. Pauline Baker: *Microcomputer Graphics: Techniques and Applications,* Prentice-Hall, Englewood Cliffs, N.J., 1983.

Rodgers, David F., and J. Alan Adams: *The Mathematical Elements of Computer Graphics,* McGraw-Hill, New York, 1976.

EXERCISES

1. We saw that the parametric form of the line segment from (x_1, y_1) to (x_2, y_2) is

$$x = ux_2 + (1 - u)x_1$$
$$y = uy_2 + (1 - u)y_1$$

Verify that in the specific case of $(x_1, y_1) = (0, 0)$ and $(x_2, y_2) = (12, 8)$, the parametric equations simplify to $x = 12u$, $y = 8u$. Also verify that when $u = 0$ and $u = 1$, we get the endpoints of the line segment.

2. Continue Exercise 1 by plotting the points $(0, 0)$ and $(12, 8)$ and drawing the connecting line segment on graph paper. Verify that for $u = \frac{1}{2}, \frac{1}{3}$, and $\frac{1}{4}$, we get points on the segment from $(0, 0)$ to $(12, 8)$.

3. What happens if in Exercise 2 we choose $u = 1.25$? Is the resulting point still on the line segment from $(0, 0)$ to $(12, 8)$? Is it on the line that is extended from this line segment?

4. Consider the line segment from $(1, 1)$ to $(12, 8)$. What is the parametric form of this segment? As was done in Exercise 1, verify that $u = 0$ and $u = 1$ lead to the endpoints of the line segment.

5. Plot the line segment of Exercise 4 with enough care to determine if the points resulting from $u = \frac{1}{2}$ and $\frac{1}{4}$ are on the line segment.

6. When filling the polygon described by Table 20.1, how will it be known when to activate or deactivate the sides of the polygon?

7. Follow the procedure described in this chapter to fill the polygon with vertices $(0, 1)$, $(1, 2)$, $(2, 1)$, and $(1, 0)$.

  a. Start by labeling the sides of the polygon in a manner similar to that given in Figure 20.9.

  b. Construct a table similar to Table 20.1 with the labeled sides, endpoints (make sure you list these in the proper order), and activity status for each side.

  c. What will be the activity status of each side when $1 < y < 2$? How about when $0 < y < 1$?

  d. When $y = \frac{3}{2}$, what line segment(s) will be drawn (give the endpoints) as a result of the filling algorithm?

8. Do parts **a, b, c,** and **d** of Exercise 7 for the polygon with vertices $(0, 1)$, $(1, 2)$, $(2, 1)$, $(3, 2)$, $(4, 1)$, and $(2, 0)$.

9. Give a transformation that will move a picture

  a. 4 units to the right

  b. 2 units down

10. Give a transformation that will move a picture 3 units to the left and $\frac{1}{2}$ unit up.

11. Give a transformation that will move a picture diagonally (on a 45° angle from the positive x-axis) 8 units away. You will need the Pythagorean property of right triangles for this problem.

12. Give a sequence of transformations that will change the rectangle with vertices (1, 1), (2, 1), (2, 3), and (1, 3) into the rectangle with vertices (3, −1), (4, −1), (4, 1), and (3, 1).

13. Give a sequence of transformations that will change the rectangle with vertices (1, 1), (2, 1), (2, 3), and (1, 3) into the rectangle with vertices (3, −1), (4, −1), (4, 3), and (3, 3).

14. Give a single transformation that combines the sequence of transformations obtained in Exercise 13.

15. Give a sequence of transformations that will change the rectangle with vertices (1, 1), (2, 1), (2, 3), and (1, 3) into the rectangle with vertices (3, −1), (3.5, −1), (3.5, 3), and (3, 3).

16. Give a single transformation that combines the sequence of transformations obtained in Exercise 15.

17. Consider the scaling transformations described in Figures 20.14a, 20.14b, and 20.15. How is the area of the object (in all three cases a rectangle) affected by each of these transformations?

18. In a more general setting than in Exercise 17, consider a rectangle with one of its vertices at (0, 0) and with sides parallel to the axes. How will the area of such a rectangle change if the scaling transformation $X = rx$, $Y = y$ is applied. How will the area be affected if the transformation $X = x$, $Y = sy$ is applied?

19. Suppose the transformation $X = rx$, $Y = sy$ is applied to the type of rectangle described in Exercise 18. How will this affect the area of the rectangle?

20. Which of the transformations described in this chapter leave the area of a two-dimensional object intact and which ones change the area?

21. What happens if a negative factor r is used in a scaling transformation? What effect is produced if $r < −1$? If $r > −1$?

22. Draw the triangle with vertices (2, 1), (5, 1), and (3, 2) together with its reflection about the line $y = x$. What are the vertices of the reflected triangle?

23. In Chapter 13 glide reflection was defined as a translation along a line followed by a reflection in that line. Give the translation and reflection needed for moving an object 3 units to the right and then reflecting it in the x-axis.

24. With a drawing show the triangle with vertices (2, 1), (5, 1), and (3, 2) together with its glide reflection described in Exercise 23.

25. Give a single transformation that combines the translation and reflection described in Exercise 23.

◆ Answers to Odd-Numbered Exercises ◆

Chapter 1

1. A:1; B:3; C:3; D:3; and E:0. The graph shows that geographically E is isolated, perhaps on an island or a different continent than the other cities.

3. Graph (b).

5.

The valences are A:2; B:3; C:4; D:1; and E:2. The real-world consequences are that if you are at one city, you may or may not be able to travel to another particular city.

7. The supervisor is not satisfied because all of the edges are not traveled upon by the postal worker. The worker is unhappy because the end of the worker's route wasn't the same point as that where the worker began. The original job description is unrealistic because there is no Euler circuit in the graph.

9. Both graphs (b) and (c) have Euler circuits:

(b)

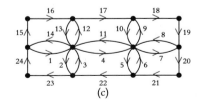

(c)

The valences of all of the vertices in (a) are odd, which makes it impossible to have an Euler circuit there.

11. Remove the vertical edge in the middle.

13.

15.

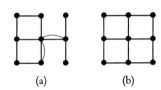

(a)   (b)

17.

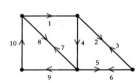

19.

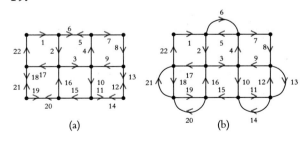

(a)   (b)

21.

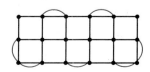

23.

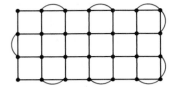

25.

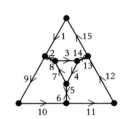

27.

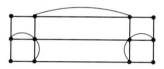

29.

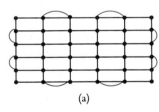

(a)

(b)

31.

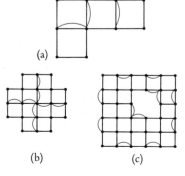

(a)

(b)   (c)

33. Yes, because every vertex has even valence and each street is represented by two edges.

35. Suppose there are k edges in the graph G. Then the sum of the valences of vertices of G is $2k$, since each edge meets two vertices. But $2k$ is an even number, and so is the sum S of the valences of all of the even vertices. The sum of the valences of the odd vertices is $2k - S$, also an even number. The only way to get a sum of odd numbers to equal an even number is if there are an even number of those odd numbers. Thus n, the number of odd vertices, must be even.

Chapter 2

1. a. $X_1, X_6, X_5, X_2, X_3, X_4, X_1$.
  b. $X_1, X_6, X_7, X_8, X_9, X_{10}, X_{11}, X_{12}, X_5, X_4,$ X_3, X_2, X_1.
  c. $X_1, X_4, X_5, X_8, X_9, X_6, X_7, X_2, X_3, X_1$.
  d. $X_1, X_2, X_5, X_8, X_3, X_4, X_7, X_6, X_1$.
  e. $X_1, X_9, X_8, X_7, X_6, X_5, X_4, X_3, X_2, X_1$.

3. Other Hamiltonian circuits include *ABIGDCEFHA* and *ABDCEFGIHA*.

5. a. Add edge AB.   **b.** Add edge X_1X_7.

7. a. There is no Hamiltonian circuit. Any Hamiltonian circuit would have to use edges X_1X_5, X_1X_2, and X_1X_4. This would force X_1 to be revisited.
  b. There is no Hamiltonian circuit. Any Hamiltonian circuit would have to use edges X_1X_2, X_2X_3, $X_{10}X_{11}$, and $X_{11}X_{12}$. To visit X_6 and X_7 would require a revisit of X_2 or X_{11}.

9. a. For any $m \geq 2$ and $n \geq 1$, the graph has a Hamiltonian circuit.
  b. If either m or n is odd, the graph has a Hamiltonian circuit. If both m and n are even, the graph has no Hamiltonian circuit.
    A real-world application would be to design an efficient route to check that the traffic con-

trol equipment at each vertex was in proper working order.

11. a. No Euler circuit; Hamiltonian circuit.
 b. Euler circuit; Hamiltonian circuit.
 c. No Euler circuit; Hamiltonian circuit.
 d. Euler circuit; no Hamiltonian circuit.

13. Examples include inspection of traffic control devices at corners and placing new hour stickers on mailboxes located at street corners.

15. $9(9)(9)(9)(9) = 59,049$.

17. a. $26(26)(26)(10)(10)(10) - (26)(26)(26) = (26)^3 (10^3 - 1)$.

19. $5! = 120$, $6! = 720$, $7! = 5040$, $8! = 40,320$, $9! = 362,880$, $10! = 3,628,800$. The number of TSP tours in a 10-vertex complete graph is 181,440.

21. She should follow the route FMCRF (or FRCMF), which takes 32 minutes.

23. a. *ACBDA* (nearest neighbor). *ADBCA* or *ACBDA* (sorted edges).
 b. *ABCDA* (nearest neighbor). *ABDCA* (sorted edges).
 c. *ADBCEA* (nearest neighbor). *ADCBEA* (sorted edges).

25. A traveling salesman problem.

27. The complete graph shown has a different nearest-neighbor tour that starts at *A* (*AEDBCA*), a sorted-edges tour (*AEDCBA*), and a cheaper tour (*ADBECA*).

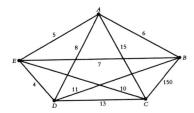

29. Graphs (b) and (d) are trees.

31.

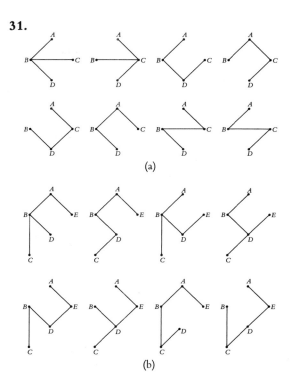

33. The wiggled edges in the figure constitute a spanning tree.

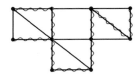

35. In drawing the graph model, do not include edges between two vertices where there is a hill of more than 800 feet between the towns represented by the vertices.

37. Yes. Change all the weights to negative numbers and apply Kruskal's algorithm. The resulting tree works, and the maximum cost is the negative of the answer you get. If the numbers on the edges represent subsidies for using the edges, one might be interested in finding a maximum-cost spanning tree.

39. a. True. **b.** False. **c.** True. **d.** False.

41. Two different trees with the same cost are shown:

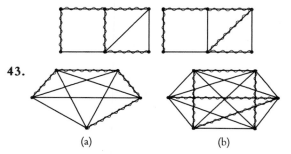

43.

(a)   (b)

45. a. Critical path: $T_3T_6T_7$. Earliest completion time is 31.
  b. There are two critical paths: $T_1 T_3 T_6 T_8$ and $T_1 T_3 T_5 T_7$. Earliest completion time is 35.
  c. Critical path: $T_2 T_5 T_7$. The earliest completion time is 29.

47. The critical path is $T_1 T_5 T_7$. Hence, if T_1, T_5, or T_7 are shortened, the earliest completion time will decrease, while if T_2, T_3, T_4, or T_6 are shortened, the earliest completion time will not decrease. If T_5 is shortened to 7, the earliest completion time is 28 since $T_1T_4T_7$ is now the critical path.

49. One possibility is (times in minutes):

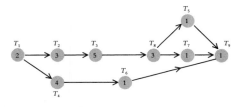

The earliest completion time is 16 minutes.

51.

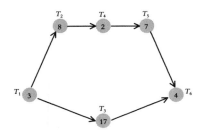

Chapter 3

1. a. Operating room schedules, doctor schedules, emergency room staffing schedules, etc.
  b. Schedules for the trains or buses and their crews, etc.
  c. Scheduling runway use, reservation agents, food service for planes, etc.
  d. Schedules for each mechanic, radiator repair, etc.
  e. Schedules for drilling, welding, etc.
  f. Schedules for washing clothes, cleaning rooms, dusting, etc.
  g. Schedules for bus run, recess duty, etc.
  h. Day, night, afternoon shift schedules, etc.
  i. Firefighter shift schedules, schedules for checking if equipment on trucks is in repair, etc.

3. a. Processor 1: T_1, T_6, idle 13 to 15; T_5, T_7, T_{11}, idle 34 to 38; T_{10}. Processor 2: T_2, T_9, idle 21 to 27; T_8, idle 38 to 45. Processor 3: T_3, T_4, idle 15 to 45.
  b. Processor 1: T_1, T_3, T_4, T_6, T_8, idle 39 to 40; T_{10}. Processor 2: T_2, T_5, T_7, T_9, T_{11}, idle 40 to 47.

5. Such criteria include: decreasing length of the times of the tasks, order of size of financial gains when each task is finished, and increasing length of the times of the tasks.

7. a. The critical path, which has length 17, is $T_1T_2T_3$.
  b. T_1, T_4, T_5, T_2, T_6, T_7, T_3 is the list to be used. The one processor would have the tasks scheduled on it: T_1, T_4, T_5, T_2, T_6, T_7, T_3.
  c. T_6, T_1, T_7, T_2, T_4, T_3, T_5 would be the list. The resulting schedule on one processor would be: T_1, T_7, T_4, T_2, T_5, T_6, T_3.
  d. No idle time. Their completion times are the same.
  e. Earlier completion of tasks giving rise to cash payments.

9. a. Identical machines, typists who type the same number of words per minute, etc.

b. Runways at an airport in different directions, humans with different levels of skills, etc.

11. a. Task times: $T_1 = 3, T_2 = 3, T_3 = 2, T_4 = 3,$ $T_5 = 3,$ $T_6 = 4,$ $T_7 = 5,$ $T_8 = 3,$ $T_9 = 2,$ $T_{10} = 1, T_{11} = 1,$ and $T_{12} = 3.$ This schedule would be produced from the list: $T_1, T_3, T_2,$ $T_5, T_4, T_6, T_7, T_8, T_{11}, T_{12}, T_9, T_{10}.$
b. Task times: $T_1 = 3, T_2 = 3, T_3 = 3, T_4 = 2,$ $T_5 = 2,$ $T_6 = 4,$ $T_7 = 3,$ $T_8 = 5,$ $T_9 = 8,$ $T_{10} = 4, T_{11} = 7, T_{12} = 9, T_{13} = 3.$ This schedule would be produced from the list: $T_1, T_5,$ $T_7, T_4, T_3, T_6, T_{11}, T_8, T_{12}, T_9, T_2, T_{10}, T_{13}.$

13. a. One reasonable possibility is (time in min):

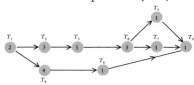

The earliest completion time is 16.
b. The decreasing time list is: $T_3, T_4, T_2, T_8,$ $T_1, T_5, T_6, T_7, T_9.$ The schedule is: Processor 1: $T_1, T_4, T_3, T_8, T_5, T_9$; Processor 2: idle 0 to 2, $T_2, T_6,$ idle 6 to 14, $T_7,$ idle 15 to 16.

15. The task times total to 34. Since (34/3) rounded up is 12, the earliest completion time is 12.

17. a. One such rule, admittedly artificial, could be: if by keeping a machine voluntarily idle, there is a longer task that becomes ready one time unit later, keep the machine idle.
b. If a longer task becomes ready at a certain time than the remaining time on a task currently scheduled, schedule the longer task and reschedule the interrupted task later. An assumption must be made whether or not an interrupted task must be resumed on the machine it was originally scheduled on or can be rescheduled on any machine that becomes free.

19. a. The tasks are scheduled on the machines as follows: Processor 1: 12, 13, 45, 34, 63, 43, 16, idle 226 to 298; Processor 2: 23, 24, 23, 53, 25, 74, 76; Processor 3: 32, 23, 14, 21, 18, 47, 23, 43, 16, idle 237 to 298.

b. The tasks are scheduled on the machines as follows: Processor 1: 12, 24, 14, 34, 25, 23, 16, 16, 76, idle 183 to 240; Processor 2: 23, 23, 21, 63, 43, idle 173 to 240; Processor 3: 32, 23, 53, 74, idle 182 to 240; Processor 4: 13, 45, 18, 47, 43, idle 166 to 240.
c. The decreasing time list is: 76, 74, 63, 53, 47, 45, 43, 43, 34, 32, 25, 24, 23, 23, 23, 23, 21, 18, 16, 16, 14, 13, 12.
The tasks are scheduled on three machines as follows: Processor 1: 76, 45, 43, 24, 23, 18, 16, 13; Processor 2: 74, 47, 34, 32, 23, 21, 14, 12, idle 257 to 258; Processor 3: 63, 53, 43, 25, 23, 23, 16, idle 246 to 258.
The tasks are scheduled on four machines as follows: Processor 1: 76, 43, 24, 23, 16, idle 182 to 194; Processor 2: 74, 43, 25, 23, 16, 13; Processor 3: 63, 45, 32, 23, 18, 12, idle 193 to 194; Processor 4: 53, 47, 34, 23, 21, 14, idle 192 to 194.
d. The new decreasing time list is: 84, 82, 71, 61, 55, 45, 43, 43, 34, 32, 25, 24, 23, 23, 23, 23, 21, 18, 16, 16, 14, 13, 12.
The tasks are scheduled as follows: Processor 1: 84, 45, 43, 25, 23, 23, 16, 12; Processor 2: 82, 55, 34, 32, 23, 18, 14, 13; Processor 3: 71, 61, 43, 24, 23, 21, 16, idle 259 to 271.

21. Examples include times to insert different chips into a circuit board, scheduling classes at a college, and putting dust jackets onto books of different sizes as part of a book manufacturing process.

23. Examples include jobs in a video tape copying shop, data entry tasks in a computer system, scheduling nonemergency operations in an operating room. These situations may have tasks with different priorities, but there is no physical reason for the tasks not to be independent, as would be the case with putting on a roof before a house had walls erected.

25. a. Each task heads a path of length equal to the time to do that task.
b. (1) The worst finish time is $(2 - \frac{1}{3})(450) = 750$. (2) The worst finish time, if the decreasing-time list is used, is $[\frac{4}{3} - 1/(3)(3)](450) = 550$.

27. The times to photocopy the manuscripts, in decreasing order, are: 120, 96, 96, 88, 80, 76, 64, 64, 60, 60, 56, 48, 40, 32. Packing these in bins of size 120 yields: Bin 1: 120; Bin 2: 96; Bin 3: 96; Bin 4: 88, 32; Bin 5: 80, 40; Bin 6: 76; Bin 7: 64, 56; Bin 8: 64, 48; Bin 9: 60, 60. Nine photocopy machines are needed to finish within 2 minutes.

29. a. (1) The schedule with four secretaries is as follows: Processor 1: 25, 36, 15, 15, 19, 15, 27; Processor 2: 18, 32, 18, 31, 30, 18; Processor 3: 13, 30, 17, 12, 18, 16, 16, 16, 14; Processor 4: 19, 12, 25, 26, 18, 12, 24, 9.

The schedule with five secretaries is as follows: Processor 1: 25, 25, 31, 12, 16, 14; Processor 2: 18, 12, 17, 12, 15, 30, 9; Processor 3: 13, 32, 26, 16, 15, 18; Processor 4: 19, 36, 18, 19, 24; Processor 5: 30, 18, 15, 18, 16, 27.

(2) The decreasing time list is: 36, 32, 31, 30, 30, 27, 26, 25, 25, 24, 19, 19, 18, 18, 18, 18, 18, 17, 16, 16, 16, 15, 15, 15, 14, 13, 12, 12, 12, 9.

The schedule using this list on four processors would be: Processor 1: 36, 25, 19, 18, 17, 16, 13, 9; Processor 2: 32, 26, 25, 18, 16, 15, 12; Processor 3: 31, 27, 24, 18, 16, 15, 12, 12; Processor 4: 30, 30, 19, 18, 18, 15, 14.

The schedule using this list on five processors would be: Processor 1: 36, 24, 18, 16, 14, 12; Processor 2: 32, 25, 18, 18, 15, 12; Processor 3: 31, 25, 19, 18, 15, 9; Processor 4: 30, 27, 18, 17, 15, 13; Processor 5: 30, 26, 19, 16, 16, 12.

(3) The five-processor decreasing time schedule is optimal (time 120), but the four decreasing time schedule is not. One can see this since when the task of length 17 scheduled on processor 1 and the task of length 18 on processor 3 are interchanged, the completion time is reduced to 154 from 155 for the four-processor decreasing time schedule.

b. As a bin packing problem, each bin will have a capacity of 60. Using the decreasing list we obtain the following packings:

(1) (First-fit decreasing): Bin 1: 36, 24; Bin 2: 32, 27; Bin 3: 31, 26; Bin 4: 30, 30; Bin 5: 25, 25, 9; Bin 6: 19, 19, 18; Bin 7: 18, 18, 18; Bin 8: 18, 17, 16; Bin 9: 16, 16, 15, 13; Bin 10: 15, 15, 14, 12; Bin 11: 12, 12.

(2) (Next-fit decreasing): Bin 1: 36; Bin 2: 32; Bin 3: 31; Bin 4: 30, 30; Bin 5: 27, 26; Bin 6: 25, 25; Bin 7: 24, 19; Bin 8: 19, 18, 18; Bin 9: 18, 18, 18; Bin 10: 17, 16, 16; Bin 11: 16, 15, 15; Bin 12: 15, 14, 13, 12; Bin 13: 12, 12, 9.

(Best-fit decreasing): Bin 1: 36, 24; Bin 2: 32, 26; Bin 3: 31, 27; Bin 4: 30, 30; Bin 5: 25, 25; Bin 6: 19, 19, 18; Bin 7: 18, 18, 18; Bin 8: 18, 17, 16; Bin 9: 16, 16, 15, 12; Bin 10: 15, 15, 14, 13; Bin 11: 12, 12, 9.

(3) Since the total weight of all the objects is 596, a minimum of 10 bins is required. However, since there are no small weights, there is no way to achieve 10 bins, and in fact, 11 bins is optimal.

31. a. Using the next-fit algorithm, the bins are filled as follows: Bin 1: 12, 15; Bin 2: 16, 12; Bin 3: 9, 11, 15; Bin 4: 17, 12; Bin 5: 14, 17; Bin 6: 18; Bin 7: 19; Bin 8: 21; Bin 9: 31; Bin 10: 7, 21; Bin 11: 9, 23; Bin 12: 24; Bin 13: 15, 16; Bin 14: 12, 9, 8; Bin 15: 27; Bin 16: 22; Bin 17: 18.

b. The decreasing list is: 31, 27, 24, 23, 22, 21, 21, 19, 18, 18, 17, 17, 16, 16, 15, 15, 15, 14, 12, 12, 12, 12, 11, 9, 9, 9, 8, 7.

The next-fit decreasing schedule is: Bin 1: 31; Bin 2: 27; Bin 3: 24; Bin 4: 23; Bin 5: 22; Bin 6: 21; Bin 7: 21; Bin 8: 19; Bin 9: 18, 18; Bin 10: 17, 17; Bin 11: 16, 16; Bin 12: 15, 15; Bin 13: 15, 14; Bin 14: 12, 12, 12; Bin 15: 12, 11, 9; Bin 16: 9, 9, 8, 7.

c. The best-fit schedule using the original list is: Bin 1: 12, 15, 9; Bin 2: 16, 12; Bin 3: 11, 15; Bin 4: 17, 12; Bin 5: 14, 17; Bin 6: 18, 7; Bin 7:

19, 9; Bin 8: 21, 15; Bin 9: 31, Bin 10: 21, 9; Bin 11: 23, 8; Bin 12: 24; Bin 13: 16, 12; Bin 14: 27; Bin 15: 22; Bin 16: 18.

d. The best-fit decreasing schedule would be: Bin 1: 31; Bin 2: 27, 9; Bin 3: 24, 12; Bin 4: 23, 12; Bin 5: 22, 12; Bin 6: 21, 15; Bin 7: 21, 15; Bin 8: 19, 17; Bin 9: 18, 18; Bin 10: 17, 16; Bin 11: 16, 12, 8; Bin 12: 15, 14, 7; Bin 13: 11, 9, 9.

33. The bins have a capacity of 120. (First-fit): Bin 1: 63, 32, 11; Bin 2: 19, 24, 64; Bin 3: 87, 27; Bin 4: 36, 42; Bin 5: 63. This schedule would take five station breaks, however, the total time for the breaks is under 8 minutes.

The decreasing list is: 87, 64, 63, 63, 42, 36, 32, 27, 24, 19, 11. (First-fit decreasing): Bin 1: 87, 32; Bin 2: 64, 42, 11; Bin 3: 63, 36, 19; Bin 4: 63, 27, 24. This solution uses only four station breaks.

35. The sum of the integers from 1 to n is $n(n + 1)/2$. Hence, the sum of the numbers from 1 to 20 is 210. Since each weight occurs twice, the weights total 420. Hence, at least 17 bins are required since the capacity of each bin is 25. For each weight occurring three times, at least 26 bins are required.

37. Such a heuristic will fill many bins to capacity, but the computation to find numbers summing to exactly W may be very time consuming.

39. a. Packing boxes of the same height into crates; packing want ads into a newspaper page.
b. We assume, without loss of generality $p \geq q$. One heuristic, similar to first-fit, orders the rectangles $p \times q$ as in a dictionary (i.e., $p \times q$ listed prior to $r \times s$ if $p > r$ or $p = r$ and $q \geq s$. It then puts the rectangles in place in layers in a first-fit manner; that is, do not put a rectangle into a second layer, until all positions on the first layer are filled. However, extra room in the first layer is "wasted."
c. The problem of packing rectangles of width 1 in an $m \times 1$ rectangle is a special case of the two-dimensional problem, equivalent to the bin-packing problem we have discussed.
d. Two 1×10 rectangles cannot be packed into a 5×4 rectangle, even though there would be an area of 20 in this rectangle.

41. There is an example of a bin-packing problem for which a given list takes a certain number of bins, and when an item is deleted from the list, more bins are required. In this example, the deleted item is not first in the list.

Chapter 4

1.

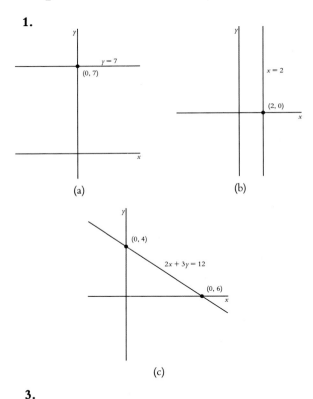

3.

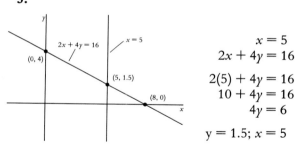

$$x = 5$$
$$2x + 4y = 16$$
$$2(5) + 4y = 16$$
$$10 + 4y = 16$$
$$4y = 6$$
$$y = 1.5; \ x = 5$$

5.

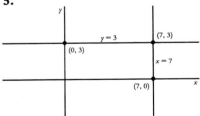

$$y = 3; \; x = 7$$

7.

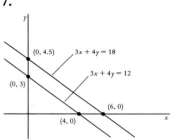

Lines do not intersect.

9.

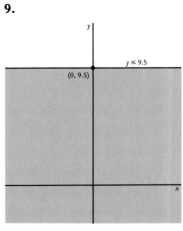

11.

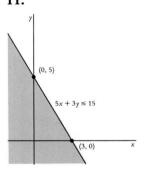

13.

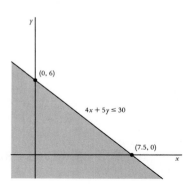

15.

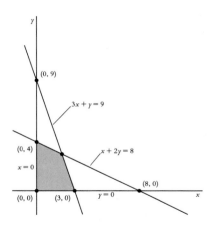

17.

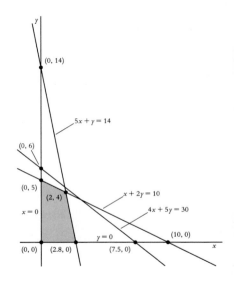

19. a. No. **b.** Yes. **c.** Yes. **d.** No. **e.** No.

21. a. No. **b.** Yes. **c.** Yes. **d.** No. **e.** Yes.

23. $3(20) + 4(25) = 60 + 100 = 160$; $3(0) + 4(50)$ $= 0 + 200 = 200$; $3(30) + 4(10) = 90 + 40 = 130$.

25. $4x + 2y \leq 28$.

27. $6x + 4y \leq 240$.

29. a. $(0, 50) = 0$ gallons cranapple and 50 gallons appleberry.
$(50, 25) = 50$ gallons cranapple and 25 gallons appleberry.
$(200/3, 0) = 200/3$ gallons cranapple and 0 gallons appleberry.
$(0, 0) = 0$ gallons of each.
b. $(50, 25)$; $(0, 50)$; $(200/3, 0)$.

31.

Mixture chart	Wheat (800 lb)	Sugar (40 lb)	Profit
Hefties (x boxes)	$\frac{1}{2}$ lb	0 lb	10¢/box
Sweeties (y boxes)	$\frac{4}{10}$ lb	$\frac{1}{10}$ lb	13¢/box

Constraint inequalities

$$\tfrac{1}{2}x + \tfrac{4}{10}y \leq 800$$
$$\tfrac{1}{10}y \leq 40$$
$$x \geq 0 \text{ and } y \geq 0$$

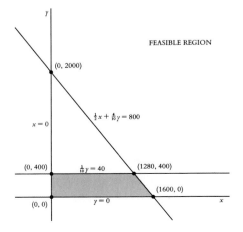

Profit

$$\text{Profit} = 0.10x + 0.13y$$

At $(0,0)$, profit $= \$0$.
At $(0, 400)$, profit $= \$52$.
At $(1280, 400)$, profit $= \$182$.
At $(1600, 0)$, profit $= \$160$.

$(1280, 400)$ yields the largest profit.

33.

Mixture chart	Milk (720 qt)	Flavoring (30 qt)	Profit
Regular milk (x qt)	1 qt	0 qt	30¢/qt
Chocolate milk (y qt)	0.9 qt	0.1 qt	50¢/qt

Constraint inequalities

$$x + 0.9y \leq 720$$
$$0.1y \leq 30$$
$$x \geq 0 \text{ and } y \geq 0$$

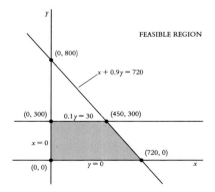

Profit

$$\text{Profit} = 0.30x + 0.50y$$

At $(0, 0)$, profit $= \$0$.
At $(0, 300)$, profit $= \$150$.
At $(450, 300)$, profit $= \$285$.
At $(720, 0)$, profit $= \$216$.

$(450, 300)$ yields the largest profit.

35.

Mixture chart	Beef (500 lb)	Pork (300 lb)	Grain filler (100 lb)	Profit
Beef hot dogs (x packages)	1 lb	0 lb	0 lb	80¢/package
Regular hot dogs (y packages)	$\frac{1}{4}$ lb	$\frac{1}{2}$ lb	$\frac{1}{4}$ lb	70¢/package

Constraint inequalities

$$x + \tfrac{1}{4}y \le 500$$
$$\tfrac{1}{2}y \le 300$$
$$\tfrac{1}{4}y \le 400$$
$$x \ge 0 \text{ and } y \ge 0$$

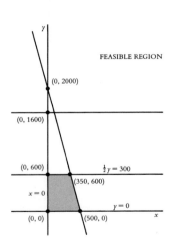

Profit

Profit $= 0.80x + 0.70y$

At (0, 0), profit $=$ \$0.
At (0, 600), profit $=$ \$420.
At (350, 600), profit $=$ \$700.
At (500, 0), profit $=$ \$400.

(350, 600) yields the largest profit.

37.

Mixture chart	High octane (500 gal)	Low octane (600 gal)	Profit
Premium gas (x gal)	$\frac{1}{2}$ gal	$\frac{1}{2}$ gal	40¢/gal
Regular gas (y gal)	$\frac{1}{4}$ gal	$\frac{3}{4}$ gal	30¢/gal

Constraint inequalities

$$\tfrac{1}{2}x + \tfrac{1}{4}y \le 500$$
$$\tfrac{1}{2}x + \tfrac{3}{4}y \le 600$$
$$x \ge 0 \text{ and } y \ge 0$$

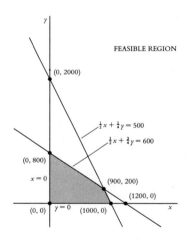

Profit

Profit $= 0.40x + 0.30y$

At (0, 0), profit $=$ \$0.
At (0, 800), profit $=$ \$240.
At (900, 200), profit $=$ \$420.
At (1000, 0), profit $=$ \$400.

(900, 200) yields the largest profit.

39.

Mixture chart	Nurse (6,250 min)	Doctor (11,000 min)	Laboratory (5,000 min)	Profit
Routine visit (x times)	10 min	5 min	5 min	$30/ visit
Comprehensive visit (y times)	5 min	25 min	10 min	$50/ visit

Constraint inequalities

$$10x + 5y \le 6,250$$
$$5x + 25y \le 11,000$$
$$5x + 10y \le 5,000$$
$$x \ge 0 \text{ and } y \ge 0$$

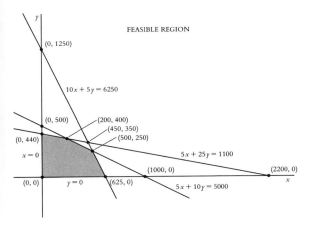

Profit

Profit $= 30x + 50y$

At $(0, 0)$, profit $= \$0$.
At $(0, 440)$, profit $= \$22,000$.
At $(200, 400)$, profit $= \$26,000$.
At $(500, 250)$, profit $= \$27,500$.
At $(625, 0)$, profit $= \$18,750$.

$(500, 250)$ yields the largest profit.

41. The old profit formula of $3x + 4y$ becomes $6x + 8y$. The point that is optimal does not change because the lines that correspond to the old profit formula are parallel to the lines that correspond to the new profit formula.

43. The new best point would correspond to the old best point, but both x and y coordinates would double, so the profit would double.

Chapter 5

1. *Population:* All registered voters in the Second Congressional District. *Sample:* The 800 voters interviewed.

3. *Population:* All chips of this type made by the supplier, including future production. *Sample:* The 400,000 chips inspected.

5. Probably lower, because only households with phones were surveyed. Black households are poorer on the average than white households and so are more likely to lack a telephone. (More households have television sets than have telephones.)

7. a. RDD omits households without telephones (over 7% of U.S. households). **b.** Voter lists omit persons not registered to vote. Young adults and poor people are less likely to be registered than older or richer people.

9. If labels 00 to 29 are assigned to the 30 students in alphabetical order, the sample consists of 21 = Pirelli, 01 = Aspin, and 19 = Olds.

11. a. The answers obtained will depend on the parts of the table used. In the long run, we expect an average of 2 of the 5 tickets to go to women. **b.** A rough empirical answer is based on how many of your 20 samples included cases in which no women received tickets. (In fact, the probability that no tickets go to women is about 0.056. It is somewhat unlikely that no women will receive tickets.)

13. a. All 380 faculty members. **b.** Label the faculty members with three-digit labels, such as 000 to 379. Then look at three-digit groups in Table 5.1 to choose the sample. **c.** With labels 000 to 379, the first five members of the sample are the faculty members labeled 157, 274, 296, 001, and 338.

15. Ann Landers's poll relies on voluntary response. It attracts readers with strong feelings, especially those with negative feelings toward their children. The random sample gives everyone the same chance, so it is much more trustworthy. The voluntary response poll gives the result 70% "no" when the truth about the population is close to 90% "yes." Such polls give *no* useful information about anyone except the actual respondents.

17. 46% of the sample believes in life on other planets. (A recent opinion poll found this result.) We can be confident that between 43% and 49% of all adults believe in extraterrestrial life.

19. The effect (if any) of the tea is confounded by the effect of visits and conversation with college students. The visits alone might make the residents more cheerful.

21. The unemployment rate is most strongly influenced by general economic conditions; it might have been still higher without the training program. The effect (if any) of the program on unemployment is confounded by the stronger effect of economic conditions.

23. Subjects who do not receive the drug should receive a placebo in order to avoid confounding the placebo effect with the effect of the drug. In the absence of a reason to do otherwise, it is best to assign equal numbers of subjects to each treatment. The design is:

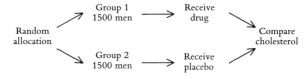

25. Because the diagnosis of mild heart attacks is somewhat subjective, it is best if the diagnosing physicians are blind in both experiments. The study in Exercise 23 should be double-blind; in Exercise 24c the subjects know whether or not they are in an exercise program, so they cannot be blind.

27. a. A direct comparison will eliminate the possibility that unusual nerve responses are due to something else in the diet or environment of the rats. **b.** The design is similar to Figure 5.4, with 10 rats in each group and diets with and without DDT as the treatments. If the rats are labeled 00 to 19, the DDT group contains rats 07, 10, 05, 00, 15, and so on. The randomization is tedious because half the rats must be chosen. The remaining 10 form the control group.

29. The outline is similar to Figure 5.4, with 10 subjects in each group and surgery and placebo as the treatments. Labeling the subjects 00 to 19, group 1 consists of 03, 18, 07, 10, 19, 08, 04, 09, and 00. The remaining subjects form group 2.

31. There is variation among subjects in their response to the treatments and random variation in the results of the assignment to groups. Averaging this variation over a large number of subjects assures that there is very little systematic difference between the two groups, whereas two groups of 10 might by bad luck be quite different.

33. a. The factors, or experimental variables, are type of corn (normal or floury-2) and protein level (12%, 16%, or 20%). **b.** The experimental units are 60 chicks; these are divided at random into six groups of 10 chicks each; each group is fed one of the six diets; weight gains after 21 days are measured and compared. The outline is similar to Figure 5.5, but with six groups rather than three.

35. Here is a 3 × 3 Latin square, formed by starting with *ABC* and sliding each row one character to the left:

A B C
B C A
C A B

Here is a 4 × 4 Latin square formed in the same way:

```
A  B  C  D
B  C  D  A
C  D  A  B
D  A  B  C
```

Chapter 6

1. There are no outliers or other unusual features.

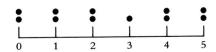

3. The distribution is roughly symmetric, with center near noon (12 hours from midnight). There are no outliers or gaps.

5. a. Your histogram will depend a bit on your choice of cells. Here is one choice.

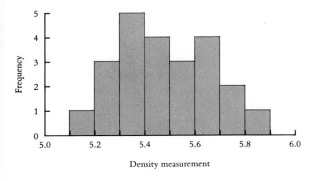

Density measurement

b. The distribution is roughly symmetric, with no outliers or other unusual features. (With only 23 observations, we cannot insist on a close approach to exact symmetry.)

7. a. $\bar{x} = 2.45$, $M = 2$. **b.** $Q_1 = 1$, $Q_3 = 4$.

9. a. The 1968 result, 43.4%, is a low outlier. In 1968 a third-party candidate drew over 13% of the vote. **b.** $M = 53.9\%$. **c.** $Q_3 = 58.8\%$ so that the 1964 (Johnson defeats Goldwater), 1972 (Nixon defeats McGovern), and 1984 (Reagan defeats Mondale) elections were landslides.

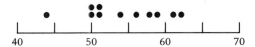

11. a. There are no outliers or other unusual features. **b.** $\bar{x} = 35.3$, $M = 35.5$. c$Q_1 = 32.5$, $Q_3 = 39.5$.

13. $M = 5.46$, $Q_1 = 5.34$, and $Q_3 = 5.63$; the five-number summary is 5.10, 5.34, 5.46, 5.63, 5.85. In a symmetric distribution, the two quartiles will fall about the same distance from the median, as will the two extremes. In this case, the quartiles are 0.17 and 0.12 from the median and the extreme observations are 0.39 and 0.36 from the median.

15. a. Here is a histogram, using classes of length $500.

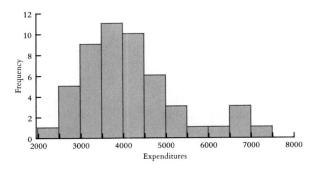

Expenditures

b. The distribution is skewed to the right, but there are no outliers.

17. $\bar{x}$ = \$4190.90, M = \$3994. The mean is larger than the median because of the right skewness of the distribution.

19. The quartiles are Q_1 = \$3553 and Q_3 = \$4871. Therefore, IQR = \$1518. Adding $1.5 \times IQR$ to Q_3 gives \$7148. The largest observation (Alaska's \$7038) is less than this value, confirming that there are no outliers in the long right tail of the distribution.

21. Because a few million-dollar houses pull up the mean but not the median, the mean is the larger of the two numbers.

23. A single high outlier is enough. For the data 1, 1, 2, 3, 3, 4, 28, the third quartile is 4 and the mean is $42/7 = 6$.

25. The five-number summary of these data is \$2410, \$3353, \$3994, \$4871, \$7038. Here is the boxplot.

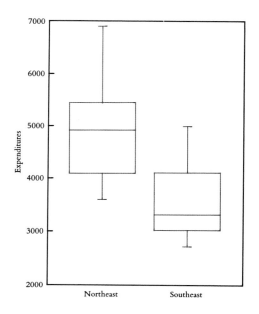

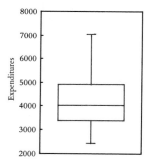

27. The five-number summary for the northeast is \$3616, \$4122, \$4949, \$5456, \$6910; for the southeast the five-number summary is \$2752, \$3007, \$3355, \$4267, \$4994. The side-by-side boxplots show clearly that public school expenditures per pupil are much higher in the northeast. Indeed, the first quartile for the northeast is close to the third quartile for the southeast. In addition, the boxplots show that the spread among states is also greater in the northeast.

29. The five-number summary for males is 46.9, 47.4, 51.9, 62.0, 62.9; for females, it is 33.1, 38.2, 42.0, 49.55, 54.6. The generally higher lean body mass of males is apparent in the boxplots. The extremes in the male distribution do not extend far beyond the quartiles, but this is not surprising in a boxplot of only $n = 7$ observations.

31. Because both variables have the same units, it is best to use the same scale on both axes. U.S. prices, the explanatory variable, are plotted horizontally. There is a generally linear pattern, with many items costing much more in Japan. Cantaloupe falls outside the linear pattern; it costs much more in Japan than the overall pattern would suggest. Note that movie prices do fall in the overall linear pattern, although they are among the highest of those given in both countries.

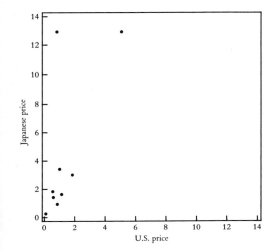

33. For $x = 3300$, $y = 4557$ pounds per square inch.

35. a. Higher. **b.** For $x = 422$, the prediction is $y = 461.7$. The error is 4.3 points. **c.** Hawaii's math score is higher (or its verbal score lower) than the general relationship suggests. Over 60% of Hawaii's population consists of Asians and Pacific islanders, so it is possible that less use of English may lower verbal scores.

37. a. There is a clear linear pattern. There are no pronounced outliers, although the first observation ($x = 7.2$, $y = 1.56$) lies a bit above the overall pattern. **b.** Use the "up and over" graphical method as in Figure 6.8b; the y obtained depends on the line drawn.

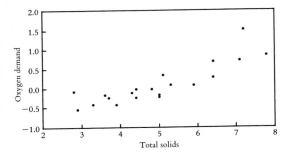

39. a. The explanatory variable x is powerboat registrations. There is an increasing and roughly linear overall pattern, but the points are not tightly clustered about a straight line. There are no clear outliers (1983 is closest to being an outlier). **b.** Use the "up-and-over" graphical method as in Figure 6.8b; y will depend on the line drawn.

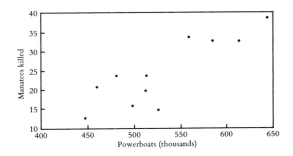

41. The basic sums are: $\Sigma x = 1328$, $\Sigma y = 2317$, $\Sigma x^2 = 241,196$, and $\Sigma xy = 392,217$. The least-squares regression line is $y = 58.588 + 1.304x$.

43. The basic sums are: $\Sigma x = 99.3$, $\Sigma y = 2.4$, $\Sigma x^2 = 533.17$, and $\Sigma xy = 24.15$. The least-squares regression line is $y = -1.393 + 0.3047x$. For $x = 4$, the prediction is $y = -0.174$.

Chapter 7

1, 3, 5. No answers provided for these exercises.

7. S can be taken to be all positive whole numbers, or you can choose a reasonable maximum, for example, S contains the whole numbers from 1 to 30.

9. $S = \{0, 1, 2, 3, \ldots\}$ (all nonnegative whole numbers) is the simplest choice because we do not have to decide a largest possible amount.

11. Models A and B are legitimate, because all probabilities are between 0 and 1 and their sum is 1. Model C has negative entries, so is not legitimate. Model D has entries with sum 1.5, so is not legitimate.

15. In a shuffled deck, all 13 possible outcomes are equally likely, so each has probability $\frac{1}{13}$. That is Model A. Model B is legitimate and would be correct if all the face cards were queens. Model C is not legitimate because the sum of the probabilities is greater than 1.

Outcome	Model A	Model B	Model C
Ace	$\frac{1}{13}$	$\frac{1}{13}$	$\frac{1}{10}$
King	$\frac{1}{13}$	0	$\frac{1}{10}$
Queen	$\frac{1}{13}$	$\frac{3}{13}$	$\frac{1}{10}$
Jack	$\frac{1}{13}$	0	$\frac{1}{10}$
Ten	$\frac{1}{13}$	$\frac{1}{13}$	$\frac{1}{10}$
Nine	$\frac{1}{13}$	$\frac{1}{13}$	$\frac{1}{10}$
Eight	$\frac{1}{13}$	$\frac{1}{13}$	$\frac{1}{10}$
Seven	$\frac{1}{13}$	$\frac{1}{13}$	$\frac{1}{10}$
Six	$\frac{1}{13}$	$\frac{1}{13}$	$\frac{1}{10}$
Five	$\frac{1}{13}$	$\frac{1}{13}$	$\frac{1}{10}$
Four	$\frac{1}{13}$	$\frac{1}{13}$	$\frac{1}{10}$
Three	$\frac{1}{13}$	$\frac{1}{13}$	$\frac{1}{10}$
Two	$\frac{1}{13}$	$\frac{1}{13}$	$\frac{1}{10}$

17. The probability that either A or B occurs is found by adding the probabilities of all outcomes that are either in A or in B. Because no outcome is in both A and B, there is no double counting if we first add the probabilities of outcomes in A and then separately add the probabilities of outcomes in B. So adding $P(A)$ and $P(B)$ is the same as adding the probabilities of all outcomes in either A or B.

19. The model is

Sum	2	3	4	5	6	7	8	9	10	11	12
Probability	$\frac{1}{36}$	$\frac{2}{36}$	$\frac{3}{36}$	$\frac{4}{36}$	$\frac{5}{36}$	$\frac{6}{36}$	$\frac{5}{36}$	$\frac{4}{36}$	$\frac{3}{36}$	$\frac{2}{36}$	$\frac{1}{36}$

a. $\frac{8}{36}$.   **b.** $\frac{21}{36}$.

21. Repeats allowed: $(20 \times 20 \times 20)/(26 \times 26 \times 26) = 0.455$. No repeats allowed: $(20 \times 19 \times 18)/(26 \times 25 \times 24) = 0.438$.

23. No x:
   $(35 \times 35 \times 35)/(36 \times 36 \times 36) = 0.919$
No digits:
   $(26 \times 26 \times 26)/(36 \times 36 \times 36) = 0.377$

25. The possibilities are *ags, asg, gas, gsa, sag, sga*, of which "gas" and "sag" are English words. The probability is $\frac{2}{6} = 0.33$.

27. $\frac{21}{6} = 3.5$.

29. $\frac{12}{8} = 1.5$.

31. **a.** The probabilities are all between 0 and 1 and have sum 1.   **b.** $A = \{9, 10, 11, 12\}$, $P(A) = 0.931$.   **c.** 11.251.

33. **a.** $\frac{12}{38} = 0.316$.   **b.** $-\frac{2}{38} = \$0.053$.

35. Optional group exercise. No answer provided.

37. Draw a normal curve, then mark the axis so that the center (mean) is at 69 and the change-of-curvature points are at 66.5 and 71.5. The curve should reach the horizontal axis at about 61.5 and 76.5.

39. The first quartile is 67.325 inches, and the third quartile is 70.675 inches.

41. **a.** The median (same as the mean) is 10%.   **b.** 9.6% to 10.4%.   **c.** 9.866% to 10.134%.

43. **a.** 50%.   **b.** 2.5% (half of 5%).

45. **a.** 234 and 298 days.   **b.** Shorter than 234 days.

47. 245.5 days or shorter.

49. About 4.9% ($7/\sqrt{2}$).

51. 200 students.

Chapter 8

1. 64.5 is a statistic, 63 a parameter.

3. Both are parameters.

5. Mean = 35%, standard deviation = 3.37%.

7. Mean = 3.0, standard deviation = 0.23.

9. 1.18%, 1.26%, 1.29%, 1.26%, 1.18%.

11. If we repeated Gallup's sampling process, we might get a different answer. But 95% of all samples will give an answer that is within $\pm 3\%$ of the percent of all adults who jog. Gallup announces the margin of error to allow for this variation and indicate how accurate his result will usually be.

13. 59.3% ± 8.02%. Because the interval falls entirely above 50%, we are 95% confident that more than half of all visitors are in favor.

15. a. 2.53%. **b.** Gallup does not use a simple random sample, but rather a complex multistage sampling design. The margin of error for 95% confidence is somewhat larger for this sampling design than for a simple random sample of the same size.

17. Only **c.** The margin of error includes only the random sampling error described by the sampling distribution of the statistic.

19. The margin of error would be greater than ± 3 points. Smaller sample sizes result in wider intervals for the same level of confidence. (Because the poll did not use a simple random sample, we cannot calculate the margin of error.)

21. a. By the 68-95-99.7 rule, 68% of the observations in any normal distribution fall within ± 1 standard deviation of the mean. For large samples, the sample proportion $\hat{p}$ follows approximately a normal distribution with mean p and standard deviation $\sqrt{p(100-p)/n}$. So the unknown p is within this standard deviation of the observed $\hat{p}$ in 68% of all samples in the long run. Replacing the unknown p by $\hat{p}$ in the standard deviation changes its numerical value very little. So the unknown p falls in the interval given in 68% of all samples. That's what 68% confidence means. **b.** A 99.7% confidence interval is

$$\hat{p} \pm 3 \sqrt{\frac{\hat{p}(100-\hat{p})}{n}}$$

23. 59.3% ± 4.01%. The 68% confidence interval is half as wide as the 95% confidence interval (one standard deviation rather than two) because a smaller margin of error is sufficient if we allow lower confidence in the result.

25. a. 18 ± 12 points. **b.** 18 ± 2.4 points.

27. 3.4137 ± 0.00115 grams.

29. 36.9 ± 1.74 points.

31. 0.7505 ± 0.00032 inch.

33. a. $\bar{x} \pm 1.28\sigma/\sqrt{n}$. **b.** 36.9 ± 1.11 points.

35. Center line = 5 grams, control limits = 5 ± 0.00173 grams.

37. The control charts in Exercises 37 to 39 have a center line (drawn solid) at 101.5 and control limits (drawn dashed) at 101.2 and 101.8. Here there are no points outside the limits and no run of 8 or more on the same side of the center line.

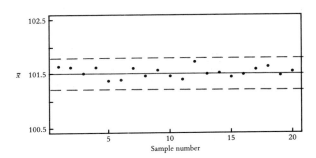

39. Samples 15, 18, and 19 fall above the upper control limit. The last six points lie above the center line. There is a clear upward drift in the plot.

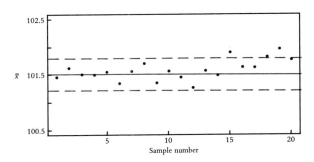

41. The percents of students who smoke for each parent condition are: both parents smoke, 22.5%; one parent smokes, 18.6%; neither parent smokes, 13.9%. There is a clear association between parent smoking and student smoking, with students of smoking parents being more likely to smoke themselves.

43. a. The two-way table is

	Bill	Will
Hits	120	130
Outs	380	370
At bats	500	500

Bill gets a hit .240 of his times at bat, while Will's average is .260. So Will has the higher batting average. **b.** Bill hits better against right-handers (.400 versus .300) and also against left-handers (.200 versus .100). **c.** Both players hit much better against right-handers; Will bats against right-handers much more often than does Bill. So even though Bill does better against both types of pitchers than Will, Will has a higher average overall. We should choose Bill for our team.

Chapter 9

1. a. $3! = 6$. **b.** $4! = 24$. **c.** $n!$.

3. a. First (Not a) (60%) beats a (40%), and then b (75%) beats c (25%). **b.** First b (75%) beats c (25%), and then b (60%) beats a (40%). **c.** b wins with 3 (20%) + 3 (15%) + 2 (40%) + 2 (25%) = 2.35 points over 2.00 for a to 1.65 for c. **d.** b (60%) beats a (40%) and b (75%) beats c (25%). **e.** There are no possibilities for strategic voting in cases **a** and **c** (for secret or simultaneous voting). In case **b**, A may vote for c on the first ballot in an attempt to eliminate b, but this will fail if C votes for b on the first ballot (as in this case when described in the text). In case **d**, A could create a cycle (tie) in voting for c in the pairwise comparison between b and c.

5. a. First E (7 votes) beats H (3), and then D (7) wins over E (3). **b.** First D (7 votes) beats E (3), and then H (6) wins over D (4). **c.** Yes, in case **a**. They vote for H on the first (and second) ballots and their second choice H wins. No, in case **b**. **d.** With sincere voting, D wins (21 points) over E (20) and H (19). **e.** No.

7. a. V with 6 firsts. **b.** B and V tied with 27 points. **c.** V (8 votes) beats B (5) in the runoff.

d. B (9 votes) beats S (4) in the runoff. **e.** S is eliminated first, and V beats B. **f.** No. V beats B beats S beats V.

9. a. (1) Chinese; (2) Italian; (3) Mexican. **b.** Mexican. **c.** (1) Mexican; (2) Mexican and Chinese (tied); (3) Chinese. **d.** No. Set $z = 0$. Italian wins if $4x + 2y > 5x$ and $4x + 2y > 2x + 9y$. This implies the contradiction $4y > 2x > 7y$.

11. a. A (14 points) ties C (14), B (12). **b.** C (9 points) wins over A (8) and B (7). **c.** (1) C ($14\frac{1}{2}$), A (14), B ($11\frac{1}{2}$). (2) C ($9\frac{1}{2}$), A (8), B ($6\frac{1}{2}$). **d.** (1) A (16), B (13), C (11). (2) A (9), B (8), C (7).

13. a. A. **b.** B. **c.** There is no difference on the ranking of the nominees.

15. E.

17. p/q is not in lowest terms if both p and q are even integers.

Chapter 10

1. a. 2. **b.** 3. **c.** 12. **d.** 6. **e.** 8. **f.** 4. **g.** 4. **h.** Infinity. **i.** 32. **j.** 2. **k.** Infinity.

3. a. (1) $\{1\}$, $\{1, 2\}$. (2) $\{1\}$. (3) $\{2\}$, 0. (4) 2. (5) $\{1\}$, $\{1, 2\}$. **b.** (1) $\{1, 2\}$, $\{1, 3\}$, $\{2, 3\}$, $\{1, 2, 3\}$. (2) $\{1, 2\}$, $\{1, 3\}$, $\{2, 3\}$. (3) $\{1\}$, $\{2\}$, $\{3\}$, 0. (4) None. (5) See (1). **c.** See **b**. **d.** (1) $\{1, 2\}$, $\{1, 3\}$, $\{1, 2, 3\}$. (2) $\{1, 2\}$, $\{1, 3\}$. (3) $\{2, 3\}$, $\{1\}$, $\{2\}$, $\{3\}$, 0. (4) None. (5) $\{1, 2\}$, $\{1, 3\}$, $\{1, 2, 3\}$, $\{2, 3\}$, $\{1\}$. **e.** (1) $\{1, 2\}$, $\{1, 3\}$, $\{2, 3\}$, $\{1, 2, 3\}$, $\{1, 2, 4\}$, $\{1, 3, 4\}$, $\{2, 3, 4\}$, $\{1, 2, 3, 4\}$. (2) $\{1, 2\}$, $\{1, 3\}$, $\{2, 3\}$. (3) $\{1, 4\}$, $\{2, 4\}$, $\{3, 4\}$, $\{1\}$, $\{2\}$, $\{3\}$, $\{4\}$, 0. (4) 0. (5) See (1). **f.** See part **e**. **g.** See part **e**. **h.** (1) Any coalition with 3, 4, or 5 players. (2) Any coalition with 3 players. (3) Any coalition with 2, 1, or 0 players. (4) None. (5) See (1). **i.** (1) Any superset of those in (2). (2) $\{1, 2\}$, $\{1, 3\}$, $\{1, 4\}$, $\{2, 3, 4\}$. (3) $\{2, 3\}$, $\{2, 4\}$, $\{3, 4\}$, $\{1\}$, $\{2\}$, $\{3\}$, $\{4\}$, 0. (4) None. (5) See (1). **j.** (1) Any superset of those in (2). (2) $\{1, 2\}$, $\{1, 3\}$, $\{2, 3, 4\}$. (3) $\{2, 3\}$, $\{1, 4\}$, $\{2, 4\}$, $\{3, 4\}$, $\{1\}$, $\{2\}$, $\{3\}$, $\{4\}$, 0. (4) None. (5) $\{2, 3\}$ and the subsets in

(1). **k.** (1) Any superset of those in (2). (2) {H1, H2}, {H1, NH}, {H2, NH}. (3) Any subset not in (1). (4) OB, GC, LB. (5) See (1). **l.** (1) Any superset of those in (2). (2) {H1, H2}, {H1, OB}, {H2, OB}. (3) Any subset not in (1). (4) NH, GC, LB. (5) See (1). **m.** (1) Any superset of those in (2). (2) {1, 2}, {1, 3}, {1, 4}, {1, 5}, {2, 3, 4, 5}. (3) Any subset not in (1). (4) None. (5) See 1.

5. a. $(1, 0)$. **b.** $(1, 1, 1)$. **c.** $(1, 1, 1)$. **d.** $(3, 1, 1)$. **e.** $(1, 1, 1, 0)$. **f.** $(1, 1, 1, 0)$. **g.** $(1, 1, 1, 0)$. **h.** $(1, 1, 1, 1, 1)$. **i.** $(3, 1, 1, 1)$. **j.** $(5, 3, 3, 1)$. **k.** $(1, 1, 1, 0, 0, 0)$. **l.** $(1, 1, 0, 1, 0, 0)$. **m.** $(7, 1, 1, 1, 1)$.

7. a. {1, 2}, {1, 3, 4}. **b.** [6 : 4, 2, 1, 1]. **c.** (5, 3, 1, 1).

9. A minimal winning coalition consists of the five permanent members plus four others: $5.7 + 4.1 = 39$.

11. Note that each member is in some minimal winning coalitions and is thus pivotal.

13. a and **b.** (These rather long lists are omitted.) **c.** Case A: (40, 40, 40, 24, 24, 15, 15, 15). Case B: (49, 49, 49, 21, 21, 21, 21, 21). Case C: (56, 56, 56, 6, 6, 6, 6, 6). **d.** [9 : 4, 4, 4, 1, 1, 1, 1, 1] or [71 : 35, 35, 35, 8, 8, 8, 8, 8]. **e.** No; each of the five borough presidents has equal power.

15. {1, 2} and {3, 4} win implies $(w_1 + w_2) + (w_3 + w_4) \geq q$, whereas {1, 3} and {2, 4} lose implies $(w_1 + w_3) + (w_2 + w_4) < q$, a contradiction.

17. Let A represent any Atlantic province and P any prairie province. Coalitions of the form (O, Q, BC, A, A, P) win, whereas those of the form $(O, Q, BC, P, *P, *P)$ and (O, Q, BC, A, A, A) lose. Summing the positive weights for 6 of the former, and for 2 and 4, respectively, of the latter two leads to a contradiction.

19. Three. There is only one minimal winning coalition in each case: Either {1, 2}, {1}, or {2}.

21. There are "too many" minimal winning coalitions, or "too many" veto power coalitions. Discuss.

23. Answers vary.

25. Countries and answers vary.

Chapter 11

1. Answers vary. Check with your school.

3. Answers vary. Yes, if 1.9996 . . . is rounded up to 2.000.

5. Stirnweiss, .30854 > .30846.

7. Johnson, .3290 > .3286.

9. a. Pistons 47.1%, Lakers 41.1%. **b.** Pistons 86.7%, Lakers 85.4%.

11.

Pistons			Lakers		
Aguirre	29:50	30	Green	37:42	38
Edwards	32:21	32	Worthy	46:38	47
Laimbeer	42:14	42	Thompson	43:45	44
Dumars	39:52	40	E. Johnson	39:50	40
Thomas	44:41	45	Scott	32:39	33
Sailey	23:07	23	Cooper	32:33	32
V. Johnson	14:00	14	Drew	18:31	18
Rodman	38:55	39	Divac	13:22	13
Totals	265:00	265	Totals	265:00	265

13. These long divisions can be done by means of a calculator.

15. Sheila receives the Frisbee and gives Jean 69 cents.

17. Mary receives the house and the car and gives John $43,831.25.

19. a. A receives the farm plus $7,333.33; B receives $132,333.33; and C receives the house and sculpture and pays $139,666.66. **b.** A receives the farm plus $80,000; B receives $120,000; and C receives the house and sculpture and pays $200,000. **c.** A receives the farm plus $78,500; B receives $99,250; and C receives house and sculpture and pays $177,750.

21. Sheila receives the Frisbee and gives Jean 75 cents.

23. Mary receives the house and car, and gives John $45,600.

25. a. *A* receives the farm plus $17,333.33; *B* receives $142,333.33; and *C* receives the house and the sculpture and pays $149,666.66. **b.** *A* receives the farm plus $78,500; *B* receives $128,100; and *C* receives the house and sculpture and pays $206,600. **c.** *A* receives the farm plus $78,500; *B* receives $106,750; and *C* receives the house and sculpture and pays $185,250.

27. Answers vary. (The assumptions needed for a conclusion need not be unique.)

29. Answers vary.

31. a. All of the apportionment methods given in the text give: Arts 2, Science 2, and Business 1. **b.** Arts 2, Science 1, and Business 2. **c.** Yes. Science

33. a. Answers vary as to which is "best."
  b. (1) Hamilton   88, 2, 2, 1, 1, 1, 1, 1, 1, 1, 1
     (2) Jefferson   90, 1, 1, 1, 1, 1, 1, 1, 1, 1, 1
     (3) Webster-
         Willcox   90, 1, 1, 1, 1, 1, 1, 1, 1, 1, 1
     (4) Adams   80, 2, 2, 2, 2, 2, 2, 2, 2, 2, 2
  c. The 90 in the Jefferson and Webster-Willcox methods violates the upper quota of 88. The 80 in Adams method violates the lower quota of 87.

35. (1) Hamilton   10, 9, 7, 5, 4, 1
     (2) Jefferson   11, 9, 7, 5, 3, 1
     (3) Webster-Willcox   10, 9, 8, 5, 3, 1
     (4) Adams   10, 9, 7, 5, 3, 2

37. (1) Hamilton   2, 2, 1
     (2) Jefferson   3, 2, 0
     (3) Webster-Willcox   2, 2, 1
     (4) Adams   2, 2, 1

39. (1) Hamilton   3, 1, 1
     (2) Jefferson   4, 1, 0
     (3) Webster-Willcox   3, 1, 1
     (4) Adams   3, 1, 1

41. Proofs vary. An important observation is the following. The Hamilton method rounds the larger fractional parts $|a_i - q_i|$ upward and the smaller ones downward. Any alternate rounding (that preserves the quota condition and apportionment sum) must round a larger fractional part down and a smaller one up. This cannot decrease the sum of these two terms. An apportionment that violates the quota condition replaces some fractional part $|a_i - q_i|$ by a term greater than 1.

Chapter 12

1. a. and **b.** Saddle point at row 1, column 2, and value 5. **c.** Bad strategies: row 2, column 1.

3. a. No saddle point. **c.** No bad strategies.

5. a. No saddle point. **c.** None.

7. a. Saddle point at row 3, column 3, and value −20. **c.** Rows 1 and 2 and columns 1 and 2.

9. Batter's optimal mixed strategy is $(\frac{5}{6}, \frac{1}{6})$, pitcher's is $(\frac{2}{3}, \frac{1}{3})$, and value is .267.

11. Saddle point at knuckleball and .250.

13. Offense $(\frac{5}{8}, \frac{3}{8})$, defense $(\frac{3}{4}, \frac{1}{4})$, and value 0.575.

15. a.

		Officer does Not patrol	Patrols
You park in	Street	0	−40
	Lot	−32	−16

b. You $(\frac{2}{7}, \frac{5}{7})$, officer $(\frac{3}{7}, \frac{4}{7})$, and value −$22.86. **c.** It is unlikely that the officer's payoffs are the (proportionally) opposite of yours. **d.** Use some available random device. Discuss.

17. $\dfrac{1}{2}\begin{pmatrix} .4 \\ .3 \\ 0 \end{pmatrix} + \dfrac{1}{2}\begin{pmatrix} 0 \\ .2 \\ .4 \end{pmatrix} = \begin{pmatrix} .20 \\ .25 \\ .20 \end{pmatrix} \leq \begin{pmatrix} .2 \\ .4 \\ .3 \end{pmatrix}$

19. *MD* is best in Exercise 17.

21. From Exercise 18, .160 < .240.

23. Player II plays T and wins $\frac{1}{2}$ on average.

25. Player I plays $(0, \frac{3}{5}, \frac{2}{5})$, player II plays $(\frac{4}{5}, \frac{1}{5})$, and the value is $\frac{12}{5}$.

27. Saddle point at row 1, column 1, and value 6.

29. $(\frac{2}{5}, 0, \frac{3}{5})$, $(\frac{2}{5}, \frac{3}{5})$, $-(\frac{1}{5})$.

31. $(0, \frac{2}{3}, 0, 0, \frac{1}{3})$, $(\frac{1}{3}, \frac{2}{3})$, $\frac{5}{3}$.

33. Player I plays $(0, 1)$, Player II plays $(1 - p, p)$ where $\frac{1}{2} \le p \le 1$, and the value is 2.

35. $(0, \frac{1}{2}, \frac{1}{2})$ or $(\frac{1}{3}, 0, \frac{2}{3})$ or anything between these two, $(\frac{1}{2}, \frac{1}{2})$, and 0.

37. $(\frac{4}{5}, \frac{1}{5})$ and value \$420,000. See parts **a, b**, and **d** in Exercise 38 for alternate approaches.

39. Row 1, column 1 gives the equilibrium point $(5, 5)$, which is the overall best payoff. The point $(2, 2)$ is an inferior equilibrium point.

41. $(4, 0)$, $(0, 0)$, and $(0, 4)$ are all in equilibrium. [It would be better if they could flip a coin to decide between $(4, 0)$ and $(0, 4)$.]

43. Answers vary.

45. a. Player 2 avoids "call" because "fold" dominates it. **b.** $(\frac{1}{3}, \frac{2}{3}, 0)$, $(\frac{2}{3}, 0, \frac{1}{3})$, and $-(\frac{1}{12})$. **c.** Player 2 since the value is nearest. **d.** Yes; Player I bets first while holding L with probability $\frac{2}{3}$. Player II raises while holding L with probability $\frac{1}{3}$.

Chapter 13

1. a. 1. **b.** 3; 9 times as large. **c.** 4; 24 square inches. **d.** 3.06. **e.** The 4 by 6 prints are almost twice as expensive per square inch of paper. **f.** 79 cents; \$1.46.

3. a. $\frac{1}{87} \approx 0.0115$. **b.** It is $(\frac{1}{87})^3 \approx 0.00000152$ times as large. **c.** $\frac{1}{48}$.

5. a. G. **b.** G. **c.** J.

7. $A'(-1, 2)$, $B'(0, -2.5)$, $C'(2.5, 3)$.

9. All true except part **a**, which is false unless $k = 1$.

11. All true except for part **a**.

13. a. The new altar would have a volume 8 times as large — not "8 times greater than" or "8 times larger than," and definitely not twice as large, as the old altar. **b.** $\sqrt[3]{2} \approx 1.26$.

15. a. Both the width and the height of the buildings are proportional to the cost, so that Warren, with a cost about 1.5 times as much as South Mountain, has a building whose area on the page is about $(1.5)^2 = 2.25$ times as large, and whose implied volume is $(1.5)^3 = 3.4$ times as large. **b.** Simply monstrous! The line for 27.5 mpg, which is about $1\frac{1}{2}$ times as much as 18 mpg, is about 9 times as long as the line for 18 mpg. **c.** The picture shows the dollar bill shrinking in both length and width, even though the value shrinks only once. To use area to reflect the purchasing power of the dollar, the 1978 dollar should have about twice the area shown (and the other depictions also adjusted accordingly).

17. a. 147,840. **b.** 2.06 million. **c.** 1.86×10^{12}.

19. a. 0.00013. **b.** We assume that all parts of the scale model are made of the same materials as the real locomotive. **c.** 0.27 lb. **d.** 0.12. **e.** 0.00012.

21. \$3.74/lb.

23. 185 meters, or 607 feet.

25. 9 ft 3 in to 11 ft 9 in (in modern times there have been men over 9 ft tall); 282 cm to 358 cm.

27. a. 900 lb per cu ft. **b.** Almost twice as dense. **c.** Since 230 lb of compost is supposed to add about 5%, the original should be about 230 lb divided by 0.05, or 4600 lb. So the revised quotation should say that the mineral soil weighs about 4500 lb.

29. a. 400,000 lb. **b.** 28.

31. $\sqrt{12} \times 20$ mph $= 69$ mph.

33. It has disproportionately large wings compared to geometric scaling up of a bird, hence lower wing loading and lower minimum flying speed. Also, in part it glides rather than flies.

35. Weighing only half as much as an adult, it would need only half the wing area; with both length and width of the wing growing proportionally, wing area (and hence weight) scales as the square of wingspan. Hence, wingspan scales as the square root of weight. The half-weight dinosaur would need a wingspan $\sqrt{\frac{1}{2}}$ times as large, or 25 ft.

37. Human infants may be only a foot long at birth, barely twice Lilliputian size. Other mammals — either as infants (e.g., pandas) or as adults (e.g., mice) — are smaller than Lilliputians. So Lilliputians are not ruled out by area-volume considerations.

39. Answer will vary.

Chapter 14

1. a. $1080.00; 8.000%. **b.** $1080.00; 8.000%. **c.** $1082.43; 8.243%. **d.** $1083.29; 8.329%.

3. $8104.03.

5. a. $1.28. **b.** $0.78. **c.** $4437.05. **d.** $3476.55.

7. a. 23.1, 11.56, 7.7 years, all close to the predictions. The number 72 has the convenience of being evenly divisible by many small numbers. **b.** "rule of 22": Divide $100r$ into 110 to get 22 years.

9. There were 120 payments of $100 each, for a total of $12,000. The formula applies with $n = 120$ and x unknown, where $x - 1$ is the monthly rate of interest. There is no way to solve exactly the resulting equation:

$$\frac{x^{121} - 1}{x - 1} = 377.47$$

Use of a computer graphing program, a calculator equation solver, or successive guessing gives $x = 1.016467$, for an annual interest rate of $12(0.016467) = 0.1976 = 19.76\%$. The effective annual yield is $(1.016467)^{12} - 1 = 0.2165$, or 21.65%.

11. 3.80 billion.

13. 3.72 billion.

15. 16.4 million.

17. a. 51 years. **b.** 42 years. **c.** 25 years.

19. The projections assume that the growth rate will be about 4.8% per year.

21. 15.7 million.

23. a. 11,400. **b.** 72,000.

25. About 40,000 years.

27. 50.

29. a. 34.4, 54.7, 77.0, 92.9, 98.8, 62.2, 68.4, 69.8, 70.0, 70.0. **b.** 16.3, 26.2, 40.2, 58.9, 80.7, 102.1, 119.5, 131.8, 140.2, 146.6. **c.** 28.1, 69.9, 124.8, 114.9, 133.5, 126.3, 142.6, 137.3, 151.8, 148.1. **d.** For $k = 0.7$, the population is increasing; it approaches an upward-sloping trend line from below. For $k = 2$ the population oscillates about an upward-sloping trend line. In neither case is there an upper bound to the size of the population.

31. Equilibrium population size 25, maximum sustainable yield 7 for an initial population of 10.

33. The cost of harvesting the first individual is $20/[10 + 5(20)] = 0.182$; for the second, $20/[10 + 5(19)] = 0.190$; for the third, $20/[10 + 5(18)] = 0.200$; for the fourth, $20/[10 + 5(17)] = 0.211$; for the fifth, $20/[10 + 5(16)] = 0.222$. The total cost is 1.005.

35. About $250.

37. The sustainable yield for an initial population of 20 is 4, so the total yearly revenue is $32. The costs of harvesting are the sum of the costs for harvesting one individual from populations of size 24, 23, 22, and 21. From the graph we guess the cost of harvesting from a population of size 24 is $3, and from one of size 21 is $2.50, and for all four together it's about $11. The resulting net revenue is $21.

39. The net revenue is $0.

Chapter 15

1. No. Equiangular triangles have the same shape but not necessarily the same size.

3. Yes, SAS.

5. $\angle A$ and $\angle D$, $\angle ABC$ and $\angle DCB$, $\angle BAC$ and $\angle CDB$, AB and DC, and BC and BC, AC and DB.

7. 10 cm.

9. 24 ft.

11. 754 ft; 344 ft.

13. 40 ft.

15. 100 ft.

17. 3.2.

19. Yes. The triangles $\triangle ADE$ and $\triangle ABC$ are similar by ASA; note that $AE = EC$.

21. $m(\angle 2) + m(\angle 3)$.

23. 44.

25. $m(\angle D) = 53\frac{1}{3}°$, $m(\angle E) = m(\angle F) = 63\frac{1}{3}°$.

27. 32°.

29. 17°.

31. 46,250 km, for an error of $(46,250 - 39,375) = 6875$ km, which is 17.5% too much.

33. $\tan 0.25° = RM/ER$, so $RM = ER \tan 0.25° = 238,857 \text{ mi} \times 0.00436 = 1041$ mi.

35. a. 500 miles is the same fraction of the earth's circumference of 25,000 miles — one-fiftieth — that the angle $\angle AOB$ is of 360°, so the angle measures 7.2°. **b.** 86.4°. **c.** Because the tangents HA and HB make right angles with OA and OB, the angles ABH and BAH each measure $90° - 86.4° = 3.6°$. **d.** We add the measures of angles ABH and BAH to the measures of angles MAH and MBH to find the measures of angles MAB and MBA, then subtract those values from 180° to get the measure of angle AMB. **e.** Think of the moon as the center of a circle of radius $MA = MB$. Then the distance from A to B along the arc of this circle is approximately the same as the distance along the circumference of the earth, or 500 miles. We have that 500 miles is to the circumference of this circle as the measure of angle AMB is to 360°, from which we can determine the circumference of this circle and hence its radius — the distance from the earth to the moon.

37. $VS = ES \times \sin 47° = 93,000,000 \text{ mi} \times 0.73135 = 68$ million miles.

39. a. Let L be the distance to the moon and l be its radius. Then the circumference of the moon's orbit is $2\pi L = 360(4l) = 360(4)(1080)$, so $L = 248,000$ miles. **b.** 42,524.05 min, of which 125 min is $1/340.19$; 360 times this gives 1.06° for the angular measure of the diameter of the moon. Our new estimate of the distance L to the moon comes from $2\pi L = 340.19(4l) = 340.19(4)(1080)$, which gives 234,000 miles. (The distance of the moon varies between 221,463 and 238,857 miles; during the eclipse of January 4, 1992, the moon will be near the farther distance, and the eclipse will not be total but *annular,* meaning that the moon's disk will not completely cover the sun's.)

Chapter 16

1. a. 32 ft. **b.** The stone hits the ground when $y = 0$, or $-16t^2 + 48t + 64 = -16(t^2 - 3t - 4) = -16(t - 4)(t + 1) = 0$. The only possibilities are $t = 4$ and $t = -1$. (If it had not been so easy to come up with the factors $t - 4$ and $t + 1$ by guesswork, we would have had to use the quadratic formula, a computer graphing program, or else trial-and-error approximation.) The stone hits the ground 4 seconds after it is thrown upward. **c.** The solution $t = -1$ in part **b** tells us that the stone would have the same trajectory (from the building height on) if it had been launched from ground level at $t = -1$. The symmetrical time between $t = -1$ and $t = 4$ is $t = 1.5$. Putting this value into the equation gives $y = 100$ ft.

3. a–f. Constructed figure will be the answer.

5. a. $(0, 0)$. **b.** $(0, \frac{3}{2})$. **c.** $x = 0$. **d.** -6 or 6. **e.** $\frac{2}{3}$.

7. a–e. Constructed figure will be the answer.

9. $(x + 4)^2 + (y - 3)^2 = 9$.

11. $(x - 5)^2 + (y + 6)^2 = 80$.

13. $x^2/5^2 + y^2/4^2 = 1$.

15. We find $a + c = 252{,}710$ and $a - c = 221{,}643$, so $a = 237{,}177$, $c = 15{,}554$, and $b = 236{,}666$. The equation is $x^2/237{,}177^2 + y^2/236{,}666^2 = 1$.

17. The moon's orbit takes 39,343.2 minutes, while Sputnik's took 88, so Sputnik (which has long since reentered the atmosphere and burned up) took 447 times as long. Taking this ratio to the two-thirds power gives the ratio of their average distances from the earth; 58.5. So Sputnik was $(1/58.5)238{,}857 = 4{,}083$ miles from the center of the earth, or $4083 - 3963.5 = 120$ miles above the surface of the earth.

19. The moon takes 27.322 times as long to orbit. Taking this to the two-thirds power gives that the moon must be 9.0714 times as far away, so that the distance of the satellite is $238{,}857/9.0714 = 26{,}331$ miles. Subtracting the radius of the earth, the satellite must be $26{,}331 - 3963.5 = 22{,}367$ miles above the surface of the earth.

21. Constructed figure will be the answer.

23. Hint: There are two cases, depending on which focus the point is closer to, leading to the two equations (written as one):

$$\sqrt{(x - (-c))^2 + (y - 0)^2} - \sqrt{(x - c)^2 + (y - 0)^2} = \pm 2a$$

25. The light from a star is reflected from a parabolic mirror and directed toward the focus of the parabola F_P. The light is intercepted by a hyperbolic mirror having the same focus $F_{H_1} - F_P$ as the parabolic mirror. The light is then reflected from the hyperbolic mirror through the other focus of the hyperbola F_{H_2} and reflected from an elliptical mirror having focus $F_{E_1} = F_{H_2}$. The light is finally reflected to an eyepiece (or camera) located at the second focus of the ellipse F_{E_2}. [From "Contemporary conic sections," by Charles G. Moore, in *Consortium: The Newsletter of the Consortium for Mathematics and Its Applications*, no. 20 (November 1986), p. 4.]

27. Two equal times correspond to two line segments P_1P_2 and Q_1Q_2 of equal length. The triangles ΔSP_1P_2 and ΔSQ_1Q_2 have equal bases and equal altitudes, hence equal areas.

29. The square root of the ratio of the masses, $\sqrt{330{,}000} = 574$ gives the ratio of the distances. So the object must be $\frac{1}{574}$ $(92{,}956{,}000) = 162{,}000$ miles from the center of the sun (a location inside the sun).

31. 1.924×10^{13} miles; 3.26 light-years.

33. a. Diameter of the moon: 2000 mi. Circumference of the earth: 22,000 mi. Circumference of the moon: 6240 mi. **b.** 100 mi. **c.** Here BT is perpendicular to (forms a 90° angle with) CB, because a tangent to a circle is perpendicular to the radius of the circle through the point of tangency. **d.** Let r be the radius of the moon. Then

$$\begin{aligned}
AT &= CT - CA \\
&= CT - r \\
&= \sqrt{(CB)^2 + (BT)^2} - r \\
&= \sqrt{r^2 + \tfrac{2r^2}{20}} - r \\
&= r\sqrt{\tfrac{404}{400}} - r \\
&= 0.0050r \\
&= 5.0 \text{ mi}
\end{aligned}$$

35. Each of the interior angles of the triangle has the same measure as one of the three angles making up the straight angle (180°) at C.

37. a. Draw the line segments AC and BD, to form two pairs of triangles. The first pair, ADB and BCA, are congruent by SAS (two sides and the included angle of one triangle are equal to two sides and included angle of the other). Hence, $AC = BD$. The second pair, ACD and BDC, are congruent by SSS (the three sides of one triangle are equal to the three sides of the other). Hence, the corresponding angles C and D are equal. **b.** The sum of the measures of the interior angles of the two triangles ABD nd BCD is equal to the sum of the measures of the four angles A, B, C, and D. The two angles C and D are acute, so this sum is less than $360°$. Consequently, the angles of one or both of the triangles must add to less than $180°$.

39. a. $p = 3, q = 3$. **b.** Cube: $pf = qV = 24$. Tetrahedron: $pf = qV = 12$. Octahedron: $pf = qV = 24$. **c.** Twice the number of edges. **d.** $V = 20$, $E = 30$. **e.** $V = 12$, $E = 30$.

Chapter 17

1. 144 pairs.

3. a. 1, 1, 3, 5, 11, 21, 43, 85, 171, 341, 683, 1365. **b.** $B_n = B_{n-1} + 2B_{n-2}$. **c.** 1, 3, 1.667, 2.2, 1.909, 2.048, 1.977, 2.012, 1.994, 2.003, 1.999. **d.** $x = 2, -1$; we discard the -1 root. **e.** $B_n = [2^n - (-1)^n/3$.

5. a. 1; 1. **b.** 5. **c.** 233.

7. a. 9. **b.** 16.

9. F_{61} and F_{62} both end in 1.

11. Use the Pythagorean Theorem.

13. First, since $AB = AE$, the angles ABE and AEB are equal. Since the interior angles of a regular pentagon measure $108°$ and the angles of a triangle add up to $180°$, each of the angles ABE and AEB must measure $36°$. Hence angle BDE must measure $108° - 36° = 72°$; by symmetry, angle BDE also measures $72°$, as do angles FBD and FDB. Second, each of angles FAE and FEA also measures $180° - 108° - 72°$. Triangles FAE and BDE have two angles and the included side equal, so they are congruent. Hence AF has length x. Third, triangles FBD and BDE are similar, as they both have angles

of $72°$, $72°$, and $36°$. Hence, their corresponding sides are proportional, which gives

$$\frac{x+1}{x} = \frac{x}{1}$$

or $x + 1 - x^2$, or $x^2 - x - 1 = 0$, whose only positive solution is ϕ.

15. All except (e) have a single vertical line of symmetry through the center of the figure. Figures (b) and (d) have a single horizontal line of symmetry through the center of the figure.

17. All are true.

19. a. B, C, D, E, H, I, K, O, X. **b.** A, H, I, M, O, T, U, V, W, X, Y. **c.** H, I, N, O, S, X, Z. **d.** H, I, M, O, X.

21. a. MOM, WOW (both either horizontally or vertically); MUd and bUM reflect into each other. **b.** pod rotates into itself; MOM and WOW rotate into each other. **c.** Here are some possibilities:

NOW NO

SWIMS

ON MON

CHECK BOOK BOX

OX HIDE

23. For all parts, translations. **a.** Reflection in vertical lines through the centers of the A's or between them. **b.** Reflection in the horizontal midline. **c.** Reflection in the horizontal midline, reflections in vertical lines through the centers of the X's or between them, $180°$ rotation around the centers of the X's or the midpoints between them, glide reflections. **d.** None other than translations.

25. *11, 1m, m1, 12, 1g, mg, mm.*

27. *mm, 1g, mg, 12, mm* (perhaps), *1m, mg, 11.*

29. *cm, pmg, p4m, p4g.*

31. Exterior angle: $360°/8 = 45°$. Interior angle: $180° -$ interior angle $= 135°$.

33.

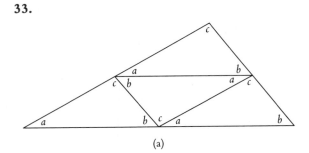

(a)

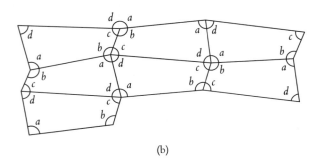

(b)

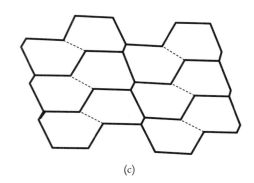

(c)

35.

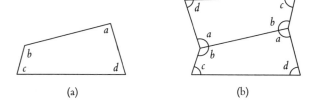

(a)   (b)

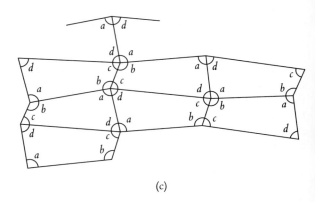

(c)

37. Project. Individual answers will vary depending on computer program used.

39. a. *ABAABABA.* **b.** Each *B* comes from an *A* in the preceding month, which must lie immediately to the *B*'s left; so each *B* is preceded directly by an *A.* **c.** The two leftmost *A*'s would have had to come from two *B*'s in a row in the preceding month. **d.** Verify that the sequence for the fourth month, *ABA*, follows the rule. Then, assuming that the rule holds for all previous months, consider last month and what happens as it transforms into the current month's sequence. Last month's sequence consists of a first part that is the sequence from 2 months ago; this first part, we know from our assumption, transforms into last month's sequence. The second part of last month's sequence is the sequence from 3 months ago, which we know from our assumption transforms into the sequence from 2 months ago. So the current month's sequence consists of last month's sequence followed by the sequence from 2 months ago, as claimed.

Chapter 18

1. No. Explanation: taken literally, the first four steps (wet hair, apply shampoo, lather, rinse) will be repeated indefinitely.

3. The output of the modified algorithm will be

4

5

6

5. 1   1   1

   2   4   8

. . .

. . .

. . .

20 400 8000

7. 1

2

4

9. Statements with line numbers 10, 20, and 60 are executed once each. Statements with line numbers 30, 40, and 50 are executed 7 times each.

11. Replace the statement with line number 60 with 60 print A, B, C, D, SUM

13. 10   input N

20   SUM = 0

30   I = 1

40   input A

50   SUM = SUM + A

60   I = I + 1

70   if I ≤ N go to 40

80   print SUM

90   end

15. Modify the solution of Exercise 10 by replacing lines 50 and 60 with:

50   AVG = (A + B + C + D)/4

60   print AVG

17. Statements with line numbers 10, 20, 30, 80, 90, and 100 are executed once each. All other statements are executed N times each. This answer corresponds to the solution of Exercise 16 given above.

19. 10   input A

20   input B

30   input C

40   MAX = A

50   if B > MAX then MAX = B

60   if C > MAX then MAX = C

70   print MAX

80   end

21. Statements at lines 10, 20, 30, 40, 90, and 100 are executed once each. All other statements are executed N − 1 times each. This answer corresponds to the solution of Exercise 20 given above.

23. 10   BALANCE = 1000

20   MONTH = 1

30   BALANCE = BALANCE * (1 + 6/1200)

40   print BALANCE

50   MONTH = MONTH + 1

60   if MONTH ≤ 24 go to 30

70   end

25.

Data element inserted	Number of data movements	Number of comparisons
2nd	3	1
3rd	3	2
4th	2	1
5th	3	2
6th	3	2
7th	6	5
8th	3	2
9th	8	7
10th	3	2
Total	34	24

27. The number of data movements is 9 and the number of comparisons is 6.

Chapter 19

1. 86 in base 2 positional notation is 1010110.
It is stored as 0101 0110.

3. The integer is 39, and it is stored as 0010 0111.

5.   3419 will be represented by 0000 1101 0101 1011.
  -3419 will be represented by 1000 1101 0101 1011.
  17,842 will be represented by 0100 0101 1011 0010.
  $-17,842$ will be represented by 1100 0101 1011 0010.

7.   3419 will be represented by 0000 1101 0101 1011.
  -3419 will be represented by 1111 0010 1010 0100.
  17,842 will be represented by 0100 0101 1011 0010.
  $-17,842$ will be represented by 1011 1010 0100 1101.

9.   3419 will be represented by 0000 1101 0101 1011.
  -3419 will be represented by 1111 0010 1010 0101.
  17,842 will be represented by 0100 0101 1011 0010.
  $-17,842$ will be represented by 1011 1010 0100 1110.

11.

Bit configuration	Value represented in sign-magnitude
0000	0
0001	1
0010	2
0011	3
0100	4
0101	5
0110	6
0111	7
1000	-0 (or simply 0)
1001	-1
1010	-2
1011	-3
1100	-4
1101	-5
1110	-6
1111	-7

13. Largest is 7; smallest is -7.

15. 734 is represented by 0010 1101 1110 and 1081 is represented by 0100 0011 1001. Therefore,

$$
\begin{array}{ll}
0001\ 1111\ 000 & \text{carry} \\
0010\ 1101\ 1110 & \\
\underline{0100\ 0011\ 1001} & \\
0111\ 0001\ 0111 & \text{sum}
\end{array}
$$

17. An integer is divisible by 2^k if and only if the rightmost k bits in its sign-magnitude representation are all 0.

19. The sequence of bits for Jane would be 0100 1010 0110 0001 0110 1110 0110 0101.

21. 51.

23. Answer will vary.

25.

A	NOT A
0	1
1	0

27.

A	B	A OR B	NOT $(A$ OR $B)$
1	1	1	0
1	0	0	0
0	1	1	0
0	0	0	1

NOT A	NOT B	(NOT A) AND (NOT B)
0	0	0
0	1	0
1	0	0
1	1	1

29. The array method requires 1,370,000 bytes to store the data, and the use of pointers will require an additional 40,000 bytes of memory. The increase is 2.9%.

31. The initial arrangement will be

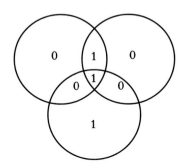

In the case of the first single-bit data error, we have

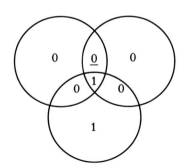

In the second case (1110), we have

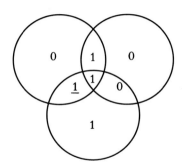

In the third case (1000), we have

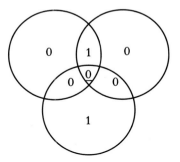

In the fourth case (1011), we have

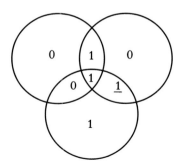

In all cases, the only way we can place an even number of 1s in all the circles is to reverse the underlined digit.

Chapter 20

1. Substitution of $x_1 = 0$, $y_1 = 0$, $x_2 = 12$, and $y_2 = 8$ in the parametric equations yields $x = 12u$ and $y = 8u$. When $u = 0$, we get $(0, 0)$; when $u = 1$, we get $(12, 8)$.

3.

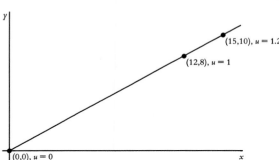

5.

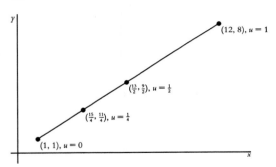

7. a.

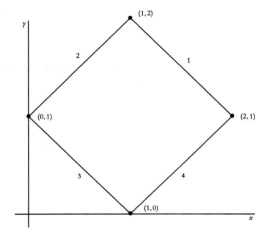

b.

Side	Endpoint	Endpoint	Activity status
1	(1, 2)	(2, 1)	Off
2	(1, 2)	(0, 1)	Off
3	(0, 1)	(1, 0)	Off
4	(2, 1)	(1, 0)	Off

c. When $1 < y < 2$, the activity status of sides 1 and 2 will be on and the activity status of sides 3 and 4 will be off. When $0 < y < 1$, the activity status of sides 1 and 2 will be off and the activity status of sides 3 and 4 will be on. **d.** When $y = \frac{3}{2}$, sides 1 and 2 will be active with points $(\frac{1}{2}, \frac{3}{2})$ and $(\frac{3}{2}, \frac{3}{2})$ on these sides. So the segment from $(\frac{1}{2}, \frac{3}{2})$ to $(\frac{3}{2}, \frac{3}{2})$ will be drawn.

9. a. $X = x + 4, Y = y$ **b.** $X = x, Y = y - 2$.

11. $X = x + 4\sqrt{2}, Y = y + 4\sqrt{2}$.

13. The following transformations applied in the given order:

$$X = x - 1, Y = y - 1$$
$$X = x, Y = 2y$$
$$X = x + 3, Y = y - 1$$

15. The following transformations applied in the given order:

$$X = x - 1, Y = y - 1$$
$$X = \tfrac{1}{2}x, Y = 2y$$
$$X = x + 3, Y = y - 1$$

17. In the case of Figure 20.14a, the area is doubled; in the case of Figure 20.14b, it is halved; in the case of Figure 20.15, it is tripled.

19. The area will change by a factor of rs.

21. Where $r < 0, X = rx, Y = y$ produces a scaling by a factor of $|r|$ in the horizontal direction followed by a reflection about the y axis. If $r < -1$, the scaling becomes a stretching; if $r > -1$, it is a shrinking.

23. Translation: $X = x + 3, \ Y = y$. Reflection: $X = x, Y = -y$.

25. $X = x + 3, Y = -y$.

◆ Index ◆